Grzimek's ANIMAL LIFE ENCY
D0901375

Volume 1

LOWER ANIMALS

Volume 2

INSECTS

Volume 3

MOLLUSKS AND ECHINODERMS

Volume 4

FISHES I

Volume 5

FISHES II AND AMPHIBIA

Volume 6

REPTILES

Volume 7

BIRDS I

Volume 8

BIRDS II

Volume 9

BIRDS III

Volume 10

MAMMALS I

Volume 11

MAMMALS II

Volume 12

MAMMALS III

Volume 13

MAMMALS IV

Grzimek's ANIMAL LIFE ENCYCLOPEDIA

Editor-in-Chief

Dr. Dr. h.c. Bernhard Grzimek

Professor, Justus Liebig University of Giessen
Director, Frankfurt Zoological Garden, Germany
Trustee, Tanzanian National Parks, Tanzania

89499

VAN NOSTRAND REINHOLD COMPANY

New York Cincinnati Toronto London Melbourne

ILLINOIS PRAIRIE DISTRICT LIBRARY

First published in paperback in 1984

Copyright © 1968 Kindler Verlag A.G. Zurich

Library of Congress Catalog Card Number 79-183178

ISBN 0-442-23041-9

All rights reserved. No part of this work covered by the copyright hereon may be reproduced or used in any form or by any means—graphic, electronic, or mechanical, including photocopying, recording, taping, or information storage and retrieval systems—without written permission of the publisher.

Printed in Federal Republic of Germany

Van Nostrand Reinhold Company Inc.
135 West 50th Street
New York, New York 10020

Van Nostrand Reinhold Company Limited
Molly Millars Lane
Wokingham, Berkshire RG11 2PY, England

Van Nostrand Reinhold
480 Latrobe Street
Melbourne, Victoria 3000, Australia

Macmillan of Canada
Division of Gage Publishing Limited
164 Commander Boulevard
Agincourt, Ontario M1S 3C7 Canada

16 15 14 13 12 11 10 9 8 7 6 5 4 3 2 1

91.03
R2
vol 7
copy 1

EDITORS AND CONTRIBUTORS

Editor-in-Chief
DR. DR. H.C. BERNHARD GRZIMEK
Professor, Justus Liebig University of Giessen, Germany
Director, Frankfurt Zoological Garden, Germany
Trustee, Tanzanian National Parks, Tanzania

DR. MICHAEL ABS Curator, Ruhr University	BOCHUM, GERMANY
DR. SALIM ALI Bombay Natural History Society	BOMBAY, INDIA
DR. RUDOLF ALTEVOGT Professor, Zoological Institute, University of Münster	MÜNSTER, GERMANY
DR. RENATE ANGERMANN Curator, Institute of Zoology, Humboldt University	BERLIN, GERMANY
EDWARD A. ARMSTRONG Cambridge University	CAMBRIDGE, ENGLAND
DR. PETER AX Professor, Second Zoological Institute and Museum, University of Göttingen	GÖTTINGEN, GERMANY
DR. FRANZ BACHMAIER Zoological Collection of the State of Bavaria	MUNICH, GERMANY
DR. PEDRU BANARESCU Academy of the Roumanian Socialist Republic, Trajan Savulescu Institute of Biology	BUCHAREST, RUMANIA
DR. A. G. BANNIKOW Professor, Institute of Veterinary Medicine	MOSCOW, U.S.S.R.
DR. HILDE BAUMGÄRTNER Zoological Collection of the State of Bavaria	MUNICH, GERMANY
C. W. BENSON Department of Zoology, Cambridge University	CAMBRIDGE, ENGLAND
DR. ANDREW BERGER Chairman, Department of Zoology, University of Hawaii	HONOLULU, HAWAII, U.S.A.
DR. J. BERLIOZ National Museum of Natural History	PARIS, FRANCE
DR. RUDOLF BERNDT Director, Institute for Population Ecology, Heligoland Ornithological Station	BRAUNSCHWEIG, GERMANY
DIETER BLUME Instructor of Biology, Freiherr-vom-Stein School	GLADENBACH, GERMANY
DR. MAXIMILIAN BOECKER Zoological Research Institute and A. Koenig Museum	BONN, GERMANY
DR. CARL-HEINZ BRANDES Curator and Director, The Aquarium, Overseas Museum	BREMEN, GERMANY
DR. DONALD G. BROADLEY Curator, Umtali Museum	UMTALI, RHODESIA
DR. HEINZ BRÜLL Director; Game, Forest, and Fields Research Station	HARTENHOLM, GERMANY
DR. HERBERT BRUNS Director, Institute of Zoology and the Protection of Life	SCHLANGENBAD, GERMANY
HANS BUB Heligoland Ornithological Station	WILHELMSHAVEN, GERMANY
A. H. CHRISHOLM	SYDNEY, AUSTRALIA
HERBERT THOMAS CONDON Curator of Birds, South Australian Museum	ADELAIDE, AUSTRALIA
DR. EBERHARD CURIO Director, Laboratory of Ethology, Ruhr University	BOCHUM, GERMANY

DR. SERGE DAAN	Laboratory of Animal Physiology, University of Amsterdam	AMSTERDAM, THE NETHERLANDS
DR. HEINRICH DATHE	Professor and Director, Animal Park and Zoological Research Station, German Academy of Sciences	BERLIN, GERMANY
DR. WOLFGANG DIERL	Zoological Collection of the State of Bavaria	MUNICH, GERMANY
DR. FRITZ DIETERLEN	Zoological Research Institute, A. Koenig Museum	BONN, GERMANY
DR. ROLF DIRCKSEN	Professor, Pedagogical Institute	BIELEFELD, GERMANY
JOSEF DONNER	Instructor of Biology	KATZELSDORF, AUSTRIA
DR. JEAN DORST	Professor, National Museum of Natural History	PARIS, FRANCE
DR. GERTI DÜCKER	Professor and Chief Curator, Zoological Institute, University of Münster	MÜNSTER, GERMANY
DR. MICHAEL DZWILLO	Zoological Institute and Museum, University of Hamburg	HAMBURG, GERMANY
DR. IRENÄUS EIBL-EIBESFELDT	Professor and Director, Institute of Human Ethology, Max Planck Institute for Behavioral Physiology	PERCHA/STARNBERG, GERMANY
DR. MARTIN EISENTRAUT	Professor and Director, Zoological Research Institute and A. Koenig Museum	BONN, GERMANY
DR. EBERHARD ERNST	Swiss Tropical Institute	BASEL, SWITZERLAND
R. D. ETCHECOPAR	Director, National Museum of Natural History	PARIS, FRANCE
DR. R. A. FALLA	Director, Dominion Museum	WELLINGTON, NEW ZEALAND
DR. HUBERT FECHTER	Curator, Lower Animals, Zoological Collection of the State of Bavaria	MUNICH, GERMANY
DR. WALTER FIEDLER	Docent, University of Vienna, and Director, Schönbrunn Zoo	VIENNA, AUSTRIA
WOLFGANG FISCHER	Inspector of Animals, Animal Park	BERLIN, GERMANY
DR. C. A. FLEMING	Geological Survey Department of Scientific and Industrial Research	LOWER HUTT, NEW ZEALAND
DR. HANS FRÄDRICH	Zoological Garden	BERLIN, GERMANY
DR. HANS-ALBRECHT FREYE	Professor and Director, Biological Institute of the Medical School	HALLE A.D.S., GERMANY
GÜNTHER E. FREYTAG	Former Director, Reptile and Amphibian Collection, Museum of Cultural History in Magdeburg	BERLIN, GERMANY
DR. HERBERT FRIEDMANN	Director, Los Angeles County Museum of Natural History	LOS ANGELES, CALIFORNIA, U.S.A.
DR. H. FRIEDRICH	Professor, Overseas Museum	BREMEN, GERMANY
DR. JAN FRIJLINK	Zoological Laboratory, University of Amsterdam	AMSTERDAM, THE NETHERLANDS
DR. DR. H.C. KARL VON FRISCH	Professor Emeritus and former Director, Zoological Institute, University of Munich	MUNICH, GERMANY
DR. H. J. FRITH	C.S.I.R.O. Research Institute	CANBERRA, AUSTRALIA
DR. ION E. FUHN	Academy of the Roumanian Socialist Republic, Trajan Savulescu Institute of Biology	BUCHAREST, RUMANIA
DR. CARL GANS	Professor, Department of Biology, State University of New York at Buffalo	BUFFALO, NEW YORK, U.S.A.
DR. RUDOLF GEIGY	Professor and Director, Swiss Tropical Institute	BASEL, SWITZERLAND

DR. JACQUES GERY — ST. GENIES, FRANCE

DR. WOLFGANG GEWALT
Director, Animal Park — DUISBURG, GERMANY

DR. DR. H.C. DR. H.C. VIKTOR GOERTTLER
Professor Emeritus, University of Jena — JENA, GERMANY

DR. FRIEDRICH GOETHE
Director, Institute of Ornithology, Heligoland Ornithological Station — WILHELMSHAVEN, GERMANY

DR. ULRICH F. GRUBER
Herpetological Section, Zoological Research Institute and A. Koenig Museum — BONN, GERMANY

DR. H. R. HAEFELFINGER
Museum of Natural History — BASEL, SWITZERLAND

DR. THEODOR HALTENORTH
Director, Mammalology, Zoological Collection of the State of Bavaria — MUNICH, GERMANY

BARBARA HARRISSON
Sarawak Museum, Kuching, Borneo — ITHACA, NEW YORK, U.S.A.

DR. FRANCOIS HAVERSCHMIDT
President, High Court (retired) — PARAMARIBO, SURINAM

DR. HEINZ HECK
Director, Catskill Game Farm — CATSKILL, NEW YORK, U.S.A.

DR. LUTZ HECK
Professor (retired), and Director, Zoological Garden, Berlin — WIESBADEN, GERMANY

DR. DR. H.C. HEINI HEDIGER
Director, Zoological Garden — ZURICH, SWITZERLAND

DR. DIETRICH HEINEMANN
Director, Zoological Garden, Münster — DÖRNIGHEIM, GERMANY

DR. HELMUT HEMMER
Institute for Physiological Zoology, University of Mainz — MAINZ, GERMANY

DR. W. G. HEPTNER
Professor, Zoological Museum, University of Moscow — MOSCOW, U.S.S.R.

DR. KONRAD HERTER
Professor Emeritus and Director (retired), Zoological Institute, Free University of Berlin — BERLIN, GERMANY

DR. HANS RUDOLF HEUSSER
Zoological Museum, University of Zurich — ZURICH, SWITZERLAND

DR. EMIL OTTO HÖHN
Associate Professor of Physiology, University of Alberta — EDMONTON, CANADA

DR. W. HOHORST
Professor and Director, Parasitological Institute, Farbwerke Hoechst A.G. — FRANKFURT-HÖCHST, GERMANY

DR. FOLKHART HÜCKINGHAUS
Director, Senckenbergische Anatomy, University of Frankfurt a.M. — FRANKFURT A.M., GERMANY

FRANCOIS HÜE
National Museum of Natural History — PARIS, FRANCE

DR. K. IMMELMANN
Professor, Zoological Institute, Technical University of Braunschweig — BRAUNSCHWEIG, GERMANY

DR. JUNICHIRO ITANI
Kyoto University — KYOTO, JAPAN

DR. RICHARD F. JOHNSTON
Professor of Zoology, University of Kansas — LAWRENCE, KANSAS, U.S.A.

OTTO JOST
Oberstudienrat, Freiherr-vom-Stein Gymnasium — FULDA, GERMANY

DR. PAUL KÄHSBAUER
Curator, Fishes, Museum of Natural History — VIENNA, AUSTRIA

DR. LUDWIG KARBE
Zoological State Institute and Museum — HAMBURG, GERMANY

DR. N. N. KARTASCHEW
Docent, Department of Biology, Lomonossow State University — MOSCOW, U.S.S.R.

DR. WERNER KÄSTLE
Oberstudienrat, Gisela Gymnasium — MUNICH, GERMANY

DR. REINHARD KAUFMANN
Field Station of the Tropical Institute, Justus Liebig University, Giessen, Germany — SANTA MARTA, COLOMBIA

DR. MASAO KAWAI
Primate Research Institute, Kyoto University — KYOTO, JAPAN

DR. ERNST F. KILIAN
Professor, Giessen University and Catedratico Universidad Austral, Valdivia-Chile — GIESSEN, GERMANY

DR. RAGNAR KINZELBACH
Institute for General Zoology, University of Mainz — MAINZ, GERMANY

DR. HEINRICH KIRCHNER
Landwirtschaftsrat (retired) — BAD OLDESLOE, GERMANY

DR. ROSL KIRCHSHOFER
Zoological Garden, University of Frankfurt a.M. — FRANKFURT A.M., GERMANY

DR. WOLFGANG KLAUSEWITZ
Curator, Senckenberg Nature Museum and Research Institute — FRANKFURT A.M., GERMANY

DR. KONRAD KLEMMER
Curator, Senckenberg Nature Museum and Research Institute — FRANKFURT A.M., GERMANY

DR. ERICH KLINGHAMMER
Laboratory of Ethology, Purdue University — LAFAYETTE, INDIANA, U.S.A.

DR. HEINZ-GEORG KLÖS
Professor and Director, Zoological Garden — BERLIN, GERMANY

URSULA KLÖS
Zoological Garden — BERLIN, GERMANY

DR. OTTO KOEHLER
Professor Emeritus, Zoological Institute, University of Freiburg — FREIBURG I. BR., GERMANY

DR. KURT KOLAR
Institute of Ethology, Austrian Academy of Sciences — VIENNA, AUSTRIA

DR. CLAUS KÖNIG
State Ornithological Station of Baden-Württemberg — LUDWIGSBURG, GERMANY

DR. ADRIAAN KORTLANDT
Zoological Laboratory, University of Amsterdam — AMSTERDAM, THE NETHERLANDS

DR. HELMUT KRAFT
Professor and Scientific Councillor, Medical Animal Clinic, University of Munich — MUNICH, GERMANY

DR. HELMUT KRAMER
Zoological Research Institute and A. Koenig Museum — BONN, GERMANY

DR. FRANZ KRAPP
Zoological Institute, University of Freiburg — FREIBURG, SWITZERLAND

DR. OTTO KRAUS
Professor, University of Hamburg, and Director, Zoological Institute and Museum — HAMBURG, GERMANY

DR. DR. HANS KRIEG
Professor and First Director (retired), Scientific Collections of the State of Bavaria — MUNICH, GERMANY

DR. HEINRICH KÜHL
Federal Research Institute for Fisheries, Cuxhaven Laboratory — CUXHAVEN, GERMANY

DR. OSKAR KUHN
Professor, formerly University Halle/Saale — MUNICH, GERMANY

DR. HANS KUMERLOEVE
First Director (retired), State Scientific Museum, Vienna — MUNICH, GERMANY

DR. NAGAMICHI KURODA
Yamashina Ornithological Institute, Shibuya-Ku — TOKYO, JAPAN

DR. FRED KURT
Zoological Museum of Zurich University, Smithsonian Elephant Survey — COLOMBO, CEYLON

DR. WERNER LADIGES
Professor and Chief Curator, Zoological Institute and Museum, University of Hamburg — HAMBURG, GERMANY

LESLIE LAIDLAW
Department of Animal Sciences, Purdue University — LAFAYETTE, INDIANA, U.S.A.

DR. ERNST M. LANG
Director, Zoological Garden — BASEL, SWITZERLAND

DR. ALFREDO LANGGUTH
Department of Zoology, Faculty of Humanities and Sciences, University of the Republic — MONTEVIDEO, URUGUAY

LEO LEHTONEN
Science Writer — HELSINKI, FINLAND

BERND LEISLER
Second Zoological Institute, University of Vienna — VIENNA, AUSTRIA

DR. KURT LILLELUND
Professor and Director, Institute for Hydrobiology and Fishery Sciences, University of Hamburg — HAMBURG, GERMANY

R. LIVERSIDGE
Alexander MacGregor Memorial Museum — KIMBERLEY, SOUTH AFRICA

DR. DR. KONRAD LORENZ
Professor and Director, Max Planck Institute for Behavioral Physiology — SEEWIESEN/OBB., GERMANY

DR. DR. MARTIN LÜHMANN
Federal Research Institute for the Breeding of Small Animals — CELLE, GERMANY

DR. JOHANNES LÜTTSCHWAGER
Oberstudienrat (retired) — HEIDELBERG, GERMANY

DR. WOLFGANG MAKATSCH — BAUTZEN, GERMANY

DR. HUBERT MARKL
Professor and Director, Zoological Institute, Technical University of Darmstadt — DARMSTADT, GERMANY

BASIL J. MARLOW, B.SC. (HONS)
Curator, Australian Museum — SYDNEY, AUSTRALIA

DR. THEODOR MEBS
Instructor of Biology — WEISSENHAUS/OSTSEE, GERMANY

DR. GERLOF FOKKO MEES
Curator of Birds, Rijks Museum of Natural History — LEIDEN, THE NETHERLANDS

HERMANN MEINKEN
Director, Fish Identification Institute, V.D.A. — BREMEN, GERMANY

DR. WILHELM MEISE
Chief Curator, Zoological Institute and Museum, University of Hamburg — HAMBURG, GERMANY

DR. JOACHIM MESSTORFF
Field Station of the Federal Fisheries Research Institute — BREMERHAVEN, GERMANY

DR. MARIAN MLYNARSKI
Professor, Polish Academy of Sciences, Institute for Systematic and Experimental Zoology — CRACOW, POLAND

DR. WALBURGA MOELLER
Nature Museum — HAMBURG, GERMANY

DR. H.C. ERNA MOHR
Curator (retired), Zoological State Institute and Museum — HAMBURG, GERMANY

DR. KARL-HEINZ MOLL — WAREN/MÜRITZ, GERMANY

DR. DETLEV MÜLLER-USING
Professor, Institute for Game Management, University of Göttingen — HANNOVERSCH-MÜNDEN, GERMANY

WERNER MÜNSTER
Instructor of Biology — EBERSBACH, GERMANY

DR. JOACHIM MÜNZING
Altona Museum — HAMBURG, GERMANY

DR. WILBERT NEUGEBAUER
Wilhelma Zoo — STUTTGART-BAD CANNSTATT, GERMANY

DR. IAN NEWTON
Senior Scientific Officer, The Nature Conservancy — EDINBURGH, SCOTLAND

DR. JÜRGEN NICOLAI
Max Planck Institute for Behavioral Physiology — SEEWIESEN/OBB., GERMANY

DR. GÜNTHER NIETHAMMER
Professor, Zoological Research Institute and A. Koenig Museum — BONN, GERMANY

DR. BERNHARD NIEVERGELT
Zoological Museum, University of Zurich — ZURICH, SWITZERLAND

DR. C. C. OLROG
Institut Miguel Lillo San Miguel de Tucuman — TUCUMAN, ARGENTINA

ALWIN PEDERSEN
Mammal Research and Arctic Explorer — HOLTE, DENMARK

DR. DIETER STEFAN PETERS
Nature Museum and Senckenberg Research Institute — FRANKFURT A.M., GERMANY

DR. NICOLAUS PETERS
Scientific Councillor and Docent, Institute of Hydrobiology and Fisheries, University of Hamburg — HAMBURG, GERMANY

DR. HANS-GÜNTER PETZOLD
Assistant Director, Zoological Garden — BERLIN, GERMANY

DR. RUDOLF PIECHOCKI Docent, Zoological Institute, University of Halle	HALLE A.D.S., GERMANY
DR. IVO POGLAYEN-NEUWALL Director, Zoological Garden	LOUISVILLE, KENTUCKY, U.S.A.
DR. EGON POPP Zoological Collection of the State of Bavaria	MUNICH, GERMANY
DR. DR. H.C. ADOLF PORTMANN Professor Emeritus, Zoological Institute, University of Basel	BASEL, SWITZERLAND
HANS PSENNER Professor and Director, Alpine Zoo	INNSBRUCK, AUSTRIA
DR. HEINZ-SIBURD RAETHEL Oberveterinärrat	BERLIN, GERMANY
DR. URS H. RAHM Professor, Museum of Natural History	BASEL, SWITZERLAND
DR. WERNER RATHMAYER Biology Institute, University of Konstanz	KONSTANZ, GERMANY
WALTER REINHARD Biologist	BADEN-BADEN, GERMANY
DR. H. H. REINSCH Federal Fisheries Research Institute	BREMERHAVEN, GERMANY
DR. BERNHARD RENSCH Professor Emeritus, Zoological Institute, University of Münster	MÜNSTER, GERMANY
DR. VERNON REYNOLDS Docent, Department of Sociology, University of Bristol	BRISTOL, ENGLAND
DR. RUPERT RIEDL Professor, Department of Zoology, University of North Carolina	CHAPEL HILL, NORTH CAROLINA, U.S.A.
DR. PETER RIETSCHEL Professor (retired), Zoological Institute, University of Frankfurt a.M.	FRANKFURT A.M., GERMANY
DR. SIEGFRIED RIETSCHEL Docent, University of Frankfurt; Curator, Nature Museum and Research Institute Senckenberg	FRANKFURT A.M., GERMANY
HERBERT RINGLEBEN Institute of Ornithology, Heligoland Ornithological Station	WILHELMSHAVEN, GERMANY
DR. K. ROHDE Institute for General Zoology, Ruhr University	BOCHUM, GERMANY
DR. PETER RÖBEN Academic Councillor, Zoological Institute, Heidelberg University	HEIDELBERG, GERMANY
DR. ANTON E. M. DE ROO Royal Museum of Central Africa	TERVUREN, SOUTH AFRICA
DR. HUBERT SAINT GIRONS Research Director, Center for National Scientific Research	BRUNOY (ESSONNE), FRANCE
DR. LUITFRIED VON SALVINI-PLAWEN First Zoological Institute, University of Vienna	VIENNA, AUSTRIA
DR. KURT SANFT Oberstudienrat, Diesterweg-Gymnasium	BERLIN, GERMANY
DR. E. G. FRANZ SAUER Professor, Zoological Research Institute and A. Koenig Museum, University of Bonn	BONN, GERMANY
DR. ELEONORE M. SAUER Zoological Research Institute and A. Koenig Museum, University of Bonn	BONN, GERMANY
DR. ERNST SCHÄFER Curator, State Museum of Lower Saxony	HANNOVER, GERMANY
DR. FRIEDRICH SCHALLER Professor and Chairman, First Zoological Institute, University of Vienna	VIENNA, AUSTRIA
DR. GEORGE B. SCHALLER Serengeti Research Institute, Michael Grzimek Laboratory	SERONERA, TANZANIA
DR. GEORG SCHEER Chief Curator and Director, Zoological Institute, State Museum of Hesse	DARMSTADT, GERMANY
DR. CHRISTOPH SCHERPNER Zoological Garden	FRANKFURT A.M., GERMANY

DR. HERBERT SCHIFTER Bird Collection, Museum of Natural History	VIENNA, AUSTRIA
DR. MARCO SCHNITTER Zoological Museum, Zurich University	ZURICH, SWITZERLAND
DR. KURT SCHUBERT Federal Fisheries Research Institute	HAMBURG, GERMANY
EUGEN SCHUHMACHER Director, Animals Films, I.U.C.N.	MUNICH, GERMANY
DR. THOMAS SCHULTZE-WESTRUM Zoological Institute, University of Munich	MUNICH, GERMANY
DR. ERNST SCHÜT Professor and Director (retired), State Museum of Natural History	STUTTGART, GERMANY
DR. LESTER L. SHORT, JR. Associate Curator, American Museum of Natural History	NEW YORK, NEW YORK, U.S.A.
DR. HELMUT SICK National Museum	RIO DE JANEIRO, BRAZIL
DR. ALEXANDER F. SKUTCH Professor of Ornithology, University of Costa Rica	SAN ISIDRO DEL GENERAL, COSTA RICA
DR. EVERHARD J. SLIJPER Professor, Zoological Laboratory, University of Amsterdam	AMSTERDAM, THE NETHERLANDS
BERTRAM E. SMYTHIES Curator (retired), Division of Forestry Management, Sarawak-Malaysia	ESTEPONA, SPAIN
DR. KENNETH E. STAGER Chief Curator, Los Angeles County Museum of Natural History	LOS ANGELES, CALIFORNIA, U.S.A.
DR. H.C. GEORG H. W. STEIN Professor, Curator of Mammals, Institute of Zoology and Zoological Museum, Humboldt University	BERLIN, GERMANY
DR. JOACHIM STEINBACHER Curator, Nature Museum and Senckenberg Research Institute	FRANKFURT A.M., GERMANY
DR. BERNARD STONEHOUSE Canterbury University	CHRISTCHURCH, NEW ZEALAND
DR. RICHARD ZUR STRASSEN Curator, Nature Museum and Senckenberg Research Institute	FRANKFURT A.M., GERMANY
DR. ADELHEID STUDER-THIERSCH Zoological Garden	BASEL, SWITZERLAND
DR. ERNST SUTTER Museum of Natural History	BASEL, SWITZERLAND
DR. FRITZ TEROFAL Director, Fish Collection, Zoological Collection of the State of Bavaria	MUNICH, GERMANY
DR. G. F. VAN TETS Wildlife Research	CANBERRA, AUSTRALIA
ELLEN THALER-KOTTEK Institute of Zoology, University of Innsbruck	INNSBRUCK, AUSTRIA
DR. ERICH THENIUS Professor and Director, Institute of Paleontology, University of Vienna	VIENNA, AUSTRIA
DR. NIKO TINBERGEN Professor of Animal Behavior, Department of Zoology, Oxford University	OXFORD, ENGLAND
ALEXANDER TSURIKOV Lecturer, University of Munich	MUNICH, GERMANY
DR. WOLFGANG VILLWOCK Zoological Institute and Museum, University of Hamburg	HAMBURG, GERMANY
ZDENEK VOGEL Director, Suchdol Herpetological Station	PRAGUE, CZECHOSLOVAKIA
DIETER VOGT	SCHORNDORF, GERMANY
DR. JIRI VOLF Zoological Garden	PRAGUE, CZECHOSLOVAKIA
OTTO WADEWITZ	LEIPZIG, GERMANY

DR. HELMUT O. WAGNER
Director (retired), Overseas Museum, Bremen — MEXICO CITY, MEXICO

DR. FRITZ WALTHER
Professor, Texas A & M University — COLLEGE STATION, TEXAS, U.S.A.

JOHN WARHAM
Zoology Department, Canterbury University — CHRISTCHURCH, NEW ZEALAND

DR. SHERWOOD L. WASHBURN
University of California at Berkeley — BERKELEY, CALIFORNIA, U.S.A.

EBERHARD WAWRA
First Zoological Institute, University of Vienna — VIENNA, AUSTRIA

DR. INGRID WEIGEL
Zoological Collection of the State of Bavaria — MUNICH, GERMANY

DR. B. WEISCHER
Institute of Nematode Research, Federal Biological Institute — MÜNSTER/WESTFALEN, GERMANY

HERBERT WENDT
Author, Natural History — BADEN-BADEN, GERMANY

DR. HEINZ WERMUTH
Chief Curator, State Nature Museum, Stuttgart — LUDWIGSBURG, GERMANY

DR. WOLFGANG VON WESTERNHAGEN — PREETZ/HOLSTEIN, GERMANY

DR. ALEXANDER WETMORE
United States National Museum, Smithsonian Institution — WASHINGTON, D.C., U.S.A.

DR. DIETRICH E. WILCKE — RÖTTGEN, GERMANY

DR. HELMUT WILKENS
Professor and Director, Institute of Anatomy, School of Veterinary Medicine — HANNOVER, GERMANY

DR. MICHAEL L. WOLFE
Utah State University — UTAH, U.S.A.

HANS EDMUND WOLTERS
Zoological Research Institute and A. Koenig Museum — BONN, GERMANY

DR. ARNFRID WÜNSCHMANN
Research Associate, Zoological Garden — BERLIN, GERMANY

DR. WALTER WÜST
Instructor, Wilhelms Gymnasium — MUNICH, GERMANY

DR. HEINZ WUNDT
Zoological Collection of the State of Bavaria — MUNICH, GERMANY

DR. CLAUS-DIETER ZANDER
Zoological Institute and Museum, University of Hamburg — HAMBURG, GERMANY

DR. DR. FRITZ ZUMPT
Director, Entomology and Parasitology, South African Institute for Medical Research — JOHANNESBURG, SOUTH AFRICA

DR. RICHARD L. ZUSI
Curator of Birds, United States National Museum, Smithsonian Institution — WASHINGTON, D.C., U.S.A.

Volume 7

BIRDS I

Edited by:

BERNHARD GRZIMEK
WILHELM MEISE
GÜNTHER NIETHAMMER
JOACHIM STEINBACHER
ERICH THENIUS

ENGLISH EDITION

GENERAL EDITOR:
George M. Narita

SCIENTIFIC EDITOR:
Erich Klinghammer

TRANSLATORS:
E. Otto Höhn
Erich Klinghammer

ASSISTANT EDITORS:
Jeanine L. Grau
Frances A. Wilke

PRODUCTION DIRECTOR:
James V. Leone

EDITORIAL ASSISTANTS:
Detlef K. Onderka
Karen Boikess

ART DIRECTOR:
Lorraine K. Hohman

INDEX:
Suzanne C. Klinghammer

CONTENTS

For a more complete listing
of animal names, see systematic classification or the index.

1 The Birds

The class of birds, by W. Meise

Next to mammals, birds are the class of animals which have particularly attracted human interest and sympathy. Their ability to escape and to fly enables them to carry on their activities before all beholders and to show themselves more openly than other animals of similar size. Most people, therefore, find it easier to observe birds than the equally common but more hidden and secretive small mammals.

Distinguishing characteristics, measurements, and weights

BIRDS (Class Aves) are feathered, warm-blooded vertebrates normally capable of flight. The total length (L) is measured from the apex of the beak to the end of the central tailfeathers with the bird lying on its back; it varies from 6 cm in hummingbirds to 235 cm in the pearl peacock and up to 600 cm in phoenix fowl, a Japanese race of the domestic fowl. The L from the tip of the beak to the tip of the foot is about 300 cm in the ostrich or 400 cm in the extinct giant moa. The wing length from the bend of the closed wing to the tip of the longest primary is 2.5 cm in the bee hummingbird, and reaches 85 cm in the condor. The wingspread, measured from one wing tip across the back to the other wing tip, is 7.5 cm in the bee hummingbird to 340 cm in the wandering albatross, or 500 cm in the extinct Nevada giant condor. The weight at the time of hatching from the egg is 0.19 g in the bee hummingbird to 1000 g in the ostrich *(Struthio camelus)*, or 6500 g in the extinct ostriches of Madagascar. The weight of fully grown birds ranges from 1.6 g in hummingbirds to 144,000 g in the ostrich or 450,000 g in the extinct ostriches of Madagascar. For the structure, see the color plates on pages 35 and 36.

Skin and skin structures

There are few cutaneous glands, but there is generally a preen gland. The skin and its derivatives are light and, as with reptiles, there are scales on the legs and a horny sheath for the beak which is occasionally divided into segments. Both the scales and the feathers correspond to reptilian scales and are formed from the epidermis and dermis.

Skeleton and muscles

The bones contain airspaces. The anterior segment of the skull is shortened, the temporal fossa is reduced in size, the wall between the orbits is thinned out, and the orbits are enlarged. The brain case is large, and the skull bones are generally fused without sutures in adulthood. The attachment of the upper mandible, with the upper beak, to the brain case is mobile (only in birds), since the nasal bones behind the ramphotheca are pliable. The lower mandible is so connected through the mobile quadrate to the skull and upper mandible that the latter is raised when the former is lowered (double mandibular joint). The skull structure, including the palate, is reptile-like: there is only one occipital condyle (mammals have two) and the lower mandible consists of several bones (in mammals there is only one on each side). The neck is very mobile, and there are eleven to twenty-five cervical vertebrae (mammals have seven) without epiphyses; the lower ones often have cervical ribs. The rump is rigid, and there are three to ten thoracic vertebrae, which are sometimes partially fused. There is no mobile lumbar region, but ten to twenty-two sacral vertebrae are fused with one another and the pelvis. There are eleven to thirteen caudal vertebrae with the last five to six being fused into the pygostyle; the latter is absent in ratite birds and a few others. The clavicles are generally fused into a furcula. The ischia is directed toward the rear as in the dinosaur group Ornithischia (Vol. VI). The fingers are reduced to two or three, and the toes to two or four. The first toe is opposable. There are two, three, four, or five metatarsals in the first to fourth toes. The tarsal joint between the two rows of tarsals is like that in reptiles, forming a tibio-tarsus and a tarso-metatarsus. There is a muscle on the inner side of the thigh; the ambiens (otherwise only present in reptiles) is a weak inward rotator of the thigh, and is often reduced or absent. The sternum in flying birds is like a keel and the major and minor pectoral muscles are the largest of the birds' muscles.

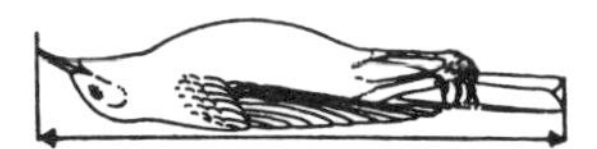

Fig. 1-1. The total length (L) is measured on the dead, stretched-out bird.

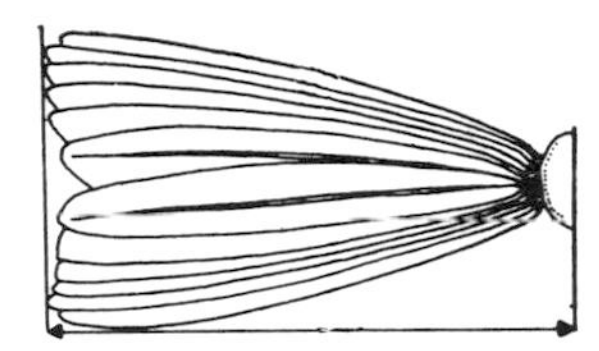

...measurement of the tail length from the point of exit through the skin of the central tail feathers to the tips of the tailfeathers (generally the central ones).

...measurement of the wing length (WL).

Nutrition, digestion, and elimination

Birds require much food, and digestion is rapid. The liver, particularly its right lobe, is large, and the cloaca is tripartite. The urinary bladder exists only in the American rhea, *Rhea americana* (not derived from the ureter); the urine flows through the two ureters into the mid-segment of the cloaca. The urodaeum is there mixed with feces (an exception is the ostrich). Nitrogen is excreted as uric acid, as in reptiles. There is additional salt excretion in many marine birds through the nasal glands (see tube-nosed swimmers, chap. 6). Indigestible food particles are regurgitated as pellets and contain insect remains, fish bones, feathers, hairs, and bones.

Respiration and circulation

The respiratory system has continuous air capillaries without an alveolar formation. A system of air sacs is connected to the bronchi. There is a syrinx (found only in birds). The heart is relatively large and is completely divided (see mammals, Vol. X). The pulse is rapid, and the red blood cells (erythrocytes) have nuclei (as in all vertebrates other than mammals).

The senses and the brain

Vision is the most highly developed sense. The eyes are large, almost always having a pecten, and always with a scleral ring. There are auditory and vestibular senses. The olfactory sense is rudimentary (exceptions are kiwis, Apterygiformes, and turkey vultures, *Cathartes aura*). The sense of taste is poorly developed. The brain is large (reptile-like), with a large cerebellum.

Reproduction

The testes are below the anterior edge of the kidneys and are smooth and much reduced (to one-thousandth of the maximal weight) outside the breeding season. The right ovary and oviduct are vestigial. The egg, as in reptiles, has a yolk. The egg white and shell membrane, as in many reptiles, has a calcareous shell. Only one egg develops in the oviduct at one time; it is almost always laid before another egg follicle ruptures. The embryo has many reptilian features (long tail, five toe limbs). The inner embryonic membrane, the amnion, and the allantois are egg membranes, as in reptiles (see Vol. VI). The embryo generally has an egg tooth.

Distribution and classification

Birds are distributed over all continents and oceans. The number of species which have become extinct since the Jurrassic (180 million years ago) is not known; about eight hundred have been described. Today 8700 species, in round figures, survive, which is twenty percent of all vertebrates or one to two percent of all animal species. There are two subclasses: 1. ANCESTRAL BIRDS (Archaeornithes), and 2. MODERN BIRDS (Neornithes).

Birds are lightly built flying machines

Birds must primarily be considered flyers. A bird is recognized by its feathers; no other animals have such structures. Another avian characteristic is the great difference between the wings and the legs. Birds are easily distinguished from other flying vertebrates. While the naked young of bats (flying mammals) have five toes, birds have only four, three, or two. They are clearly distinguished from flying saurians and other reptiles by their plumage. Apart from that, all birds can be recognized at a glance, even in the absence of the tail and plumage; they stand on two legs, and the long neck and beak form an unmistakeable appearance.

Air forces acting on the avian wing

All characteristics of birds are related to their ancestors' transformation into flying animals. Flight is an avian characteristic which man has only been able to imitate after much exertion with the help of machines. How does a bird fly? Even with extended wings, a bird will immediately drop in still air. The air current which results then passes along the wing. It finds different conditions on the upper wing surface than below the wing, for the convex upper surface causes a more rapid motion than if the wing were flat. As in a column of vehicles, the distances separating individual vehicles increase as the speed of the column increases, so the density of air particles is decreased as their speed of motion is increased. Thus, above the wing a zone of air of reduced density, a suction, is produced which lifts the wing. At the point of greatest convexity, a little behind the anterior edge of the wing, this suction is maximal and decreases towards the rear.

On the underside of the wing, the reverse takes place. The air stream is slowed through the upward convexity; instead of a suction, there is a pressure which also raises the wing. This pressure is, however, only about one third of the suction force acting on the upper surface of a wing with a normal cross section. The force resulting from suction and pressure acts at approximately right angles to the direction of motion. Its effect can be more accurately estimated if one conceives of it as acting at the pressure point of the wing. In a bird which is descending gradually with its head forward, this force acts towards the front and upwards. One can separate it into a vertical upward and a horizontal forward component (see fig. 1-2). Thus the bird glides forward relatively slowly on a slanted path. This forward movement through the air varies considerably in different groups. A gull naturally glides farther than a young greenfinch falling out of the nest. The best gliders among the birds perform about as well as the best glider planes. They glide a horizontal distance of about twenty times their initial height. Their "gliding score" is thus 1:20. In the young greenfinch, however, the ratio is only about 1:1. Thus, with its wings extended, a bird can change a downward fall into a downward and forward glide. It can even, in spite of a continuous rate of fall, rise or stay at the same height (soaring) according to its wing shape and wing load, if there are adequate upward currents in the air. In this case, gliding becomes soaring, as shown by gulls in the updraft following a moving ship or by hawks soaring in the thermal updrafts over forest clearings. In thunderclouds the updraft may reach a rate of six meters per second, so that a bird with a rate of fall of one meter per second gains five meters per second. It avoids being blown upwards by more or less closing its wings; the load per unit area and the rate of fall are thereby considerably increased.

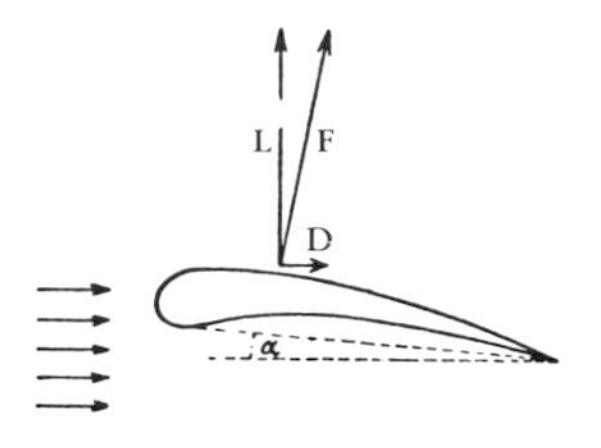

Fig. 1-2. When a bird wing meets an air current with its leading edge, a relatively small angle of attack (α) creates a sufficient force (F) consisting of a lift (L) and a drag (D).

A bird body is streamlined; air passes around it without forming eddies.

Birds smaller than jackdaws do not soar. This has not been satisfactorily explained so far. If small birds get into strong updrafts, they are, of course, blown upwards, but they do not seek out updrafts. Good soarers, however, know how to move from one updraft to another. They utilize an updraft to save energy, rising in circles and then gliding to the next one, possibly on the other side of a valley. Soaring birds congregate in such updrafts or in thermal currents. In their more extended travels, they avoid areas deficient in updrafts, as is seen, for example, in the case of migrating white storks which fly around the Mediterranean,

Gliding in wind strata

Another group of flyers can move best in strong winds which blow over the great oceans; their name, procellariids (in German, it is Sturmvögel, which means storm birds), is well deserved. It is difficult for the human observer to understand that an albatross, gliding down at an angle with the wind from behind from a height of about twenty meters, can only regain its height by turning into the wind. It then rises without flapping its wings. Once having arrived at the top, an albatross immediately begins with a new turn in its flight spiral and can, in so doing, proceed in any direction it chooses. This astonishing ascent is probably due to the headwind. The albatross rises into an ever-increasing air flow, and this current passes over the long, narrow wings to create a forward force which has upward and forward components. Such "dynamic soaring" is performed only by the larger members of the order Tubinares. Some of them thus circle the earth over the southern oceans.

Flying by beating the wings

These fabulous gliding and soaring abilities are used far less by flying birds than is flight with wing beats. The effect of wing beats is best seen when one examines a hummingbird, filmed in slow motion, "hanging" in still air in front of a flower. In order to produce a vertical force sufficient to hold a bird weighing two to four grams, thus neutralizing the effect of gravity on this hanging body, twenty-five to eighty wing movements per second are required. As in gliding, so in flapping flight the current of motion passes over and under the wing; the force is exerted at thirty wing beats per second for one sixtieth of a second on the upper side of the down-beating wing, and in the next sixtieth second on the under surface of the meanwhile completely reversed wing. In other words, the hummingbird hangs almost vertically in front of the blossom and, in the plane taken through its shoulders, each wing moves with the upper surface up well in front of the chest, and immediately thereafter with the lower surface completely reversed. We can understand this hovering of the hummingbird better if we move the arm with the back of the hand up forward and bring it back with the palm up.

Fig. 1-3. Like a helicopter, a hummingbird hangs in the air with its body almost immobile while its wings whirr back and forth twenty-five to eighty times per second, thus giving the bird the necessary lift. Thus hummingbirds can take their food from flowers.

The hovering hummingbird can well be compared to a helicopter which remains in place in the air. Only its rotors turn continuously, while the two wings are, as it were, two opposite propellers which reverse after a mere half turn. The peculiarity of hummingbirds' wings lies in the fact that they are almost like boards, so that during both the downbeat and the upbeat of the wing, upward-acting forces arise. Correspondingly, the muscle which raises the wing, the supracoracoideus, weighs from a third to over half as much as the pectoralis in hummingbirds, while in most other birds it weighs only a tenth to a twentieth as much, although it weighs one eighth as much in pigeons and doves, and between one quarter and one third as much in gallinaceous birds.

Flight with forward thrust

Hovering is much rarer in the bird world than is normal flapping flight. Besides sufficient uplift, a forward drive is also important here. There is a forward push which is characteristic of bird flight, much as is the case with propeller aircraft. If there is a wind, hoverers also need a forward push. The axis on their pair of wings must then be so directed that not only a loss of height but also a lateral drift is compensated. The horizontally-acting force must cancel the wind. Thus, if a hummingbird wants to leave a blossom, it can readily fly backwards.

Horizontal flight

In distance flight, the wing moves in space not only to and fro as in hovering, but in a continuous wave motion. If one affixes a light to each wing tip and one to the body, one may see how the wings precede the body in the downbeat but lag behind it on the upbeat. The tip of the wing moves fastest of all; thus, in the downbeat, the wing is raised so much against the current from the front that the force is as large as possible and is exerted forward. The angle of attack (between the flight direction and the wing chord) is generally between five and thirty degrees. Long wings, because of the greater speed of the wing tips, result in more forward propulsion than do short wings; thus, for example, falcons fly faster than sparrowhawks.

How artificial birds fly

The behavioral physiologist Erich von Holst has constructed artificial birds, and with their help has examined bird flight in detail. "The wing tip," he writes, "moves, while the body glides forward in a straight line, in an up and down wave motion, and the wing is not merely raised and lowered like a stiff plate but adapts its position at any time to this wave track in the finest detail; thus in the downbeat its front edge slants down somewhat while it is raised during the upbeat."

Everything depends on the accuracy of these components. When we watch a crow or gull flying close by, we hardly notice any of this; we merely see the up and down movement, but not the rotation of the wing along its long axis. Only slow motion cinematography makes these components of the motion quite clear. It is therefore not correct to say: "The bird holds itself in the air because, in the downbeat, it compresses air beneath itself and rests on this air cushion while on the upbeat it makes its wings smaller and brings them back up as soon as possible." On the contrary: "As the wing on the down- and upbeat accurately follows a wave track and glides through the surrounding liquid (air is considered a dilute liquid), it can create a force in the air on the upbeat which carries its weight in the same way as on the downbeat."

"If one watches the flight of a gull or stork carefully, one will notice that the bird's body glides forward in a straight line as if it were fixed in the sky by an invisible thread," as von Holst puts it. "It does not rise during the wing downbeat nor does it fall during the wing upbeat. The same is true for my artificial birds." The elegance of the flight motion and the aesthetic pleasure one derives from watching it is a direct consequence of this accurately adapted wing motion.

With some exaggeration, one can say that in flapping flight, many large birds glide or soar with little motion of the arm (secondary) portion of the wing, while the hand (primary) portion with its stronger motion is responsible for forward motion. So as to not lose altitude, most birds bend the hand portion of the wing down during the upbeat,

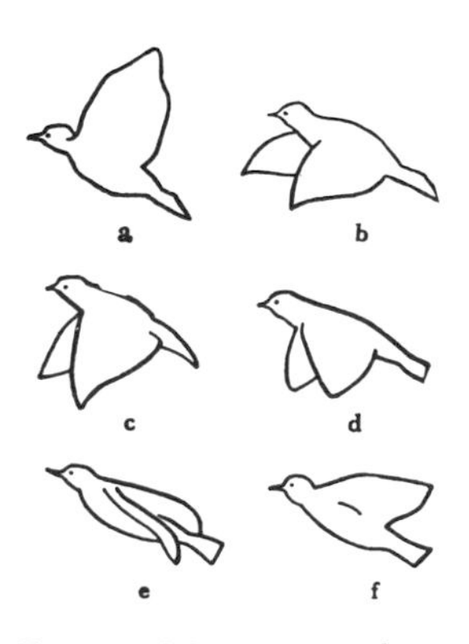

Fig. 1-4. Movement of a canary in flight: a-c: downbeat. d-f: upbeat.

to the rear or in both directions toward the arm portion, and this, as it were, "neutralizes" it, or the wing tips open like a venetian blind by rotation of the individual primaries. At this time at least, the arm portion of the wing still produces an uplift. This ceases when, on the upbeat, the whole wing is drawn towards the body, as is the case with many small birds. The wing beat is then, however, so fast that a drop cannot be noticed.

Undulating flight

Besides normal flapping flight and hovering (flying in place), an undulating flight is also found in birds. A flurry of wing beats then raises the flight path which drops again during the following pause. The amplitudes of these waves are very deep in small birds which show a low load per unit wing load, such as wagtails.

Digitation of the wing tips

Many species of birds, even the very smallest which do not have particularly pointed wings, spread out the primaries into finger-like projections during flight. This avoids the danger of the formation of eddies which are particularly likely to arise at the wing tip, or of going into a spin. Air passes through the slots between the primaries and the slot between the wing and the alula at high speed, and thus brings energy into the boundary layer on the upper wing surface. This simple method has been imitated in the flaps of planes; it increases lift and forward push without causing a dangerous flow over the upper wing surface.

Directional changes in flight

It was once believed that the tails of birds played an important role in changes of direction during flight. However, we now know that here, too, the action of the wings is much more important. Turning to the right or left can be brought about by speeding up the wing beats on the outer side of the curve. As the uplift is simultaneously increased, the outer side is raised so that the feet, as in a cyclist in a curve, point outwards and downwards. Sometimes the turn is enhanced as well by the slower action of the inner wing, a partial spreading of the tail on the inner side, and a shifting of the tail and feet.

While in undulating flight, the bird's rising and falling are due to interruptions of the wing beats; this is usually achieved through shifts of the plane of the wing beats to the front or the rear. The force of the air then acts either more on the front or on the rear, the center of gravity swings under the point of attack, and the body is tilted toward the front and up or down; thus it rises or falls. Here too, tail movements upwards on the rise or downwards on the drop may also be involved. In many aerial display flights, birds make a complete loop. Some can even rotate about their long axis, but birds are generally careful to prevent rolling by regulating their flight by the degree of opening of the two wings.

Take-off, landing, and landing on water

Every pilot knows that take off and landing offer particular difficulties. At take off, the bird must reach a speed which provides a sufficient upward suction as soon as possible so as to prevent falling to the ground. In landing on land or water, the speed which always increases during a descent must be reduced. A vertically descending bird, such as the cock of certain bustard species during his courtship flights, spreads the wings and tail only when just above the ground. Air resistance then acts as a brake, the rest of the impact being taken up by the feet. In other modes of braking, the wings, particularly the secondaries, are often placed vertically to the flight direction, as in hovering when it is necessary to offset forward thrust, but even more often during landing. When the wing is at ninety degrees to the flight direction, it causes so much resistance that the propulsion which has brought the bird almost to its landing place is reduced and finally stops. A few forward beats, somewhat comparable to oar strokes in water, may further aid the braking action. In such cases, a fall would be inevitable if the bird did not estimate its flight path so that it would land on its feet at just that point. If, however, there is still some forward momentum or a strong wind from behind, the bird tips over forward on landing, though it may be able to compensate for this forward motion by running a few steps, swimming a short distance, or grasping its perch.

When the young purple heron (*Ardea purpurea*, see Chap. 8) hatches, it looks almost like a primeval reptile.

Flight generally requires considerable energy expenditure. For most species capable of flight, flying is vital; in a few groups, however, it plays only a minor role. Thus the flight to a branch in the evening protects most gallinaceous birds from nocturnal ground predators. During the day, escape by flight is often the last chance of avoiding a predator. Flights to feeding areas and in search of food over extensive areas give birds the opportunity to nourish themselves adequately. More than five percent of avian species regularly catch their food on the wing, even when, as in many flycatchers, they leave their perches for only very short pursuit flights. Wings are used most extensively when parents with a nest full of young must supply them with insects for weeks. It has been estimated that a common swift (*Apus apus apus*) with two young flies one thousand kilometers per day, but a great tit (*Paus major*) with ten young flies only one hundred kilometers. Both repeatedly fly away from the vicinity of the young, whose nest is in one place and quite endangered. Many flights, however, are associated with reproduction even when there are as yet no young, such as during courtship. This is evident, among other things, in the wing differences between male and female lapwings (*Vanellus vanellus*), and we also see it in many magnificent aerial displays and courtship flights.

Flying capabilities

The achievements attained in flight vary considerably in different species and groups. Common swifts fly for twelve to fourteen hours per day at speeds from about 65 kilometers per hour in their normal hunting flight to about 145 kilometers per hour in very fast flight. Speeds range between about 40 (in redstarts) and 200 kilometers per hour (in falcons). With a wind from behind or in dives (swoops),

swift-flying species may reach speeds of up to three hundred kilometers per hour. In their daily lives, birds generally fly steeply upwards; sparrows and some other species even do so vertically. They generally do not rise above one hundred meters; many, however, can climb to a height of four thousand meters, and, for geese, 8800 meters have been calculated.

Residents, wanderers, and migrants

The ability to fly enables many birds to carry out the greatest migrations known in the animal world. Without these regularly repeated migrations, a considerable number of species would not have been able to inhabit areas which lack food during certain seasons. The more we penetrate into such unfavorable parts of the world, the more species we encounter which, as migrants or partial migrants (species in which only part of the population migrates), temporarily leave the area. The shortest distances are involved when birds merely move from the upper parts of rivers which are subject to freezing to the lower parts which remain open, as is the case with kingfishers, or when, as in many other birds, they move from high mountains into valleys. Wanderers, like many tits and chickadees, remain within a relatively small area. The wanderers may vary between strict residents, like nuthatches, to true migrants which leave the breeding area completely during the unfavorable season. Although the proportion of migrants increases the more unfavorable the climate and feeding possibilities become, I know of no areas which are totally deserted by all birds even in the severest season.

Where birds migrate to

The wintering area of a migrant can, among birds that migrate over long distances, be completely separate from the breeding area. In the intervening regions, such birds are known only as birds of passage; thus the spotted redshark *(Tringa erythropus),* which migrates from its breeding grounds in northern Eurasia through Central Europe, winters only in southern Europe or even further south. Birds of passage are more likely to winter occasionally in areas in between than are species which breed there but normally migrate away from there, since they are already some distance from their cold home countries. More striking is the great number of birds of passage, for example among shorebirds, which spend the summer south of their breeding range and thus remain there through the breeding season without breeding. Such non-breeding birds are either not yet sexually mature or they encounter unfavorable conditions to complete their migration, or they may simply skip a breeding season.

The archaeopteryx (*Archaeopteryx lithographica,* see Fig. 1-27) of the Jurassic Period had a long-feather-bearing tail. The tail-feathers of recent birds, however, arise from a short tail, the pygostyle. The wings of archaeopteryx had claw-bearing fingers suitable for climbing.

Of the 152 migrant species in Germany, according to Rudolf Berndt, three winter in southern Asia, fifty-two in the Mediterranean area, eleven in northern Africa from Senegal to Abyssinia, twenty-four in South Africa as far south as Natal, and thirty-one down to the Cape Province; seven more species cannot be so assigned. As Berndt further states, migrants from north and east Asia winter mainly in the Indo-Malayan area and partly even in the Australo-Polynesian area. Certain

89499
ILLINOIS PRAIRIE DISTRICT LIBRARY

East Asiatic breeders migrate to East Africa. "Of North American migrants," writes Berndt, "only a few reach Tierra del Fuego, but, nevertheless, thirty species reach Chile. On the other hand, species breeding in southern South America migrate north into the tropical zone, but not beyond twenty-three degrees north latitude. About twenty-two species migrate from the Antarctic and adjacent areas as far as the southwest coast of Australia. Certain shore birds, terns, skuas, and tube-nosed swimmers migrate from the Arctic to the Antarctic and a few sea birds of the extreme south, such as the Wilson's storm petrel *(Oceanites oceanicus)* and the greater and sooty shearwaters *(Puffinus gravis* and *Puffinus griseus),* migrate to the northernmost seas. In Africa about twenty species leave the Cape Province for the north; some may go as far north as the northern Congo. In subtropical Australia, there are over twenty species which migrate as far as Celebes, Java, and, in small numbers, Borneo and Sumatra."

There are also migrations within the tropics which depend on the changing rainy and dry seasons. The subspecies of certain bird species may winter somewhat separately, just as they breed separately, but, for example, up to five subspecies of the yellow wagtail *(Motacilla flava)* may winter in the same area.

We particularly notice those birds of passage which leave their breeding grounds early and then remain in an intermediate area before they finally move on to their wintering grounds. The intermediate area may lie in the direction of the final goal; thus, for example, East German and Baltic starlings *(Sturnus vulgairis)* rest in Hamburg, Germany, on their way to Britain. They arrive there in the months of June and July and are unpopular among the orchard owners as "cherry thieves." Many species, however, choose an intermediate area which lies off the direct course to their wintering grounds. Thus British shelducks *(Tadorna tadorna)* wander into the Helgoland Bay before moving on to Belgium and northern France.

Direction and routes of migration

The westward migration of starlings from East Germany toward England is somewhat unusual, while Central European migrants move principally southwest during their fall migration and northeast during their spring migration. Rarely do they move directly to the south in the fall; occasionally they move to the southeast. Even an eastern or east-southeastern direction of the fall migration may occur, as is the case in the red-breasted flycatcher *(Ficedula parva parva)* which winters in India. All these directions are more or less rigidly followed, though least so in a number of sea birds which follow shore lines to avoid overland flights. Thus, terns in general follow shore lines; this enables them to fish in open water, and so on their way from Europe to South Africa, they take a detour to the west via Portugal and Senegal. The barn swallow *(Hirundo rustica),* however, as a land bird not dependent on water, follows a true southern course from Europe to South Africa. However, land birds may also follow coast lines, like the Kurische Nehrung, a narrow strip of land bounded on one side by the Baltic Sea and on the other by an inland waterway which it forms, off East Prussia, which gathers all migrants that want to avoid crossing the Baltic.

Apart from mass migration routes, landmarks are also utilized by migrants. One of these is the small island of Helgoland off the German coast; many other islands which have become staging areas for the rarest birds of passage are therefore famous in connection with birds of passage. Thus important bird observatories have been built there. Direct observation and the results of banding have shown that most members of a species move in much the same direction between their breeding and wintering areas. They cross intermediate areas on a wide front. In the case of certain European migrants which wander around the Mediterranean in a southwestward or southeastward direction, the migratory pathways approach one another in a somewhat funnel-shaped manner in the south. More rarely, a narrow front is followed right from the start. The best example of this is the European crane *(Grus grus grus).* Wide and narrow front migration routes may be followed by the same species; thus the white stork *(Ciconia ciconia ciconia)* departs on a wide front from its breeding range and later follows the narrow Nile Valley.

The return flight to the breeding area may follow the same pathways or quite different routes may be taken. The returning white stork simply follows the same route in the opposite direction, although beyond the Bosporus its routes no longer form a funnel but rather a fan. On the other hand, the red-backed shrike *(Lanius collurio collurio),* which follows the Nile southwards, migrates northwards over Arabia, thus following quite a different route in the spring. Its routes form a counterclockwise loop. A clockwise loop is followed by the European black-throated divers (Arctic loon, *Gavia arctica*) which only visit the lower Elbe and adjacent areas to the east in the spring, but in the fall move over eastern Russia and western Siberia to the Black Sea.

Daily and total distances covered during migration

As is true of their daily flights, many birds cover extraordinary distances during their seasonal migrations. The total annual migratory route of the common swift is about 12,000 kilometers, while that of the white stork is 20,000 and that of the Arctic tern *(Sterna paradisaea),* which probably has the longest migratory path, is 35,000–40,000 kilometers. When such great distances are involved, flight without stopovers is, of course, impossible. Probably the longest distances without stopovers are covered by migratory American golden plovers *(Pluvialis dominica).* One subspecies migrates from Alaska across the Pacific to Hawaii, and another from Nova Scotia across the sea to Argentina in about forty-eight hours. Since plovers cannot settle on the sea, they must cover 3300 to 4000 kilometers without rest. In general, however, bird migrations involve daily flights of 200 to 800 kilometers; moreover, they do not move every day of the migratory season. Resting periods are inserted, more in the fall than in spring. The birds may rest for merely part of a day, but sometimes they do so for several weeks. Bird watchers thus have opportunities to observe rarities on passage over a more or less extended period of time. The average daily flight distance is much reduced by rest periods; on the other hand, it must be increased when headwinds carry the bird back, so that more wing beats are required to cover the same distance. Migratory speed corresponds in general to ordinary flight speeds. Among the slow migrants are swallows, crows, and finches, which reach forty-four to fifty-two kilometers per hour; ducks and shore birds are fast migrants, reaching ninety or more kilometers per hour.

The effect of weather on migration

Moderate tail winds are utilized by migrants and so influence the actual speed of migration, although a few small birds seem to favor a head wind. In strong winds and during fog, migration ceases; when sudden fogs or marked cold occur, there are, particularly during the spring migration, great catastrophes to which many birds often fall victim. Sidewinds and storms carry many birds into countries far off their normal route, or even into other continents. Thus 500–1000 lapwings *(Vanellus vanellus)*, which left the British Isles with an air speed of seventy kilometers per hour and a tail wind of ninety kilometers per hour, landed about 3500 kilometers away in Newfoundland after twenty-four hours. Many a "vagrant" has, however, left its route not as result of such wind drift, but rather because it has made an error of navigation or has travelled for a while as a passenger on a ship or train and then flown on, or has been chased on by people.

Birds which depart late in the year and return early are particularly endangered. They are threatened by spells of bad weather and often have no opportunity to find food under a snow cover, or they may be unable to rest on freezing waters. They react in a more "nervous" manner to signs of favorable or unfavorable weather. Thus they may let themselves be tempted into a too-early return migration by a warm front, or take off at the onset of a cold front in the fall or spring in a "winter exodus" to the south. The starlings, skylarks *(Alauda arvensis)*, and wild geese belong to these "weather birds." Students of bird migration designate these as migrants influenced by external factors, in contrast to so-called "instinct birds," those migrants, such as common swifts, cuckoos *(Cuculus canorus)*, and white storks, which migrate according to the calendar date, depending on internal cues.

When birds migrate and how long they stay in their breeding areas

The migratory season is by no means always determined by the available food supply at the onset of the migration; in extreme migrants, it is determined by the day length. Many species leave as many weeks after the summer solstice as they arrived before it. Thus the common swift spends only a very short time in Europe; nevertheless, the region where it hatches must be designated as its home. Many breeding populations spend only three months in Central Europe, while others may, as a rule, remain eight to nine months. An intermediate position is taken by the white stork, among others. The normal bird migrations in Central Europe, excluding intermediary migrations, last for at least eight months of the year. From January until May or June, migrants arrive in this area; they leave from the end of July to November. The members of a particular species may often arrive at a given locality within a few days of one another, particularly as a result of a "pile up" if they have been held up by bad weather on their homeward flight. More often, however, the period of arrival of a species extends over several weeks, and that of departure is even longer. Thus the barn swallow takes three months, from early March to early June, to settle its European breeding range. If one draws a line through localities for which the arrival date is the same, one finds that this species advances at a rate of forty kilometers per day. When southern Europe is being occupied, the birds wintering in the Cape of Good Hope region will not depart for another two months. Many must travel to Britain, and so they fly over swallows which are already breeding in southern Europe.

Banded birds

The fact that so much is known today about the dates during which the various species and populations occupy their breeding habitats, as well as winter and passage areas, is due to banding. Though this procedure has provided extensive data on certain species, continuous checks are needed since migration times, routes, and breeding ranges may change. Banding does not merely aid in the understanding of bird migration. Birds provided with a colored band are readily recognized in the wild and observed as individuals, and so banding contributes to the solution of many other ornithological problems as well.

Migration by day and night

Many species are "day migrants," such as the white stork, pigeons and finches. They migrate mainly during the first three to four hours after sunrise, more rarely in the late afternoon and evening. Those that soar on thermal updrafts can take advantage of these only late in the day. The European cuckoo, bitterns, and many small passerines only start after sunset for a flight of several hours. They are noctural migrants which do not become active again until about two A.M. Herons, ducks, cranes, larks, thrushes, starlings, and many other groups migrate both day and night. Some can remain aloft day and night continuously, as can be shown by the daily travel records of birds which cross the open seas.

Sociability during migration

The sequence of individual migrants which do not form flocks, such as birds of prey passing through the Kurische Nehrung, the narrow strip of land off East Prussia, is a memorable experience for an observer. Much more impressive, however, are days of mass migration when hundreds of thousands of birds continuously pass through a locality, so that accurate counting is impossible and numbers can only be estimated. Such mass movements have impressed man since ancient times. The swarms often consist of only one species, as is the case with starlings and cranes, although in other cases, several species migrate and also rest together. Sometimes members of a species are separated according to age and sex; then males arrive a few weeks before the females on the breeding grounds and occupy territories. In the fall, adult shorebirds generally migrate before the young of that year; in other groups the reverse is the case.

The altitude during migration

Migrating swallows, when flying into a strong wind, fly as low as possible, seeking protection behind ground elevations. When crossing mountain ridges or mountain passes, small birds also fly low above the ground. But this may change even within a species; radar observations have frequently shown migrants beyond visual range at heights of one thousand to fifteen hundred meters. Without binoculars, one can just detect buteos but not small birds at this altitude. However, with binoculars, night observations of small birds passing in front of the moon are possible. Many seabirds, such as auks and penguins, migrate by swimming rather than flying. Very rarely are migratory movements made on the ground or by hopping from branch to branch.

Invasions

Occasionally birds move into areas which they do not normally frequent, or they occur in certain localities in unusually large numbers, or they migrate even from areas where they are normally residents. Such "invasions" into Central Europe were undertaken until 1913 by the central Asiatic Pallas' sand grouse *(Syrrhaptes paradoxus).* The fact that since then they have not revisted Central Europe suggests that there has been no population surplus and no serious food shortage in their Asiatic home. The Bohemian waxwing *(Bombycilla garrulus garrulus),* however, reaches Central Europe annually in small numbers and migrates homeward in spring. Slender-billed nutcrackers *(Nucifraga caryocatactes macrorhynchos)* do not visit Central Europe from Siberia quite so often, and they generally do not return. Even a few Central European birds in great numbers often visit areas that are particularly rich in food, as is the case with long-eared owls *(Asio otus)* and short-eared owls *(Asio flammeus flammeus),* as well as certain gulls which may visit cities. Crossbills congregate to breed in localities with good fir and pine cone crops; among Central European birds they are the least tied to specific breeding areas and are therefore called "nomads" or "gypsies."

The origin of migration

Since zones of different temperatures surround the earth, the bases for regular bird migrations have probably existed since the development of birds, during the earth's middle age. It is known with certainty that they have existed since the Miocene (middle Tertiary, about

Fig. 1-5. Rump contour feather of a pheasant *(Phasianus colchicus)* (left) and fluff feather (plumula) of an adult bird (right).

twenty-five million years ago), for at that period the earth became markedly cooler. However, for the present state of bird migration, the glacial and post-glacial periods have probably played a major role. Many species which had to leave certain areas during the glacial period returned during the post-glacial period. The migration routes of certain species reflect the routes along which areas formerly covered by ice were recolonized. Thus Alaskan yellow wagtails fly to Asia in the fall; on the other hand, sandhill cranes *(Grus canadensis canadensis)*, which nest in east Siberia, migrate to North America. How long will this continue? When will these species discover the more easily-reached areas on the continent in which they nest? Even now, certain changes in bird migration can be shown to have taken place. Favorable winter conditions in cities have prevented many black-headed gulls *(Larus ridibundus ridibundus)* from leaving Central Europe during the winter as they formerly did. The serin finch *(Serinus canaria serinus)* and the collared dove *(Streptopelia decaocto decaocto)* are examples of the reverse, where migratory behavior has appeared anew.

What controls migration?
Migratory readiness

What influences the extraordinary performances of migrants? First, a migratory readiness in the widest sense must take place; evidently it often depends on day length, as has been shown experimentally. Increasing day length acts, via the hypothalamus, on the hypophysis (pituitary gland), which in turn stimulates other hormone-producing glands, e.g. the thyroid. The beginning of the migration period, after the conclusion of the moult, is almost always linked with a change in metabolism, which in a few days results in the formation of an appreciable fat deposit. When all these conditions have been fulfilled, the readiness to migrate in a narrower sense, the "migratory mood," can be induced by weather influences or other stimuli. Caged birds also show a migratory mood; thus European robins *(Erithacus rubecula rubecula)* show wing fluttering at night so that they must be protected by a soft roof lining from damaging themselves. "Instinctive migrants" need only a short period of migratory readiness before they get into the migratory mood, as they begin their movements within a relatively short time. On the other hand, many "weather birds" wait for months for the releaser, e.g. a few cold days, and they must remain in migratory readiness throughout this time. The basis for the formation of migratory readiness is always herditary, in all cases. Thus certain European warblers (Sylviidae) reared in isolation from contact with others of their species still show migratory restlessness; the young European cuckoo migrates to the southeast and Africa long before its parents do. Furthermore, it does so at night.

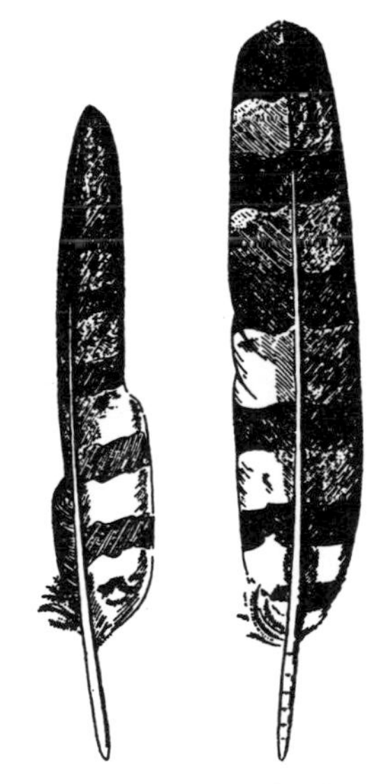

Fig. 1-6. Primary feather (left) and tail feather (right) of the Euasian sparrowhawk *(Accipiter nisus nisus)*

How do migrants find their way?

Hence, the ability to find their way to the wintering area must be inborn. The migrants' abilities to find their way have often been ascribed to a special sense. Today we assume that if a southeasterly or southwesterly direction of migration is inherited, the bird has the

The anatomy of a bird (Domestic pigeon, *Columba livia*):

A. Viscera

Ventral view after removal of the breast muscles, sternum, and ventral body wall. (A = Artery, V = Vein, M = Muscle, Mm = Muscles)

I. Respiratory organs:
1. Trachea; 2. Tracheal muscle; 3. Lung (Pulmo); 4. Thoracic and abdominal air sacs.

II. Digestive organs (see fig. 1-15):
5. Gullet or Esophagus; 6. Crop (Ingluvies); 7. Gizzard (Pars muscularis); 8. Duodenum; 9. Small intestine; 10. Liver (Hepar); 11. Pancreas; 12. Cloacal opening.

III. Heart and circulatory organs (see fig. 1-8):
13. Heart (Cor); 14. Arterial arch; 15. Common carotid artery; 16. Subclavian artery, for breast and wing; 17. Wing vein (V. cutenea ulnaris), suitable for blood withdrawal.

IV. Musculature:
18. Neck muscles (Mm. cervicales); 19. Flight musculature (Mm. pectorales); 20. M. triceps; 21. M. biceps; 22. Tensores patagii; 23. M. extensor carpi radialis; 24. Deep M. pronator teres; 25. Superficial M. pronator teres; 26. M. extensor pollicis longus; 27. M. flexor digitorum; 28. M. flexor carpi ulnaris; 29. M. flexor metacarpi posterior; 30. Muscles of the hand; 31. M. sartorius; 32. M. intermedius; 33. M. medialis; 34. Mm. adductores; 35. M. semimembranaceus; 36. M. gastrocnemius; 37. M. soleus; 38. M. tibialis anterior.

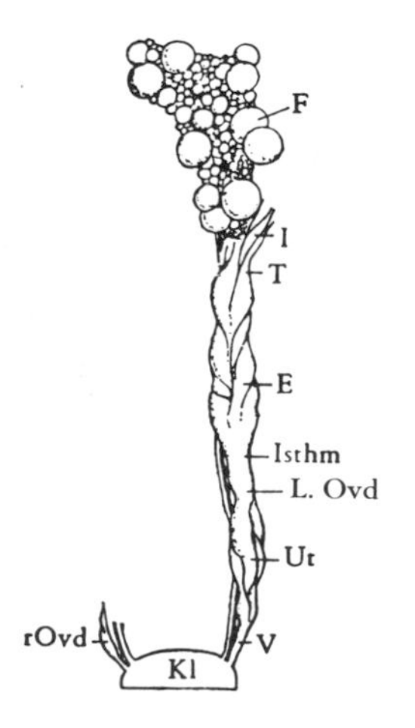

Fig. 1-7. Female reproductive organs of a bird: F—egg follicle of the left ovary; 1. Ovd.—left oviduct with infundibulum (I); T—Tube; M—magnum; (Isthm)—isthmus; Ut.—uterus (shell gland); V—vagina; r. Ovd.—right oviduct, atrophic in all birds; Cl.—cloaca.

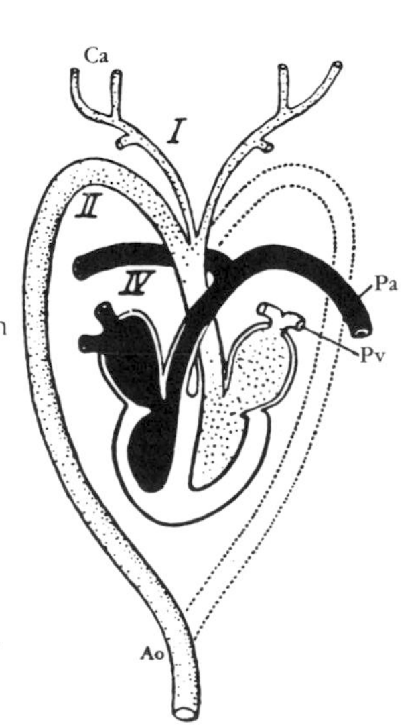

Fig. 1-8. Schematic representation of the heart and principal vessels of a bird. Black: venous blood (deficient in oxygen); A.—aorta; C.a.—common carotid artery; P.a.—pulmonary artery; P.v.—pulmonary vein; I, II, and IV—first, second, and fourth right arterial arch; the second left arterial arch is not developed in birds and is shown here in dotted outline.

A
5
18
6
22
23
26
15
5
2
15
24
25
29
30
14
19
21
16
3
3
28
13
17
20
10
10
31
4
32
33
4
7
34
35
8
1
9
8
11
36
37
12
38
GK

B
1
2
3
4
5
5'
5''
5'''
6
6'
7
7'
9
9'
9''
9'''
10
11
11'
12
13
14
15
16
17
18
19
20
21
22
23
24
25
26
27
28
29
30
31
32
33
34
35
36
37
38
GK
C
1
2
3
4
5
6
8
9
10
11
12
13
14
15
16
17
18
19
20
21
22
23
24
25
26
28
29
30
31
32
33
34
35
36
37
38
39
39'
40
41
41'
42
43
44
44'
45
46
47
47'
48
49
50
51
52
53
54

The anatomy of a bird (domestic pigeon):

B. Skeleton of one side seen from the left:

1. Skull; 2. Eye socket orbit; 3. Upper mandible; 4. Lower mandible; 5-5″. Cervical vertebrae: 5. First cervical vertebra (Atlas), 5′. Second cervical vertebra (Axis), 5″. Tenth cervical vertebra, 5‴. Cervical rib; 6-6′. Thoracic vertebrae, 6″. Crista ventralis; 7-7′. Ribs (Costae), 7″. Ucinate process; 8. Sternal Ribs (Ossa sternocostalia); 9. Sternum, 9′. Keel (Crista sterni), 9″. Incisure sterni (Sternal notch) closed to a foramen by a ligament, 9‴. Pectoral process; 10. Lumbar and sacral vertebrae which are, moreover, fused to the last thoracic and first sacral vertebra (0s lumbosacrale or Synsacrum); 11. Caudal vertebrae, 11′. Pygostyle from the fusion of several lowermost caudal vertebrae; 12. Furculum (clavicles united to a fork); 13. Coracoid; 14. Scapula; 15. Humerus; 16. Radius; 17. Ulna; 18-19: Carpal bones, 18. Radial carpal bone; 19. Ulnar carpal bone; 20-21. Metacarpals: 20. Metacarpal of digit-II; 21. Metacarpal of digit III; 22. First digit I phalanx; 23. Second digit, two phalanges; 24. Third digit, one phalanx, 25. Ilium; 26. Ischium; 27. Pubis; 28. Sciatic foramen for exit of sciatic nerve; 29. Foramen obturatum; 30. Femur; 31. Patella; 32. Tibia fused with some of the proximal tarsals to a Tibiotarsus; 33. Fibula; 34. Metatarsus (fused metatarsals); 35. First digit, two phalanges; 36. Second digit, three phalanges; 37. Third digit, four phalanges; 38. Fourth digit, five phalanges.

C. External features:

1-7. Beak: 1. Lower mandible; 2. Lower beak; 3. Cleft of beak; 4. Upper mandible; 5. Ridge; 6. Nostril; 7. Cere; 8. Lores; 9. Eye ring; 10. Ear coverts; 11-20. Dorsal parts: 11-13. Top of head; 11. Forehead; 12. Crown; 13. Back of head; 14-16. Upper neck: 14. Nape; 15. Lower back of head; 17-19. Back: 17. Upper back; 18. Lower back; 19. Rump; 20. Upper tail coverts; 21. Tail feathers (retrices); 22. Outer tail feathers; 23. Side of neck; 24-25. Cheek; 26. Angle of the lips; 27-38. Ventral parts: 27. Chin; 28. Throat; 29-30. Crop; 31. Upper breast; 32. Lower breast; 33. Side; 34. Abdomen; 35. Cloacal region; 36. Caudal region; 37. Lower tail coverts; 38. Leg feathers; 39-48. Wing: 39-39′. Primaries; 40. Secondaries; 41. Alula; 42-42″. Tertials; 43-43′. Greater wing coverts; 44-44′. Middle coverts; 45. Lesser coverts; 46. Scapulars; 47-47′. Thumb-feathers; 48. Wing butt; 49-54. Foot: 49. Hock; 50. Tarsus; 51. First toe; 52. Second toe; 53. Third toe; 54. Fourth toe.

ability to determine these directions at the beginning of its flight. At least starlings and domestic pigeons can be trained to fly in specific directions. Beyond this, the bird must have additional abilities in order to reach its goal. Thus birds transported in closed boxes into various directions during the breeding season have nevertheless found their way back to their nests. Perhaps the annual return to the breeding grounds after migration is linked to these abilities; however, it could also be learned on the way back along a familiar route.

In many cases, birds orient themselves by the position of the sun. But at least the starling, once it has been trained to fly in one direction, continues on it even if the sun has moved on in the meantime. It determines the proper direction, much like we find south by means of a watch. If we move the apparent position of the sun by moving its point of reference, then the bird is deceived and, in combination with its "inner clock" (inborn time sense), chooses the direction which is correct for the wrong sun position, which is actually the wrong direction.

When a bird has set out in the correct direction, it must still maintain the correct route. It finds its way by certain land marks or by course determinations en route. Where there is a bend in the route another "innate cue" must be present so that the proper direction is maintained. How this is done is not yet clear. Possibly when the sun cannot be clearly seen, the bird responds to the earth's magnetic field and thus determines the proper direction if it has drifted off course or after it has been transported to a strange place. During fog, migration ceases; from this it may be concluded that vision plays a major role, but this does not exclude other possible factors. We still do not know how a red-backed shrike was able to return from Marseille, France, where it had never been, to its nest 1200 kilometers away in Berlin within eleven to thirteen days. Though migrants have often been seen to make circling search flights, birds transported some distance away may take off in the right direction just like pigeons. It seems as if the bird can determine the direction toward home "unconsciously." For us, however, it looks as if the bird were able to determine latitude and longitude and to calculate its direction accordingly. Birds which migrate in flocks have it easier, since it is quite possible that among them are some which know the route and its land marks and hence act as guides. Birds which migrate at night do not take the moon but rather the fixed stars, which are comparable to the sun, as reference points. But this star navigation must still be more thoroughly investigated before it can be regarded as proven.

The bird's body—a light structure

A flying bird can save energy in forward propulsion and lift by being as light as possible. With full plumage, its specific gravity is about 0.6, although it is more in many diving birds; in plucked birds it is about 0.8 to 0.9. Thus a bird is lighter than water but heavier than air. No greater contrast can be imagined than that between the light skin

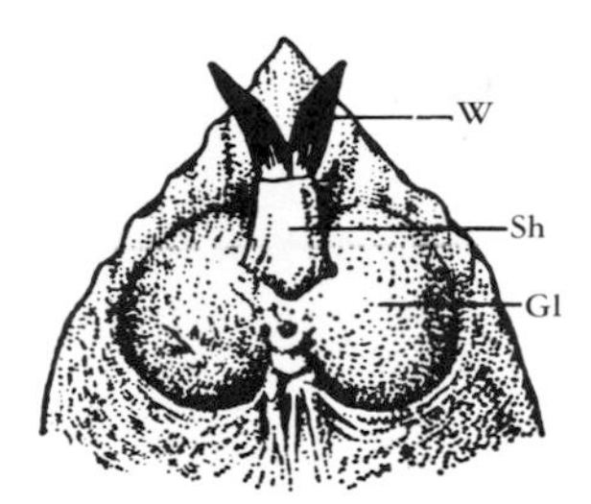

Fig. 1-9. Preen gland of hoopoe *(Upupa epops epops)*: W: wick of preen gland from which the bird takes the oil with the beak. Gl: Preen gland; Sh: preen gland shaft.

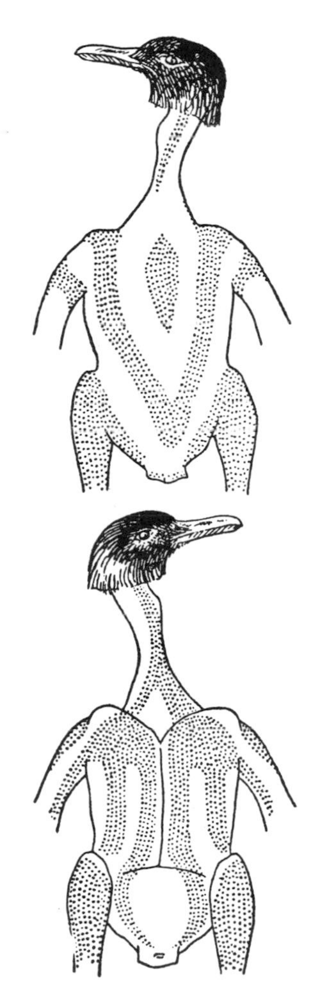

Fig. 1-10. Feather tracts (upper and lower side) of a black-headed gull *(Larus ridibundus ridibundus)*

of a bird and the armor-plated skin of its nearest present day relative, the crocodile. Bird skins lack sweat and sebaceous glands. On the other hand, only birds have a preen gland, a paired sac which accumulates a waxy secretion. The oily secretion often emerges through a feather wick and is distributed over the plumage with the beak; it protects the feathers from dessication and particularly from becoming wet. The molecules of this preen oil have interlocking side chains that make possible the formation of a continuous oil film which covers the feathers. In herons and in many species which inhabit very dry areas, the preen gland is lacking; in others it is very small. A powder down then often replaces the preen oil, which arises when the tips of constantly growing feather tufts in various parts of the plumage on various parts of the body decompose into very fine pieces of keratin.

The plumage is the bird's most important "invention." It forms a light body covering which holds heat even when, as in the ostrich, it is loose and soft. Flying birds with ostrich-like feathers are, however, inconceivable, because the surface of such soft plumage would absorb too much energy. In fact, the best fliers, the falcons, have the hardest feather surfaces. But their soft plumage also contains a great deal of air; being "dead material" consisting of keratin, it is insensitive to cold. By means of the erection or flattening of the feathers by muscles that are attached to the feathers, the amount of air held under the plumage can be regulated.

Rather surprisingly, this light feather cover is, in most birds, distributed unequally and often sparingly over the body. Contour feathers grow from the feather tracts (pterylae), which give the bird its characteristic external shape and which include large feathers of the wing and tail as well as the small feathers. In between, there are areas (apteria) without contour feathers which are, however, normally overlapped by the feathers of adjacent feather tracts. Only the screamers have, for unknown reasons, feathers all over; the non-flying ratites and penguins also have feather papillae from which contour feathers grow over the entire skin.

The small feathers of the plumage protect the body from loss of heat and from becoming wet. They also give flying birds a smooth surface and the rounded form, without projections, of a streamlined body which needs to overcome only minimal air resistance. The feet are generally hidden in the belly plumage during flight. Many swift-flying shorebirds, however, show that this is not always necessary. The basic unit of the plumage is the individual feather. The number of feathers in particular groups of birds is quite variable. Thus, hummingbirds have about a thousand; swans have twenty to thirty thousand feathers with many particularly on the neck. Penguins are similarly well endowed with feathers, although their feathers are only small scale-like structures. Every feather grows from a papilla, a small finger-like structure which, coming out of a lower layer, forms a small cone on the skin. When one examines a plucked chicken, the papillae are evident. A keratin cylinder surrounded by a thin feather sheath is pushed out of the papilla; it is the origin (Anlage) of the feather. Within the cylinder there is a prolongation of the dermis, the pulp. It contains blood vessels which serve to transport all materials necessary for the formation of the feather. On each side of this cylinder a thick keratin tube is formed, which consists of the shaft (rachis) of the main feather and, in many species, a secondary shaft as well. Ridges extend from both barbs, with strips in between which later disappear. These barbs (rami) do not branch out at right angles to the shaft, but rather parallel towards the tip (so that practically the whole cylinder is involved in feather formation). By the time these barbs (rami) have been formed, the feather sheath at the tip has long since ruptured. If it ruptures late, as in the European cuckoo, for a time the bird looks like a hedgehog. Gradually the two shafts with their many side branches become exposed. The second or aftershaft, which lies nearer the body, is generally small; sometimes it is absent. Only in the emu *(Dromaius novaehollandiae)* are both shafts equally well developed. The aftershaft always carries only loose barbs. This is also true of the down feathers, of the almost shaftless nest down, and of the most basal portion of most feathers. Loose branches are particularly extensive in pigeon feathers; therefore pigeons do not carry separate down feathers beneath the contour feathers and their features have no secondary shafts.

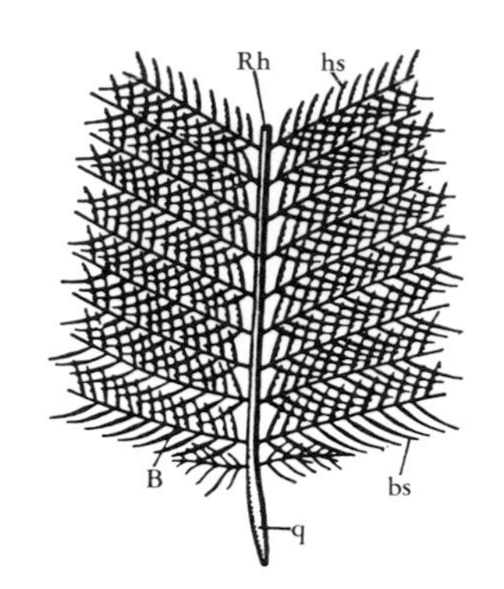

Fig. 1-11. Structure of a feather: The hollow quill (q) passes into the marrow-filled feather shaft, rhachis (Rh). From the shaft the barbs (B) pass out laterally. These barbs are connected by a series of barbules (crosswise structures) (bs) which in turn have small hooks (hs). These catch the margins of the adjoining barbules and thus hold the barbs in a continuous sheet. The figure is schematized and enlarged.

The barbs of the mainshaft are almost always interconnected by small barbules or radii. Along the shaft of a crane's primary feather there is a row of about one thousand barbs on each side with a total of up to six hundred thousand barbules. These barbules form, in the closed part of the feather vane, two rows that are different in appearance and function. Those directed toward the base of the feather (proximal barbules) have a brim on their upper surface forming a ridge of attachment for the hooklets of the distal barbules and often have notches. The barbules on the side of the barbs, which point towards the feather tip (distal barbules), on the other hand, are provided with a few hooklets which engage in the brim of the proximal barbule and hold them tightly in spite of possible movement. Thus, hundreds of barbules with hooklets and brims make the feather into a closed, elastic plane consisting of two vanes. In the wing and tail, two adjacent feathers resist separation as the lower feather sends brush barbules upward which attach to the feather lying above; thus the surface is held together.

Bristles are quite simple feathers (vibrissae) which are formed particularly about the base of the beak and about the eye as eyelashes, but which may, however, also be found elsewhere. They serve as tactile organs but also as grids against airborne particles by guarding the nostrils or ear openings.

Color of the plumage

The calamus, the hollow basal portion of the shaft and the air cells of the other parts of the shaft and the vanes, make the feathers into an extremely lightly built cover for the bird's body. The plumage, however, carries most of the bird's colors. The numerous hues of color which physicists show and name in their color charts are inadequate to describe all the colors which actually occur on bird feathers.

Pigment and structure colors

As in other animals, melanins are the most widely distributed pigments. They are poorly soluble protein derivatives, black to light brown grains of round, elongated, or lens-shaped in form. When the black eumelanin granules lie close together, black colors arise; when they are thinly scattered, the color effect is a pale gray. Similarly, the brown pheomelanin, depending on the density of distribution, gives rise to deep brown, reddish-brown or pale reddish to yellowish colors. Closely related birds in warm moist areas often look blackish, while those in cold moist areas are more reddish-brown, those in warm countries are gray or "desert-colored," and those in dry cold areas are whitish since pigments are lacking there. This phenomenon, called "Gloger's rule," evidently depends on chemical reactions of the melanins, but we still know little about it.

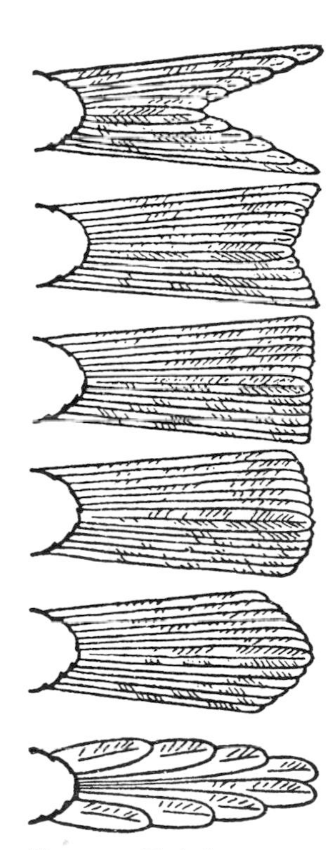

Fig. 1-12. Tail shapes: forked, indented, square, rounded, wedged, scalloped.
Designation of the most important tail shapes of birds. The shape of the tail is frequently an aid in identification.

The other pigments of the bird feather are not shaped granules when seen under the microscope, but are diffusely distributed. Among them the fat soluble carotinoids (lipochromes) and their derivatives form especially yellow and red colors; more rarely one finds blue or violet lipochrome pigments. Certain other pigments occur particularly in parrots, birds of paradise, and turacos (see vols. VIII and IX).

In addition to the color from pigments, structural colors play a great role in the appearance of bird feathers. They depend on the fine structure of the keratin. Blue feather color, for example, almost always arises as follows: of the many wavelengths of visible light which enter the feather, the long red waves reach the background of the feather, which is dark due to pigments, and are absorbed there. The shorter waves of blue light strike small air vesicles in the keratin and are refracted out again. Incidentally, the blue of the sky arises quite similarly in front of the dark background of space. If the air vesicles are larger, the longer waves are also reflected back and the feathers look white. Feathers with "blue structure" which have a yellow pigment look green; if they contain a red pigment, they look violet or purple.

Iridescent colors occur in hummingbirds and many other birds; they too are structural colors. They are explained in the chapter on butterflies in Vol. II. The sheen or the velvety, dull effect of many feather colors is also dependent on the fine structure of the keratin mass and its surface.

Many birds in captivity acquire unnatural colors after the moult. This is because certain food materials, especially carotenoids, are often lacking. Thus, captive flamingos used to grow white instead of deep red feathers. With the addition of carotenoid-rich food supplements, it is now possible to ensure that even in zoos they can grow beautiful red feathers after the moult. European goldfinches *(Carduelis carduelis)* develop black instead of the natural brown coloration if they are fed much hemp. It was often stated that fully formed feathers could still change their color, but this could only be proved in a few cases. Thus, dark feathertips extensively exposed to sun can turn brown due to the bleaching of melanin. Some birds color their white feathers yellowish to reddish with preen oil, such as the pelican's yellow throat patch and the pink color of many gulls. Others color themselves with iron-rich water, such as the bearded vulture *(Gypaëtus barbatus),* or with moorland soil as in certain cranes (see Vol. VIII). Albinos which, as a consequence of an inherited factor, form no pigments, occur in all birds, as they do in other animals, similar to melanistic individuals as a result of too much melanin.

The overall plumage pattern arises because of the colors on the individual feathers. It is amazing how adjacent primaries, each marked at a different point with a black band, form a continuous straight band on the spread wing of the Egyptian plover *(Pluvialis aegypticus).* The overall effect which such patterns can have is seen, for example, in the erect and spread tail of the peacock *(Pavo cristatus).* Such magnificent plumages are often found only in males and are at least partly attributable to selection by the female (sexual selection). Often, however, they are probably merely "uniforms" which enable the females to distinguish conspecific males from those of other species. Quite as surprising as the display plumage is the effect of protective coloration on the observer, where an incubating bird blends into the background because of its disruptive pattern. Such a pattern, with light brown and darker brown stripes on the back, is the least conspicuous on ground covered with low vegetation. Larks and sparrows, for example, are perfectly camouflaged, but so are many others, such as the females of many gallinaceous birds and ducks. Some birds or the young even imitate the coloration of other species. This so-called mimesis is discussed in detail in Vol. VII under cuckoos.

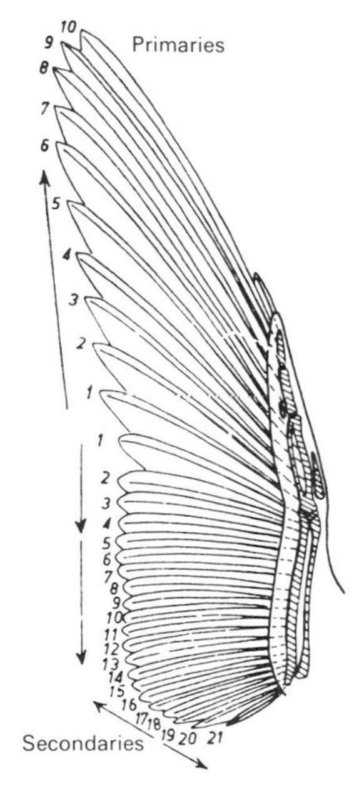

Fig. 1-13. The order in which the primaries are moulted differs in different groups of birds. Most frequently the moult of the primaries begins (see arrows) with the most proximal feather (1) and continues to the most distal feather. The order of moult of the secondaries is shown here in the wing of the Manx shearwater *(Puffinus puffinus),* which is more complex (see arrows).

The fulmar *(Fulmarus glacialis)* is one of the few species which grows only a single plumage; thus after hatching, age cannot be determined by its plumage. Generally, however, the nest down is succeeded by a juvenile plumage which partly consists of broad feathers. The down feathers are pushed up on the tips of the emerging contour feathers. The juvenile plumage is often less intensely colored than that of the adult; the tail feathers are often more pointed and narrower. The juvenile plumage is generally followed in the fall by the first plumage which is either worn for one year or only in the non-breeding season. In species which moult once a year, a new plumage is grown in each successive year; in those which moult twice a year, breeding and non-breeding plumages alternate annually. Sometimes the plumage changes its color or pattern until the fourth breeding season, as in the large gulls. Many birds of prey take even longer to acquire the adult plumage.

The moult

The dropping and replacement of feathers is called the moult. In many birds of prey and owls it lasts one year or longer. Mostly, however, it is completed in one to two months. It is probably induced by the pituitary gland (hypophysis) via the thyroid gland. Feathers removed by plucking, or those which are suddenly dropped when the bird is frightened, are also replaced. Moulting birds generally have a raised body temperature; although the plumage to be replaced represents only five per cent of the body weight, it evidently requires much energy to replace it. Certain fish hunters, such as penguins, starve during this time on land and lose about half of their body weight. But other birds also appear to eat less during the moult than normally, particularly those which shed all their flight feathers at once and thus become temporarily flightless. Many species, particularly those which wear a striking breeding plumage, moult twice a year. Only rarely, however, is the whole plumage moulted; generally there is only a partial moult of the contour feathers before the onset of spring.

Skin appendages and skin sacs

In many birds we find skin appendages and skin sacs which can be evaginated, particularly on the head. Often these can be inflated, in the turkey *(Meleagris gallopavo)* by lymph, in the prairie chicken *(Tympanuchus cupido)* by air. Other skin structures are the horny sheath of the bill (ramphotheca) found in all birds, the horny head or frontal shield of many groups, and the horny scales of the legs and toes. The claws found on the fingers (on the wings) in a few birds play hardly any role except in the nestling hoatzin (*Opisthocomus hoazin;* see Vol. VIII); claws on the toes are much more important. They are laterally compressed in many birds, but in the mound birds they are flat, in the eagle owl *(Bubo bubo bubo)* they are very long and pointed, and in the ratites they are often hoof-like. Sometimes the claw of the middle toe has a horny comb on its inner side; herons and nightjars use this to clean the plumage of the head and neck.

The skeleton

A flying body needs rigidity, particularly where the moving wings are connected to the spine. The shoulder joint of birds is therefore equipped with particularly strong ligaments and joint cavities. The pectoral girdle too, in spite of all its distensibility, forms a relatively stable structure. This is provided by the uncinate processes in most birds, which is the fusion of several thoracic vertebrae and the large sternum. Since birds only walk on two legs, the pelvis is also strengthened by the fusion of several or even many vertebrae.

The bony supports of the body and the muscles are more strongly fused with one another in birds than in reptiles. Nevertheless, the skeleton of birds is surprisingly light in comparison with that of other vertebrates. In flying birds, it amounts to only ten per cent of the total weight. This framework is so light because the bones are hollow; this has proceeded much further in birds than in any other vertebrate. Even the most distal segments of the toes are hollow in the toucans and hornbills; air spaces even extend into them. On the other hand, the flightless penguins have almost no air spaces in the bones. Wherever great demands are made on hollow bones they are strengthened by a system of bony braces; thus the common swift shows a spiral structure in its humerus which presumably offers particular resistance to twisting strain on this bone. Nevertheless, many birds which fly well, such as the gulls, have humeri which are filled with marrow, and have no air spaces.

A number of birds like toucans and hornbills have enlarged beaks, but these beaks and their bony framework, except that of the helmeted hornbill *(Rhinoplax vigil),* are always light and have a spongy bone framework. Furrows on the bill, the frontal bosses, and the bosses on the beak and on the crown are generally composed only of keratin, but bony cones covered by a horny skin occur on the skull of some species of the Cracidae, on the middle hand in screamers, and on the legs in pheasants. Bony humps, as in the mute swan *(Cygnus olor),* are often joined in birds. Combs and other appendages consist in part of skin and connective tissue, in part of keratin. The single condyle at the rear edge of the foramen magnum rotates in the pan of the atlas; the dentate process of the second cervical vertebra (axis or epistropheus) fits into its floor.

It is important to condense the weight of a flying body within a small space. This alone insures a favorable aerodynamic shape. With its light head and lighter wing tips and tail, the bird is more stable than were the flying saurians with their armoured heads, the archaeopteryx with its long tail spine, or other animals with heavy hand or leg muscles. In birds such muscles rest on the sternum or their main mass is attached to the humerus or the thigh; they act more distally on the bones only via the tendons. This body compactness has persisted even in birds which have lost the ability to fly, including the largest ones, although in the ostrich and cassowary the foot is relatively heavy.

The thorax hangs from the spine (see Color plate, p. 36). Its lateral boundaries are formed by three to nine ribs; the large, variably-shaped sternum forms a ventral wall and thus a cavity which encloses the heart and other organs. The center of gravity of the entire bird is just above the sternum. Birds are the only vertebrates which have processes on the ribs which may reach from the posterior edge of one rib beyond the forward edge of the next; these uncinate processes are absent only in the tinamous and screamers. Here parts of the important external intercostal muscles are inserted, enlarging the angle between the upper and ventral segments of the rib, which are linked by a joint. This lowers the sternum and may laterally widen the thorax. The thorax can also be anteriorly widened when the pectoral girdle moves. On each side, a corocoid connects anteriorly with a joint right and left, and leads from here to the front and up to the shoulder joint, forming the anterior edge of its fossa. The furcula also arises near this joint cavity, but generally unites with that of the other side. This is also the case with our clavicles; in gliding birds it is often joined to the anterior edge of the sternum and the keel. In other birds (not just flightless ones), it may be vestigial or completely absent. The narrow scapula begins at the front as part of the shoulder joint fossa and reaches back over the dorsal segments of the ribs as far as the pelvis.

The forelimbs

In the wings, the skin (progatagium) extends between the upper and lower arm, and the metapatagium extends between the lower arm and the hand. When the wing is fully extended, this skin can lie up against the bone and effect a thickening of the anterior edge of the wing. The humerus with a longish condyle fits into the shoulder joint. Thus it can especially move forward, upward, and back. Some groups of birds have long ridges on this bone which enlarge the areas of insertion for the breast muscles. The two bones of the lower arm (radius and ulna) become displaced, relative to one another, as the wing is extended and the ulna is moved more outward. Thus the hand is pushed outward by the ulna, and is also stretched. In certain gliding birds, there are even additional bones in the angles of the elbow joint and at the wrist joint to ensure that the arm will remain extended without muscular exertion: the glider needs only to move some muscles to pull these sesamoid bones out, which makes the folding of the wing possible. The Procelaridae especially have these small additional bones and therefore they hardly change the wing position in their hour-long gliding flight.

The root of the hand consists of only two small bones (radiale and ulnare); some of the carpals become part of the metacarpus during development. The original metacarpal of the first digit is seen only during development, while that of the second digit forms externally on the fully developed metacarpus as a small process and carries a distinct phalanx. In the young there is often a claw at its tip, which is rare in the adult. The third digit is rare; the fourth was found only in the archaeopteryx, which provided it with a claw and was movable only in this bird. In other birds, the third and fourth digits can only be moved together. The fifth digit is always absent. The metacarpus supports the inner five primaries; the outer four to five are located on the third and fourth digits.

The legs

While the front limbs have been converted into flying tools, the legs differ less from those of other higher vertebrates. The ilium, pubis, and ischium meet in an almost hemispherical acetabulum (socket) in which the head of the femur is held. There is an opening in the socket between the ischium and the ilium. In spite of this ball-bearing-like arrangement, bird legs cannot perform a complete rotation, for bony processes (antitrochanter) behind the fossa prevent, in conjunction with another bone process (trochanter) on the femur, a lateral raising of the leg. The thigh almost always runs forward and down, corresponding somewhat to the position in man when he is seated, but it projects only a little out of the body wall. Hence the knee is almost always covered with feathers or hidden by the skin of the abdomen, so that the uninitiated might take the visible uppermost part of the leg as a backward-directed thigh. In reality only the lower leg with the tibia and the fibula, which is smaller and much reduced in its lower part, emerge from the plumage.

When the tibia and, along with it, the toes are moved at the knee joint, the fibula acts like a door spring to turn the leg back as soon as the muscle pull stops. Through incorporation of the proximal tarsals, the tibia is lengthened into a tibio-tarsus and thus ends further down. The joint which follows is the ankle or intertarsal joint. It is followed by a single bone, the tarsometatarsus, which is derived from the distal tarsals and all the metatarsals; it ends in two to four condyles to which the toes are attached. The first toe, which almost always points to the rear and is fixed in the forward position only in a few swifts, is often absent. In certain kingfishers the second toe has disappeared. When the fourth as well as the first toe are in a fixed position to the back, one speaks of a zygodactilous foot; such a climbing foot is best known in the parrots, owls, cuckoos and woodpeckers. When the fourth toe can be turned to the rear as in the osprey *(Pandion halietus),* the bird then has four pairs of pincers for grasping.

The muscles

Among the muscles of birds, the flight muscles are rather unique. The pectoral lies mainly laterally to the keel of the sternum and sends its tendon to the anterior and lateral part of the humerus; its opponent, the supracoracoideus, lies beneath it on the sternum. Since it sends its tendon through the foramen triosseum to the upper side of the humerus, it raises the upper arm. In order to make the lower arm form an angle with the upper, a large part of the great delta muscle (brachialis) and the biceps are used; the triceps are used to extend it. In addition to the other red coarse-fibered muscles, many birds also have "white" muscles. In many gallinaceous birds they make sudden flight movements possible, but they are easily fatigued.

Diving and swimming

The wings are also used by birds for locomotion under water. The diving of many species can quite correctly be called "flying under water." Penguins move by using the wings and not the feet, even when on the surface of the water. The auks and shearwaters are excellent wing divers. Early stages of this capacity, however, are found in the dippers of mountain streams. Since water is much denser than air, the wings of wing divers are quite small or are not fully opened underwater. Penguins have developed regular fins since they no longer need their wings for flying.

Diving birds that propel themselves with their feet instead of their wings have a somewhat higher specific gravity than most other birds; their bones contain less air and their plumage can be pressed fairly closely against the body in order to reduce the air cushion beneath. All have webs or lobes on the toes and push the water to the rear and upwards in order to move forward and down. Diving ducks, mergansers, and coots are not quite as good divers as are cormorants, anhingas, marine diving ducks (scoters), loons, and grebes. In some of these groups the wings take part in steering under water. Most of them rarely dive for longer than a minute or deeper than ten meters. Only loons, crested grebes, and penguins dive thirty meters, and sixty meters have even been ascribed to loons; they are said to be able to survive fifteen minutes under water. The behavior of the plunge divers, such as ospreys, gannets, brown pelicans, and terns is different. They utilize the force generated by the fall through the air to dive into the water; gannets can dive up to twenty meters in this fashion. Others, like tropic birds and gulls, plunge down until just above the water, then spread out the wings to brake their fall and with their beaks seize the prey they have spotted. They can usually do this without getting their plumage wet.

Short-legged birds use their legs as little as possible for locomotion on land. Thus nightjars, mousebirds, swifts, hummingbirds and most cuckoos and coraciiformes fly for even the shortest distances. Loons, with their legs attached far to the rear, cannot walk upright like penguins; their laborious movements on land resembles sliding on the abdomen.

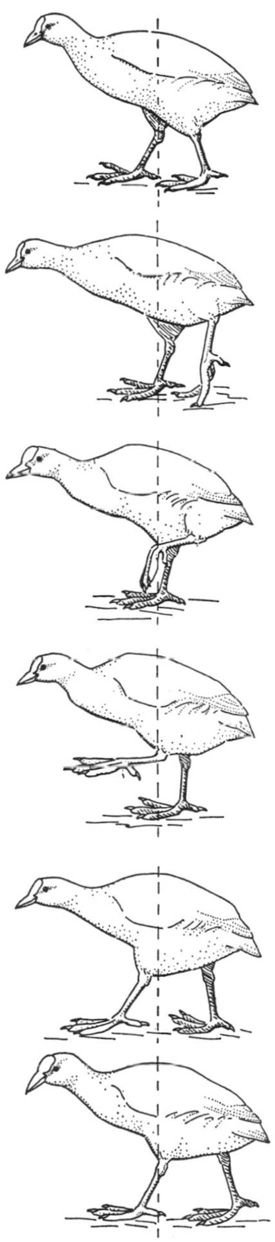

Most birds, however, regularly move by using their feet. They walk, run, crouch, step, or stalk by alternately placing one foot in front of the other. The fewer toes they have, the more swiftly they can do this. Thus the ostrich (a swift runner) is the only bird with only two toes. In ancestral birds, walking must have been very important for the standardization of the leg, because a similar basic structure has evolved considerably, compared to reptiles. The ankle joint was, as it were, moved back, which may be of importance for balancing the bird in the usual horizontal position of the body to maintain the center of gravity over the point of support.

There are differences in walking. In pronounced walkers, the hind toe is inserted high or is absent; the toes are short. There are, however, fast walking birds which have long toes like the water rail *(Rallus aquaticus)*, a long hind claw like the larks, or very long claws like the jacanas.

Fig. 1-14. Walking in the coot *(Fulica atra)*. The dotted line shows the position of the center of gravity in the various phases of movement.

Such birds generally walk in swamps, sand, or on the water.

In walking, the bird extends the toes of the foot which is behind, thereby pushing its body forward, and then supports it with the same which by this time is in front of the body; the other foot meanwhile supports the body as a supporting leg and becomes the stepping leg with the next step.

Locomotion on the ground

If both feet are extended simultaneously, the bird takes a jump. Jumps of up to three meters in length are known in the roadrunner *(Geococcyx californianus)*. The yellow-fronted New Zealand parakeet *(Cyanoramphus auriceps)* also leaps about one meter, but the movements of tree-inhabiting parrots consist of a grasping from one branch to the next in which even the beak may be used. Some species, like the blackbird *(Turdus merula)*, only hop; others, like the house sparrow *(Passer domesticus)*, can walk as well as hop.

The bird stands on the ground between hops. Its toes do not lie quite flat but are often somewhat raised. In typical tree birds, the digit next to the tarsometatarsus is relatively short; this enables the bird to grasp either thin or thick twigs better than is possible for walking birds with relatively long toes.

Climbing

In a number of birds, climbing is the preferred form of locomotion. In woodpeckers and tree creepers, it is an upward hopping on the tree bark; wall creepers similarly hop upwards on rocks. Nuthatches can hop head up or down on trees and rocks. All these climbers, as well as the pygmy parrots, have short tarsi and thus keep their center of gravity close to the trunk, even when they brace themselves with the hard feathershafts of their tails as do woodpeckers, pygmy parrots, and tree creepers. Woodpeckers and nuthatches especially hold themselves with the laterally far-reaching claw of the fourth toe on the trunk, somewhat in the manner of a man using climbing irons.

All these movements can occur in the most varied combinations. Albatrosses fly an extraordinary amount of the time, although they get their food from the sea while swimming. Many birds of prey fly when hunting and in between sit on the ground or on branches.

Flightless birds

Many birds, though quite capable of flight, are almost always seen on the ground; some reluctant fliers are tinamous, partridges, and lyrebirds. In no less than ten living and four extinct bird families, the capacity for flight of all or some species has been lost. Their wing and flight muscles have retrogressed. Thus the ratites and penguins have become flightless, as did the extinct giant cranes *(Diatryma* and *Phororhacos)*. Also, several species of cormorants, ducks, rails, auks, grebes, and pigeons are flightless. There are even a few passerine birds, quite capable of flight, which never or almost never do so.

Nutrition and digestive organs

As with leg shape and the two-leggedness, many peculiarities of the digestive organs are compressed as closely as possible to the center of gravity. The sharp cutting edges of the beak which are built of light

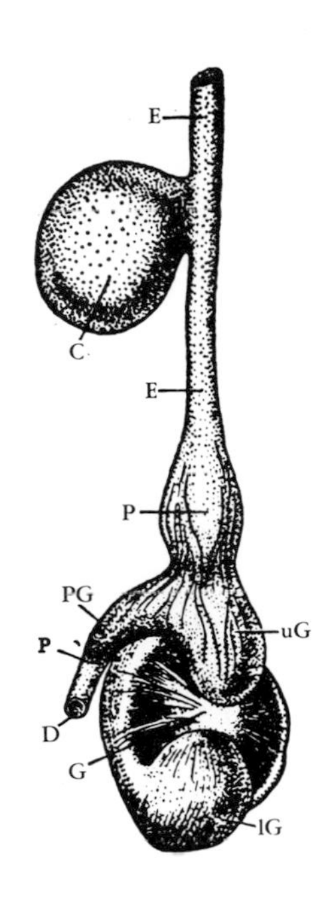

Fig. 1-15. Anterior part of the peacock's gut: E—Esophagus or gullet; C—Crop; P—Proventriculus; G—Gizzard; uG—upper, IG—lower, and PG—Pyloric region of gizzard; D—beginning of the duodenum.

keratin replace heavy teeth. Ingested food is generally moved close to the center of gravity; only light food is carried for any length of time in the beak to be brought to the young; heavy lumps of food are carried with the feet, as birds of prey do. Those birds which must take up larger amounts of food at a time are no particular exception to this rule. They have a crop, rarely more than one, as a dilatation of the gullet in which food is softened or stored, but in the plant feeders the long intestine counterbalances the weight of the crop; the birds of prey shorten the crop-intestine axis, usually by an upright posture. A glandular stomach has developed from the gullet only in birds; here pepsin and hydrochloric acid initiate protein digestion. In the adjacent muscular gizzard, hard leaves, grasses, grains, and even insects are squashed and ground between plates which can squeeze and rub the contents; often the bird takes grit to facilitate the process. Fish and flesh eaters do not need such muscles and, in many fruit eaters, the gizzard has retrogressed to varying degrees. The ileum of grain and cellulose eaters is rather long; in fish eaters it is narrow, and in the Magellan's penguin *(Spheniscus magellanicus)* it may be up to nine meters long. It is shorter but still narrow in flesh eaters, but hardly equals the body length in birds which live on soft insects. Fruit eaters, such as fruit pigeons, have a short and often very wide gut. The openings of the two caeca are found where the intestine begins; they are generally short but only rarely are they reduced to one or totally absent. In some groups, such as ostriches, fowl, and ducks, they may be long and can suddenly expel foul-smelling contents. In the caeca there are, in such cases, more cellulose-digesting bacteria than in the rest of the gut. The gut contents pass through the rectum into the cloaca from the large intestine, which in the ostrich is quite long (14.4 meters). Its innermost segment, the coprodaeum, reabsorbs water. This can be so efficient that some birds, such as larks, do not need to drink at all.

Food and water

Although most birds are light weight animals, their food requirements are substantial. They expend a considerable amount of energy on their high body temperature; this is particularly true in taking flight, when their metabolism is increased ten times. A bird needs between one sixth to twice as much food per day as its body weight. Those living on fish, which contain much water, need more than those living on a nutrient-rich meat diet. Birds can tolerate thirst only with difficulty; hence they undertake long flights to find water. In winter in the polar regions and on high mountains, snow may be eaten to supply water. Penguins and some other marine birds meet their water requirements by salt water. All birds conserve water since they lose none in perspiration. Instead, a cooling of the blood is achieved only by panting when the need arises.

Choice of and obtaining food

The choice of food varies from group to group among birds. Many take a mixed diet, while others are complete vegetarians or meat eaters.

Within the same species, too, the diet varies according to availability or individual preference; there are tawny owls *(Strix aluco aluco)* which specialize in eating fish. In general, however, any food that is suitable for a species is utilized in proportion to its availability. For example, in Eurasian sparrowhawks, domestic sparrows and skylarks take first place in its menu. Generally birds have to search for their food. The common heron *(Ardea cinerea cinerea)* flies up to thirty kilometers for this purpose, or merely descends from its resting place on a tree to stand and wait for approaching prey in the water, which it then seizes with lightning speed. Pelicans cooperate in fishing, while skuas are "parasites" which rob other birds of their food. The white stork in search of food walks through wet meadows or sedge in shallow water. Such search-hunters distribute themselves all over the land and the upper layers of water. Gallinaceous birds scratch on the ground, and starlings and snipes bore into the ground with their beaks. Honeyeaters dip into flowers; woodpeckers search for insects under the bark with their tongues.

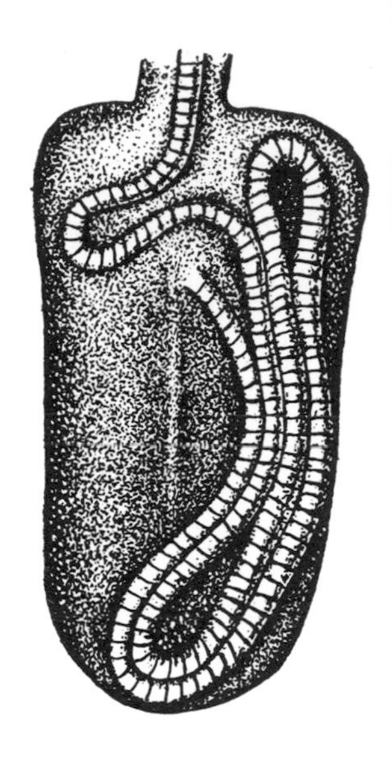

Fig. 1-16. The 120 cm long trachea of the magpie goose *(Anseranas semipalmata)* forms such a loop outside the sternum between the skin and the breast muscles.

Nuthatches, jays, and tits (chickadees) hide some of their food and are able to find it again; and shrikes store insects by impaling them on thorns that are visible from a distance (see Vol. IX). Other birds first treat their food in some way before they consume it. Finches remove the hulls of grains, hawfinches crack pits, and flycatchers remove the wings of butterflies which they have seized in the air with one bite. Eagle owls and birds of prey have developed special killing techniques: they strike their prey in the back of the head, which leads to instant death. Raptors, shrikes, and crows then hold the prey with a foot, tear off pieces, and also pull out feathers. A regular dissection of the food, however, is rare; in parrots the beak, in conjunction with the tongue, is a highly developed organ for grasping and taking in food.

There is probably nothing in nature that is edible which is not eaten by birds; many even eat wood, bones, or feces. This way they have been able to utilize all areas of the earth, the dry land, the water, and the air. They penetrate only as far as the root zone of plants into soil, however; in the water, they may dive from ten to an occasional sixty meters. Thus the coexistence of more than a hundred species within a few kilometers has become possible.

Lungs, respiration, air sacs

The respiratory organs are light but very effective. The lung is below the thoracic vertebrae and adjacent parts of the ribs, as well as between the ribs; thus it utilizes the respired air more efficiently than do the lungs of other vertebrates. Nowhere does the air find dead ends; it runs from the bifurcation of the trachea through the two main bronchi and then through thinner dorsal and lateral bronchi to the inferior bronchi and the finest capillaries. The walls of the air capillaries are also the walls of the lung capillaries, so that a rapid gas exchange is guaranteed. Air flows on the ventral side of the lung from the bronchi into the air sacs. During expiration, air returns, at least in part, through the same passage and in so doing again delivers oxygen; finally the blood contains waste products made up of carbon dioxide and water vapor from the blood.

Birds breathe much faster than cold-blooded reptiles. At rest, a domestic pigeon breathes twenty-nine times per minute; during fast walking, 180 times; in flight, 450 times. With smaller birds, such as hummingbirds, the number of respirations, even when they are at rest, is 250 per minute, and at fifty wing beats per second they breathe 3000 times.

When the rib cage moves while the bird is breathing, the air sacs are enlarged or compressed. The five pairs of air sacs surround almost all organs and often reach the vertebrae, the limb bones, and the skin. The air here provides an enlargement of body volume and thus contributes to its lift; it reduces the specific gravity of the bones, cushions the inner organs, cools the body interior, particularly in flight, and protects the skin when the bird plunges into water. In submerged birds, the oxygen contained in the oxyhemoglobin and oxymyoglobin of the muscles is important; it is displaced by carbon dioxide which birds tolerate well. Besides this, glycogen supplies energy and the circulation primarily supplies the central nervous system.

Voice and vocal organs

The larynx, with muscles through which movements of annular cartilages can cause hissing and spitting, lies behind the glottis in the trachea. The tracheal rings, normally cartilaginous, may be ossified in some birds, such as cranes. The syrinx is at the bifurcation of the trachea. Here, the normally soft skin between the tracheal and first bronchial rings is enlarged to the inner or outer membranes. Some birds have inner as well as outer membranes. This causes the air stream to vibrate during expiration since the membranes are tensed by the slightly higher pressure in the interclavicular air sac. The membranes act like reeds which vibrate in the passing air stream. The air waves caused in this manner produce a tone of a certain pitch, generally rich in overtones with a frequency between 70 and 7000 cycles per second. The longer the trachea is, the deeper the tone, as in the trombone.

Every species of a certain locality has its own voice in which timbre, pitch, and loudness, or at least a repertoire of musical pieces, calls, and occasional instrumental sounds, are involved. Melody as well as tempo and rhythm play a role. Almost all species can be recognized by their voice. A bird can sometimes be deceived by sound imitations produced by humans or by animals. It can also recognize conspecific birds, or even certain individual birds, by the voice.

Warm-bloodedness

In contrast to their reptilian ancestors, birds, like mammals, are warm-blooded. Therefore, they need an external source of heat, even as embryos or in the egg, since the embryo cannot produce enough heat on its own. An exception to this are the mound builders; their eggs were removed from the nest after fourteen days and were kept at 30°C, and they still hatched. These birds, therefore, produce more heat than others, even while still in the egg, since they have very large yolks. Most birds are also unable to maintain their body heat immediately after hatching. Even three-day-old down-covered Adelie penguins *(Pygoscelis adeliae)* die within a short time if they are not brooded by one of the parents. It is not until they are five days old that they can maintain their temperature and are truly homoiothermal.

Not only do such cold-blooded (poikilothermal) youngsters undergo a lowering of the blood temperature, but many adult birds do so also, reminding us of their reptile-like ancestors. This temperature drop was first detected in swallows, which, however, can endure such periods of cold only briefly while on migration; they then become rigid and cling to one another. The only truly hibernating bird known thus far which becomes dormant for longer periods is the whip-poor-will *(Phaleanoptilus nuttallii)*, a nightjar of the southerwestern United States. A bird of this species spent three months clinging to a rock cleft in a state of rigidity; its body temperature fell from 42°C to less than 20°C. Since then, blood temperatures of 20°C have been recorded at night in a common swift nestling and of 25°C in a mousebird. Hummingbirds have also been found which spend the night in a state of rigidity. The usual blood temperature, however, is 41°C to 43°C; in certain waterbirds it is 38°C to 40°C, and in small birds like hummingbirds whose energy consumption is extraordinary, it is 44°C, occasionally even reaching 54°C.

Heart and circulation

Although birds have acquired their warm-bloodedness independently of mammals, the bodily changes involved are nevertheless similar in both vertebrate classes. Like the mammalian heart, the relatively large bird heart (0.2 to 2.8% of the body weight) is completely divided into right and left chambers, the atria and ventricles. The descending aorta curves to the right, not to the left as in mammals. The number of red blood cells is very high, one and one-half to seven and one-half million per cubic millimeter of blood, which amounts to 3 to 10% of the body weight.

The renal portal artery system of reptiles, which is absent in mammals, has been preserved in the birds' circulation; it carries the blood laden with wastes from the posterior end of the body to the kidneys before it reaches the inferior vena cava and right atrium. Warm-bloodedness is also favored by the lung structure and the quantity of food eaten by birds. In birds, warm-bloodedness raises the basal metabolism; the pulse rate increases from 100 to 1000 and is highest in small birds. The blood pressure is also higher than in mammals. These adaptations have enabled birds to settle inhospitable areas and to leave them again in the most unfavorable seasons.

Much emotion, little sense

A fast-flying animal must not only have a high metabolism, it must also be in a position to react very quickly and appropriately when obstacles and dangers suddenly appear. Conscious deliberation would take too long in most cases, and even simple associations of memories and experiences would often take too much time and be too uncertain. Hence it became necessary for birds to adapt to swift reaction without thinking, as it were. In the evolution of the brain, birds have taken a completely different path than mammals. Within the mammals, the ability to learn has been further improved with each step, while the rigid inborn responsiveness to stimuli has become correspondingly reduced. Birds, however, have perfected the ability to respond very quickly to the highest degree, and they are able to respond to specific simple stimuli with biologically meaningful behavior. This property makes birds almost ideal experimental animals for ethologists; much of what we know about inborn behavior in animals and man is due to investigations made on birds.

Oskar Heinroth has characterized the innate behavior responses of birds as follows: birds are "...emotional animals to the highest degree, with many inborn drives but little reason." Much more than the innate behavior patterns in mammals, one can recognize that certain releasing stimuli are linked with biologically meaningful responses in bird behavior. Birds stand in much the same relationship to these responses as a key does to a lock. For example, a young peregrin falcon *(Falco peregrinus)* reared without contact with conspecific birds had become quite friendly with a young sparrow reared in the same room. But when the sparrow flew for the first time, the falcon rushed at its friend and killed it. The perception of the small flying "something" no longer had any relationship to the picture of the "friend" which depended on experience, but merely became the key which, as it were, "opened a lock in the falcon's brain" which released the inborn prey-catching behavior; it is, as ethologists say, an innate releasing mechanism (IRM).

In many cases, the object which releases an innate response is itself not inborn. Thus a newly hatched greylag gosling *(Anser anser)* does not know its parents. The first object which it meets after hatching is accepted as the parent by the gosling; only this object can release the inborn following behavior of the gosling from then on. In nature, the first object seen by the greylag chick will always be the parents, but for a gosling hatched in an incubator, a person whom it first sees is thereafter father and mother, as Konrad Lorenz describes in Chapter Eleven. This process is called imprinting; it takes place early in life, during a short, sensitive period, after which the object that has been learned is usually not forgotten. This process was first discovered in birds, although it has since been reported in mammals as well.

Displacement and vacuum activities

Quite often a bird is in a situation in which it perceives stimuli which release antagonistic behavior. The approach of an enemy on the ground may trigger the most rapid flight; simultaneously, however, the sight of the unprotected nest may stimulate the bird to incubate with equal intensity. In such conflict situations, it may happen that the bird shows quite a different form of behavior not connected with either of the two stimuli, a so-called displacement activity. A bird which, at the sight of a strong opponent, vascillates between the urge to attack and to escape often shows displacement pecking on the ground, as if it were looking for food. In comparable situations, people often scratch their heads in displacement, a form of behavior which belongs in the functional system of body comfort movements. The bird which sees an enemy on the ground and wavers between the urge to incubate and to escape, often begins to show a distraction display (injury feigning); it staggers about and slowly moves from the nest, dragging its wings. Thus it gradually decoys the enemy behind itself until the eggs are safe. This injury feigning is definitely not a deliberate pretense, but an innate response which presumably began as a displacement activity.

Fig. 1-17. "Pseudo-sleeping" as a displacement activity in the oystercatcher.

At times certain innate behavior patterns occur in the absence of any detectable releasing stimuli; they go off "on their own" or, as it were, in a vacuum: hence the term vacuum activities. An example was described by Lorenz in captive starlings, which occasionally go through the motions of watching a flying insect, catching, killing, and finally swallowing it, in the absence of any insect. Young honey buzzards *(Pernis apivorus)* still in the nest may perform the typical movements of digging out a wasp's nest. Vacuum actions being thus, by definition, not related to relevant environmental factors, they achieve no result, except possibly that of "exercising" the particular motor patterns involved.

Fig. 1-18. Distraction display, which occurs in many birds, appears like a feigning of an injury. Top: A ringed plover *(Charadrius dubius curonicus)*; Bottom: ringed plover *(Charadrius hiaticula hiaticula)*.

Many peculiarities of innate behavior will be dealt with in connection with the particular orders, families, and species of birds. The behavior of birds is almost always simpler and easier to interpret than that of most mammals. But although the innate modes of behavior are rigid, birds are certainly capable of learning; thus they adapt to the specific conditions of their own environment. For many birds, there is a period during which they must learn new song patterns in addition to the inborn basic forms of their own specific song. Thus, through imitation of certain characteristics of the adults' song, regular dialects may originate which are passed on from generation to generation in the same area.

Other innate modes of behavior are by no means completely rigid at all times. When crows with a drive to build a nest cannot find a tree, they may nest on the ground, and, if necessary, use nesting materials other than those prescribed for them by inborn mechanisms. In such cases it is not yet known whether actual changes in behavior are involved or whether the innate mode of behavior has several additional modes of behavior available. In any case, birds are able to perfect innate capacities, such as flying, through experience.

Memory and learning

The capacity for true learning is not unequally distributed only among bird species, but even within one species some birds learn more easily than others. Certain birds learn to repeat words and short sentences, to imitate the whistling of melodies, and, in nature, to mimic songs of other species. They learn to recognize certain conspecifics and people individually, they observe the number of eggs in their nests, and in a short time, they get to know their eggs "personally" so that strange ones cannot be substituted for their own. Birds also become familiar with favorable feeding places and generally have a good memory for localities. Many can find hidden food, as well as the previous year's nest. Cockatoos readily memorize the place in a wire mesh where a hole permits them to leave the cage; jackdaws are not as adept at this, and domestic chickens cannot do it at all.

Insight and thinking

No insight is required in hiding food. But when a raven *(Corvus corax)* hides a piece of food under a carpet and immediately stops when an observer is present, one might think a certain degree of insight is involved. Many psychologists consider such behavior a conditioned reflex. The same assumption is made when a gray parrot *(Psittacus erythacus)* says "good-bye" as soon as its owner picks up his hat. Generally, however, even birds which can speak very well will chatter randomly without making much sense. Parrots and starlings are exceptionally good talkers; bullfinches *(Pyrrhula pyrrhula)* and mynahs are very good at imitating whistling. The language of birds does not reflect the use of concepts, but one can ascribe to birds a kind of prelinguistic thinking, as Otto Koehler discusses in the parrot chapter (Vol. VIII). In some species of birds we find the use of tools, which is amazing at first. For example, certain bower birds (Ptilonorhynchidae) use fibers and pieces of bark as paint brushes to paint their display bowers; one of the Darwin finches *(Camarhynchus pallidus)* uses a cactus spine to dig out insects. But these abilities do not presuppose a higher level of intelligence; like nest building, they are a part of the innate repertoire.

On the one hand, birds can be trained to respond to a specific number of objects or a specific cue direction; on the other hand, they act as if bound rigidly to specific forms of behavior. We must therefore ask whether or not such behavior has been preceded by periods of learning in which the bird has found what is correct by trial and error.

Trial and error

If several conspecific birds, unknown to one another, are placed in the same cage, each one will try to first chase the others away from the food. After several hours or days, a peck order has developed in which bird no. 1 usually chases no. 2 away from the food, and no. 2 chases no. 3, and so on. Less often, as with the ring-necked pheasant, no. 3

may chase off no. 1. By trial and error, a bird learns the better food, and which trees to peck at; for example, a woodpecker learns to prefer telephone poles when they are drier than healthy trees. In training cormorants to fish or raptors to catch and deliver prey (as in falconry), man is the dispenser of immediate reward or punishment; thus by trial and error, learning can be successfully accomplished. This learning, however, depends upon individual abilities. Among cormorants, certain strains have been bred which learn very well and as a result are expensive.

The language of birds

In order to determine whether specific body postures constitute part of a bird's "language," one must establish to what extent conspecifics or other species understand them. In response to warning calls, the young of the common curlew *(Numenius arquata arquata)* squat; in response to other calls, the young of precocial birds follow; in response to the alarm call of a captured starling, even when the call is tape recorded, starlings fly out of a cherry tree. When mallard ducks *(Anas platyrhynchos)* move their beaks up and down, they convey with such "intention movements" that they will soon fly off together and thus they remain together; they arouse the same mood in other conspecifics. Whenever a bird with outstretched and lowered head, and sometimes lowered wings and ruffled plumage, assumes a defensive posture which looks as if it were about to attack, this is understood at once by the real attacker; this posture alone often helps to prevent fights. Like many actions, it is a ceremony or ritual which often prevents unnecessary expenditure of energy. The display behavior or showing-off of many birds, such as the tail display of the peacock, attracts females and keeps most males at a distance. Often parts of the plumage or the skin, which are normally hidden, are displayed. A certain rigidity is characteristic of such displays which occasionally occur even in females when they wear a bright nuptial plumage. In such cases, however, the female, when meeting a male, immediately stops her display. Thus unnecessary and possibly harmful fights with males are avoided.

Fig. 1-19. The white stork scratches itself as shown.

The black-winged stilt *(Himantopus himantopus)* does so over the lowered wing.

Expressive postures and actions are thus "understood" within a species without the complete behavior being performed and without an overt "intent to communicate" being apparent. Thus social behavior within a species has particular modes of expression at its disposal which are understood only by its members. Thus parents use their own signals to communicate with their young in the nest; others are used among the young, and still others are used by the young toward their parents. Members of the pair, companions within a flock, and individual birds or flocks all communicate with other conspecifics. Hence a bird recognizes a member of its own species by visual and audible signs as well as by its behavior. Frequently birds even understand the warning calls of other species, such as that of the jay *(Garrulus glan-*

Anting

darius). The innate reactions to enemies or prey are also understood by conspecifics.

Evidently the bird does not understand the significance of this species language any more than it does the meaning of its other forms of behavior. Picking up grains or fetching building material, therefore, does not seem to take place "consciously"; neither does the particularly interesting "anting" behavior which has so far been observed in about 160 species of birds. These birds rub ants into their plumage with their beaks for periods lasting up to forty-five minutes; less frequently they stand with fluffed-up feathers on a swarming ant heap. When in captivity, they accept mealworms, mustard, and the like, as substitutes. Presumably formic acid or the sight of ants and other insects triggers this behavior. Anting is purposive; originally birds probably killed certain external parasites, such as mallophoga, in their plumage by this method. There are quite a number of other such self-directed actions, from oiling the plumage from the preen gland to cleaning and ordering the plumage with the beak, foot, or toe. In scratching, birds bring up the foot either directly, i.e. under the wing, or indirectly, i.e., over the somewhat lowered wing. When sunbathing, a bird ruffles its plumage. Many birds bathe only in dust, but most of them bathe in water, using several methods. The movements which are performed, such as feather erection and shaking and sleeking the plumage, may, besides their actual function, be valuable in communication with others of the species, with partners, young, or enemies.

In many of the species which are discussed in subsequent chapters, forms of behavior will be considered which should be regarded as mere parts of the total behavior repertoire of the species, its behavior inventory or ethogram. Any number of questions concerning the behavior are presented to the ethologist. For example, in the display call of the goldeneye duck *(Bucephala),* the bird extends its neck, lays its head on its back, flings it to the front, and calls. The ethologist asks, "Why? How was this behavior released?" He asks about its significance or its function. "How do members of the species react or what else is achieved by the behavior? From where and when in its evolution did the bird acquire this behavior? When does it first appear in the life of the individual?" The call seems to have gradually developed through the evolutionary transformation of the trachea; the peculiar shape of the trachea in male goldeneyes compels this forced posture with the head laid back, if a sound is to be produced at all. In the display behavior of other species, components of nest building behavior or of the tending of the young may be "built in." Often the male "symbolically" presents nest material to the female or feeds her, as if she were a helpless youngster. Components of feeding behavior involving the use of fish are part of the courtship display of terns (see Vol. VIII).

The social behavior of most birds is highly developed, and is less so only in a few species. To these few belong certain species of woodpeckers, in which even the members of a pair avoid one another as much as possible; in their case, in the words of Heinroth, male and female are permanently at war and seem to find it "horrible that a second bird is needed for reproduction." Most birds can lose their fear of enemies, including man, in a process which is inherited, as is the case in domestic turkeys as compared with wild ones, and in the case of birds that live on islands where enemies are few, such as the Galapagos Islands. Presemably out of "fear," crows and many small birds flock together and, crying loudly, perform real or sham attacks (the so-called mobbing) on predatory animals such as raptors, owls, cuckoos, or cats. Even man may be attacked by a goshawk or a tawny owl if he approaches the nest. When birds temporarily live in families, the members of the family maintain body contact in only a few groups, e.g., the whiteye. Snapping at the same piece of food is best avoided if each bird remains a certain distance from the other (individual distance); in small birds this is at least ten centimeters.

Social behavior

Social behavior is, of course, most pronounced in colonial birds which often breed together in hundreds of thousands or even millions. Penguins, while breeding, perform swimming games and social games. Otherwise, games are played most frequently in young birds. The "aerial games" of adults are generally a part of reproductive behavior.

There are phenomena in bird life which strike us as unsocial or antisocial, including brood parasitism, the robbing of eggs, young, or nests, taking of food, and the robbing of nest material. Apart from this, various forms of mutual aid which characterize the social life of birds are also most remarkable. Large aggregations of birds gathered for sleeping, breeding, or feeding are less likely to be surprised by enemies, since there are always a few birds awake and alert. The members of a jackdaw colony collectively drive off attacking enemies; pelicans collaborate in fishing. Young are often not only fed by strange members of the species, but even "adopted." Mutual preening is also a part of the social life of birds.

Personality

Apart from the conspicuous differences in the behavior of the species, there are also less obvious individual differences in the behavior of members of a species. One individual conspecific bird may be more fearful, another more aggressive, one more quiet, another more excitable, one more and one less "intelligent." Birds are thus by no means "automata" which exclusively follow the hereditarily predetermined modes of behavior of the species. Therefore, pair formation among birds in captivity depends not only on the will and the health of the birds, but often on their individual preferences as well.

Eyes and vision

In the life of birds, more so than in mammals, automatic responses to releasers are of primary importance in controlling their behavior. For this the bird needs distance receptors which deliver information

Fig. 1-20. From the behavior of the common curlew: threat posture of the male.

Threat during a conflict over rank.

Incipient breast rotation in the course of pair formation.

Copulation ceremony.

for the release of adaptive behavior to appropriate areas of the brain. With one exception, the kiwis, birds are visual animals. In general structure, their eyes are very similar to those of reptiles. Here, in contrast to other parts of the body, nature has not spared any weight. Excluding muscles and the optic nerve, the eyes of a tawny owl weigh 1/32, and those of a peacock about 3/100, of the body weight; small birds comparatively have the largest eyes. They are quite rigidly fixed in orbits which they fill; thus the weight of the six eye muscles was eliminated. The bony partition between the orbits often contains large holes. The head and neck move the eyes; some eye movements, however, are not noticed since the eye ring moves with the eye so that the latter looks rigid. An exception to this is the bearded vulture. Owls, because of their relatively large and light permeable cornea and their very small retina, have eyes with the greatest light concentration. In the tawny owl the eyes take up one third of the head. These eyes are directed forward; their visual fields overlap considerably and their axes together form an angle of about ninety degrees, while in most other birds the axes form much wider angles, up to 180 degrees. For this reason, owls, when looking about, must turn their heads around. They can see sixty to seventy degrees of their visual field with both eyes; in many other birds overlap is only six to ten degrees. In an exceptional case, the black-footed penguin *(Spheniscus demersus),* each eye receives the picture of only one side. Such eyes, which permit all-around vision, are particularly important for ground birds which are endangered from all sides.

Of the two eyelids, birds usually close only the lower, but not the upper lid. In addition, birds can draw a third eyelid, the nictitating membrane, in front of the eye; it cleans the cornea from the inner upper to the outer lower edge and in some diving birds presumably protects the eye. The inside of the cornea is often supported by ten to eighteen bony plates which, for example in the eagle owl, form a tubular protection for the long eyes. The lens, unlike in mammals, is flattened by the relaxation of muscle tension and made convex by pressure. In this way birds achieve accommodation, which is poor in most nocturnal owls but very good in divers. The iris may be almost any color and in many species is of a different color according to age and sex. It has a continuation on all sides, the choroid, from which the pecten projects like a fan into the vitreous body. The kiwis are again an exception. Here the pecten consists of two to eighteen folds and acts like a venetian blind, thus improving the ability to detect movement. The retina contains rods and cones throughout; it is thus capable of forming sharp images right up to the edge. Cones contain yellow, reddish, or greenish oil droplets and are individually connected to the brain; they are the receptors for color vision. The rods, which are interconnected in groups of thousands, are used in light-dark vision. Besides that, there are areas of maximal visual activity, the fovea. In

F. Reimann

North Sea Coast in Winter:

In addition to coastal birds which remain with us all year (residents), one can see many winter visitors from the north along the North Sea coast. □ Loons: 1. Red-throated loon *(Gavia stellata)*. □ Tube-nosed Swimmers: 2. Fulmar *(Fulmarus glacialis)*. □ Pelecaniformes: 3. Cormorant *(Phalacrocorax carbo)*. 4. Northern Gannet *(Morus bassanus)*. □ Geese: 5. Brent Goose *(Branta bernicla)*. 6. Tufted Duck *(Aythya fuligula)*. 7. Greater Scaup *(Aythya marila)*. 8. Goosander or American Merganser *(Mergus merganser)*. 9. Smew *(Mergus albellus)*. □ Raptors: 10. Gyrfalcon *(Falco rusticolas)*. □ Shorebirds and Gulls (see Vol. VIII): 11. Dunlin *(Calidris alpina)*. 12. Turnstone *(Arenaria interpres)*. 13. Kittiwake *(Rissa tridactyla)*. 14. Greater Black-backed Gull *(Larus marinus)*. 15. Caspian Tern *(Hydroprogne caspia)*. 16. Razor-billed Auk *(Alca torda)*. 17. Guillemot *(Uria aalge)*. 18. Black Guillemot *(Cepphus grylle)*.

a few bird groups, there are two fovea fields, a lateral one for light rays which come from objects close to the beak (in owls the only one), and a medial one which helps each eye perceive rays entering the eye along the line of sight. This medial visual field is often band-shaped and thus is suitable for the discovery of enemies coming from the horizon.

Because of these peculiarities, the visual capacities of birds are hardly comparable with those of other animals. Nevertheless, birds smaller than the domestic pigeon or the hobby *(Falco subbuteo subbuteo)* probably see less sharply than man, but see a much larger area equally in focus at one time. They can also probably perceive and process twice as many images per second as man, up to 150, which is important in swift flight. Larger birds, however, which have more visual cells (which are about the same size in all species), generally see better than humans. What an eagle sees with its eyes, we can only recognize with a six power telescope.

Auditory and vestibular senses

Birds not only see well, they also have excellent hearing. Their ears can perceive sounds varying in pitch from forty to almost thirty thousands cycles per second. In this regard they are comparable to man, who can perceive sound frequencies from sixteen to twenty thousand cycles per second. Their auditory acuity is, however, less than that of man, with the exception of owls, which can also localize sounds best. The auditory organ of birds (the lagena) lies in a slightly curved bony tube. The vibrations of the eardrum are transmitted by a single bone, the calumella, to the legena. "Feather ears" found in many birds do not belong to the external ear but to the plumage. The sense of equilibrium, which relays information about the changing position of the body in space, depends on the semicircular canals and other parts of the membranous labyrinth.

Other senses

Though most birds have nostrils, their olfactory mucosa is usually small, except in kiwis, the turkey vulture, and perhaps the king vulture *(Sarcoramphus papa).* Their sense of taste is not well developed either. Thus domestic chickens can only distinguish salt and acid and possibly a mixture of sweet and bitter. Tactile organs are found in great numbers on the skin, the buccal cavity, on the beak, and in the muscles of the lower leg and lower arm. Receptors for other body sensations such as pain and temperature are also located here. They signal, among other things, whether the feathers should be erected or pressed to the body, and they may be responsible for the detection of radar. Researchers continue to look for a "magnetic sense" which may help birds orient themselves. There is some evidence to support its existence; however, there is no convincing demonstration yet of a special sense.

The brain and the nervous system

With a light flying body, the bird must, as much as possible, keep the connections from the central nervous system to the brain at a minimum. There is also some saving of weight in the brain itself since liquid-filled spaces within the skull are absent. Nevertheless, in proportion to the body weight the brain of birds is five to twenty times larger than in reptiles. "Thus by weight, it equals that of most mammals," as J. Schwartzkopf writes in Berndt-Meise's *Natural History of Birds.* "Its structure, however, is closer to that of reptiles. Compared to them, it is primarily the cerebrum and cerebellum which have become enlarged. Within the class of birds, ostriches, gallinaceous birds, and pigeons have the lowest relative brain weights, and parrots the highest. Among related species, the larger members have relatively the largest brains. The large parrots are 'smarter' than the smaller species, and the raven is the largest and also the most intelligent passerine bird."

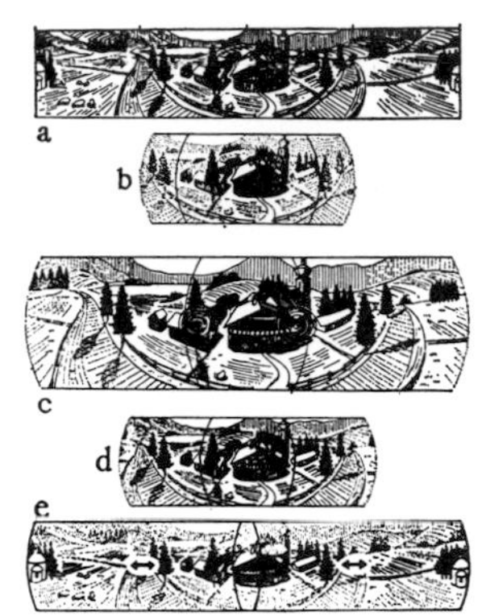

Fig. 1-21. How a bird sees its environment: Of the panorama (a) the human eye perceives, only a segment of barely 90 degrees (b), the visual fields of both eyes overlap considerably (indicated by the central portion of the sketch enclosed by curved lines). The eyes of a bird of prey (c) stand more to the side of the head, the overlap of the two visual fields is less, and the segment of the panorama perceived is wider. The eyes of owls (d) are directed more to the front; their visual field is intermediate between that of man and that of raptors. The greatest visual angle (360 degrees) is shown by many shore birds (e); the visual fields of their eyes overlap not only in front but also behind them (the chapel appearing at the extreme right and left lies exactly behind the bird).

While the optic lobes and the cerebellum of birds are large, the diencephalon is not prominent. All parts of the brain, as Schwartzkopf puts it, "are closely packed together in the tight bird skull; the large eyes press against the brain, so that its basis is directed forward and often even forward and up, particularly in the Scolopacidae." The movements of various parts of the body are controlled and modified according to information received from the vestibular sense in the cerebellum, which is made up of folds. The impulses then are transmitted to the tegmental motor nucleus which extends from the metencephalon to the midbrain. They are combined there with impulses from the tegmentum of the midbrain and the cerebrum, which are then passed on to the spinal cord. Since a falcon, after removal of the cerebrum, still pursues flying objects, the image of a sparrow must be transmitted to a structure or a region in the midbrain; thus the appropriate movements are released.

What is then the function of the cerebrum? When a falcon with an excised cerebrum captures a sparrow, it does not know what to do with it. It lacks the hunger motivation. The cerebrum of birds has no folds but has a smooth cortex; here the perceptions coming from each eye are projected to the right and left hemispheres respectively. But from here there is no connection to the anterior horn of the spinal cord, as in mammals, but only one leading to its beginning. The cerebrum receives impulses from the diencephalon and transmits them to the midbrain, cerebellum, medulla, and the anterior part of the spinal cord. If the midbrain area is stimulated by electrodes, different moods and reactions can be elicited, such as flight, display, or eating. The different moods evidently are not restricted to sharply defined brain centers.

Hormones, the chemical messengers

In addition to the nervous system, which might be thought of as a telephone system, there are the hormones which are messengers of a sort. They are made in glands, secreted into the blood, and transported throughout the body to their site of action. They particularly affect the autonomic nervous system, and they are effective only at specific sites. The anterior lobe of the pituitary gland secretes numerous hormones. In wydah birds, one hormone stimulates the formation of the nuptial plumage, another the drive to incubate and in pigeons the formation of the crop "milk." Most, however, deliver trigger responses in other glands. The thyrotrophic hormone stimulates the thyroid gland and the gonadotrophins stimulate the sex glands. The thyroxin of the thyroid gland is, among other things, important for increasing the metabolism for release of migratory restlessness and for moulting. The male hormone, testosterone, stimulates the formation of the male nuptial plumage in the ruff *(Philomachus pugnax),* but not in the domestic cock, where the male plumage is neutral. Here a female sex hormore is responsible for the fact that a genetic female does not look and crow like a cock. In old age, when the production of the female sex hormone diminishes, or after disease of the ovaries, hens may show cock feathering. If male and female sex hormones are combined, hermaphroditism may occasionally occur in birds.

Sex organs

In their sex organs as well, birds conserve weight. Although they have two ovaries, as all vertebrates do, only the left one is developed. Whenever, occasionally, a right ovary is present, as is frequent in the Eurasian sparrowhawk and sometimes in domestic hens, a functional right oviduct is still missing. The egg containing the yolk with the embryonic streak erupts from an enlarged egg follicle; it passes via the funnel of the oviduct which is much enlarged at the time of reproduction, and there it is surrounded by layers of albumen (egg white), as well as by two shell membranes and the shell. At the large end the two shell membranes are separated by an air chamber, which can be seen when opening a hen's egg. The yolk sphere always remains surrounded by albumen, for it is suspended by the spiral chalaza, which runs from pole to pole as dense, twisted albumen. This ensures buffering and an ability of the yolk with its sensitive ovum or embryo to rotate. The egg shell consists of calcium carbonate and generally has pores, canaliculi, a sort of artificial membrane, and color determined by pigments.

As in reptiles, in birds the passages for feces, urine, and secretions from the sex organs empty into the cloaca. In the midportion of the cloaca the male has lateral openings for the ureters and the openings of the sperm ducts, while the female has a single opening, the vagina, in the corresponding position. Only in a few groups of birds, such as ducks and geese, does the male have a copulatory organ; it is generally spiral in form so that it can reach the female sex opening which lies somewhat to the left in the cloaca. In most species where males have no penis, the cloacas are pressed together during copulation, so the semen can reach the vagina in this manner. After the breeding season the primary sex organs of males as well as females retrogress.

The bird egg

Birds are the only vertebrate class in which there are no species which give birth to live young. The oviduct at any given time contains but one egg; a female bird can therefore lay only one egg per day (only in domestic hens is a second egg sometimes laid within a day). This too saves weight. When the egg appears at the cloaca, as it does in the domestic hen twenty to twenty-four hours after the rupture of the follicle, the pointed end first emerges, but the egg may still be turned before laying. Thus, twenty to thirty per cent of chicken eggs are laid with the large end first. The egg of the smallest bird (hummingbird) weights a quarter gram; that of the largest, the extinct elephant birds (Aepyornithidae), weighed just about ten kilograms. The eggs can be white or colored and from an almost spherical to a cylindrical shape. Some birds lay eggs which weigh twenty-five percent of the body weight; the clutch laid within a few days often exceeds the bird's body weight.

External sex differences

In many birds, such as cormorants or crows, males and females cannot be externally distinguished. Only from the response of one animal to the display behavior of the other can they recognize whether they are of the same or of opposite sexes. In many species, however, the sexes look strikingly different. In these the partner can be identified at a distance. Males are generally larger and more colorful than females. Apart from such differences in color and pattern, there are differences in feather structure and, in the males, inflatable skin sacs, combs or other processes on the crown or beak, appendages on the side of the beak or throat or in the middle of the throat, spurs on the leg, and other features.

Seasonal differences

Often the nuptial plumage of the male or of both sexes is not present for the entire year. Thus European robins *(Erithacus rubecula)* always have a red throat, but that of blue throats *(Luscinia svecica)* is only blue in the breeding season. The change from one plumage to another can be brought about in different ways: by growing additional feathers, rubbing off of feather tips, a partial moult, a complete moult of the contour feathers or even a complete moult; furthermore, it may also occur by a color change of the beak, the growth of warts, the formation of a beak appendage, or other changes.

These changes are generally completed at the beginning of the breeding season. The breeding season, in the widest sense, follows a rest period. The bird comes into a breeding condition with an enlargement of the sex organs. When the young become independent, the breeding season ends. In Central Europe, birds generally breed in spring and summer, and only occasionally in winter. The time of breeding is basically determined by the availability of adequate food during the growth period of the young. Since the breeding seasons of the various species of a particular region can begin at different times, the external releasing factors (increase of day length, temperature, and small animal and plant life) must have different threshold values. In the tropics there are dry season, rainy season, and year-round

▷

Fig. 1-22. The male common redstart, which has occupied a territory within a nest hole, shows the nest hole to a female which has entered his territory. This striking hole-showing behavior is inborn; the action itself, however, is not rigid, but may be performed quite variably. Three variations of hole-showing are illustrated.

Breeding season

breeders; in the latter category, individual birds or pairs probably have long non-breeding periods as well. The duration of the breeding season varies from about two to fifteen months; it is shortest in auks, longest in king penguins *(Aptenodytes),* although in the large albatrosses it lasts a year. Albatrosses can, of course, only have one brood a year and then they rest a year; king penguins solve these problems in a different manner (see Chapter 5).

We generally find in many bird species that several broods follow one another each year without pause. Often they overlap, i.e. new eggs are already laid again while the parents must still care for the young of the first brood. In many Central European species, a pair may raise two to three broods per season; house sparrows and pigeons even manage five broods a year.

Breeding places, territories

A bird selects a specific locality in the breeding season. Brood parasites also do this, as well as species which migrate outside the breeding season. Many birds establish the breeding place as their own territory, and defend it against members of their own species, at least those of the same sex. In insect-eating birds particularly, the ownership of an area probably guarantees the success of the brood more than if many nests of conspecifics existed in the same place, thus necessitating long flights to obtain food. In some colonial breeders, however, such as fulmars and common herons, long flights to feeding grounds are quite common.

Bird territories often comprise almost the entire space occupied by a pair in the breeding season; this is true with the nuthatch *(Sitta europea).* Penguins, tube-nosed swimmers, pelecaniformes, gulls, auks, and other colonial breeders which feed at sea and therefore have their breeding space confined to a small area on the coast, often defend only the immediate area surrounding their nests, as far as the beak can reach. In hoatzins, anis, song shrikes (Cracticidae), and a few other species, small groups maintain a common territory. Only rarely is a bird non-territorial, such as the nutcracker.

Often the territorial boundaries change somewhat during the breeding season, but they are well known to neighbors. If a member of the species approaches the boundary, then an interchange of innate patterns of behavior and postures usually prevents serious fights and the intruder gives way to the territory owner. The territorial song of most song birds and the corresponding calls of many other male birds indicate to conspecifics, "Here is a male with territory." The widely audible song warns off males of the species in time, while it attracts unmated females. In many species the bird's appearance or its displays are understood as claims to a territory. By the occupation of territories by the males before the start of the breeding season, the breeding pairs distribute themselves according to the most favorable food and nest site conditions over a district.

In addition to breeding territories, there are also display and food territories, and territories claimed by young birds; many birds occupy a different territory in the winter than they do during the breeding season. The females of certain species also behave like territory owners and so chase strange females away.

With most birds pair formation begins only after the males have occupied the breeding territories. The females of many species arrive in the breeding area after the males and thus have the opportunity to select a male who looks the most impressive, sings the best, or has the most adantageous territory. We do not know whether female birds really make their choice this way, however. When the sexes are similarly colored, pair formation is often preceded by mutual threat displays.

Pair formation

At the time of pair formation, most birds are already in a reproductive mood. But there are also species which already court in the fall or winter, and many ducks have a regular period of engagement lasting several months before they come into the reproductive mood. Pairs stay together at least until the time of egg-laying, and very often until the young are independent. A re-mating with another partner may occur for the next brood. Many birds, however, mate for the season and the pair remains together after the first brood until the young of the next brood reared that same year are also independent. In other species, pairs remain together for several years, and in geese and some other species, pairs are for life.

Monogamy and polygamy

There are also species which live in polygamy, where several females are paired with one male (polygyny) or others where several males mate with one female (polyandry). Complete lack of pair formation also occurs. In such species, the sexes come together only for copulation; this holds for the black grouse *(Lyrurus tetrix),* the woodcock *(Scolopax rusticola),* the ruff, and the lyrebirds (Menuridae), as well as many other species.

Monogamy is, however, by far the most common form of mating in the bird world. Particularly in insect-eating species, the rearing of the young is evidently better assured when two parents care for the young than if the entire job had to be borne by one adult.

The many innate forms of behavior which lead to pair formation, bringing the sexual behavior of both partners to the same level, copulation, and keeping the pair together are generally "ritualized," as ethologists call it. Nest construction is also innate. The nest site is selected by the male, the female, or by both partners together. The nest may be an unaltered piece of ground, the bare floor of a tree cavity, a hollow made by a few rotations of the body, or a work of art constructed after weeks of work; between these there are all kinds of transitions. Most birds build their own nest, althought some take over nests of other species. Still others, like the Eurasian cuckoo, are brood

Nest-building

parasites; they lay their eggs in the nests of other species and leave incubation and rearing to the "foster parents." Only in the largest penguins is there no nest; they carry their egg about on their feet.

Small birds normally build a new nest for each brood; this is important since their nests are heavily infested by parasitic mites and insects. Larger birds often use the same nest for several successive years. It is generally repaired for each brood and often these repairs continue during breeding.

Egg-laying

Once the nest is ready, generally only a few days pass before the female lays the first egg; in some species, however, the interval is longer, sometimes even many weeks. Occasionally the commencement of laying depends on a certain ambient temperature being reached and maintained. In such cases, a whole population of birds may lay its eggs almost simultaneously; in the slender-billed shearwater *(Puffinus tenuirostris),* egg-laying is completed within twelve days. Most birds lay early in the morning. Depending on the species the clutch consists of one to eighteen eggs; clutch size also varies somewhat within a species, depending on locality or food supply. If a nest contains more than eighteen eggs, they have almost certainly been laid by more than one female.

Incubation

All bird eggs need warmth for their development, which is normally supplied by incubation. Areas of bare skin develop on the abdomen of most incubating birds; these are the brood patches in which the blood vessels dilate, thus providing a particularly rich blood supply to the skin and a consequent sufficient supply of heat. The bird squats over the eggs in such a way that their upper surfaces contact the brood patches. The eggs are generally protected along the sides and downward by the nest lining from the cool outside. This is, however, not always the case; in many birds the eggs lie on the bare ground. In guillemots they even lie on rocks over which sea water may spill.

In many species male and female incubate in alternation. The nest-relief takes place in these species accompanied by innate ceremonies. The relieving partner may bring nest materials and pass them to the other, or it may touch it on the back and thus encourage it to stand up. In some birds only the female incubates; in others, only the male. In these cases, longer or shorter intervals occur during incubation since the birds must leave the nest to feed. Toward the end of incubation, birds sit more tightly on the eggs than at the beginning; they are more reluctant to leave the eggs when disturbed.

Most birds rotate their eggs from time to time. Thus the center of gravity of the egg always comes to lie near the bottom so that the embryo and later the young lie towards the top. In a few cases it is necessary to protect the eggs from ambient heat; this is done by the bird standing over them and shading the eggs with the wings, or by burying them. Some species, like the greylag goose *(Anser anser),* roll back eggs which have rolled out of the nest, while others uncover eggs which have been covered by blowing sand.

The shortest period of incubation is found in the Eurasian cuckoo and in white-eyes with ten and a half days, the longest in the royal albatross *(Diomedea epomophora)* with eighty days. The normally mobile bird is tied to its nest during the incubation and nestling periods of its young; it is therefore exposed to enemies which know its locality. Hence in general the greater the potential danger to the clutch, the shorter the incubation period. Seabirds on small islands where there are no egg-eating mammals and large birds of prey, and which are capable of defending themselves can afford a long period of incubation and slow development of the young. Most of our European song birds, on the other hand, incubate only twelve to fourteen days and then feed the young for another twelve to fourteen days in the nest. Hole breeders have, on the average, longer incubation and nesting periods than birds breeding in the open; only woodpeckers are an unexplained exception. Within the same group, larger species as a rule incubate longer than smaller ones.

Embryonic development

Bird eggs, like reptile eggs, contain much yolk. The embryonic disc rests on the large yolk sphere, which is part of the ovum. In it the nucleus of the egg unites with the nucleus of the sperm after fertilization. The first divisions, up to a stage of development of several hundred cells and the formation of the inner and outer embryogenic layers, still take place within the maternal body. Then development is interrupted until the onset of incubation. The embryo and later the fetus develop from the embryonic disc. During this development the yolk is used up. It does not reach the body through a connection between the yolk sac and the embryonic gut, but via the blood vessels of the yolk sac. The albumen (generally called the egg white), or at least its remainder, is pressed into the amniotic sac toward the end of development, and from there it enters the beak to the fetal gut as the chick's first meal. Often part of the yolk, sometimes a third of the original mass, remains under the abdominal wall of the young bird and is only absorbed after hatching. Many precocial young only take their first food after two to three days.

The body of the fetus absorbs calcium from the egg shell and builds it into its skeleton during development. This changes the egg shell; it permits the passage of carbon dioxide and oxygen and the release of water vapor. The egg membranes are the same as in reptiles (Vol. VI), and embryonic development also proceeds much as it does in reptiles. At first the bird embryo seems to consist primarily of brain, heart and tail. This long reptilian tail, however, soon retrogresses. In front of and behind the heart, two fin-like pairs of limbs extend, which then form arms and two temporarily five-toed legs. The little arms eventually become wings. But when the embryo, after several days,

▷

Fig. 1-23. A—Neogaea (Neotropical region) A/B —Sonorea (Neotropical-Nearctic transitional area) Ba—Holarctis (Nearctic and Palearctic regions) Bb—African or Ethiopian region; Bc—Madagascan region.

becomes a fetus, it has neither a cloacal opening nor a mouth. These two openings develop later. In the last days before hatching the feather buds grow; in many altricial birds they submerge again below the skin so that the little ones are either completely or almost nude at hatching. Their eyes and ears are generally still closed as well.

A few days before hatching, most fetuses grow a calcarious projection on the upper mandible of the beak, the egg tooth. Young woodpeckers also have such a tooth on the lower mandible. Shortly after this structure is ready, it is used to open the air chamber of the egg and the young bird can then peep within the egg, sometimes three days before hatching. In large birds hatching may extend over two to three days. Most birds hatch in the morning; the parents help their young out of the egg only in rare cases.

A bird hatches, by K. Lorenz

How a greylag goose chick hatches from the egg is described by the well-known ethologist, Konrad Lorenz. "Important things must be happening in the goose egg. If one places one's ear to it, one hears within a crackling and a soft rumbling, and then one hears a soft, flutelike 'peep.' Not until an hour later does the egg have a hole, and in this hole one sees the first part that is to be seen of the new bird—the tip of the beak with its egg tooth. The movement of the head, with which the egg tooth is pressed from within against the egg shell, not only effects the cracking open of the shell, but it also brings about a movement of the rolled-up little bird, which thus rotates a little each time about the long axis of the egg. The egg tooth thus moves in a 'parallel circle' within the egg shell and makes a series of cracks along this line until when the circle is completed, the whole blunt end of the egg can be pushed off in one piece by a stretching of the neck.

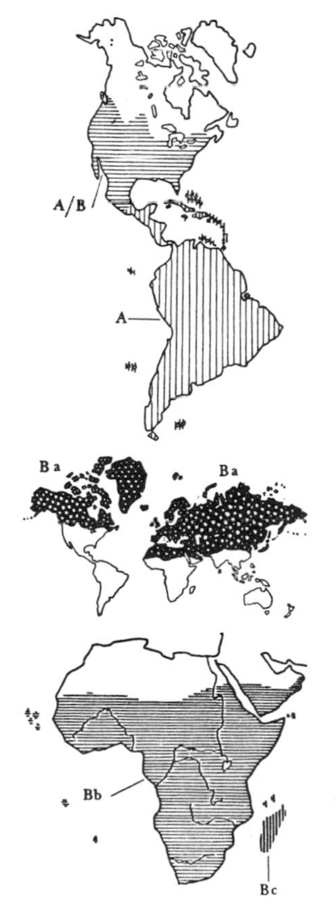

"Slowly and with difficulty, the long neck, which cannot yet carry the heavy head, becomes free. The nape still remains stiffly bent down in the embryonic position which it has held since its beginning. Hours are required until the joints stretch and become pliable; the muscles strengthen and the vestibular organs in the inner ear function to maintain balance so that for the gosling there is, for the first time, an up and down, and it can now hold its head free and upright.

"The wet 'something' which emerges from the shell looks unbelievably ugly and quite pitiful and, more than anything, wetter than it really is. For if one touches it, it feels only moist. The impression that the plumage is wet and stuck together arises because each downy feather is still tightly folded and enclosed in a thin membrane. In this form it is no thicker than a hair. All these feather-hairs are stuck together in strands by the albumen-rich liquid within the egg so that they occupy a minimum of space. When these feather membranes dry, they disintegrate into dust and liberate the enclosed downy feathers. These, therefore, do not really dry out; they are already dry when they were covered by capsules and thus were protected against the fluids in the egg. The bursting of the feather membrane is, of course, facilitated by the movement of the newly-hatched youngster and is accelerated as it rubs itself against its siblings and against the abdominal plumage of the brooding mother. When this friction is absent, as in an incubator-hatched gosling, the feather membranes remain intact longer than usual. In such a case, one can perform a surprising little bit of magic. One takes the bird in one hand, a lightly greased ball of cotton wool in the other and with it rubs softly against the feathers over the youngster. The brittle feather membranes break into very fine particles resembling hair scales; the gosling, however, changes in a magical way. Where the cotton wool passes a thick growth of airy, fine, golden yellow down appears and in a few seconds one has, instead of the naked, moist, gummed-up little monster, a sweet round ball of down which is at least twice as large as before."

Growth, distribution, and environment, by W. Meise

The egg tooth generally drops off soon after hatching; in one species of penguin it remains, however, for forty-two days. The young of some species are as large as their parents after only ten days; in the ostrich and king penguin, however, growth takes a whole year. The emperor penguin *(Aptenodytes forsteri)*, however, which is twice as heavy as its slower growing cousin, must be full-grown in five months when the hard Antarctic winter begins.

Rearing of the young

After hatching, the parents almost always brood (keep the young warm), feed, and care for the young. They keep them clean, guide and defend them, warn them of danger, and often distract enemies away from the young by fluttering their wings "helplessly." Substituted strange young are often cared for, but there are species which are not so readily deceived by brood parasites (cf. cuckoo, Vol. VIII and weaver birds, Vol. IX). Rearing behavior and the behavior of young are so variable among birds that a general treatment is here dispensed with.

Fledging

The fully fledged young of altricial birds generally leave the nest in the morning, almost always without the urging of the parents. Many species leave the nest before they can fly well; their initial clumsiness readily leads many people to the assumption that the young must first "learn" to fly. This is, however, not the case; the capacity to fly matures almost simultaneously with development of the primaries. To investigate this, ornithologists took some nestlings from their nest shortly before fledging, reared them artificially, and prevented them from flying. They waited until the others left the nest and could fly well. Then they released their experimental birds. It was shown that those which had been prevented from flying flew at once, just as well as their siblings which had had the opportunity to "learn" to fly. Generally, there are so many variations of the altricial and precocial type that details are considered under the individual species.

Birds, like butterflies, belong to the few animal groups in which the female has only one sex chromosome in every cell nucleus, while the

▷

Fig. 1-24. Bd—Indomalayan or Oriental region. B/C—Wallacea (Indomalayan Papuan transitional area). (Ba, Bb, Bc, and Bd together form the Arctogaea or Megagea) Ca—Papuan region. Cb—Polynesian region; the greatest part of the islands of the middle and southern Pacific (not shown here) belong to this region. Cc—Australian region, Cd—New Zealand region (Ca, Cb, Cc, and Cd together form the Notogaea). D—Antarctic region.

Heredity

male has two. In mammals, including man, the situation is reversed. In birds, it is the chromosome distribution in the egg nucleus and not in the sperm which determines the sex of the embryo. Further details are given in the chapter on heredity in Vol. I.

Duration of life

The maximum age which birds can attain varies greatly. Hundred-year-old parrots certainly belong in the realm of the fable; nevertheless, a raven in captivity reached an age of sixty-nine years. Banding of free living birds has so far shown an age of thirty-one years in the herring gull *(Larus argentatus)*, twenty-four years in the common heron, twenty-one years in the common swift, twenty years in the mallard duck, and eleven years in the European robin.

Environment

The size of an individual bird's range can, depending on its species, vary a great deal. Many nuthatches cover only a few hundred meters throughout their lives, whereas the range of the wandering albatross *(Diomedea exlulans)* encircles the earth. The data on the distribution in the text and the maps always shows, as is common in ornithology, the breeding range of the species or subspecies involved. If, for a particular reason, the wintering or non-breeding summering areas or the routes of migration or distribution as vagrants are also given, then it is specially indicated.

Bird distribution

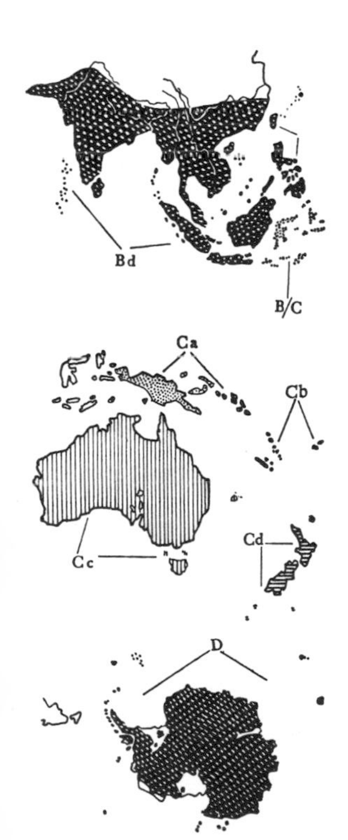

The distribution of birds over the earth's surface is intimately related to its geological history, with the appearance and disappearance of habitats, and with climactic changes and the formation of mountains. Thus in the course of millions of years, bird species have evolved and become extinct, they have evolved into new species or emigrated, and they have spread over new territories or become confined to narrow residual areas. Beyond this, many birds species have been exterminated by man in recent history, or have fallen prey to changes in their environment which he caused. This inglorious beginning was made in the seventeenth century by mariners who massacred the flightless dodos (Raphidae) on Mauritius and Réunion. Since then the decrease and extermination of birds has proceeded irresistibly in spite of all efforts to preserve these animals which man cannot recreate. The conversion of moors and waste land into useful agricultural land displaces birds that depend on the former environment; outright persecution eliminates portions of other species. Furthermore, the introduction of foreign species into new areas can also have disastrous effects on the native bird population of an area.

Even today there are, however, species which are still enlarging their range. The serin finch and fulmar are slowly but steadily invading new areas. The collared dove *(Streptopelia decaocto)* and the cattle egret *(Ardeola ibis)* are rapidly invading areas in which they were unknown until recently. Particularly in the northern regions, various bird species are enlarging their area of distribution; this is presumably because the climate in the north is generally becoming warmer.

Some species, for example the golden plover *(Pluvialis apricaria)*, inhabit only very specific habitats (or biotopes, from the Greek bios = life, and topos = place) within their range. Such species may be called stenoecious (from the Greek stenos = narrow, and oikos = house). Other species, like the Eurasian cuckoo, are not selective in the choice of their habitats; they are called euryoecious (Greek eurus = wide). But even euryoecious birds do not inhabit all areas within their range with equal readiness. Such preference and avoidance may vary within a species from one country to another. Thus, in continental Europe the cormorant *(Phalacrocorax carbo)* nests on high trees, while the British subspecies breeds on cliffs; phalaropes breed in fresh water areas but spend the non-breeding season at sea.

Habitat and Biocenosis

The environment of a bird (and of any animal species) consists not only of the ground and the flora of its habitat, such as desert, prairie or deciduous forest, its biotope, but also of all the other animals and plants which live in the same habitat and with which the bird is linked in a network of mutal interdependence. This community of all organisms inhabiting a biotope in a region is called a biocenosis (from Greek bios = life, koinos = in common). When about twenty bird species inhabit the same piece of woodland and thus are members of a biocenosis, the competition between them is much less strong than one might expect at first sight. Every species makes somewhat different demands on its environment, using a certain "ecological niche," a particular part of the environment. Competition within a species depends on its population density, which is often reduced by the establishment of territories (discussed earlier in this chapter).

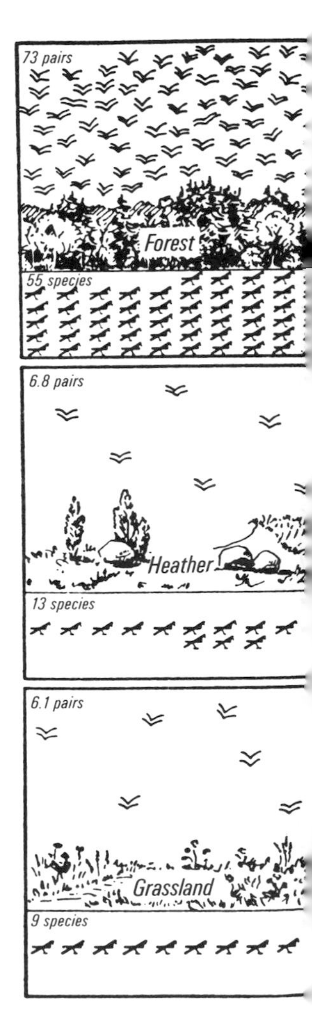

Fig. 1-25 and 1-26. The density of bird populations in six different habitats of the Lüneburg Heather region in Germany.

The facts of animal geography (the science of the distribution of species over the earth) and ecology (the science of the relationships of living things to their environment) apply in principle to all animals, but they have been particularly well studied in birds. Bird censuses are generally made not just for single species, but rather, the birds are grouped according to their habitats, called bird communities. Thus in a large part of the Old World the birds of inland waters are counted at regular intervals, providing information for carrying out bird protection programs. The breeding pairs of all five to fifteen species of the European coastal seaboard colonies are regularly counted. In such a census, it was discovered that the number of snowy owls *(Nyctea scandiaca)* decreased suddenly every three to four years. Moreover, this always happened after a "crash" of the lemming population which always follows a time of overpopulation.

Ornithologists pay special attention to the fluctuations of numbers in bird populations and their causes. The rate of survival of eggs, young, and adults varies widely in different species. Thus it was found that only ten young fledge from one hundred oyster catcher *(Haematopus ostralegus)* eggs, but a hundred common buzzard *(Buteo buteo)* eggs resulted in seventy-four fledged young. Sexually mature adults survive from only eight to eighteen percent of the eggs. Once a yellow-eyed penguin *(Megadyptes antipodes)* reaches adulthood it has a ninety percent probability of living into the next year; the adult barn swallow has only a forty percent chance. The mean life expectancy of yellow-eyed penguins is seven years; in common swifts, five to six years, and in small to middle-sized song birds, it is only one to two years. In order to balance these high losses, each pair of these birds must lay five to fifteen eggs per year, but the yellow-eyed penguin and swift manage with an average of two eggs a year.

Only a few of the diseases and causes of death of birds will be mentioned here. Many birds starve to death in unfavorable weather or because winds have blown them into areas lacking in food; it is not unusual for weak altricial young birds to starve because the parents feed only the more vigorously begging, stronger siblings. Not uncommonly, birds die accidentally as a result of collisions with electrical wires, lighthouses, or glass panes. Others become victims of oil pollution, i.e. oil drifting on water from oil tankers which have suffered an accident or from oil tanks cleaned out at sea by irresponsible captains. Pesticides used in agriculture and forestry, particularly DDT, poison not only insects but also their natural predators, the insectivorous birds, in masses. Naturally there are many diseases and causes of death not attributable to man: poisoning by certain plants, disturbances in the balance of vitamins or hormones, diseases due to parasites, bacteria, and viruses. Particularly important is ornithosis, formerly called psittacosis, which is a virus disease of the respiratory pathways. A considerable proportion of not only wild birds but also of domestic fowl and cage birds are carriers of the organism, although they do not suffer from the disease. Among bacterial diseases in birds, paratyphoid due to salmonella and botulism in ducks are particularly dangerous since they can also be conveyed to man. Looking at animal parasites, in addition to protozoans (particularly coccidia), worms of all types, mites, ticks, bugs, fleas and flies (particularly Mallophaga) play a role. This last group of insects comprises more than 3000 species which live exclusively as parasites on birds (see Vol. II).

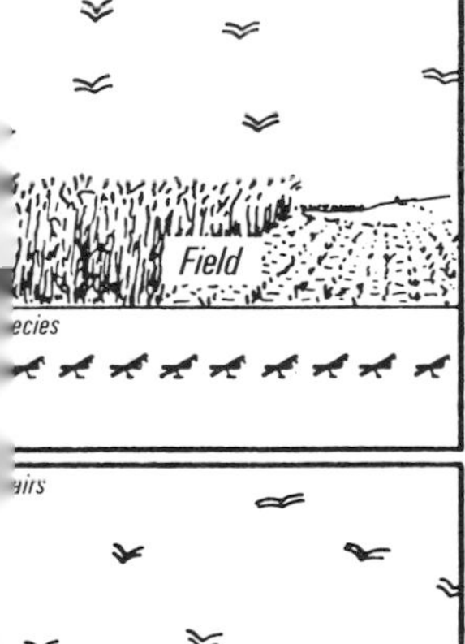

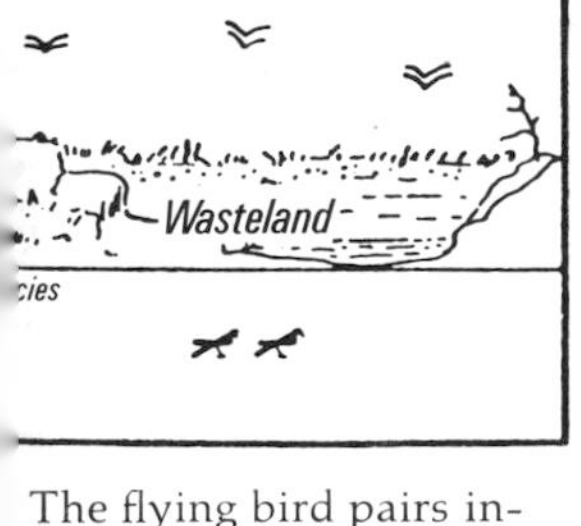

The flying bird pairs indicate breeding pairs per square kilometer; the birds shown below the line represent the number of species.

For many decades ornithologists have studied not only bird migration, but also the increase of bird populations by means of banding and other techniques. Some populations fluctuate strongly. After cold winters, kingfishers *(Alcedo atthis)* and green woodpeckers *(Picus viridis)* almost disappear, but soon recover their former numbers. The losses in such recovering populations are evidently less than among species with denser populations. In many species, the census has resulted in a count of all breeding pairs or individuals of a population. In Steller's albatross *(Diomedea albatrus)* this number is between one and ten, in the whooping crane *(Grus americana)* between ten and one hundred, in the Hawaiian goose *(Branta sandvicensis)*, including captive birds, between one hundred and one thousand, in other species between one and ten thousand, and so forth. The most common free-living species of land birds are probably the house (English) sparrow and the red-billed dioch *(Quelea quelea),* while among seabirds, they are the slender-billed shearwater and the guanay cormorant *(Phalacrocorax bougainvillei).* All are without doubt exceeded in number by domestic fowl, of which there are probably several billion.

Population density

The density, i.e., the number of birds irrespective of species per square kilometer, is lowest in the food-scarce deserts although some species may be surprisingly numerous. There are far more birds in prairies or steppes and still more in woodlands. The highest density, according to number of species and individuals, is reached in riparian woodland in which fresh water, swamp, and grassland birds, woodland birds as well as birds characteristic of isolated trees, can live. Thus, in Central European meadows there are 0.06 pairs (of nine species) per hectare, in moorland, 0.27 pairs (of twelve species), in heather woodland, 0.73 pairs (of fifty-five species), in the pine forests of Brandenburg, Germany, 1.07 pairs, but in the mixed oak and hornbeam forests of northwest Germany, ten pairs nest per hectare. Population density is particularly high where there is abundant food available beyond the actual habitat, or where man has provided additional nest sites and artificial feeding stations. Thus in the bushes and trees of the Frankfurt Zoo and in the nest boxes provided, fifteen pairs of birds breed per hectare; in small woodland sanctuaries there may be fifty, and on the North Sea island or Norderoog, there are even five hundred pairs per hectare which utilize the surrounding feeding areas of the sea. For the whole surface of the earth, one to two breeding pairs per hectare probably represent a relatively high population density. This amounts to a world population of ten to thirty billion free-living pairs; on the average there are presumably about twenty billion. Cage and other captive birds should be added to this, as well as domestic birds.

The causes of phylogenetic development are the same in the whole animal world; they are discussed in Vol. I. They are generally valid and for man's understanding of himself, important insights were gained in large part through observations of birds. Ornithologists also discovered that the subspecies of a particular species are generally larger in colder than in warmer climates (Bergman's rule), and that in cold areas they have shorter beaks and feet than in warm countries (Allen's rule). Both are associated with the heat economy. Larger bodies have, in proportion to their weight, a smaller surface and therefore lose less heat; shorter beaks and feet also lose less heat. Bergman's and Allen's rules are, however, also valid for warm-blooded mammals; in cold areas they have especially shorter ears and tails than in warmer ones. Gloger's rule, according to which darker races live in humid and lighter colored races live in dry areas, is probably a result of selection; it too is valid for members of other animal groups, such as mammals and butterflies.

The principles governing the formation of subspecies and species are the same for birds as for all other animals. Certain peculiarities will be considered, among others, with the Galapagos finches (Darwin's finches) and with the carrion crow *(Corvus corone corone)* (both in Vol. IX). A species is, as defined by Meise,"a group of individuals or representative populations in space, which as far as environmental barriers permit, interbreed more or less freely, but which are almost always reproductively isolated from other groups (other species) under natural conditions."

Opinions about bird classification, i.e., about the placing of birds into related groups, differ widely. In part this is because fossil evidence is particularly scarce for birds. In addition, the bases for judging the relationships of different bird groups with respect to structure and function of the body (e.g., its morphology, anatomy, and physiology), to behavior and geographical distribution, are often not clear. A comparison of the different systems is given in the section on systematics at the end of this book.

Subclass: primitive birds (Archaeornithes), by E. Thenius

Because of the unambiguous body characteristics of birds, there is never any doubt about presently living birds belonging to this vertebrate class; even species with secondary retrogression of the wings can be recognized at once as birds. It may, therefore, seem surprising that with certain primeval forms known from a few skeletal remains, doubts have been expressed as whether they should be regarded as reptiles or birds, since these fossil forms combine reptilian as well as bird characteristics, somelike as the mammal-like reptiles combine characteristics of both reptiles and mammals.

These fossil forms are known from the earth's Mesozoic Period, the Upper Jurrassic; they lived about 150 million years ago. They are the primeval birds (subclass Archaeornithes) with the famous *Archaeopteryx lithographica,* which is illustrated on the color plate on page 26. It has the following characteristics: a skull with teeth in the upper, intermediate, and lower mandible, and bi-concave vertebrae. All thoracic vertebrae are movable, and the pelvis is connected with only six (nonfused) vertebrae. The ilium and pubis are not joined together. There is a caudal spine of twenty to twenty-one free vertebrae; all three fingers are free (metacarpals, not fused) with claws and the tarsometatarsus (not quite fused). The abdominal ribs are present. The back of the skull is small and the brain is still very reptilian with a small cerebrum and cerebellum. The tibia and fibula are not fused together.

The recently deceased Czech paleontologist, Josef Augusta, writes about archaeopteryx: "There is no more famous paleontological find than this! There is none more significant, for its discovery falls into the period when the struggles for the acceptance of the new ideas [about evolution. Ed.], which meant such a tremendous step forward for biology, were beginning."

The first discovery of archaeopteryx was made by H. von Meyer in 1861. It was the imprint of a single feather, found in the lithographic slate formation of Solnhofen in the Altmühl valley of Franconia in Germany. It is understandable that experts doubted the credibility of this discovery. How could a bird feather have gotten into the stone layers of the earth's Mesozoic Period? But shortly afterwards an almost complete skeleton with a long bony tail, which, as clear impressions showed, was feathered on both sides and had equally clear impressions of the primaries and secondaries on the forelimbs, was found in the same strata. Another skeleton was found in 1877 at Eichstätt, Germany. This so-called Berlin specimen is preserved even better and shows additional characteristics not known in present-day birds, such as jaws with teeth, fingers with claws, and a reptile brain.

In view of the peculiar combination of reptilian and bird characteristics, it is understandable that this primitive bird caused some conflict among scientists. Was it a reptile with "bird" wings, a bird with reptilian features, or a link between reptile and bird? Since then only one other poorly preserved skeleton was found in 1956 at Solnhofen. Scientists are, however, now agreed that *Archaeopteryx lithographica* is to be regarded as a bird which, corresponding to its old geological age and reptilian descent, still shows several reptilian characters. After all, this also applies to the geologically oldest mammals which also still show reptilian features. However, the fossil discoveries clearly show that birds evolved from another reptile group of the earth's Mesozoic Period than did the mammals. The ancestors of birds would be expected to be in the same reptilian line, the Archosauria, from which the crocodiles, the long extinct dinosaurs (see Vol. VI), and the flying reptiles (Pterodactyls) evolved. As discussed in Vol. VI, the reptiles developed their own stem of flying forms (the flying saurians), which, like the bats among mammals, also had a skin membrane for flight. The flying reptiles thus cannot be considered as ancestral forms of birds, although they lived on the shores of the Jurrasic Sea at the same time as did archaeopteryx.

Many unanswered questions which are responsible for different conceptions still exist about the origin and evolution of birds. Some investigators think that birds arose from bipedally walking dinosaurs. Others assume the ancestors of birds were tree-inhabiting reptiles which, instead of having scales, developed feathers as their body cover. These feathers, after adequate enlargement, are believed to have made flight possible, fluttering flight having been preceded by a stage of gliding. Whether the origin of the plumage is connected with the acquisition of a steady body temperature, like the hair in mammals, is a matter of speculation.

Fig. 1-27. Hand and Lower Arm Skeleton of Birds: A—ulna, B—radius, D—"thumb"—2nd digit, F2—third digit, often wrongly designated second, F3—fourth digit, often wrongly designated third, Hw—carpal bones, Mh—metacarpals

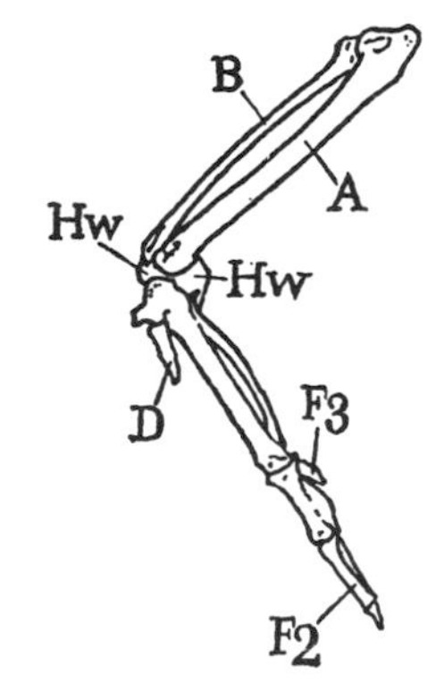

Muscovy duck *(Cairina moschata)*. The second digit is free, but the third and fourth metacarpal are fused, as in all living birds (exception, see Fig. 1-28).

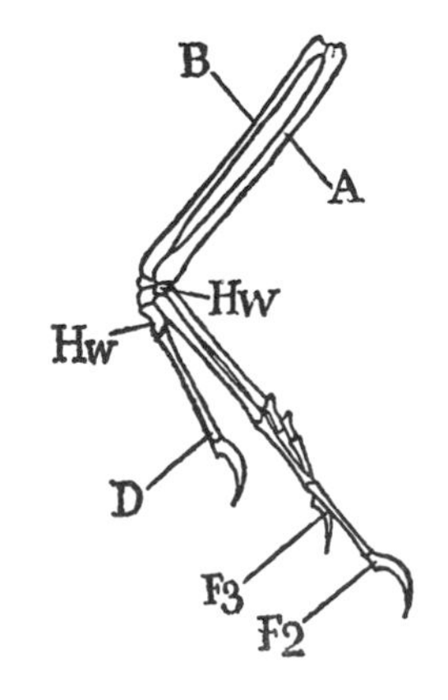

Fig. 1-28. Archaeopteryx. The digits are not fused together, and claws are still present on all fingers.

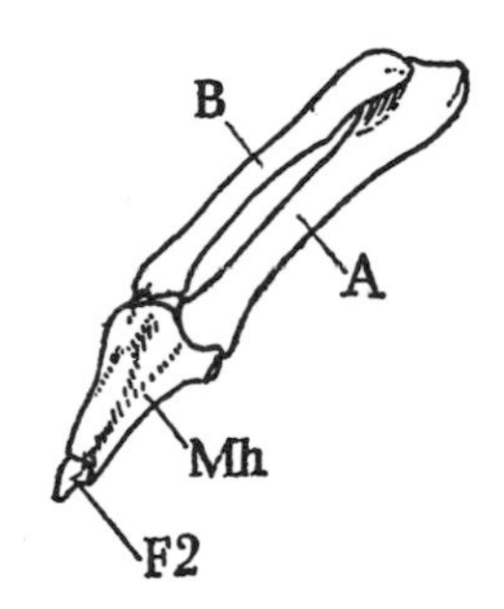

Fig. 1-29. Cassowary: The hand of the flightless wing is retrogressed.

A tree-living phase of development of so-called pre-birds (Proaves) is probably because, in bipedally walking reptiles which live only on the ground, not only the forelimbs but also the big toe are more and more retrogressed. Although until now earlier primeval birds fossils have been lacking, there can be no doubt that evolution towards a bird type began very early, perhaps in the Triassic Period, about two hundred million years ago. Reptilian remains of this period are known (the Thecodontia) which are possible ancestral forms.

The environment of primitive birds was presumably the forests of the Upper Jurassic, in which they moved about, flying and climbing (see Color plate, p. 26). From the structure of their forelimbs and the brain, which we know from an impression of the brain case, one may conclude that their capacity for flight was much poorer than that of the flying saurians which lived at the same time. The fact that, nevertheless, the bird type has won out against the latter and could develop into the many forms of present-day birds is presumably linked to the formation of the plumage.

Subclass: modern birds (Neornithes), by W. Meise

Distinguishing characteristics

All living birds as well as those known from the Cretaceous Period and later geological strata belong to the MODERN BIRDS (subclass Neornithes). With the exception of the Cretaceous divers (Hesperornithiformes see Fig. 1-28), birds are toothless, with the vertebrae anteriorly and posteriorly jointed in saddle form or posteriorly hollowed out. Some thoracic vertebrae are fused. The pelvis is formed into "synsacrum" by the fusion of several vertebrae; the caudal part of the spinal column is shortened and has fan-like tail feathers and a pygostyle. The ilia and pubes are fused, and the third and fourth fingers are free, although the metacarpals are fused. At most only two fingers have claws; the tarsometatarsus is fused at the latest shortly after hatching. There are no abdominal ribs. The back of the skull as well as the mid- and fore-brain are large. The tibia and fibula are fused in the upper region. The distribution of birds is world-wide.

Today there are about 8600 bird species with about 30,000–35,000 subspecies (geographical races), which are divided into about 1600–2400 genera and about 140 families. The classification system used here divides the living birds into twenty-six orders (see the chapter on classification in the back).

Order of primitive divers (Hesperornithiformes), by E. Thenius

Only a few bird species are known so far from the Cretaceous Period; among them are the primitive divers (Hesperornithiformes) which, in contrast to all other recent birds, are toothed. Though some investigators place them close to the loons and grebes, we prefer to separate the toothed birds into the subclass Odontognathae, distinct from all other recent birds.

The best-known species of which there are complete skeletal remains is *Hesperornis regalis* of the late Cretaceous Period of North America (about eighty million years ago). This flightless swimmer and diver lived on the shores of the Upper Cretaceous Sea and was presumably a fish eater. The marked adaptations of its skeleton are similar to those of present-day grebes and loons, but exceed them. This bird had a standing height of about one meter. Only the humerus is present in the forearm. The sternum has no keel since there is no flight musculature. In this and the skull structure, this bird resembles the ostriches (Struthioniformes) more than present-day divers. The longer hind legs show the same relative proportions as found in grebes and loons. The caudal spine is short but does not yet form the pygostyle characteristic of modern birds.

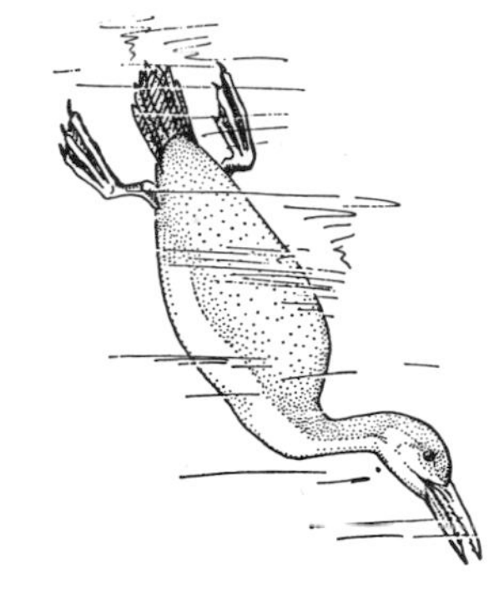

Fig. 1-30. The toothed divers of the Cretaceous period moved in and under the water like present-day loons (see chap. 4).

Birds do not make good fossils. The few primeval birds which have been found only give glimpses of the former number of forms. Hence all speculations about phylogenetic relationships within this vertebrate class are much less certain than is true of reptiles or mammals.

2 Tinamous

Order: tinamous, by A. F. Skutch

In older natural history books, the description of the birds generally began with the ratites, including the rheas, ostriches, cassowaries, kiwis, and the extinct moas and elephant-birds of Madagascar. Since these large flightless birds have no keel on the sternum, which is otherwise characteristic of birds (see Color plate, p. 36), they are known as flat-chested birds (ratites) in contrast to all the other modern birds, which are keel-breasted (Carinatae). Today we know that the ratites descended from flying keel-breasted ancestors. The tinamous (Tinamidae) are a still-living primitive bird family which are close to the ancestral group of the ratites and particularly to the nandus (rheas).

Distinguishing characteristics

The tinamous used to be classified with gallinaceous birds, because externally they resemble guinea fowl. Because of their relationship to the ratites, they are today considered a separate order: Tinamiformes. The L is 20-53 cm, and the weight is 450-2300 g. They are ground birds with a compact form, a slender neck, a relatively narrow head and a rather short beak which is slender and curves lightly downward. Their wings are short and hence their capacity for flight is poor. Their feet are strong; there are only three well developed forward toes and the hind toe is located in a high position, being either retrogressed or absent. The tail is very short, and in a few species it is hidden under the tail coverts; this and the abundant rump feathering cause the rounded body shape. Powder down and a preen gland are present. In contrast to gallinaceous birds, they do not scratch for food (see Chapter 17). The copulatory organ is present. The incubation period is from sixteen days in the slaty-breasted tinamous *(Crypturellus boucardi)* to twenty-two or more days in the ornate tinamou. The ♂ alone incubates the one to sixteen eggs, which generally come from several ♀♀ and the ♂ also leads the young. The plumage is inconspicuous, although the crown feathers of many species can be raised as a crest. The ♂♂ and ♀♀ have similar plumage, or the ♀♀ may be somewhat brighter in color and often larger than the ♂♂.

They live in tropical parts of America; in the north they live only slightly beyond the Tropic of Cancer (Zimttao in northwest Mexico), but in the south they are widely distributed in the temperate zone (Argentina, Chile). They inhabit varied habitats: rainforest, thickets, bushland, tree steppes and treeless grassland up to an altitude of 5000 m in the Andes. As far as is known, none are migratory.

Fig. 2-1. The approximately forty-three species of tinamous live in these areas of Central and South America.

There are nine genera with forty-three (or forty-five) species, divided by some investigators into two subfamilies: A. The WOODLAND TINAMOUS (Tinaminae) have connected nostrils in the middle of the beak or before it. There are three genera, which are distinguished by the back of their tarsometatarsus: 1. The genus ROUGH TAOS *(Tinamus)* contains the following species: the GREAT TINAMOU *(Tinamus major)* and GRAY TINAMOU (*Tinamus tao;* see Color plate, p. 87). 2. SCALY TAOS *(Nothocercus)*: BONAPARTE'S TINAMOU *(Nothocercus bonapartei)* and the BLACK-CAPPED TINAMOU (*Nothocercus nigracapillus;* see p. 87). 3. The genus SMOOTH TINAMOUS *(Crypturellus)* has many species, among them the LITTLE TINAMOU *(Crypturellus soui)*, the VARIEGATED TINAMOU *(Crypturellus variegatus)*, the BRUSHLAND TINAMOU *(Crypturellus cinnamomeus)*, the SLATY-BREASTED TINAMOU *(Crypturellus boucardi)*, and TATAUPA *(Crypturellus tataupa)*.

B. The STEPPE-TINAMOUS (Rhynchotinae) has nostrils which are connected at the base of the beak. There are six genera: 1. The RED-WINGED TINAMOU with one species, the RED-WINGED TINAMOU (*Rhynchotus rufescens;* see Color plate, p. 87). The L is 42 cm. This bird is hunted in the pampas as a "partridge substitute." 2. the CRESTED TINAMOUS *(Eudromia)*; among them is the MARTINETA TINAMOU (*Eudromia elegans;* see Color plate, p. 87), which has no hind toe. 3. The PUNA TINAMOU (*Tinamotis pentlandii;* see Color plate, p. 87) has no hind toe. 4. The PARTRIDGE TINAMOUS *(Nothoprocta)* has many species; among them are the ORNATE TINAMOU *(Nothoprocta ornata)* and *Nothoprocta cinerascens.* 5. The QUAIL TINAMOUS *(Nothura)* has many species; among them is the SPOTTED TINAMOU (*Nothura maculosa;* see Color plate, p. 87); and 6. The PEACOCK TINAMOUS with only one species *(Taoniscus nanus)*. They are found in southeast Brazil and northeast Argentina. The upper tail coverts are extended into a train.

Tinamous are ground birds

Walking or running, the tinamous move almost exclusively on the ground. When approached by man, they hide in thick ground cover or steal away unobserved. When they are hard-pressed in open country, they often crawl into holes which another animal has dug. Some species are very reluctant to fly. But when surprised by a larger animal or when followed too closely, they rise suddenly with loud, frightening wing beats, and often they call. They disappear swiftly and alight in thick vegetation. Before the surprised hunter can bring up his gun, they have disappeared. The effectiveness of this escape behavior is evident from the fact that one of the largest species, the great tao, can still be found in forests in which men and dogs have exterminated all other large birds.

The sudden flight, however, can also be disadvantageous. W. H. Hudson long ago reported that a spotted tinamou which is unexpectedly hunted by a rider on horseback in grassland may fly up with such an uncontrollable force that it sometimes flies into a fence or the wall of a house and fatally injures itself. He also tells of a bird which rose in a strong wind to a great height and then, tumbling down, fell to the ground and was killed on impact. From these observations it was concluded that the tinamous had insufficient control over their movements when suddenly flying upward. Today we are more careful with such judgments. Whatever may happen on the pampa, most tinamous fly so skillfully that they avoid accidents. Animals that live in dense forests and thickets do not easily hit tree trunks and branches; such unadaptive behavior would soon be eliminated by natural selection.

The initial burst of wing-fluttering is often followed by a long glide and again by renewed wing beats. Although tinamous generally cover longer distances on foot rather than in flight, D. A. Lancaster observed a brushland tinamou *(Nothoprocta cinerascens)* which regularly flew 200 meters from its nest to its feeding place. In other species as well, flights of 200 to 1500 meters have been observed.

The food of the tinamous

Tinamous eat mainly small fruits and seeds. They pick them from the ground or from the plants which they can reach from the ground; sometimes they jump up about ten centimeters to reach a particularly tempting fruit. Seeds with wing-like appendages, which make swallowing difficult, are beaten against the ground or vigorously shaken in order to remove it. They also eat opening buds, tender leaves, blossoms, and even roots. For variety they catch insects and their larvae, worms, and, in moist places, mollusks. They swallow small animals whole; they first peck at larger ones, then shake them or beat them against the ground. When searching for food, they scatter fallen leaves and other ground cover aside with their beaks, but they do not scratch with their feet. When searching for worms or larvae in moist places, brushland tinamous and probably other species fling the earth aside with their beaks, digging two to three centimeters deep in this manner. Like many other birds, they swallow small stones and grains of sand. The species which live in deserts drink either very rarely or not at all.

Some species sleep in trees, while others sleep on the ground. Perhaps this could be explained by the different development of the back tarsus (see rough taos and smooth taos). The American zoologist, William Beebe, discovered that smooth taos sleep on the ground, while rough taos sleep in trees where their rough leg lies on a horizontal branch. But smooth taos probably often sleep in trees also, as Ernst Schäfer observed in the little tinamou in Venezuela.

Their voice

Hearing the calls made by tinamous is one of the most unforgettable experiences one can have in tropical South America. Although the calls or songs of many species are simple, the purity and softness of their tones are attained only by a few birds. In several species the songs sound like organ tones, while others resemble flute-like whistles or trills that have a melancholy quality. The tones of the song sometimes have the same pitch, but in some species they may rise, and in others they rise and fall again. In contrast to most tinamous, the male Bonaparte's tinamou utters a rough crowing or barking call which can be heard for several kilometers through the mountain forests. In some species, males and females can be readily distinguished by their calls. Tinamous sing mainly during the breeding season; they are noisiest in the early morning and late evening. But one can also hear them at other times of the day, and often their moving tones interrupt the stillness of the night. When startled tinamous fly off or follow one another, they shriek or crow hoarsely.

Only the males incubate

In Patagonia, the Martineta tinamous form strings of six to thirty or more birds. But most adult tinamous live singly except during the reproductive season. Reproductive behavior differs from that of most other birds. Only the males care for the eggs and young. So far no exception to this rule is known. In the few species which have been adequately studied, the males live in polygyny and the females in polyandry. A male in breeding condition attracts two, three, or even more females by continuously calling. They all lay their eggs in the same nest and then leave their incubation to the male. The females leave in order to lay eggs into the nests of other males. When the male has raised his young or has lost the eggs, he begins to call again and attracts new hens, which then supply him with another nest full of eggs. This breeding behavior has been observed in such different species as Bonaparte's tinamou, the brushland tinamou and the slaty-breasted tinamou. The variegated tinamou, however, cares for only a single egg; furthermore the clutch of four to nine eggs incubated by the male ornate tinamou seems to come from a single female. In this mountain species, the larger and more aggressive female defends the breeding territory while in other tinamous this is done by the males.

In the ornate and in Bonaparte's tinamous, as well as in the taos and in many other species, there are about as many females as males. But in the variegated tinamou there may be four times as many males as females. Tinamous almost always nest on the ground, often in the densest herbage or between the projecting root buttresses of a large tree. In Venezuela, Bonaparte's tinamou places its nest in a crevice under a rocky ledge or on a steep slope where projecting roots hide it; in Costa Rica, it is sometimes placed on a tree stump. Many tinamous lay their eggs directly on the ground or on leaves and other parts

Fig. 2-2. A displaying female variegated tinamou.

of plants which happen to be on the chosen spot. Ornate tinamous, however, build a proper nest base out of dry earth or a mixture of earth and moss turf. On this they erect a firm structure of grass which is worked into the base so as to form a circle.

The shiny tinamou eggs are among the most beautiful natural products known. They may be green, turquoise-blue, purple, wine-red, slate gray, or a chocolate color, and they often have a purple or violet lustre. They are always uniformly colored without spots or blotches. Their shape is oval or elliptical; the two ends differ very slightly or not at all.

Incubating male tinamous sit continuously on the eggs for many hours. In most species they leave the clutch only once a day to look for food. These excursions are undertaken during mornings or in the afternoon, depending upon the weather; they last, according to the observations of several investigators, from about forty-five minutes up to four or five hours. A slaty-breasted tinamou, observed by Lancaster, left the nest only every other day at the peak of the incubation period, and sometimes sat on the eggs for forty-seven hours without interruption.

Although the eggs are conspicuous and have no protective coloration, many tinamous do not cover the eggs when leaving the nest. This is known to be true for the great tao, *Crypturellus soui*, and Bonaparte's tinamou. A brushland tinamou covers its eggs only when they are about to hatch. The ornate tinamou, on the other hand, regularly covers the eggs with feathers, which give the eggs some protection from the harsh climate in the 4000 meter high Peruvian Puna. In the warm forests of Central America, a slaty-breasted tinamou always threw leaves toward the nest when it left to feed. When it had thrown a number of leaves toward the nest, it pulled them over the eggs with its beak and then threw more leaves. It spent more time with this activity at the beginning of the incubation period than it did before hatching. This was done quite carelessly; often more than half of the eggs were left uncovered.

Incubating tinamous sit so firmly on the nest that one can approach very closely. Although they will not let themselves be touched by one's hand, an observer can sometimes touch them with the end of a long stick. If someone approaches, some species, among them the ornate tinamou and several smooth tao species, press the head and body close to the ground and raise the hind end, sometimes so much that the stumpy tail and the lower tail coverts stand almost vertically. This posture resembles that assumed by certain tinamous in courtship display and is also seen when they are alarmed while walking. It seems to serve no special purpose in the incubating cocks, since raising the hind end may expose the shiny eggs. If one approaches an incubating tinamou cock too closely, it flies off the nest and out of sight in a burst of speed. An exception is the tataupa, which flutters over the ground as if hurt and unable to fly when it is displaced from its eggs. Males of other species sometimes turn around and slowly approach the intruder with loosely held, trembling wings, but this "distraction" is not a convincing example of the "injury feigning" which has been described in so many other incubating birds and birds with young.

The chicks are precocial

Newly hatched tinamous are densely covered with long soft down which in some species is marked in dulled colors. On the first day after hatching, the male leads the young out of the nest, moving slowly and calling the young with repeated soft whistles or whining tones. Now and then he picks up an insect from the ground and moves it in his beak while uttering his soft, enticing calls. Then he lays the insect before one of the young to be picked up. On leaving the nest, the chicks of the smaller tinamous are very delicate; it is difficult to believe that they can endure the dangers of the tropical forests and thickets. But they move so skillfully and well-hidden through the dense vegetation that little is known about their lives after they leave the nest. They probably develop rapidly and soon leave their father. A slaty-breasted tinamou at twenty days differed little in color and size from the adults.

At the turn of the century, many tinamous, mainly Pampas hens, were introduced and raised as game birds in France, England, Germany, and Hungary. After this initial success, however, all attempts to settle tinamous in Europe in the wild have failed.

With their beautiful, moving songs, their magnificent eggs, and their unusual family life, tinamous create particular interest. It is unfortunate that they are so shy, which is partly because they are constantly hunted for their tasty meat. It is therefore difficult to observe them and study their behavior in detail. They are seldom seen in zoos or in South American chicken runs.

▷

Tinamous (see chap. 2): 1. Red-winged tinamou *(Rhynchotus rufescens)*; 2. Puna tinamous *(Tinamotis pentlandii)*; 3. Martineta tinamou *(Eudromia elegans)*; 4. Gray tinamou *(Tinamus tao)*; 5. Black-capped tinamou *(Nothocercus nigracapillus)*; 6. Spotted tinamou *(Nothura maculosa)*.

1
2
3
4
5
6

1
♂
a
♀
2

3
♂
♀

5
4
6
7

9
8
10
11

3 The Ratites

Order:
Struthioniformes

◁

Rheas: 1. Rhea *(Rhea americana)*; a. albino form; 2. Darwin's rhea *(Pterocnemia pennata)*.
Ostriches: 3. Massai ostrich *(Struthio camelus massaicus)*.
Moas: 4. Giant moa *(Dinornis maximus)*;
Kiwis: 5. Common kiwi *(Apteryx australis)*; 6. Little Owen's kiwi *(Apteryx owenii owenii)*; 7. Greater Owen's kiwi *(Apteryx owenii haasti)*.
Cassowaries: 8. Australian cassowary *(Casuarius casuarius)*; 9. One-wattled cassowary *(Casuarius unappendiculatus)*; 10. Bennett's cassowary *(Casuarius bennetti papuanus)*; 11. Emu *(Dromaius novaehollandiae)*, cock with young.

Birds fly; for this reason, most of their anatomic adaptations can only be understood in relation to their ability to fly. There are, however, flightless birds among which the RATITES (Order Struthioniformes) are the best known and most important group. In many of their characteristics, they are more primitive than most other living birds. Therefore, it was once thought that ratites had split off from all other birds at a time when birds had not yet "invented" flight. However, if this were so, many of the structural features of ratites would not be understandable. For example, all of them have a wing skeleton which is not fundamentally different from that of flying birds. Their wings still bear flight feathers and coverts; hence, they are degenerated wings rather than degenerated forelegs, as was the case in the bipedally walking dinosaurs (see Vol. VI). Therefore the ratites undoubtedly stem from flying ancestors and have evidently lost their ability to fly as their body size increased. This has led to considerable changes in the bones, muscles, and plumage.

Their characteristics include degenerated breast muscles, a retrogressed keel of the sternum, an almost absent wishbone (furcula), a simplified wing skeleton and musculature, flight and tail feathers which have retrogressed or have been converted to decorative plumes, strong legs, leg bones without air chambers except in the femur, no separation of pterylae and apteria, a loss of feather vanes which means that oiling the plumage is not necessary, and no preen gland.

There is a close relationship between ratites (particularly the rheas) and tinamous (see chap. 2). There are four suborders, each with one family: 1. RHEAS (Rheae); 2. OSTRICHES (Struthiones); 3. CASSOWARIES (Casuarii) and EMUS (Dromai); 4. KIWIS (Apteryges). Altogether there are six genera with ten species, as well as two extinct families, one of which is a separate suborder with ten genera and about twenty-five to thirty species. Some investigators regard each ratite family as a separate order.

Suborder: ratites, by K. Sanft

In the steppes of the South American lowlands and in the high plains of the Andes live the RHEAS OR PAMPAS OSTRICHES (Suborder Rheae, family Rheidae). In appearance and way of life they resemble the African ostriches, but they are not related to them. Fossil rheas have been found in the Upper Pleistocene of Argentina. They lived there about two million years ago; it is suspected that they were closely related to the South American tinamous.

Distinguishing characteristics

Rheas are smaller and more slender than ostriches: standing upright they reach 1.70 meters. They may weigh up to 25 kg, the head, neck, rump and thighs are feathered, and their plumage is soft and loose. There are three front toes, and the hind toe is absent. The tarsus has horizontal plates in front. The gut and particularly the caeca are very long. They are grass and leaf eaters. The urine is stored in an expansion of the cloaca and is eliminated in liquid form. The copulatory organ is extrudable. There are two genera, each with one species and seven (in part uncertain) subspecies: 1. The COMMON RHEA (*Rhea americana*; see Color plate, p. 88). The total height reaches 170 cm; the height of the back is 100 cm, the wingspan reaches 250 cm, the tarsal length is 30–37 cm, and the beak length is 9–12 cm. The ♂♂ are larger than the ♀♀. The tarsus has about twenty-two horizontal plates in front. Albinos occur. The eggs measure 135 x 95 mm and weigh 530–680 g. They are elliptical with a shiny surface and are ivory or golden-yellow with black stripe-shaped pores; the color pales with time. The young at first are yellow with black longitudinal stripes on the back; after two years they resemble the parents.

2. DARWIN'S RHEA (*Pterocnemia pennata*; see Color plate, p. 88) is smaller with a height at the back of 90 cm. The tarsus is 28–30 cm and has about eighteen horizontal plates. The eggs are 125 x 85 mm and weigh 500–550 g. They are a yellowish-green when laid and later become a pale yellow. There are two subspecies; among them is *Pterocnemia pennata garleppi* of the Andean high plain, at an altitude of between 3500 and 4000 meters.

The rhea

The home of the rheas is the grass steppes of South America; they avoid forests and mountains. Generally they live in groups of one cock and several hens within a distinct territory. After the breeding season, loose flocks of fifty or more may form. Extreme dry seasons, such as those which force Australian emus to migrate extensively, do not occur on the pampas; therefore, rheas do not wander far. Good eyes and acute hearing allow them to detect enemies from far away; their fast legs, which can take strides of up to 1.5 meters, carry them quickly out of danger. In an emergency they can escape by suddenly dodging aside. This maneuver is facilitated by the use of the wings which, for a flightless bird, are remarkably long. While running, they raise one wing and lower the other; this has a steering effect, like the rudders of an airplane, and makes such sudden changes of direction possible. This behavior can sometimes be seen in large zoo enclosures.

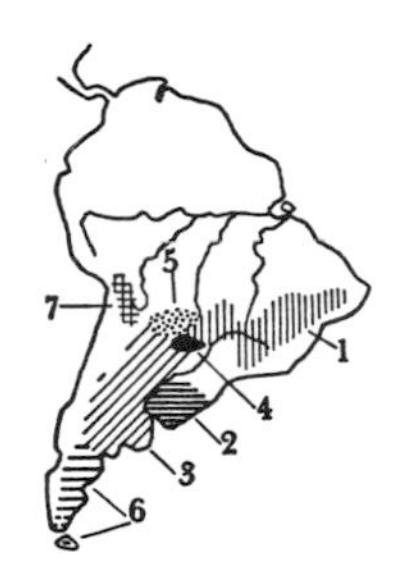

Fig. 3-1. Rhea (*Rhea americana*): 1. *Rhea americana americana;* 2. *Rhea americana intermedia;* 3. *Rhea americana albescens;* 4. *Rhea americana araneiceps;* 6-7. Darwin's rhea (*Pterocnemia pennata*); 6. *Pterocnemia pennata pennata;* 7. *Pterocnemia pennata garleppi.*

The rheas' food consists of grasses and herbs (alfalfa, clover, serradella) and insects and other small animals. Because they prefer the same food plants as sheep, they are food competitors. However, their consumption of the burr-like seeds which tangle the wool of sheep also make them useful. If they find enough juicy plants, they require very little water.

In the breeding season between September and December, earlier in the north than in the south, the cock expels all rivals from his territory. In a zoo he even expels his keeper. He only tolerates hens. The courtship display consists of his running around with his plumage erected, dodging sideways, and swinging the neck from side to side. At this time he gives his deep call, which can be heard far away, "nan-du," which is responsible for one of its names, "nandu." The cock builds the nest, a simple depression lined with a few pieces of plants; only the cock incubates. Each hen lays her eggs outside the nest at two day intervals. It was observed in zoos that the cock spread his wing under the hen, caught the egg on it, and then rolled it into the nest with his beak. Usually there are fifteen to twenty eggs in a clutch; since each hen lays ten to fifteen eggs and several hens lay for "their" cock, it hardly matters if many eggs are lost. Nests containing up to eighty eggs have been found. But the cock, during incubation, cannot cover so many and hence he cannot hatch them all. The young hatch after about forty days of incubation; they are led only by the cock. After half a year, they reach adult size, and at two to three years they are capable of reproduction. Carnivores and raptors pursue the young, but the adults have no enemies other than man. They are generally hunted only for sport since their meat is not tasty and their plumage has little value. The wing feathers can be used as dusters. The eggs, which correspond in volume to twelve chicken eggs, can be eaten. Wire fences enclosing cattle pastures increasingly reduce the habitat of rheas. They are often kept and raised in zoos. From 1955 to 1967, 273 young were hatched in the Frankfurt Zoo alone (mostly in incubators) and successfully raised.

The life of Darwin's rhea is, as far as is known, very similar to that of the common rhea. As an inhabitant of colder regions, it can, in contrast to its larger relative, become quite fat. It is rarely seen in zoos, but it has been raised successfully a few times; in 1967 fifteen young were raised in the Frankfurt Zoo.

Suborder: ostriches, by F. Sauer and E. Sauer

The AFRICAN OSTRICH (*Struthio camelus;* see Color plates, p. 88 and p. 98) is the only living representative of the suborder Struthiones, family Struthionidae. Its relatives inhabited wide areas in Asia, Europe, and Africa since the Eocene (about fifty-five million years ago). Eight extinct species all belonged to the same genus as the surviving species.

Distinguishing characteristics

Attaining three meters in height and weighing more than 150 kg, the male ostrich is the largest living bird. The head and about two-thirds of the neck are sparsely covered with short, hairlike, degenerated feathers, which make it appear naked. The skin is variably colored, depending on the subspecies. The legs are particularly strong and long. The tarsus in sexually mature ♂♂ has red horn plates, while in sexually mature ♀♀ they are black. The foot has two toes: a large, strongly clawed third toe and a weaker, generally clawless fourth (outside) toe. The first and second toes are absent. The feathers have no secondary or aftershaft. There are fifty to sixty tailfeathers. The wing has sixteen primaries, four alular and twenty to twenty-three secondary feathers. The wing feathers and rectrices have changed to decorative plumes. The copulatory organ is retractable and up to 20 cm long. There are three stomach segments with a gut up to 14 m long. The rectum is especially expanded, and the caecae are about 70 cm long. The urine is concentrated in the large cloaca but in contrast to all other living birds is secreted separately from the feces. The pubic bones are fused towards the rear, and serve to support the gut. This is only so in ostriches. The wishbone is absent, and the palate formation is different from that of other ratites. The sphenoid and palatal bones are not connected. The eggs, depending on the age of the hen, are 127 x 103 mm to 160 x 129.5 mm, and weigh 775 to 1618 g. The thickness of the shell averages 1.97 mm. The young have downs with two to three extended branches which look like spines. They reach their full height in six months. In the first year a brown spotted camouflage plumage develops, while the adult color appears in three to four years. The ARABIAN OSTRICH *(Struthio camelus syriacus)* has been extinct since 1941.

Fig. 3-2. The ostrich *(Struthio camelus)* was distributed in these areas of Africa and Asia Minor until a few decades ago. Since then it has been exterminated in many areas.
Subspecies: 1. North African ostrich *(Struthio camelus camelus)*; 2. Arabian ostrich *(Struthio camelus syriacus;* exterminated); 3. Somali ostrich *(Struthio camelus molybdophanes)*; 4. Massai ostrich *(Struthio camelus massaicus)*; 5. South African ostrich *(Struthio camelus australis)*.

The life of ostriches

Ostriches live on the open savannah, in the dry South African "bushveld" or on the wide sand plains of deserts which have hardly any plants, and also in dense bush. This adaptable grazer is at home even in steep rocky mountain country. Depending upon habitat and season, they eat various grasses, bushes, and forage on trees. Plants that store water help them through dry seasons but cannot meet all water requirements for a longer period of time. Without open water they eventually die of thirst. In the Namib desert of southwest Africa, ostriches wander regularly in search of water and they find even the smallest and most hidden water holes in the wild mountain country. They supplement their plant diet as much as possible with animal food, such as invertebrates and small vertebrates which they chase by rather ungracefully zig-zagging after them. Long periods of abundant food influence the reproductive readiness of the ostrich. Even in the most unfavorable weather, during long droughts, or during occasional local rains, the ostrich is a very adaptable and opportunistic breeder which can rear a few young in any season. Thus the ostrich population often becomes a rather mixed society of flocks, families, and individuals of all age groups which changes in composition with the season.

The social life of ostriches is one of the most complex in the animal

Fig. 3-3. Courtship display of the male ostrich.

world. In rainless periods, when wandering, and in the common grazing grounds at watering places, they often form peaceful aggregations of up to 680 birds, but the individual flocks remain recognizable. Social contacts between birds of different groups are initiated by approaching the other in a submissive posture, with lowered head and tail down. Often a family of a herd adopts the chicks or young of another. Single cocks may join together and form "schools" of half grown ostriches, with whom they wander about for days or weeks. For the communal sandbath, each flock seeks out a sandy depression.

Depending upon local circumstances and the composition of the population, ostriches live in monogamy or polygamy. Pair formation generally takes place within the large flocks. The "goading" of the old hens and the animosities between the cocks often lead to displays and "dances" of the entire flock. The most common mating pattern is polygyny, in which a cock generally lives with a "head hen" and two "auxiliary" hens. The head hen tolerates the others and all lay their eggs into a common nest. In most cases, the experienced head hen drives the auxiliary hens away from the nest as soon as they have finished laying.

In the initial phase of courtship, the cock displays by alternating wing beats in front of the flock to attract or cut out the chosen hens. He chases yearlings away with the help of his head hen. Then the birds move together to the breeding territory which, if necessary, is defended against conspecifics, or enlarged at their expense. During the actual courtship of a specific hen, the cock always drives her away from the others. Both move to a remote spot and graze for a while, their behavior becoming increasingly synchronized. The feeding evidently becomes of secondary importance and develops into a ritual which further synchronizes the behavior of the two partners. The smallest disruption in their movements leads promptly to a premature end of this preliminary display. When, however, the courtship continues undisturbed, the cock excitedly flaps the right and left wing in alternation. Both birds then slow down their steps and begin to poke their beaks into the ground on a sandy spot and pull out grasses. The cock then throws himself to the ground and stirs up the sand with tremendous wing beats; this seems to be a symbolic hollowing out of a nest bowl. Simultaneously, he turns and winds his head in a rapid spiral motion. He continually repeats his muted courtship song while the hen circles in front of or around him in submissive posture, dragging her wings. At the moment when the cock suddenly jumps up, the hen drops to the ground and the cock, beating his wings, mounts for copulation.

Breeding and raising young, by B. Grzimek

The Frankfurt Zoo is very proud of the fact that in the last few years, seventeen young ostriches were hatched there in an incubator and were successfully raised without a mother. The youngsters which, at birth, are almost the size of chickens, need much care from the very beginning. As a result, ostriches have only rarely reproduced in zoos. This may sound strange since ostrich farms are known to raise many ostriches. However, this is in the native climate of their homeland, and both parents are available for the incubation and raising of the chicks.

The male ostrich has far more to do during breeding than merely fertilizing the eggs. He is a genuine father. He scratches out the nest depression in a sandy spot, often in a dried-out river bed. Even before the eggs are going to be laid he guards the nest. Finally he sits down on it, the hen lays her eggs before his breast, and he pushes them under his body with his neck and beak. The Sauers observed in southwest Africa "that the hens came to the nest for egg-laying every other day, generally in the late afternoon, and they often laid their eggs at the same time. The greatest number laid by a single wild hen was eight. Auxiliary hens sometimes laid no more than three or four eggs." In the Nairobi National Park, four hens laid twenty-four eggs into one cock's nest. He was unable to cover all of them, and only sixteen young hatched. The male ostrich incubates from the late afternoon until early morning; the female, therefore, does not have to sit for long since she only incubates in the hottest hours of the day. There are, however, many exceptions to this rule. While in other animal species it does not matter as much if the males are predominantly shot, in ostriches this is disastrous. The excess of eggs for the remaining cocks may then result in none hatching, for to leave some and confine himself to just a few eggs is beyond the cock's abilities.

In 1960 an ostrich family group in the Nairobi National Park laid over forty eggs. Moreover, the cock had constructed the nest so that visitors could see it from the highway. Thus the cock and his hens were continuously surrounded by cars as close as two or three meters, and the animals were photographed and filmed. Since in the National Parks the animals do not know man as an enemy, the ostriches showed no alarm. One day some young lions approached the clutch, played with the eggs like balls, and scattered them quite a distance. With a laborious effort, the cock later on pushed them all back into the nest and continued to incubate. Although one might not have expected it, the eggs hatched.

When holding an ostrich egg, one may wonder how the chick could possibly get out of the shell without help. The shell is as thick as china, and we had to use a saw and hammer to open an ostrich egg. It weighs about one and a half kilograms, about as much as twenty-five to thirty chicken eggs. Ostrich eggs are good to eat; they taste almost like chicken eggs. When refrigerated, an ostrich egg keeps up to a year and remains edible. It takes about two hours to hard-boil one.

Young ostriches hatch after forty-two days of incubation; then they grow one centimeter a day. Small ostriches artificially raised as foster children in homes in Africa follow people like dogs do. If the family goes swimming, then the young ostrich swims after them like a little duck. In the Serengeti ostriches begin to incubate in September and around Christmas they walk about with their chicks.

The behavior of ostriches, by F. Sauer and E. Sauer

Behavioral interactions between chicks and parents begin a few days before hatching, with pleasant-sounding contact calls from the chicks within the eggs. Actual hatching may take hours or even days. Soon after hatching the nestlings swallow the first small stones for grinding food, and already on the first day most chicks are ready to explore the vicinity of the nest. In these first days they get to know their parents by social imprinting (a special form of learning; see chap. 12). After hatching, the chicks leave the nest with their parents. When in danger from a terrestrial enemy, such as a jackal, the adults try to decoy with conspicuous zigzag running, wing beating, and moot calls; at the appropriate time one parent calls the young, which meanwhile have cowered motionless, and leads them away. Grzimek and his co-workers met a cock ostrich with a hen and eight chicks: "A hyena attacked and was about to grab one of the chicks; the cock looked after the chicks while the hen attacked the hyena, put it to flight, and chased it for more than one kilometer. A few days later we met the same family again, but now there were only six chicks."

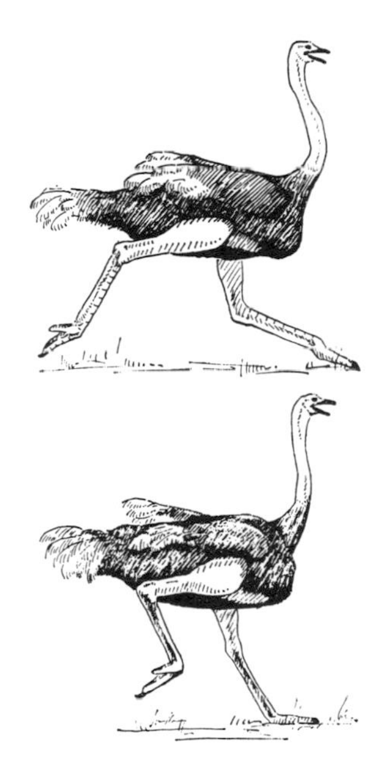
Fig. 3-4. Ostriches, when running, take steps measuring up to 3.5 meters. They can keep up a speed of 50 km per hour for over fifteen minutes, and reputedly can reach a speed of 70 km per hour.

Clutches are often destroyed by people in search of eggs as well as by hyenas and jackals; even Oryx antelopes occasionally supplement their plant diet with animal food and eat ostrich eggs.

One to three months after hatching, and sometimes much later in areas thinly populated by ostriches, the families rejoin the large flocks. It is believed that ostriches live thirty to seventy years, but there are no reliable data about the age of wild ostriches.

Excellent eyesight and acute hearing are their most important senses; they often make the ostrich an unintentional but reliable sentinel for many African grazing animals, such as zebras, antelopes, and gazelles. They are capable of producing pleasant-sounding calls, moot and harsh guttural sounds, hissing and snorting, and a far reaching territory and courtship song. The volume and quality of these calls are not inferior to those of song birds, although the vocal organ, the syrinx, is very primitive. The calling of ostrich cocks during courtship resembles the distant roaring of lions. Grzimek describes it as follows: "They blow air out of the trachea into the mouth, close the beak firmly, and press the air back into the gullet which is thereby dilated. The entrance into the stomach is tightly closed, preventing the air's entering. The entire naked red neck is expanded like a balloon and a dull but far-reaching roaring is produced which presumably tells other cocks as well as hens, 'here is my land'."

Physical capacities, by B. Grzimek

Ostriches are excellent runners. This large bird, nearly three meters high, easily takes steps of 3.5 meters high when running. If one follows an ostrich in a car, one can see on the speedometer that it can run fifty kilometers per hour for a quarter to almost half an hour without obvious exertion or signs of exhaustion. Other wild animals can only run fairly short stretches at such a high speed. Ostriches are said to be able to reach a top speed of about seventy kilometers per hour; man's is thirty. Ostriches must therefore have unbelievably efficient hearts.

In a short run, an ostrich can readily jump one and a half meters high. Hence zoo enclosures must be about two meters high and one must, in general, be very careful with the aggressive cocks. In the Hannover Zoo an ostrich, by a blow with its leg, bent a one centimeter-thick iron bar into a right angle; in the Frankfurt Zoo another one reached an attendant's back with only one toe and tore his clothes, including underwear, and moreover flung the man halfway through a wire fence. A fully grown man can ride a tame ostrich without exerting the bird very much.

A completely new discovery was made by Klaus Immelmann when he kept watch for a few nights in the ostrich house of the Frankfurt Zoo. He wanted to find out how the birds sleep. They sat with upright neck every night seven to nine hours, although the eyes were closed. In this condition they are less easily disturbed by noises or movements than when they are awake. They do, however, get up at least a dozen times to defecate and urinate. When ostriches stand during the day, they sometimes lay their "tired head" on top of a fence, or at least lean on it and then their eyes close as they stand.

Quite unknown, however, was the fact that, while sleeping, ostriches may also lay their extended necks on the ground. They do this only one to four times a night, and never for longer than one to sixteen minutes at a time. Only then is the ostrich in really deep sleep; it can be photographed with a flash and one can knock on the floor or speak fairly loudly without waking it up. The birds also like to stretch their legs, which have been in a sitting posture under their bodies, out behind themselves. All birds never fell into this deep sleep at the same time.

A very similar posture may be seen on other occasions. When an ostrich runs away, it may be that it suddenly disappears, although it has not yet reached the horizon. When one follows, one sees the bird sitting flat on the ground with its neck stretched out. This is probably the origin of the story of the ostrich hiding its head in the sand to avoid being seen. The ancient Arabs first recorded this tale and later it was repeated by the Romans and all writers after them through the centuries. Half-grown ostriches in particular like to lie down in this way. If someone approaches them, they will suddenly jump up and rush off.

▷ A magnificent Australian cassowary *(Casuarius casuarius)* in the Hellabrunn Zoo in Munich, Germany.

The economic uses of ostrich feathers, by B. Grzimek

It is unfortunate that ostriches have such beautiful, waving feathers. For the ancient Egyptians, these feathers became the symbol of justice, because they recognized that the vanes of an ostrich feather are exactly

of equal width on both sides of the shaft. Many other feathers have a narrow and a wider vane, and hence the shaft divides it "unjustly." The ancient Egyptians also had discovered that ostrich plumes make pretty decorations for people. As long as only knights in the Middle Ages decorated their helmets with them, hunting did not hurt the ostrich population. But when ostrich feathers became fashionable for women in the last century, things suddenly looked bad for the ostrich. They were exterminated in north Africa and Egypt. In Persia and Arabia they were also exterminated. In south Arabia the last ostrich disappeared about 1900. In northern Saudi Arabia the last was said to have been shot in 1933 on the border of Iraq, although according to other information two ostriches were said to have been seen and immediately killed in 1948 at the point where the borders of Iraq, Jordan, and Saudi Arabia meet.

If it were not for the ostrich farmers, the ostrich would have disappeared altogether. Otherwise those wild animals whose feathers or fur become fashionable follow a very swift downhill course. As they become rarer, the price for their fur, feathers, or whatever they supply, goes up. This induces greedy individuals to follow them into the most remote wilderness in order to capture them eventually. If however, the same species can be raised in captivity, the price slowly declines and it is no longer worthwhile to pursue the last wild ones. Thus chinchillas, nutria, silver foxes, European mink and sable survived only because someone discovered how to keep and raise them in captivity at the last moment.

The first ostrich farm was opened in 1838 in South Africa; because of the high prices of feathers, farms soon sprung up in Algeria, Sicily, and Florida. Such farms also existed in Nice in France at the time when ostrich feathers were fashionable, and Carl Hagenbeck even founded one at Hamburg-Stellingen. At the turn of the century ostrich farming was one of the largest businesses, particularly in South Africa. Even before World War I, a good breeding cock cost up to $7,500 there. About 1910, 370,000 kilograms of feathers were exported annually, while seventy years earlier it was only 1000 kilograms. The cocks' feathers are cut just above the skin, not plucked as one does with dead chickens or geese.

During World War I, no one in South Africa had time to hunt ostriches. Since ostrich feathers had meanwhile gone out of fashion and had become quite cheap, it was felt at the end of the war that there were too many ostriches, and so hunting them became legal. Businessmen hunted them from cars and often returned from a trip with 400 or 500 skins from which wallets and ladies' handbags were made. Nobody wanted the 100 to 520 kilograms of meat, particularly when such a bird was about thirty years old. The dead ostriches contaminated whole districts, for the hyenas and vultures could not consume the sudden excess.

◁
An ostrich hen *(Struthio camelus)* spreads out her large wings, forming a sunshade for her offspring. An ostrich's wings are useless for flight.

When ostrich leather is used today, most of it is not from wild birds but from South African ostrich farms which, after a temporary decline, are flourishing once again. At present 42,000 ostriches are being kept there. In Florida as well, there are a few ostrich farms; however, there the large birds are not kept for the sale of their skin but as tourist attractions.

If one wants to treat a sick ostrich in a zoo, one only needs to pull a sock over its head; it is then helpless and can be led around. In those zoos which still permit feeding, there has often been trouble because ostriches swallow the most unbelievable objects. Coins, nails, parts of horse shoes, and pocket knives have been found in dead ostriches. One had even drunk green oil paint, so its entire stomach and gut was gummed up with it.

The ostrich in historic and prehistoric times, by F. Sauer and E. Sauer

Since ancient times, the ostrich has stimulated man's thought, religion, and art, as is indicated by 5000 year old records from Mesopotamia and Egypt. For the Kalahari bushman of today, its egg is still a valuable vessel in which he keeps his scarce water; he makes beautiful jewelry for his wife and children from the egg shells. Incidentally, a few wild ostriches which were introduced in southern Australia survive there. The existence of these impressive birds is not yet in serious danger, but without protection, the danger of their extermination by man increases constantly.

Fossil bones and egg shells show that ostriches probably originated in the Eocene (about fifty-five to forty million years ago) in the Asiatic steppes as small flightless birds. In the Lower Pliocene (about twelve million years ago) they developed into gigantic forms which were distributed as far as Mongolia and, later, South Africa. The present-day ostrich is somewhat smaller and originated as a new species in the Pleistocene (about one to two million years ago); some of its early remains were found at the homesites of prehistoric man.

Suborder: elephant birds, by E. Thenius

Numerous remains of ostrich-like birds of genera such as *Mullerornis* and *Aepyornis* are known from the Quaternary Period of Madagascar. They were found in rock strata which are at most two million years old. Like many prosimian species, these elephant birds (suborder Aepyornithes, family Aepyornithidae) became extinct only recently. Besides the New Zealand moas, they are the best known giant birds; even the Carthaginians are supposed to have known about them. Flacourt, the first French governor of Madagascar, was the first to report on these giant ostriches, which make the African ostrich of today seem small in comparison. He stated that a giant bird, called Vouron Patra, was still frequently found in the southern half of the island in the mid-seventeenth century.

The first report of value to scientists did not, however, become available until the last century when a traveller by the name of Sganzin sent the sketch of one of the giant eggs to the collector, Jules Verreaux, from Madagascar in 1832. He reported that the natives used remains of such eggs as vessels. Their shell is several millimeters thick, they may be more than thirty centimeters long, and their volume is given as more than eight liters. This corresponds to more than seven ostrich eggs or more than 180 chicken eggs.

The elephant birds differ in several skeletal characteristics from the present-day continental ostrich. They form a separate branch which presumably reached Madagascar very early. This is confirmed by findings from the Lower Tertiary of North Africa, which have been called *Stromeria* and which must have been closely related to the ancestral line of the later elephant birds. They were relatively small, slender-legged forms. They suggest that the division of the ostrich-like birds must have taken place very early, probably in the earth's Middle Age. The similarity of the elephant birds to the moas, which also became exterminated in historical times, cannot be traced to a direct relationship. It is rather a consequence of flightlessness which is related to gigantic size. Within the group of living ratites, the African ostriches are the nearest relatives of the elephant birds, even though they have some features in common with the cassowaries.

Suborder: cassowaries

The CASSOWARIES (suborder Casuarii) clearly differ from the rheas and ostriches in their structure and way of life. All the feathers consist of a shaft and loose branches; there are no rectrices or preen gland and only six to seven large wing feathers. On the strongly retrogressed wing, the lower arm and hand are only half as long as the upper arm. The wishbone (furcula) and coracoid are degenerated. There is a special palatal structure, and the palatal bones and sphenoids touch one another. The deep green eggs are coarsely grained. Incubation is mostly done by the ♂♂. There are two families: 1. EMUS (Dromaiidae) with one genus *(Dromaius)* and two species; and 2. CASSOWARIES (Casuariidae) with one genus *(Casuarius)* and three species.

Family: emus, by K. Sanft and B. Grzimek

The EMUS (family Dromaiidae, genus *Dromaius*) are flightless inhabitants of the Australian bush steppes. Three subspecies which lived on coastal islands were exterminated in the last 150 years. Ancestors of today's emus lived in the Upper Pleistocene (50,000–10,000 years ago) in Australia.

Distinguishing characteristics

Externally they resemble the rhea; they are about the same size but much more compact and heavy, weighing up to 55 kg. The main shaft and secondary shaft (aftershaft) are equal in length so that every feather appears to be double. The wings are small, and are hidden by the rump plumage. There are three toes. The gut and caeca are shorter than in rheas. Their food consists of fruits and seeds. There is no preen gland. The copulatory organ is extrudable. There is only one genus with two subspecies:

1. The EMU (*Dromaius novaehollandiae;* see Color plate, p. 88) has four subspecies. The height reaches up to 180 cm and the height of the back

is 100 cm. The weight reaches 55 kg, the tarsus is 35 cm long, and the beak is 12 cm long. The eggs measure 135 x 90 mm and weigh 600 g; the surface is wrinkled. When fresh, they are dark green but become almost black with time.

2. The BLACK EMU *(Dromaius minor)* is darker and smaller. The height of the back is 80 cm; the tarsus is 28 cm long. It is extinct.

The black emu was discovered in 1802 on King Island; two of the birds reached Paris in 1804 and were kept in the private zoo of the Empress Josephine. The last of these two died in 1822 and with this the history of this interesting, insular species died. White settlers and brush fires presumably exterminated them in their home. Only a few skins and skeletons remain as valuable possessions in museums.

The killing of emus

The larger emu *(Dromaius novaehollandiae)* has also been exterminated in Tasmania and all populated areas of Australia. Emus are accused of drinking the water needed by cattle and sheep, of stamping down wheat fields and picking up large quantities of grain, and of jumping over barbed wire fences. Kangaroos, on the other hand, run blindly into fences when hunted and become entangled in them. Emus, particularly the young, consume large quantities of caterpillars and grasshoppers. The adults eat masses of the burrs which entangle sheep wool. They also eat fruit and berries; mainly, however, they live on green grass and herbs. But these are intended only for sheep in Australia. Recently the sparsely settled state of West Australia began a determined campaign against emus because the struggle had been unsuccessful for the past thirty years. The emus are to be allowed to survive only on a few hundred square kilometers in the southwestern tip of the huge state of West Australia. A fifty cent bounty was paid for every emu during the 1930's; in 1937, 37,000 emus were killed in the Northampton district alone. Five years earlier, one of the most remarkable campaigns was held near the towns of Campion and Walgoolan, for 20,000 "enemies" were said to be damaging the harvest. A company of the Royal Australian Artillery under the command of a major, together with local farmers, took to the field against the emus with two machine guns and ten thousand shells. They hoped to drive them along fences into the range of the machine guns, as had been successfully done in the northwest of the state of New South Wales. But in this "war" only twelve emus were killed because the emus were more adept at camouflage and withdrawal at the critical moment than the soldiers. In 1964 the state of West Australia still paid bounty on 14,476 emus.

Fig. 3-5. 1. One-wattled cassowary *(Casuarius unappendiculatus)*, New Guinea; 2. Bennett's cassowary *(Casuarius bennetti)*, New Guinea; 3. Australian cassowary *(Casuarius casuarius)*, New Guinea and York Peninsula; 4. Emu *(Dromaius novaehollandiae)*, Australia, extinguished in Tasmania and many parts of Australia; 5. Black emu *(Dromaius minor)*, Kangaroo and King Islands, extinct.

These large Australian ratites are said to have survived in the northern part of Western Australia and even to have multiplied. But the farmers fear that a dry year may drive them southwest to the agricultural areas in masses. They no longer wish to exterminate the birds, but to confine them; however, in dry years this might result in the same thing. A fence many hundreds of kilometers long has been built to protect the wheat and sheep farming areas of West Australia against the emus which populate the districts of Northampton and north of Hopetown.

The emu is a fast runner which can reach speeds of up to fifty kilometers per hour. Surprisingly, it also swims well and with endurance. Nearly everything we otherwise know about the life of emus comes not from Australia but from zoos, mostly European. It is very difficult to distinguish hens and cocks. For this reason alone there is often no reproduction in zoos where only two are kept. By chance these may be two cocks or two hens. It is often necessary, even for years, to exchange one for another until a zoo finally has a pair. The cock sometimes utters a call which sounds like "e-moo," and can be heard far away. Here in the Frankfurt Zoo, zoologists Ingrid and Richard Faust have studied the life of ratites for years. For this reason we have raised hundreds of South American rheas and distributed them among European zoos. By 1967, fifty-nine emus had been reared, some hatched by the cock, others in incubators. The newly-hatched emu chicks weigh 440 to 500 g. The eggs are laid between December and April. Before copulation the hen utters dull rattling sounds that sound like drum beats, and the male becomes attentive and approaches her. When two emus are paired they stand next to one another with lowered heads and bent necks. They sway their heads from side to side above the ground. Then the hen sits down, the male sits down behind her, and he shuffles up and onto her and finally grasps the skin of her nape with his beak. At the same time he utters squeaking or purring sounds, and finally he runs away while the hen remains sitting.

Reproduction in zoos and in the wild

Incubation and raising young is the male's task, as in the rhea and cassowary. The nest is a shallow depression located next to a bush. It is simply made with leaves, grass, and bark, and holds fifteen to twenty-five eggs which come from several hens. Incubation takes twenty-five to sixty days; the great variability is due to pauses during which the cock must leave to feed and drink for shorter or longer periods of time. At two to three years of age, the young are grown and capable of reproduction.

Emus are easy to keep

Emus in captivity are hardy and easy to keep. As K. Sanft has found, they were bred as early as 1830 in the London Zoo, and not much later in the Berlin Zoo's peacock island. In the Frankfurt Zoo the incubating cock used to get up and walk around between four and five o'clock in the afternoon. The hen would sit on the nest during this time and lay another egg. The Königsberg Zoo received a male emu in 1897 which was still there in 1928; his hen had arrived in 1904. There the cock, while incubating, did not eat or drink at all and apparently left the nest very rarely. While incubating or brooding, he would permit eggs or young to be removed, but when leading the young around, he was aggressive. In the Moscow Zoo, an emu cock ate no food during the fifty-two days of incubation and lost seven to eight kilograms, fifteen percent of his body weight.

We were able to leave the emu hen in the same enclosure with the cock when he was leading the young, although he occasionally hissed at her. This indicates that emus have some sort of family relationship, at least more so than the South American rheas.

I. and R. Faust fed emu chicks with large amounts of protein, particularly during the first weeks, in the form of ant eggs, chopped meat, chicken mash, chopped eggs and, of course, minced salad and other greens. Young African and South American ratites must also be given much protein if they are to be successfully raised.

Emus which have lost their fear of man, or are cornered when they are about to be caught, often deliver tremendous blows with their unbelievably strong feet. They may readily break a man's thigh or tear the muscles with their hard claws. But a tame emu kept in Syndey was satisfied by chasing men and taking their hats away.

The behavior of emus toward people who raised them suggests that they care for one another and that they know one another individually. While rheas become shy after some time and do not differentiate between those who raised them and other people, this is not the case with emus.

Family: cassowaries, by K. Sanft

The CASSOWARIES (family Casuariidae) are closely related to the emus; they inhabit the primeval forests of North Australia and New Guinea as well as some of its islands. They are known from fossils of the Upper Pleistocene (about 50,000–10,000 years ago) in Southeast Australia.

There is only one genus *(Casuarius)*. With a height at the back of up to 100 cm and a weight of 85 kg, it is the heaviest bird next to the ostrich. The legs are very strong. There are three toes, the claws of the inner toe being up to 10 cm long and straight. The feathers, like those of emus, have an aftershaft of equal length; the flight feathers are reduced to mere rods of thick keratin. On the head they have a helmet-like, horny structure. The head and neck are bare of feathers; instead some have skin folds on the neck. The species are distinguished according to the shape of the helmet and the form of the skin folds on the neck. The bare skin differs in color in the various species and subspecies, and can be bright red, yellow, blue and/or white. Males and females are similarly colored. The chicks have a yellow-brown downy plumage with dark brown longitudianl stripes, but after a few months they become uniformly brown. The eggs average 135 x 90 mm and weigh 650 g; the surface is slightly wrinkled, and the color is a shiny grass green, which later darkens somewhat. The eggs and young of the various species are almost indistinguishable. The food consists of fruit. The gut and caeca are even shorter than in the emu. There are three species:

1. AUSTRALIAN CASSOWARY (*Casuarius casuarius*; see Color plate, p. 88 and p. 97). The height of the back is 90 cm. The tarsus measures 30 cm. The helmet rises high above the head; there are two bare skin folds on the front of the neck. There are eight subspecies.

2. ONE-WATTLED CASSOWARY (*Casuarius unappendiculatus*; see Color plate, p. 88). The height of the back is 100 cm. The tarsus measures 35 cm. The helmet is flattened at the back, and there is only a small skin fold in the middle of the neck, and two folds at the base of the beak. There are four subspecies.

3. BENNET'S CASSOWARY (*Casuarius bennetti*). The height of the back is 80 cm. The tarsus measures 27 cm. The helmet is small and flattened in the front and in the back. There are no skin folds. There are seven subspecies, among them the PAPUAN CASSOWARY (*Casuarius bennetti papuanus*; see Color plate, p. 88).

The first living cassowary to reach Europe was received in Amsterdam in 1597. It was an Australian cassowary which was later presented to Emperor Rudolf II. Since then cassowaries are regularly seen in our zoos. Information about their life in the wild is still incomplete; only a few Europeans have observed them in their natural habitat. As an inhabitant of the dense primeval forest, the cassowary readily avoids an observer. With outstretched neck, it breaks through the undergrowth, and in so doing the helmet may help in parting the branches. The Australian and one-wattled cassowary live exclusively in the flatlands; only Bennett's cassowary can also be found in mountains at altitudes of 3000 meters. Swamps and broad rivers present no obstacles to cassowaries, for they swim well. They live on fallen fruit as well as small animals. They are solitary; only in the breeding season do males and females come together. The nest is a shallow depression which is lined with leaves and grass. It holds three to eight eggs, which unlike the emu's come from only a single female. The cock sits on the eggs for forty-nine to fifty-six days; he alone takes care of the young. This has been shown in the few successful broods in zoos.

The natives know the cassowary quite well. They hunt it as the only "big game" besides pigs. In many villages tame hand-raised cassowaries walk around and the children play with them. But this changes when the birds are fully grown. Then they become dangerous to man and they are kept in areas enclosed by high fences. From a standing position a cassowary can easily jump one and a half meters high. The claw of the inner toe, which is up to ten centimeters long, is a formidable weapon; one blow with it can rip open a person's abdomen and kill him. Hence the unruly captives are eventually eaten or sold to animal dealers. Gilliard reports that a Papuan can buy eight large pigs or a wife for one cassowary. A good keeper of cassowaries, therefore, need not fear that he will remain a bachelor. There has been a brisk trade with cassowaries, and those of Ceram are undoubtedly descendants of birds which were introduced there and later became wild. Papuans often bring their birds to dealers from far away, on foot or by boat, much to the chagrin of zoologists who still do not know the true distribution of the many forms.

▷
Fig. 3-6. Size of the various moa species in relation to man.

Although ostriches, rheas, and emus reproduce readily in captivity, the breeding and rearing of cassowaries is more difficult. As early as 1862, and again in 1863, one young was hatched in the London Zoo, but neither could be raised. It was not until 1957 that another young hatched in the San Diego Zoo. Incidentally, its father had lived there for thirty-one years. In 1964 the Frankfurt Zoo, and in 1965 the Dresden Zoo, both in Germany, were successful in breeding them. Unfortunately the Frankfurt young was injured by its father and died. Since cocks and hens look alike, it is not easy to obtain a breeding pair. Even when one brings a pair together they do not live in peace by any means; in zoo enclosures fights occur which are a matter of life and death. One can understand why a breeder may prefer not to endanger his birds. A cassowary, after all, costs five hundred or more dollars.

Suborder: kiwis, by R. A. Falla

Family: moas

As on Madagascar, gigantic flightless ratites also once lived on the large islands of New Zealand; they reached their peak in the Glacial Period and were able to survive almost to the present. These very plump moas (family Dinornithidae) were very similar to the much smaller kiwis. We therefore combine both into a single suborder (Apteryges), but some zoologists divide them into two orders. Moas, like kiwis, have four toes, their sense of smell is good, and their eyes are small. The clutch consists of only one or two eggs. Both families are confined exclusively to New Zealand.

Distinguishing characteristics

The total height of moas is 100 to 300 cm, and the weight reaches 250 kg. The ♀♀ are apparently larger than the ♂♂. The neck, as in emus and cassowaries, is curved, and not upright as in rheas and ostriches. The head is small, the legs are heavy, and there are no traces of wing bones. There is no pygostyle. The feathers are entirely loose with long aftershafts. The eggs are white, slightly oval, relatively thin-shelled, and weigh about 500 to 7000 g. There are nineteen species in two subfamilies and six genera; among them is the GIANT MOA (*Dinornis maximus*; see Color plate, p. 88), which is the largest, and the DWARF MOA (*Megalapteryx hectori*), which is the smallest. The ELEPHANT FOOT (*Pachyornis elephantopus*) and the PLUMPFOOT MOA (*Euryapteryx gravis*) had especially heavy feet.

The moas were able to develop into the dominant plant eaters in their habitat, mainly because of the absence of large land mammals. Hence these birds encountered neither food competitors nor large carnivores. It was not until man appeared in New Zealand that their numbers began to decline. Most species of moas became extinct between the tenth and seventeenth centuries because the Maoris pursued them for their meat, their bones, and their egg shells, and because

Two *Diornis* species.

Three species of the genus *Pachyornis*.

Megalapteryx, *Anomalopteryx*, and *Emeus*.

fires set by man altered the vegetation on the islands. Only one or two of the small species may have survived into the beginning of the nineteenth century.

The Polynesian immigrants to the island called these birds "moa." For a long time, the tradition of the Maoris was the only source of our knowledge of moas. This information was replaced by scientific information in 1839, when Richard Owen described a piece of a femur which John Rule had brought back to England. Owen's view that it was the bone of a large and probably flightless bird stimulated further research. Many more bones were found, often close together in swamps and limestone caves but also in exposed sand dunes. Nearly everything we know today about moas stems from these discoveries, which were made on the surface or at the most four meters deep in the ground. Very few fossils have been found in deeper rock layers. Hence they provide insufficient clues about their geological age. It is assumed that they are at least one million years old and that moas already lived in New Zealand before the end of the Pleistocene, at least two million years ago, but these assumptions have not been verified.

In the description of the various species there was some confusion because the bones of several species were intermingled at some sites. Occasionally bones not belonging together were ascribed to one species. Today nineteen species are recognized, but some paleantologists believe the remains belong to twenty to twenty-seven species. Pieces of skin and parts of the plumage have been found only for the DWARF MOA *(Megalapteryx hectori)*. The feathers are a purple-black with gold-yellow-brown edges. Age determination using the carbon-14 dating technique indicates that the remains examined were from the fifteenth century, but this age determination may well have to be corrected. The behavior of moas can be inferred, in part, from the nest sites found in rock caves. Crop contents and feces indicate that they ate grass, twigs, leaves, berries, and seeds.

Family: kiwis

The KIWIS (family Apterygidae) are the most peculiar of the ratites, and also differ most from the other families. Among the many birds found only in New Zealand, they are the most characteristic of these islands.

There is only one genus *(Apteryx)*. The length is 48 to 84 cm, the standing height reaches 35 cm, and the weight is 1.25 to 4 kg. The ♀♀ are larger than the ♂♂. The beak is long, pliable, and senstive to touch, and the nostrils are lateral at the tip of the beak; the eye interior has a much reduced pecten. The feathers have no aftershaft; rather, the shafts project much like coarse hair from the vanes which lack barbules and are therefore loose. There are large vibrissae around the gape and thirteen flight feathers which are only a little stronger than the other feathers. The second finger is absent. There is no tail, only a small pygostyle. The legs are strong but short, and the claws are sharp. The gizzard is weak. The caeca are long and narrow. The young are colored like the adults but with softer plumage. They inhabit wooded parts of New Zealand up into the high mountains; today they also inhabit partly cleared bush and deserted farmland. There are two readily distinguishable species:

1. COMMON KIWI (*Apteryx australis*; see Color plate, p. 88). The length is 54–55 cm and the beak is 110 to 206 mm long. The weight of the ♀ is about 3 kg and that of the ♂ is about 2 kg. There are three subspecies: SOUTHERN, NORTHERN and STEWART'S COMMON KIWI (*Apteryx australis australis*; *Apteryx australis mantelli* and *Apteryx australis lawryi*).

2. OWEN'S KIWI (*Apteryx owenii*). The beak is 75 to 160 mm. There are two subspecies: LITTLE OWEN'S KIWI or DWARF KIWI (*Apteryx owenii owenii*; see Color plate, p. 88) and GREATER OWEN'S KIWI or HAAST'S KIWI (*Apteryx owenii haasti*; see Color plate, p. 88). They were formerly regarded as two separate species; the length is 35 to 45 cm and 45 to 55 cm respectively. Both are found today on the south island; the little Owen's kiwi, at least as a subfossil, is also found on the north island.

Fig. 3-7. Original areas of distribution of kiwis (genus *Apteryx*) in New Zealand; today they are extinct in parts of their former range.

Kiwis have little in common with other ratites aside from the lack of a sternal keel and the structure of the bony palate. Superficially they do not appear to be related to any other group of birds. The missing "breast" gives them a strange outline. Their senses of smell and hearing are well developed; the eyes are small. More is known about their anatomy than about their mode of life and behavior. Kiwis are truly nocturnal and spend the day hidden in caves which are generally surrounded by dense vegetation. They search for their food at night in the same habitat. Only the weak, shrill "kee wee" or "kee kee" whistles of the male and the hoarse "kurr kurr" of the female betray their presence. They feed on insects, particularly their larvae, and also on worms and fallen berries. While probing for hidden worms and larvae in the soft forest floor, they use their long beaks in the same manner as snipes. Little is known about their life in the wild beyond the location of their nest-holes under roots, on slopes, in the ground, under grass tufts, or on rocks. It is known that they sometimes enlarge these holes in the breeding season. Fortunately kiwis can be readily kept in captivity and the reports of F. D. Robson provide at least an idea about the breeding behavior.

The northern common kiwi generally lays two eggs in late winter at an interval of several days; often it may be a week or longer until the second egg is in the nest. The kiwis of the south island, on the other hand, apparently lay only one egg. The shiny, white, thin-shelled, elongated egg weighs about 450 grams, fourteen percent of the female's weight. The smaller male incubates for fifty-seven to eighty days in a hidden nest made of fallen leaves and leaf mould. When they hatch, the young are entirely covered by a soft juvenile plumage. It appears that they stay in the nest for the first five days and do not eat during this time, but use up the remaining yolk in their body cavity.

Then they leave the nest, led by their father, and independently seek food. The father sometimes scratches the ground for them. The young grow slowly; they are said to be sexually mature in five or six years. Their solitary, nocturnal life is always unobserved. Outside the breeding season they appear to have only weak family or pair bonds. Owen's kiwis are said not to differ in their calls and way of life from the common kiwi. However, for several decades, we have no reports about them. Many flightless New Zealand birds have been threatened by the expanding human settlements, or even exterminated. Thus a rapid disappearance of the kiwi was also feared, but fortunately this has not happened to any noticeable degree. Even today kiwis live in suitable wooded districts in the north of the North Island, in the west of the South Island, and on Stewart Island. A few of the birds are killed by dogs, but for most carnivores the kiwi does not seem to be a desirable prey; perhaps this is because it defends itself by strong blows with its feet, or its odor may be unpleasant. Many kiwis are also caught in traps which are set for the KUSU (*Trichosurus,* see Vol. X), which was introduced from Australia. Today an effort is made to avoid endangering the kiwis. Strict laws have placed them under complete protection. Even zoos and scientific institutions can generally obtain only sick kiwis which are unable to live in the wild. It is therefore likely that this strangest of all ratites will be preserved for the future.

4 Grebes and Loons (Divers)

Fig. 4-1. Great crested grebe *(Podiceps cristatus)*

Loons (divers) and grebes have their feet positioned far back; as a result, they can only walk in a very clumsy manner on land, so they are more dependent on the water than are most other swimming birds. Only penguins surpass them in their diving ability. Zoologists formerly included grebes and divers in a single order, the Pygopodes. However, the two groups differ in many respects, including the number of cervical vertebrae and the special adaptations for swimming found on their toes. While the divers have webbed feet, the toes of grebes have lobes, each toe with its own characteristic type which expands the surface of the toes when they move the feet back. For these and other reasons, zoologists now believe that grebes and loons are not very closely related and that the loons probably derive from the same ancestors as the gulls. The numerous common features of the two orders are presumably the result of adaptation to similar conditions of life and are an example of convergence.

Distinguishing characteristics

Both are footdivers of fresh and saltwater but they nest only on fresh water. The L is 20–90 cm, and the weight ranges from 120 up to 4500 (exceptionally 6400) g. There are 11–12 primaries, and all are moulted simultaneously. Nests are floating or built up from the bottom in shallow water or along the shoreline. The sexes are similar, the ♂♂ being on the average larger than the ♀♀. Their distribution is worldwide, except for the Antarctic.

Order: grebes, by W. Wüst

The GREBES (Order Podicipediformes, family Podicipedidae) have a long geological history. They evolved in the Northern Hemisphere but now inhabit all continents except the Antarctic. They are from thrush to duck size; the L is 20–78 cm, and the weight is 120–1500 g. There are 17 to 21 cervical vertebrae. Some thoracic vertebrae are fused. The legs are positioned far back on the trunk, and the tarsus is laterally compressed with a sharp front edge, but on the back a double row of horny sawteeth is found, which is not known in any other group of birds. The toes have one centimeter-wide lobed membranes along one

Fig. 4-2. Red-necked grebe *(Podiceps griseigena)*

Fig. 4-3. Great crested grebe chick

side. The claw of the mid-toe resembles a fingernail and is somewhat comb-like at the tip; possibly this is used to clean the plumage. Tail feathers are small and soft (unlike most birds), and so these birds appear tailless. The patella is extended upward. The stomach is generally lined with swallowed feathers from the bird itself and from others. There are four genera with nine species:

A. GREBES *(Podiceps)* with six species. B. PIED-BILLED GREBES *(Podilymbus)* with two species: 1. ATITLAN GREBE (⊕*Podilymbus gigas*). The L is 35 cm. They are flightless and almost extinct. In 1966 there were less than 100 animals left. 2. The PIED-BILLED GREBE (*Podilymbus podiceps;* see Color plate, p. 115) has an L of 31 cm and a weight of 119 to 146 g. Its voice is very loud, resembling that of an owl or one of the American cuckoos. C. The RUNNING GREBES have only one species, the WESTERN GREBE *(Aechmophorus occidentalis).* Their largest form has an L of 55–73 cm and a weight of 892–1811 g. It breeds in colonies of hundreds or thousands of pairs on lakes in western North America. It is a resident in part of its range. Courtship displays end with a race in which both members of a pair paddle upright rapidly over the water surface. D. TITICACA GREBES have only one species (⊕*Centropelma micropterum;* see Color plate, p. 115). The L is 40 cm and the weight is 400 g. They are flightless.

Fig. 4-4. Horned grebe *(Podiceps auritus)*

Grebes move on land only when they have no other choice, such as when building nests, in order to incubate, or to get from one open water hole to another in severe frost. Although they generally escape pursuers by diving, they are not bad fliers. During migration the species which is capable of flight can cover great distances.

Fig. 4-5. Black-necked grebe *(Podiceps nigricollis)*

They are all excellent divers, although they dive neither as deeply nor for as long as the loons, generally for less than half a minute and less than seven meters deep. They live in still, fresh water and are seen at sea only outside the breeding season. The felt-like, thick, silky-soft contour feathers protect the underside against the water. In early times these feathers were used on clothing in place of fur collars.

Fig. 4-6. Little grebe *(Podiceps ruficollis)*

The nests are built of rotting plants, they float, and they are anchored to reeds or branches. A clutch consists of at least three eggs; the incubating bird always covers the eggs when leaving the nest. The eggs are at first snow-white and covered with chalky calcium carbonate, but they soon become chocolate brown on the wet plants on which they lie. The downy young are generally colorfully marked and striped, often producing a clown-like effect; right after hatching they move under the wings into the fur-like back plumage of whichever parent happens to be on the nest. Thus protected, they swim and dive with their parents weeks before they can dive themselves. In the breeding season of the second year of their lives, they usually resemble their parents.

The great crested grebe

The largest grebe in Europe is the GREAT CRESTED GREBE (*Podiceps cristatus;* see Color plates, p. 115 and pp. 255/256). The L is 48 cm and the weight is 590–1400 g. Its deep, rolling "korrr" call is known to every naturalist. On calm days, this call carries for kilometers over the smooth water of lakes, especially when the bird is in a mood to display. In the breeding plumage, as well as in the cryptic off-season plumage, the two sexes appear similar, as do all grebes. This might account for the fact that in courtship display both sexes behave similarly.

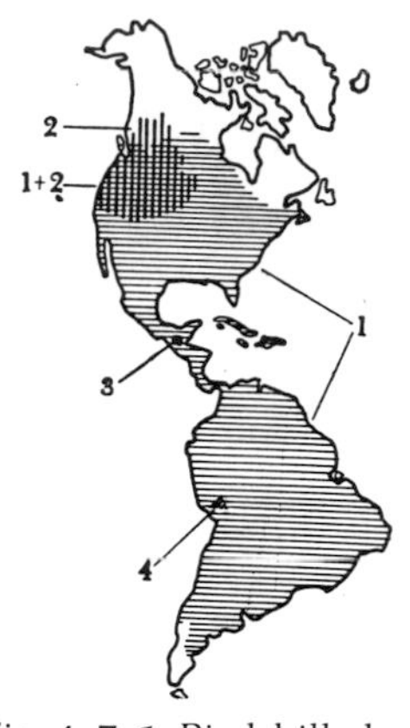

Fig. 4-7. 1. Pied-billed grebe *(Podilymbus podiceps);* 2. Western grebe *(Aechmophorus occidentalis);* 3. Atitlan grebe *(Podilymbus gigas);* 4. Titicaca grebe *(Centropelma micropterum).*

The two partners face off in the water, shaking their erected crests and ruffs; in between they seem to be preening. They offer each other nesting material and at the height of excitement they rise up in front of one another. This upright pose, called the "penguin-posture" by ethologists, was formerly confused with copulation; it is, however, a prelude to the latter and is performed on the open water. The actual copulation takes place on the hidden nest. Nests are constructed in loose stands of cattails, sedges, or horsetails, and also under bushes overhanging the shore. Unfortunately fishermen often destroy these nests.

In many countries, including West Germany, this grebe is not under legal protection at all. In spite of being hunted it has been able to survive in great numbers. In Britain in earlier times, it was so persecuted that in 1860 there were only forty-two pairs left. Because of later protection the population was back to 1240 pairs by 1931. The great crested grebe, more than the smaller grebes, feeds on fish, but these are on the average not over thirteen centimeters long and often are commercially useless surface fish. This large grebe needs about 200 grams of fish a day. They can be seen throughout the year in Central Europe, but numbers and distribution vary with the season. They never form large flocks like ducks; from late summer into spring they are, however, often found in groups of dozens. The Swiss lakes provide a main wintering area for them. Over 20,000 great crested grebes have been counted there at different times.

Fig. 4-8. Courtship ceremony of the great crested grebe: cat posture (upper); standing up in the water (middle); and penguin pose with offering of water plants (lower)

The RED-NECKED GREBE (*Podiceps griseigena;* see Color plate, p. 115), with an L of 43 cm and a weight of 450–925 g, is only slightly smaller. In its non-breeding plumage it may easily be mistaken for the great crested grebe. It breeds much more rarely than the latter in Central Europe and is particularly scarce in West and South Germany. It is certainly more easily overlooked in the breeding season than its larger relative, although one cannot miss its call. Its bawling courtship call reminds one of the squealing of young pigs or the whinnying of young foals; this has given it the common names "pig caller" or "stallion." Red-necked grebes can be seen migrating in Central Europe in all seasons, and occasionally even on the waters of city parks. Since they generally remain quiet, they are readily overlooked.

The HORNED GREBE or SLAVONIAN GREBE (*Podiceps auritus;* see Color plate, p. 115) has an L of 33 cm and weighs 320–720 g. It is intermediate in size between the great crested and the little grebe. It regularly winters in Central Europe. It is found in great numbers along the North Sea and Baltic coasts, but appears less often inland; it is thus generally seen in this region only in its inconspicuous juvenile or non-breeding plumage.

The black-necked grebe

In this plumage it greatly resembles the slightly smaller EARED or BLACK-NECKED GREBE (*Podiceps nigricollis;* the L is 30 cm and the weight is 213–450 g). Only during the last century has it penetrated most parts of Central Europe, coming from the east and southeast. That it is a "newcomer" seems to be reflected in the fact that just as suddenly as it settles in one locality, it may be gone again within a few years. It may even change its breeding place during the mating season. Horned and eared grebes quite often live in gull and tern breeding colonies. This hardly benefits the gulls, but the grebes gain from the watchfulness of the black-headed gulls. When the "sentinels" among the gulls give the alarm call at the approach of danger, the grebes also leave their nests immediately. Eared grebes often nest on fish ponds. They like abundant shore vegetation and large areas of open water, on which they spend much time even in the spring and summer, in contrast to the red-necked and little grebe. The nests stand or float in flooded meadows, sometimes singly and well hidden, but also often in larger aggregations of up to 100 nests; then they are as conspicuous as molehills. The somewhat strained high-pitched call is disyllabic; the soft second note can only be heard at close range. Several days before hatching, the young can be heard peeping in the egg. Like most swimming birds, these grebes lose all their flight feathers in the moult so suddenly that they cannot fly for at least two weeks. Before this important event, they often gather in flocks on waters with abundant food and cover where they must remain until the new flight feathers have grown. In the European sanctuary for waterbirds, the Ismaninger pond region near Munich, about 600 eared grebes were counted at one time in early August.

Fig. 4-9. Threat posture of the red-necked grebe.

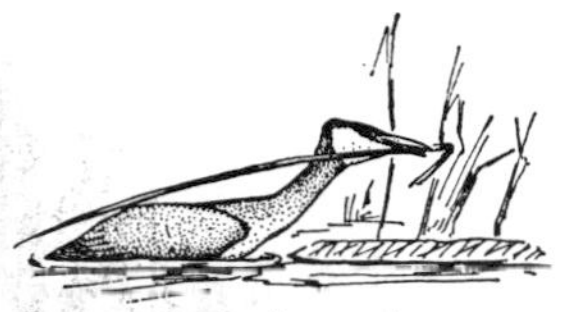

Swimming, the red-necked grebe brings building material to the nest.

The "sitting" posture in the courtship display of the red-necked grebe.

The little grebe

The smallest of the five European grebes is the LITTLE GREBE or DABCHICK (*Podiceps ruficollis;* see Color plates, p. 115 and pp. 385/386; the L is 27 cm and they weigh 105–305 g). It is mostly seen in winter, either alone or in small groups on ice-free lakes or rivers, provided the currents are not too swift. These fist-sized yellowish-brown feather balls generally carry their plumage fluffed up and when seen from behind they show their down, somewhat as a roe deer shows its whitish rump patch. These tiny divers float as easily as corks. Before diving, they press their feathers down, thus squeezing the air out of their plumage and sinking deeper. With a jump they then disappear and surface again elsewhere. In summer their melodic trilling calls may be heard. From August to April they make a similar but shorter contact call which they also use for communication when they are flying. This is the only grebe that also breeds on small ponds, even within cities.

Above all, clear water and protecting plant growth are important for it. Newly laid eggs may be found as early as March, and occasionally as late as September. The young are fledged forty-four to forty-eight days after hatching. These little divers are often seen in zoos; in the Frankfurt Zoo one can watch their diving skills through a glass pane.

By far the smallest grebe lives in tropical America. It is the LEAST GREBE (*Podiceps dominicus;* the L is 20 cm and the weight is 100–137 g). Its iris is brilliantly orange colored. The incubation period is twenty-one days.

Order: loons or divers, by L. Lehtonen

The LOONS OR DIVERS (order Gaviiformes, family Gaviidae) have an L of 58–90 cm and a weight of 1–4.5 (occasionally 6.4) kg, and are generally larger than grebes and have a thicker neck. The skull shows a deep supra-orbital groove for the nasal gland, and 14 to 15 cervical vertebrae. All thoracic vertebrae are free; the sternum is narrow and rather prolonged. There are extensive webs between the three front toes, and they have a short tail with 16 to 20 tail feathers and 11 primaries. In flight the head and neck are somewhat lowered. Copulation is carried out on land. They have two brown spotted eggs which are left uncovered when the parents leave the nest. The downy young are dark. They develop a second downy plumage from the same feather papillae which supply the first downy coat and later the small body feathers. They acquire their breeding plumage in the third year.

Distinguishing characteristics

Fig. 4-10. The penguin-dance of the red-necked grebe.

Only one genus, *Gavia,* with four species is found in the northern tundra and forest zones of the Old and New Worlds. A. Three species with a black and white checkered pattern on the wings of the breeding plumage: 1. ARCTIC LOON (Black-throated diver, *Gavia arctica;* see Color plate, p. 116); the L is 70 cm and the weight is 2–3.5 kg; the nape is gray. 2. The COMMON LOON or GREAT NORTHERN DIVER (*Gavia immer;* see Color plate, p. 116). The L is 75 cm and the weight is 4 kg; the nape and beak are black. 3. The YELLOW-BILLED LOON or WHITE-BILLED NORTHERN DIVER *(Gavia adamsii).* The L is 87 cm and the weight is 4.5 (reaching 6.4) kg. The beak is ivory-colored in adults. B. One species with small white stripes on the non-breeding plumage: 4. The RED-THROATED LOON, or DIVER (*Gavia stellata;* see Color plate, p. 116). The L is 58 cm and the weight is 1–2.4 kg.

The Arctic loon

The ARCTIC LOON *(Gavia arctica)* inhabits the deep northern clear-water lakes. During the breeding season pairs occupy territories from which they chase others of their species. If an intruder approaches, the territory-owning male calls "kooik-kookooik-kookooik," a sound which can be heard up to four to six kilometers away. The intruder then soon dives and leaves the area. By these calls these loons, which inhabit a particular lake, come to know their respective territories and neighbors before the breeding season so that boundary violations are rare. Besides the loons of this species which breed, there are other pairs which defend a territory but do not build a nest. These are non-

▷

Grebes: 1. Titicaca grebe *(Centropelma micropterum);* 2. Pied-billed grebe *(Podilymbus podiceps);* 3. Great crested grebe *(Podiceps cristatus);* 4. Horned grebe *(Podiceps auritus);* 5. Little grebe *(Podiceps ruficollis);* 6. Red-necked grebe *(Podiceps griseigena).*

3
2
1
4
5
6

1
a
2
b
3

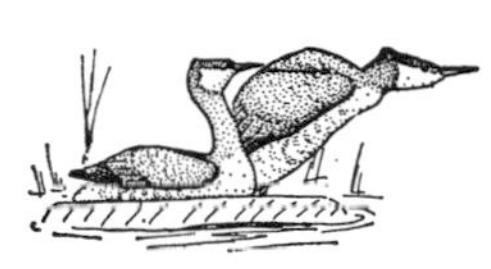

Fig. 4-11. Copulation and its after-display in the red-necked grebe.

Fig. 4-12. Arctic loon *(Gavia arctica)*

◁ Loons: 1. Arctic loon *(Gavia arctica)*; 2. Red-throated diver *(Gavia stellata)*; a. Non-breeding plumage; b. breeding plumage; 3. Common loon *(Gavia immer)*

breeding pairs, five to six years old, which generally only breed a year or two later. Arctic loons which wander around on a lake from spring to fall without a permanent home are even younger.

In the spring, Arctic loons migrate to their breeding grounds and in the fall to their wintering places. Those nesting in Siberia make a peculiar loop in their migrations. According to E. Schüz, they begin their migration in April. They move from the Black Sea to the Baltic Sea first, and then through the Finnish and Bothnian Gulfs northeastward as far as the Lena River. The fall migration begins early in August. At first the birds swim along the large rivers to the Arctic ocean, reaching its shores in September, where they gather in large numbers. Then they follow the coasts of the Arctic Ocean westward as far as the White Sea; from there they continue their journey southward over the lake regions of Eastern Karelia and the Pripjet marshes. They reach their wintering areas on the Black Sea in November and December. Occasionally a few lose their way and turn up in Central or Western Europe.

They arrive at their breeding lakes soon after the melting of the ice. After five to twelve days they begin to build their nests. The nest, in which the female lays two eggs, is usually built on the shore of an island. Near this nest the male often builds one to three sham nests. The nest is not more than 120 centimeters from the water, since these loons cannot cover much distance on land. Its legs are so close to the end of its body that they are not suitable for walking.

The female incubates for about four weeks. The young are precocial, leaving the nest on the first or second day after hatching. They swim well from the very beginning, and are able to dive when three to four days old. But the parents feed them until the fall. At first they are fed small invertebrates, but when the young are more than a month old they only eat fish, particularly smooth silvery species like small maranes *(Coregonus albula* and *Coregonus lavaretus)*, white fish *(Alburnus)*, and smelt *(Osmerus eparlanus)*. Arctic loons are skillful fish catchers, quite adapted to their aquatic life. Except for the incubating female, they spend ninety-eight to ninety-nine percent of their lives in or on the water. Although many water birds can swim as well, there are no seabirds besides the penguins which are such accomplished divers. They can dive for a distance of 500 to 800 meters and stay submerged for three to five minutes. They usually dive only just beneath the surface, but they are capable of diving thirty meters deep.

As early as June, the non-breeding birds gather on lakes in the mornings or evenings. As soon as one Arctic loon arrives at such a gathering place, others also appear there. Each arrival is greeted with a special little ceremony: all members of the flock swim with outstretched necks toward the new arrival and the birds touch beaks. This "beak salutation" is followed by a collective "circle swimming": all loons swim with outstretched necks in a circular course of one to two meters in diameter. Toward the end of the summer these flocks continually increase in number. First they are joined by those pairs which nested unsuccessfully as well as by the territory-owning but non-breeding pairs. When the migratory urge awakens, the successful parents also leave their territory for a few hours and join the flocks. In August and September these flocks are at maximal size; one can then see twenty to forty birds together on the lakes in Finland, which are most frequented by Arctic loons.

In such flocks, pair formation takes place in the late summer and fall. Spring courtship is almost absent and is replaced by group rituals in the autumn. These include circle-swimming, a fast "flutter-run" in an upright posture over the water surface, and rapid noisy diving accompanied by loud calls. Those loons one sees performing rapid swimming thrusts which carry them to and from for distances of five to ten meters are unpaired males trying to attract the attention of unattached females. Even in the fall fights connected with courtship are rare; only when two males are vying for the same female are there occasional disputes. The opponents attack one another and show an unbelievable maneuverability. One may dodge another's attack with a skillful movement, and next both may shoot up out of the water so that only their tails and feet remain under the surface. They may face one another only twenty to thirty centimeters apart with their bills pushed down against their breasts. They alternatively move forward and back, moving their feet like propellers. Such fighting males show their acrobatic skills not only on the surface of the water; they also dive so that only their wing tips show like shark fins. The struggle is close to a decision when the antagonists seize one another with their beaks, and one tries to force the other under water. The loser then escapes on the surface. If the victor follows, the other members of the flock follow too and get in between the two, thus ending the fight. The birds soon calm down and once more resume their normal swimming postures.

Fig. 4–13. 1. Common loon *(Gavia immer)*; 2. White-billed diver *(Gavia adamsii)*.

These flock rituals last until the fall migration. Although the Arctic loons have short, narrow wings, they can fly for long distances. To become airborne they need 40 to 200 meters of open water. They begin by pushing both feet forward horizontally. This is accompanied by wing beats for speed. The rate of climb is always so gradual and slow that Arctic loons must fly about one kilometer to gain an altitude of twenty meters.

Fig. 4–14. Red-throated diver *(Gavia stellata)*.

The three other species of loons have much in common with the Arctic loon in their way of life. Probably the ancestor of all loons was closely related to a small but fairly large-winged species *(Colymboides minutus)* which lived twenty-five to forty-five million years ago. Loons are probably not related to the toothed divers (genus *Hesperornis*; see fig. 1-28) of the Cretaceous Period, which became extinct more than seventy million years ago.

The common loon

The COMMON LOON (*Gavia immer*; see Color plate, p. 116) is the largest member of its genus. According to Gier, its air sacs are particularly large, enabling it to make long dives without difficulty. In the breeding season common loons occupy a fairly small territory. That of the female extends for only 200 to 300 meters, while that of the male is 500 to 1000 meters around the nest. Since these loons are strong, they have, apart from man, only few enemies. At the approach of an enemy, they escape from the nest and dive long distances. Only when the eggs are hatching is the female's incubation or brooding drive so strong that she will not flee when in danger. These loons can readily chase off polecats *(Putorius putorius)*, and they also defend themselves so vigorously against foxes and even young polar bears that the attacker withdraws in most cases. The loons' main weapons are their beaks, with which they aim at the eyes of the attacker.

The RED-THROATED LOON or DIVER (*Gavia stellata*; see Color plates, pp. 61/62 and p. 116) is the smallest of this group. Its distribution greatly coincides with that of the Arctic loon, but the Arctic loon lives on the larger lakes, while the red-throated loon primarily lives on small ponds with grass-covered banks. In contrast to other loons, it can take off from smaller lakes since it needs much less of a running start. Its wings are relative large, as large as those of the Arctic loon; the ratio of the body weight of the two species is about 60:100. Red-throated loons fly more than their relatives. There are often no fish in their breeding ponds so they have to make daily flights over many kilometers to seek prey. In recent decades the number of these small loons in southern Finland has steadily declined.

5 The Penguins

Order: penguins, by B. Stonehouse

The PENGUINS (order Sphenisciformes, family Spheniscidae) are a group of seabirds with very distinguishing characteristics. Their relationship to other orders of birds is uncertain, and therefore some ornithologists separate the penguins as a superorder or even a subclass which is distinct from all other living birds. Their closest relatives appear to be the TUBE-NOSED SWIMMERS (Procellariiformes).

Distinguishing characteristics

The morphological characteristics which distinguish penguins from other birds are a result of their adaptation to life in the water. The L is from 40–115 cm and the weight ranges from 1–30 kg (fossil forms up to 120 kg?). The long spindle-shaped body has legs which are inserted far back, so that they are most effective as oars and steering organs. The tail, as a steering rudder, is streamlined and triangular; the wings are transformed into flippers, but contain all the bones of a wing suitable for flying. The bones, however, are shortened, flattened, and tightly connected by ligaments, thus forming a rigid surface. The breast muscles (wing muscles) are large, taking up the whole front from the neck on down to the lower abdomen. The trachea is, as in Procellariiformes, divided lengthwise. The body is uniformly covered with feathers except for the brood patch; they have thick subcutaneous fat-pads. There are six genera with eighteen (according to some authors, fifteen) species, which are confined to the Southern Hemisphere.

▷

Penguins: 1. Emperor penguin *(Aptenodytes forsteri);* 2. King penguin *(Aptenodytes patagonica);* 3. Gentoo penguin *(Pygoscelis papua);* 4. Adelie penguin *(Pygoscelis adeliae);* 5. Chinstrap penguin *(Pygoscelis antarctica);* 6. Rock hopper penguin *(Eudyptes crestatus);* 7. Macaroni penguin *(Eudyptes chrysolophus).*

Although at first glance penguins resemble the AUKS and GUILLEMOTS (family Alcidae, see Vol. VIII) of the Northern Hemisphere, they are not even remotely related. They are appreciably more adapted to life in the water than the Alcidae. Some species may spend several months continuously in the water and live exclusively on food from the sea. All species drink salt and fresh water and most of them also eat snow. They move with agility on land or on ice; even mire and rubble, smooth rocks, and soft snow are no obstacles for them. Penguins endure long periods without food on land. They incubate for a long

▷▷

Emperor penguins breed on the ice edge of the Antarctic coast, and only rarely on land.

1
2
6
5
4
3
7

5
3
2
4

Penguins: 1. Peruvian penguin *(Spheniscus humboldti)*; 2. Black-footed penguin *(Spheniscus demersus)*; 3. Magellan's penguin *(Spheniscus magellanicus)*; 4. Galapagos penguin *(Spheniscus mendiculus)*; 5. Little penguin *(Eudyptula minor)*.

time and their young grow up slowly. They inhabit mainly the coasts of the Antarctic, but they are also common in the southern temperate cool zone. Northwards, one species is found as far as the Galapagos Islands which lie on the Equator; they also occur on the subtropical coasts of South America, South Africa, and Australia.

The feathers have a surprisingly uniform shape. Each single abdominal or back feather is slightly bent and has an extensive downy portion at the base, formed from the aftershaft. The tips of the feathers overlap like tiles and thus form a waterproof outer shell. The downy part of the feathers also forms a waterproof undercoat; a layer of air thus remains trapped around the body and body heat is conserved.

Swimming and diving ability

The shape of their bodies makes penguins excellent swimmers. They can remain in the water for a long time without being harmed by the cold, yet they can move quickly and with agility on land. Their adaptation to life in the water is developed to a high degree, similar to that shown by the seals among mammals. If this adaptation were to improve further it would be at the expense of their ability to live on land as well. After all, they cannot live exclusively in the water like whales or sea cows since they must breed and moult on land.

Penguins generally swim on or just below the surface of the sea; under water they reach a speed of thirty-six kilometers per hour. At higher speeds they dive alternately up and down like dolphins; this enables them to breathe at regular intervals without interrupting their forward movement and to reduce friction by "lubricating" the plumage with air bubbles. The heat-conserving air-cushion under their plumage gives their bodies a buoyancy which is not altogether favorable for prolonged diving. Only rarely do they dive for longer than two to three minutes without surfacing. The large penguins can, however, presumably search for food under water for five to one hundred minutes or longer, particularly in winter when much of the prey is found only in the depths of the sea. In general, penguins find their food at a depth of ten to twenty meters. This diving ability, although modest, allows them to seek food at depths which are beyond the capacity of most tube-nosed swimmers. This has undoubtedly contributed to an ecological separation of these two bird groups in the feeding grounds of the southern oceans. The greatest difficulty in the specialization of penguins probably was to achieve better swimming and diving ability. They had to become heavier and faster than tube-nosed swimmers and this was only possible at the expense of the ability to fly.

Moving on land

On land or on ice, penguins walk upright, although in snow they often move by sliding forward on their bellies. The rock-penguins and their relatives hop in an upright posture with both feet simultaneously; hence the English call them "rock hoppers." At the edge of the ice, penguins often pick up speed under water so that they shoot out of the waves and land securely on the ice upright on both feet.

Heat economy

Warm-blooded animals which live in cool regions or in cold water must conserve their heat. When an emperor or an adelie penguin dives into the Antarctic Ocean, it meets a water temperature which is about forty degrees centigrade lower than its body temperature; a man cannot survive for more than ten minutes under such conditions. Since penguins are relatively small, the ratio between body surface and body volume, even in a large penguin, is less advantageous than it is in any other warm-blooded vertebrate which spends much time in the water; it is therefore particularly important for penguins to counteract the heat loss at sea. This is done in part through elevation of the metabolic rate; penguins are much more lively in the water than on land and thus produce more metabolic heat, thereby causing the body temperature to remain constant. A further help is the two to three centimeter thick layer of subcutaneous fat, particularly in the polar penguins; further, their thick waterproof plumage and the air trapped beneath it form a very effective protection.

Such heat insulation is very useful in a cold climate and is essential in enabling penguins to survive in and near the Antarctic. The heat insulation of the Antarctic adelie penguin is so effective that, while they are incubating, snow can accumulate on them without melting. In warmer areas, and sometimes even in the particularly cold ones, heat insulation becomes a problem; on land penguins are always in danger of overheating, especially when they fight, run, or are very active in some way. Of the two insulating layers the fat is the lesser problem, for it is penetrated by blood vessels. These can dilate so that more blood reaches the outermost skin layer where it cools off. The protective plumage, however, allows less heat exchange. Penguins can erect the feather tips somewhat but the downy under layer still holds back much of the heat. The plumage is very resistent to disturbances by wind; even in storms with winds of sixty or more kilometers per hour it retains its smooth outer surface.

Tropical penguins have a much thinner plumage than do the polar species, and their layer of fat is also much thinner. On land they seek protection from the heat, particularly in caves, in the undergrowth, and in thick coastal forests, and they confine their activities mostly to the night. Polar penguins live in more open areas, but even they can suffer from the heat when the sunshine is intense.

Penguins have some special areas of the skin which have a good blood supply so that heat can be radiated and which, when necessary, is used for giving off excess heat. Among them are the underside of the wings, the surfaces of the feet, and the brood patch, when present. Tropical species, particularly the Peruvian penguin *(Spheniscus humboldti),* have a very sparse plumage and they even have bare spots on the face which, if necessary, contribute to heat loss.

Plumage color

At first glance penguins seem to be only black and white. Closer examination, however, shows their plumage to be surprisingly varied and colorful (see Color plates, pp. 121 to 124). The colors are brightest just after the moult, but in the course of the year the plumage becomes less impressive and at the end of the breeding season the originally black areas turn suddenly into a dirty brown, which looks rather shabby. This shabbiness increases when the birds rear their young. The moult begins soon after the young have become independent and lasts from two to five (sometimes even six) weeks. Penguins change their entire plumage at once in a so-called "catastrophic moult," but the bird does not become naked, for the new feathers push the old ones out of the skin and the latter only fall out, often in whole patches, when the new ones cover the body.

Large penguins lay only one egg; the rock hoppers on Tristan da Cunha often lay three eggs, but most species have a clutch of two eggs and only occasionally one or three. The weight of the egg ranges from one and a half to four percent of the body weight. The incubation period lasts from thirty-three to sixty-two days. The crested penguins *(Eudyptes)* lay eggs of different sizes; the first one is usually small, but the second is twenty to fifty percent larger. The first egg is often lost or is destroyed at the beginning of incubation; if that does not happen it develops normally.

The young of the same clutch hatch at the same time or, at most, within a day of each other; they are covered with a sparse downy coat and are carefully brooded until, at the age of six to ten days, they begin to regulate their own body temperatures. They keep their thick, almost woolly, down plumage until they are almost fully grown. They are fed by both parents, which regurgitate food for them; within a short time they almost reach the size of a fully grown penguin. The down is replaced by a new plumage and the young, without parental help or guidance, go into the water. Young differ from the adults by their somewhat smaller size; furthermore, the crests or patches of color of some species are generally less bright in the young. They acquire the adult plumage after the moult, which occurs at the end of their first year of life; although they are then fully grown, a few years may elapse until they breed for the first time.

Most penguins are sociable. They breed in groups or in large noisy colonies; they take to the water in flocks and seek out the feeding grounds in large swarms. One of their largest and most densely populated habitats lies at the edge of Antarctica; on several Antarctic and sub-Antarctic islands, breeding colonies of hundreds of thousands, or even of millions, of penguins have been found. In lower latitudes their numbers are generally fewer, although there are very densely populated colonies on some islands off the South African coast and on some oceanic islands with temperate climates, and even on the islands of New Zealand. Penguins in warmer areas generally nest in scattered groups between grass, tussocks, or in bushland; they dig themselves into coastal sand dunes or guano deposits, and gather in small or large caves which offer them protection from the sun's heat. The only partially sociable species also go out to feed in great flocks. It is presumably advantageous for the birds to swim in groups, for the inexperienced gain from the experience of older animals in finding their way between the breeding and feeding places.

Those features which serve to distinguish the various genera and species are on the heads and necks of penguins; thus these features are easy to see when they swim on the surface of the water.

Their contact call is a loud, monosyllabic "krohk," which sounds different in each species and is readily heard far over land and water. With this call and the visible distinguishing marks, the birds can maintain contact with others of their species, which provides them with the protection of the community. This prevents them from getting lost at sea; in areas where there is little food or where it occurs only in scattered places, it helps them to find food and protects them from enemies. On land as well, the advantages of living together in flocks are evident: a locality suitable for breeding for some penguins is likely to be equally suitable for many, and the presence of large numbers of breeding pairs surely reduces nest robbing.

The weight of individual penguins is quite variable according to the season and other factors. The average weights of healthy birds of the various species during the breeding season and time of rearing young are given below. Penguins are heaviest just before the moult, when they weigh up to fifty percent more than the average; immediately after the moult they may weigh twenty to thirty percent less than average. Body length is also variable; it is only given there to indicate size differences which are not clearly evident from weight alone. The lengths given were measured from the tip of the bill to the tip of the tail stump; hence the tail feathers were not included. In a few species, such as the adelie penguin, this would increase the figure by twenty percent. Standing height is seventy to eighty percent of the length given; thus the emperor penguin, with a length of 115 centimeters, is about eighty to ninety centimeters high, and the little blue penguin, with a length of forty centimeters, stands about thirty centimeters high.

Simpson lists seventeen species in his review of fossil penguins, which lived from the Lower Eocene to the beginning of the Miocene (between about fifty-five and twenty-five million years ago). As far as can be determined from the rather sparse data, all would have been immediately recognizable as penguins; in other words, there is no reason to assume that they differed much in shape from present-day penguins, although they resembled most other birds a little more than they do today since many had more slender tarsal bones or straighter humeri. Seven of these extinct species were considerably larger than the emperor penguins; they reached standing heights of 120 to 150 centimeters, and total lengths of 150 up to 180 centimeters. It seems likely that these larger penguins lived in a temperate rather than cold climate, in which case they needed neither a very dense plumage nor a layer of subcutaneous fat. Nevertheless, they must have been relatively heavy, perhaps weighing over a hundred kilograms, and heat exchange must have therefore presented some problems for them. Presumably they could dive very deeply. Their extinction is perhaps due to changes in the environment, and possibly to changes in the average temperatures of the sea to which they could not adapt because of their long breeding season and growth period. They disappeared at about the time when toothed whales and dolphins suddenly evolved. Perhaps the giant penguins could not compete with these mammals or became their prey.

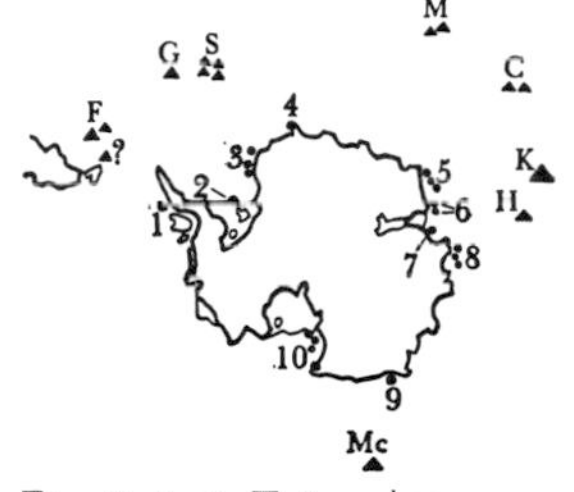

Fig. 5-1. 1. Triangles: King penguins *(Aptenodytes patagonica)*, breeding colonies: Falkland Islands (F); South Georgia (G); Northern part of New Sandwich Islands (S); Marion Island (M); Crozet Island (C); Kerguelen Island (K); Heard Island (H); Macquarie Island (Mc).
2. Dots: Emperor penguin *(Aptenodytes forsteri)*, breeding colonies: Dion Island (1); Austral Bay (2); Halley Bay, Caird Coast, and Norsel Bay (3); Lazarev (4); Kloa Point, Fold Island, and Taylor Island (5); Auster Island and Cape Denley (6); Amanda Island (7); Gauss Mountain, Harwell Island, and Shackleton Barrier (8); Pointe Géologie (9); Coulman Island, Franklin Island, Beaufort Island, and Cape Crozier (10).

Penguins feed on floating (planktonic) and swimming animals, particularly small fish, small floating crabs, and squids. Although a few species limit themselves to certain animals, as the black-footed penguin apparently does to fish or the gentoo penguin to small crabs and the largest penguins to fish and squids, nevertheless they usually take whatever is most abundant. Black-footed penguins need approximately half a kilogram of fish per day; this species takes about five million kilograms of fish per year from the South African waters during the 185 days of the breeding season. The abundance of penguins by the hundreds of thousands or millions in the Antarctic and sub-Antarctic is related to the fact that these waters are very rich in food during the summer. There are some indications that the slaughter of baleen whales in recent times has been of some benefit to the Antarctic gentoo penguins and their relatives. Since the baleen whales are consumers of plankton shrimp, particularly of the genus *Euphausia*, they are therefore food competitors of the penguins.

It was once assumed that penguins were limited in their distribution to the cool waters of the Southern Hemisphere. It was stated that those which live in lower latitudes had reached the tropics by way of the cold Benguella and Humboldt currents. However, penguins actually live in waters of very different temperatures; average sea temperatures range from twenty-three degrees at the Galapagos to minus one degree centigrade in the Antarctic. The water temperature is therefore not likely to be the most important factor in determining their world distribution.

A better hypothesis is that they prefer areas where the annual fluctuation of temperature of the sea is slight. Most species and subspecies live as adults in waters which vary in temperature only from one to five degrees throughout the year. In the Antarctic, the water temperature remains at the freezing point and only varies one to two degrees; similarly, the Humboldt current, between Valparaiso and Callao, has a temperature variation between thirteen and seventeen degrees centigrade. The Benguella current is only slightly colder, its temperatures fluctuating four to five degrees centigrade. The yearly temperature fluctuation among the islands of the cool, temperature regions of westerly winds, where many species live, is only two to four degrees centigrade. The coastal climate exhibits the same temperature fluctuations as the water, so the penguins find the same balanced climate on water as on land. Breeding places, where the annual temperature fluctuates more, are only temporarily used by penguins; after breeding, they try to find waters with more constant temperatures.

This restriction to areas with the narrowest possible temperature range is probably necessary because of the penguins' main difficulty, the maintenance of a constant body temperature. Presumably, they are unable to adapt quickly to sudden changes of the water temperature. Hence, each species is restricted to a relatively small area with a constant water temperature. Young birds are more inclined to migrate far than are adults. The latter show a definite preference for their breeding areas and only rarely move far from their home waters.

Penguins are among the most popular zoo animals, although they require much care because of their susceptibility to mainland germs. Aspergilosis, a disease of the respiratory pathways caused by a mold, is very dangerous to them. Penguins appear to thrive best in filtered air behind glass. But even long before these techniques were used, some species bred in captivity, such as the king penguins in Edinburgh around 1930.

Giant penguins

The largest living penguins are the GIANT PENGUINS (genus *Aptenodytes*). Their beaks are long, narrow, and curved down somewhat. They have vividly colored orange or golden-yellow patches on the ears and sides of the neck. There are violet, reddish-lilac, or orange plates on the lower mandible. The sexes are similar; the ♂♂ are a little larger than the ♀♀. The young are similar to the adults in the first year, but they have paler ear patches and a bluish-gray head. The distribution ranges from the Antarctic to the cool temperate zone. There are two species, the EMPEROR PENGUIN (*Aptenodytes forsteri*; See Color plates, pp. 121, 122/123), with an L of 115 cm, a feather length of 4.2 cm, an average weight of 30 kg, and a feathered tarsus; and the KING PENGUIN (*Aptenodytes patagonica*; see Color plate, p. 121) with an L of 95 cm, a feather length of 2.9 cm, and an average weight of 15 kg. The tarsus is not feathered.

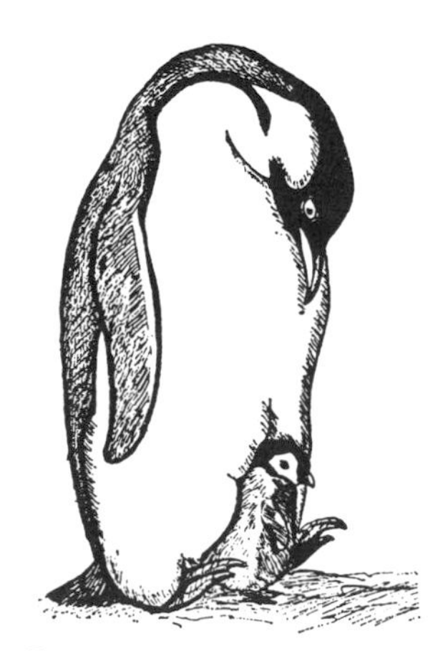

Fig. 5-2. Emperor penguin with young in its first downy plumage.

Although the emperor penguin is only a little larger than the king penguin, it weighs almost twice as much. It is a bird of the high Antarctic and has long dense plumage and very large fat stores. The king penguin, on the other hand, is a bird of the sub-Antarctic and temperate-cold zone. Its body is more slender and its plumage is less dense. During most of the year it has little subcutaneous fat. The vividly colored patches on the sides of the head of both species are shown off prominently during courtship; if they are colored dark experimentally, the bird fails to attract a partner and hence does not breed. Neither species builds a nest; instead they carry the egg on top of the feet and walk around with it.

The emperor penguin

Emperor penguins breed only on the coast of the Antarctic mainland, particularly on the sea ice. Their breeding range extends from 66° south latitude in the eastern Antarctic to 77°, which is only 1400 kilometers from the South Pole. They are rarely found outside of the Antarctic, although occasionally migrating emperor penguins, presumably young ones, have been seen near the Falkland Islands and near south New Zealand. Emperor penguins lay their eggs in the fall as soon as the sea forms an ice cover along the coast. The females return to the sea soon after laying, while the males alone incubate during the Antarctic winter. They do so standing close together in order to keep warm, remaining without food for about ninety days, sixty-two to sixty-four of which are spent in incubation. The mothers return when the young hatch; they find their partners by calling and then they alone care for the young. Meanwhile the males stay on the open sea and in fourteen to twenty-four days reach their original weight again. The young grow slowly at first. Later in early summer when food becomes more plentiful, they rapidly grow larger. After five months, at the beginning of the new year, they can go to sea. This leaves the parents time to moult before the next brood; the moult probably lasts thirty-five to forty days, which means another long period without food.

The king penguins

King penguins are birds of the sub-Antarctic and temperate-cool latitudes. In some areas they were exterminated in the last century, since their oil and feathers were in great demand. They lay their eggs in the summer. Males and females incubate for fifty-one to fifty-seven days; the young stay in the colonies through the whole winter but are then fed only about every fourteen days, so that they lose half their weight until they get more food. King penguins can raise two young every three years by laying early one year, late the next, and not at all the third. The second down plumage of the young is uniformly nutria-brown, whereas that of the emperor penguin has white sides on the head like the first down plumage of both species, which changes after one to three weeks.

Adelie penguins

The ADELIE PENGUINS and two more species form the genus *Pygoscelis,* with an L of 72.5–75 cm. The beak is short to fairly long, and the tail is long. There are no vividly colored parts of the plumage, although the beaks and feet may be yellow or orange. The sexes are very similar, and can be distinguished mainly by behavior. The young of the chinstrap penguin resemble the parents, but in others they may be distinguished by the white chin and throat. There are three species: 1. The ADELIE PENGUIN (*Pygoscelis adeliae;* see Color plate, p. 121) has an L of 70 cm, a weight of 5 kg, and a feather length of 3.6 cm. 2. The CHINSTRAP PENGUIN (*Pygoscelis antarctica;* see Color plate, p. 121) has an L of 68 cm, a weight of 4.5 kg, and a feather length of 2.9 cm. 3. The GENTOO PENGUIN (*Pygoscelis papua;* see Color plate, p. 121), including the subspecies, the NORTHERN GENTOO PENGUIN *(Pygoscelis papua papua),* has an L of 81 cm, a weight of 6.2 kg, and a feather length of 3 cm. The SOUTHERN GENTOO PENGUIN *(Pygoscelis papua ellsworthii)* has an L of 71 cm, a weight of 5.5 kg, and a feather length of 3.2 cm. The MACQUARIE GENTOO PENGUIN *(Pygoscelis papua taeniata)* has an L of 75 cm, a weight of 5.5 kg, and a feather length of 3.3 cm.

The adelic penguin

All penguins of this genus have long curved tail feathers which sweep behind them like brooms as they walk. The adelie penguin lives farthest south on the coasts of the Antarctic continent and the barren islands which surround it. It is met by all Antarctic travellers and is described with loving enthusiasm. The strange white ring of skin around the eye and the short, partly feather covered beak are its main distinguishing features.

Fig. 5-3. . . . . . .Adelie penguin *(Pygoscelis adeliae);* △ Chinstrap penguin *(Pygoscelis antarctica);* • Gentoo penguin *(Pygoscelis papua).*

The three species of this genus breed during the short Antarctic summer, and their young take to the water in February or March. Adelie penguins have been particularly well studied. In soft snow they can run faster than a man. They construct their nests of small stones which they often steal from one another. Only the males incubate, from thirty-three to thirty-eight days, and during this time they do not go into the water, and hence they fast for six weeks. The young are first covered with silvery white, and later with dirty brown down; after four weeks they are left alone for some periods of time. They then gather into "kindergartens" of up to two hundred animals. At nine weeks of age, they take to the water without their parents. In the southern colonies the young grow up more quickly and the time of fasting for the parents is shorter.

Chinstrap and gentoo penguins

Chinstrap penguins are the boldest of all penguins. They attack people with their beaks and wings. Like the adelies, they also breed in very large communities; for example, a single colony with two million chinstrap penguins was reported on the south Sandwich Islands.

The calls of the gentoo penguin sound like the braying of a donkey. The fact that this species has evolved into three subspecies is probably an indication that they are more sedentary than other penguin species. They breed in colonies, generally in small groups between plant tussocks; southern gentoos often breed in places without any vegetation at all.

Crested penguins

In the CRESTED PENGUINS (genus *Eudyptes*), long feathers form a crest above the eyes. The L is 70–72 cm. The beak is fairly strong. In most species the ♂♂ are larger with a stronger beak than the ♀♀. The young are like the parents, but with smaller crests. There are four species: 1. The MACARONI PENGUIN (*Eudyptes chrysolophus;* see Color plate, p. 121)

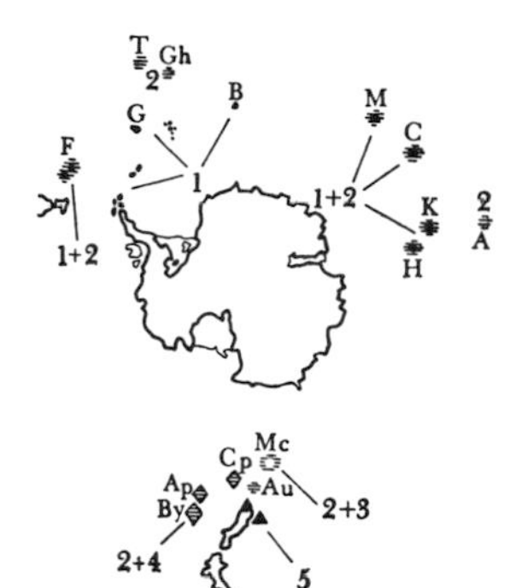

Fig. 5-4. 1. Macaroni penguin *(Eudyptes chrysolophus chrysolophus)*; 2. Rock hopper penguin *(Eudyptes crestatus)*; 3. Macquarie Island macaroni penguin *(Eudyptes chrysolophus schlegeli)*; 4. Erect-crested penguin *(Eudyptes atratus)*; 5. Fiordland crested penguin *(Eudyptes pachyrhynchus)*.

Breeding colonies: Falkland Island (F); South Georgia (G); South Sandwich Island, Bouvet Island (B); Gough Island (Gh); Tristan da Cunha (T); Crozet Island (C); Kerguelen (K); Heard (H); Macquarie (Mc); New Amsterdam (A); Campbell Island (Cp); Auckland Island (Au); Antipodes (Ap); Bounty Island (By).

has two subspecies; *Eudyptes chrysolophus chrysolophus* has an L of [70] cm, a weight of [4.2] kg, and a feather length of 2.9 cm (weights and measurements in brackets in this chapter are average values based on less than ten individuals). The MACQUARIE ISLAND MACARONI PENGUIN or ROYAL PENGUIN *(Eudyptes chrysolophus schlegeli)* has an L of [62] cm, a weight of 4.5 kg, and a feather length of 2.9 cm. The cheeks and throat are whitish. They are regarded as a separate species by some zoologists. 2. The ROCK HOPPER PENGUIN (*Eudyptes crestatus;* see Color plate, p. 121) has an L of 55 cm, a weight of 2.5 kg, and a feather length of 2.9 cm. 3. The ERECT-CRESTED PENGUIN *(Eudyptes atratus)* has an L of [67] cm, a weight of 3.6 kg, and a feather length of 2.9 cm. It has a short yellow crest, red eyes, and a brownish-red or blackish beak with some red. 4. The FIORDLAND PENGUIN *(Eudyptes pachyrhynchus)* has an L of [55] cm, a weight of [3] kg, and a feather length of 2.7 cm; there is a short yellow crest. The eyes are brown and the beak is red. The SNARES ISLAND PENGUIN *(Eudyptes robustus)* is regarded by some zoologists as a subspecies of the fiordland penguin. Both species have measurements similar to those of the erect-crested penguin.

The macaroni penguin is the most southern representative of this genus; the others live in warmer water. Using its sharp claws, the rock hopper can climb on rocks when it allows the waves to throw it on land. It lives on many islands with either temperate-warm or temperate-cold climates, and is also found in unexpectedly large numbers on Heard Island in the sub-Antarctic. The three yellow-crested forms live only in the New Zeland area. In spite of their wide distribution, the genus is very uniform and is in no way adapted to extremes temperatures. Macaroni penguins which breed with adelie penguins on the edge of the Antarctic have feathers that are no longer than those of the erect-crested penguins which live on the much warmer islands around New Zealand or the rock hoppers of Tristan da Cunha and New Amsterdam. In the erect-crested penguins, the male incubates for nineteen days, taking no food for thirty-five days. When incubating, it leans forward at a forty-five degree angle, which looks very uncomfortable to us; more rarely it incubates lying on its breast.

All crested penguins leave their breeding places in late summer or fall and spend three to five months at sea. Their migrations have not yet been thoroughly investigated, but presumably they move north with a current of constant temperature. Hence they remain all year round in water of much the same temperature. The northern species breed in late winter or early spring, enabling their young to go to sea in mid-summer. Antarctic species, on the other hand, lay eggs in the spring and their young can seek warmer waters in the fall.

The BLACK-FOOTED PENGUINS (genus *Spheniscus*) are small to medium-sized; the L is 50–71 cm. The beak is high and strong at the base with longitudinal grooves; it is used for digging. They are short-tailed and have a smooth plumage. In front of the eyes and on the chin they have red to black bare spots. Their webs are often white spotted. They rub their beaks and necks on one another in a greeting ceremony, in contrast to other penguins. Four species are distinguished by facial patterns and throat bands:

Black-footed penguins

1. The BLACK-FOOTED PENGUIN (*Spheniscus demersus;* see Color plate, p. 124) has an L of 70 cm, a weight of 2.9 kg, and feathers which measure 2.3 cm. 2. The MAGELLAN PENGUIN (*Spheniscus magellanicus;* see Color plate, p. 124) has an L of 70 cm, a weight of 4.9 kg, and a feather length of 2.4 cm. 3. The PERUVIAN or HUMBOLDT PENGUIN (*Spheniscus humboldti;* see Color plate, p. 124) has an L of 65 cm, a weight of 4.2 kg, and a feather length of 2.1 cm. 4. The GALAPAGOS PENGUIN (⊖*Spheniscus mendiculus;* see Color plate, p. 124) has an L of 53 cm, a weight of 2.2 kg, and a feather length of 2.1 cm.

Peruvian penguins breed almost solely on the islands of the Peruvian coast. Originally they nested in guano caves, but the exploitation and removal of the guano, which has made this coast so famous, has contributed to a reduction of the population of these penguins. They now breed primarily in rock caves or under the sparse cover of tropical plants. The black-footed penguins, on the islands off the South African coast, lead a similar life. The very small Galapagos penguins inhabit the warmest area of all the penguins, the islands of Narborough and Albemarle of the southwest Galapagos. They breed in holes under rocks which are very close to the water, probably in localities where the temperature is the coolest and most equitable. On land they are very confiding but, at sea, they are shy like all penguins. There are probably no more than 500 pairs of this endangered species. Their nests are occasionally found in groups of two or three in spots which offer the best protection. On the guano islands off South Africa and Peru, penguins of this genus come fairly close together for breeding, presumably because the available space is limited. In the cool, moist, and wooded coastal areas of southern Chile they breed as single pairs.

Fig. 5-5. Male and female rock hopper penguin in mutual courtship display; the feature shown here is the "rapid head shaking."

Magellan's penguins

The Magellan's penguins of Patagonia and the Falkland Islands breed in holes which they dig out in firm sand dunes or clay. When frightened, they withdraw into their holes and lie down in the hole entrance, ready to defend it. Unfortunately they are still killed in great numbers along the Straight of Magellan by the natives who manufacture gift articles from the skins.

The three northern species have very short plumage; the Peruvian penguin resembles the Antarctic chinstrap and gentoo penguins in size, but its plumage is thirty to thirty-five percent shorter and it has appreciably smaller fat deposits. Galapagos penguins breed mainly from May to July during the coolest period, whereas Peruvian penguins are reported to breed all year round and black-footed penguins twice a year, mainly in February and September. Magellan's penguins

▷

Fig. 5-7. The female (right) of the yellow-eyed penguin greets the male on his return to the nest.

breed only once a year, laying their eggs in October. All species lay one to three eggs, but most often two. They incubate for thirty-nine days, and the young go to the water for the first time when they are three months old.

Yellow-eyed penguins

The YELLOW-EYED (or YELLOW-CROWNED) PENGUIN *(Megadyptes antipodes)*, is about the size of the gentoo penguin. The L is 75 cm, the body length without the tail feathers is 66 cm, the weight is 5.2 kg, and the feather length is 2.7 cm. It has a fairly long beak. There is a black and yellow crown patch, and behind it is a golden-yellow stripe across the upper ear coverts. The nape has slightly elongated feathers. The eyes are pale yellowish-green. Both sexes are similar. The young have a less developed head pattern. They reside on New Zealand and a few nearby islands.

Fig. 5-6. 1. Galapagos penguin *(Spheniscus mendiculus)*; 2. Peruvian penguin *(Spheniscus humboldti)*; 3. Magellan's penguin *(Spheniscus magellanicus)*; 4. Black-footed penguin *(Spheniscus demersus)*.

This penguin breeds in scattered communities; the nests are generally grass-covered, located between tussocks, below rocky overhangs or bushes, in flat caves, or hidden beneath roots. The two eggs are laid in September and October. The male and female relieve one another every one to five days while incubating. L. E. Richdale observed this species for eighteen years (1936–1954) on the Otago Peninsula and furnished important data for an understanding of their behavior and changes in population. Among other things, he found that they "trumpet" with their heads raised, their backs hollowed out, and their wings held forward; then they lower their heads almost to the ground, swing them to and fro, and continue calling. Almost all females reproduce for the first time when three years old, although almost half of them attempt to do so, but unsuccessfully, when two years old. From the fourth year onward, the survivors of that age group decline in numbers by about thirteen percent each year. Richdale estimated that they live to an average age of six to seven years; only a few reach an age of twenty years. The egg-tooth is lost when the young are five to six weeks old. The moult lasts twenty-four days.

Dwarf penguins

The DWARF PENGUINS (genus *Eudyptula*) are even smaller than the Galapagos penguins. The L is 40–42 cm. The beak is moderately long and somewhat hookshaped; it is generally stronger in ♂♂ than in ♀♀. The young resemble the adults very much. They are rather shy birds which are active at night, and hence are inconspicuously colored. Generally two species with several subspecies are described, but a thorough revision of this group would be desirable: 1 The LITTLE PENGUIN (*Eudyptula minor*; see Color plate, p. 124), with a southern subspecies *(Eudyptula minor minor)*, has an L of 40 cm, a weight of 2.5 kg, and a feather length of 1.1 cm. The northern subspecies *(Eudyptula minor novaehollandiae)* has an L of 41 cm, a weight of 2.2 kg, and a feather length of 1.1 cm. The CHATHAM ISLAND LITTLE PENGUIN *(Eudyptula minor iredalei)* has an L of 39 cm, a weight of 2.1 kg, and a feather length of [0.1] cm. 2. The WHITE-FLIPPERED PENGUIN *(Eudyptula albosignata)* is re-

garded by some as a subspecies of the little penguin. The L is 40 cm, the weight is 2.4 kg, and the feather length is 1.5 cm. It has a broad white band on each flipper edge.

All dwarf penguins have very similar ways of life. Their nests are generally well hidden. They dig holes up to two meters long or use holes dug out by shearwaters, as well as those found in natural rock or earth caves, under rocks or plants. They come ashore only after sunset. To see them crossing an illuminated coastal strip to reach their nests is a tourist attraction near Melbourne. They spend the night on shore, usually in their breeding areas in which they live the whole year. At dawn they return to the sea. Only birds in courtship display, breeding, or in the moult stay on shore during the day, but they hide from the sun as much as possible. This is probably because they need a cool environment. They soon begin to suffer from the heat of the sun; shade and water are vital requirements for them.

The egg-laying season begins in the spring (February); two eggs are laid. In the northern subspecies, egg-laying extends over several months. Both males and females incubate for thirty-three to forty days. The young enter the water when they are about eight weeks old. In general, this species is very quiet; only during the breeding season do they become noisy. House owners on the coasts of New Zealand and Australia are somewhat wary of them because they like to breed under floor boards, and their nightly "conversations" when breeding are not welcomed by their "landlords."

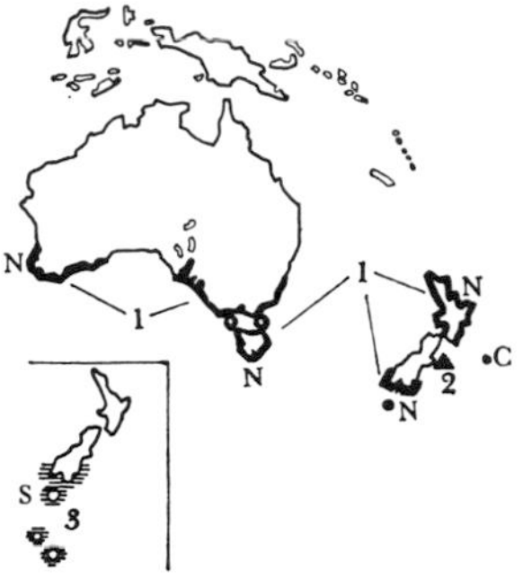

Fig. 5-8. 1. Little penguin *(Eudyptula minor);* N. Northern subspecies *(Eudyptula minor novaehollandiae);* S. Southern subspecies *(Eudyptula minor minor);* C. Chatham Island little penguin *(Eudyptula minor iredalei);* 2. White-flippered penguin *(Eudyptula albosignata);* 3. Yellow-eyed penguin *(Megadyptes antipodes).*

▷
The largest gatherings of birds are found on the coasts of seas rich in food. This swarm of seabirds of various species was photographed by Heinz Sielmann on a shore of one of the Galapagos Islands.

1
2
3
4
5

6 The Tube-nosed Swimmers

When one comes from the inland to the sea coast, one is astonished at the number of birds of the shore and tidal zone. Gulls and terns, in particular, seem to the inland observer the epitome of the light-winged fliers of the high seas, and he is likely to assume that these birds are to be found out on the oceans as well as along the coasts. However, this is not the case; gulls and terns (see Vol. VIII) are coastal birds and only rarely travel far out to sea from the coast.

Order: tube-nosed swimmers, by B. Stonehouse

The narrow-winged flying acrobats of the high seas belong to quite a different order of birds than the gulls and terns: they are the petrels, albatrosses, and their relatives, a group which ornithologists call TUBE-NOSES (order Procellariiformes) because of the peculiar shape of their nostrils.

Distinguishing characteristics

The nostrils are horny tubes which are generally found on the culmen and less often on the sides of the beak, which is straight with a hooked tip. The beak-sheath is composed of several separate horny plates. There are large nasal glands for salt excretion. The long gullet and proventriculus secrete an oil. There are fifteen cervical vertebrae, and the furcula is movable and located on the keel of the sternum. The pelvis is penguin-like, and is fused with the synsacrum. The knee joint has a projection of the tibial crest. There are three forward toes that are connected by webs, and the hind toe is dengenerated. There is a long arm-skeleton, but short secondaries. The skeleton of the hand and primaries is very long. These birds have an outstanding flight capacity and great endurance. The bones, particularly in the larger species, have large air spaces.

◁ Albatrosses: 1. Wandering albatross *(Diomedea exulans)*. 2. Waved albatross *(Diomedea irrorata)*. 3. Royal albatross *(Diomedea epomophora)*, chick. 4. Black-footed albatross *(Diomedea nigripes)*. 5. Light-mantled sooty albatross *(Phoebetria palpebrata)*.

They are inhabitants of the high seas, particularly of the Southern Hemisphere, and they only go on land to breed. They lay one egg, and have a long incubation period. The young grow very slowly. This order includes both the largest and the smallest sea birds; the L is 14–135 cm, and the weight is 20–8000 g. Fossils have been found from the Upper Eocene of France (about thirty million years ago) and in the

Miocene of North America (about twenty-five to twelve million years ago). Today there are four families: 1. The ALBATROSSES (Diomedeidae); 2. The PETRELS and SHEARWATERS (Procellariidae); 3. The STORM-PETRELS (Hydrobatidae); and 4. The DIVING-PETRELS (Pelecanoididae). There are twenty-two genera and ninety-two species.

Nasal tubes and stomach oil

The paired nose tubes may vary in form and length; they join round or oval openings with large nasal or olfactory cavities. The significance of the tubes is not known; many suggestions have been made regarding their purpose, but none are altogether convincing. Evidently the olfactory sense, otherwise of low efficiency in birds, is well developed in the tube-nosed swimmers, but this does not explain the peculiar nostrils. All members of this order fly low over the sea, so possibly the nose tubes keep spray out of the inner nasal cavities. All tube-noses also have large nasal glands which secrete a saturated salt solution, and so the tubes could possibly serve to keep this solution away from the eyes and the skin of the gape.

Another peculiarity of these birds is the flesh-colored, oil liquid which most species secrete from special cells of the proventriculus. When in danger, breeding birds and the young as well can regurgitate this stomach-oil and spray it at an aggressor for a distance of several meters. When this oil cools it solidifies into a wax-like consistency; in this form it is often found around the nests in cold regions. The birds can use this oil when preening their plumage and possibly they apply it directly from the nose tubes. It is also possible that the stomach-oil accounts for the strong musky smell, peculiar to all tube-noses, which can even persist on skins which have been in museums for over a hundred years.

Long incubation and growth period

While the tubular nostrils and the stomach-oil occur only in tube-noses, their particularly slow reproduction and the long growth period of their young are not so unique in the bird world. All Procellariiformes lay only one egg. As a rule, it is relatively large, weighing six to ten percent of the mother's body weight in large species and ten to twenty-five percent or more in the smaller ones. If the egg or the youngster perishes, species breeding in moderate and high latitudes are incapable of laying another egg as a replacement during the same breeding season. Nothing is known about laying a new egg in tropical species, but they can breed at less than yearly intervals. Incubation and the growth periods last longer in tube-noses than in all other birds of similar size. Their eggs are larger than those of gulls of similar weight, although the weight of their single egg may be exceeded by that of the gull's entire clutch of eggs. Tube-nose's eggs are incubated roughly twice as long as gull eggs of similar size, and the young need about twice as long as gull chicks of similar size to reach the fledging stage.

There are as yet no satisfactory explanations for the peculiarities of this group. The absence of terrestrial enemies on the breeding islands may have facilitated the long periods of incubation and development. A connection between clutch size, the incubation period, and the rate of growth on the one hand and the parents' ability to feed the young, on the other, is evident. The most successful way to reproduce in most tube-noses is apparently to have a single, slow-growing young well protected from heat loss by long down. Since food is rarely abundantly available in the immediate vicinity of the breeding places, the fast-growing young would be more liable to starve during the long periods between meals than those with slower metabolic and growth rates. The benefit which would arise through the competition between two or more fast-growing young in each brood is therefore not desirable for tube-nosed swimmers. The very sociable behavior of most species, which is necessary during the communal search for food far from the breeding place, as well as the long incubation and growth period, would make second or additional broods per breeding season disadvantageous. Thus the single brood per season with the long incubation and growth period, often in a secure underground nest, is altogether appropriate for these birds.

All Procellariiformes are birds of the high seas, adapted in various ways for feeding on or just beneath the surface of the water. They can spend days, weeks, and even months away from land. Two-thirds of the living species live in the Southern Hemisphere, which must be regarded as the main area of development of this order. From the temperate-cool zone of westerly winds, they have spread south as far as the coasts of the Antarctic and north over the equator as far as Arctic latitudes.

Family: albatrosses, by J. Warham

The largest members of the order are found among the ALBATROSSES (family Diomedeidae); the royal albatross and the wandering albatross have the largest wing-spread of all seabirds, from about 200 to a little over 320 cm. The nasal tubes are inconspicuous, located on each side of the beak immediately in front of its base. They are not connected with one another laterally. The flight-feathers are extraordinarily long, and the arm portion of the wing, with 27-40 secondaries, is well-developed. There are thirteen species which are easily distinguished by the shape of the horny plates of the beak; at sea they may be recognized by the pattern of the under wing as well as the color of the beak and the head. Generally only two genera are recognized, *Diomedea* and *Phoebetria:*

1. Southern forms of medium size, the "Mollymauks" that are very similar to one another with the same wing pattern include: (a) the BLACK-BROWED ALBATROSS *(Diomedea melanophris),* which has a large yellow beak and an upper mandible with a pink hook at the tip. The legs are flesh-colored. The dark gray back forms a single dark area with the blackish-brown upper wing surface. The underside of the

wing is grayish-black along the front edge and white toward the back. The tail is short and gray, and the head and the rest of the body are white. Above the eye is a stripe of dark feathers; hence, one may think they have a "sinister facial expression." (b) The YELLOW-NOSED ALBATROSS *(Diomedea chlororhynchos)* is smaller. (c) The GRAY-HEADED ALBATROSS *(Diomedea chrysostoma)* and (d) BULLER'S ALBATROSS *(Diomedea bulleri)* are of the same size as the black-browed albatross, while (e) the SHY ALBATROSS *(Diomedea caudata)* is larger and heavier.

2. The large southern forms, with a weight of 7–8 kg, bones which are largely filled with air spaces, a predominantly white plumage, black wing tips, and a whitish beak include: (a) the ROYAL ALBATROSS (*Diomedea epomophora;* see Color plate, p. 138), whose young are colored white like the adults; and (b) the WANDERING ALBATROSS (*Diomedea exulans;* see Color plate, p. 138), whose young are brown with white faces. The underside of the wing is hardly different from that of adults.

3. The northern forms include: (a) the SHORT-TAILED ALBATROSS (⊕*Diomedea albatrus*) which are entirely white, although the young are dark brown; (b) The LAYSAN ALBATROSS (*Diomedea immutabilis;* (c) The BLACK-FOOTED ALBATROSS (*Diomedea nigripes;* see Color plate, p. 138).

4. The single tropical species is the WAVED ALBATROSS (*Diomedea irrorata;* see Color plate, p. 138).

5. The SOOTY ALBATROSSES (genus *Phoebetria*) include: (a) the SOOTY ALBATROSS *(Phoebetria fusca)* and (b) the LIGHT-MANTLED SOOTY ALBATROSS (*Phoebetria palpebrata;* see Color plate, p. 138). Both species are dark brown with pointed tails. They are the most graceful flyers among the tube-noses.

Of all birds albatrosses are most adapted to the high seas. Their range extends through the band of strong air currents, where they glide up and down without any effort, using the updrafts of air which are deflected upwards by the waves. In a calm, when they would have to use their wing beats to remain airborne, most albatrosses prefer to rest by sitting on the water.

Like other seabirds, albatrosses drink sea water and eliminate much of its salt through their nasal glands, which are depressions in the skull directly above the orbits. The salt solution dripping from the bird's beak has often been mistaken for stomach-oil. Young albatrosses squirt this stomach-oil rather indiscriminately while young giant petrels and fulmars eject their oil at an aggressor with accuracy.

Squids, often of appreciable size, are the main food of albatrosses and the horny beaks of this prey are generally found in great numbers in their stomachs. Off the southern coast of Australia, the large squid *(Amplisepia verreauxi)* is evidently caught by wandering albatrosses at night as well. In addition, fish, crabs, and some plant constituents make up their food. Floating nuts, like the seeds of *Aleurites moluccana,* have been found in the stomachs of Laysan albatrosses. Seabirds are seized as well; at the breeding places one sometimes finds regurgitated pellets containing the skeletons of entire prions (small petrels of the genus *Pachyptila*); even small penguins are consumed. They can be readily caught when they gather in numbers about some larger food object, such as a dead seal. Albatrosses are considered to be scavengers. The black-browed albatross and the wandering albatross often follow ships on the southern oceans, taking up food remains which are thrown overboard. Individual birds may thus follow a ship for several days.

The importance of squids as part of their food raises the question as to whether there are species differences in this preference. Only Tickell has given information on this. He found that the gray-headed albatross takes more unicata, crabs, and amphipods than does the black-browed albatross, which consumes lampreys *(Geotria australis)* that are avoided by the gray-headed albatross. Yet both species feed mainly on fish and squids.

Albatrosses are normally silent at sea unless they are fighting for food. Then they utter various gurgling sounds. In the breeding grounds, sounds such as beak clattering, groaning calls, high-pitched whistles, a low-pitched call, and croaks accompany their various behavior patterns.

Richdale has thoroughly studied the reproduction of a number of species. In the royal albatross, the birds do not return to their breeding places until they are four to seven years old, and even then most of them do not breed successfully until they are still older. Once they have begun to breed, they may continue for many years since they have a long life. An age of thirty-six years has been established for this species. A number of Buller's albatrosses, banded as breeders on Snare Island in 1948, were still breeding nineteen years later in 1967. A return to the same nest site and partner is the rule. Males normally arrive at the breeding places before the females. Mollymauks often use the same nest several times, while the larger albatrosses usually make a new nest for each brood, which, however, is close to the old one.

The nests are either low mounds made of plants encircled by a ditch which the birds scoop out with their beaks, or occasionally cylindrical ridges of hard-trodden earth or peat. They are almost always on islands with a view of the sea. An exception is the small royal albatross colony on the mainland of New Zealand at Tairoa Heads. Most species breed in colonies; only the wandering albatross often breeds in solitary pairs, and the sooty albatross usually does so also.

Fig. 6-1. A pair of wandering albatrosses in courtship display postures.

The clutch consists of only one egg, which is not replaced if it perishes. It is white or cream-colored and often has reddish spots at its blunt end. Both male and female incubate, although the male does so at first for a fairly long time; in the royal albatross the male sits for at least four to six continuous days. The incubation period is seventy-nine days in the royal albatross and sixty-five days for both the black-footed and the sooty albatross.

The young is guarded by both parents in alternation until it is big enough to defend itself. This period lasts for five weeks in the large and about three weeks in the smaller species. The chick usually stays in the nest until fledged, but the young wandering albatross often leaves the nest and builds a new one a few meters away. Both parents feed the chick irregularly and as it gets older the intervals between feedings increase. The times when each individual parent feeds the young are quite independent of the other parent. The young wandering albatross can make its first flight at the age of 278 days, the royal albatross at 236, and the sooty albatross at about 139 days. Because of this long growth period, large albatrosses breed only every other year. Hence new pairs appear on the breeding islands to breed a few months before the young of the previous breeding season depart. But if a brood fails early in the season, the wandering albatross and the royal albatross often breed again the next year.

Fig. 6-2. Courtship display of the waved albatross (top to bottom): beak duelling; beak snapping; shoulder preening of bird on left and beak snapping by the bird on the right; "presentation" bird on the right; showing the nest site; and social preening.

Irenäus Eibl-Eibesfeldt has described the courtship of the waved albatross: "The frequently-repeated ritual begins with a dance. The male walks around the female, his neck pulled in, and sways from one side to the other with his steps synchronized in an exaggerated manner. In step with these swaying motions, they both turn their heads to the left and right so that their beaks touch the raised shoulder. This dance is followed by a beak-duel. The partners stand facing each other, stretch their necks forward, and with fast sideways head movements beat their beaks against their partner's, while nibbling with the bill at the same time. Following this, different forms of behavior may be shown, such as beak clapping in which the bird raises the head and neck, opens the beak and closes it with a loud snap. Often they both do this simultaneously. Then there follows either a beak-duel or a display in which the beaks are pointed upward and calls are uttered. Occasionally one's bill clatters rather like a white stork's. With its bill stretched forward, the partner then preens its shoulder feathers. These different movements may be repeated in any sequence. Toward the end, both birds bow to one another, pointing their bills at the ground. This is probably a symbolic indication of the nest site. Then they sit down and begin to comb one another's neck plumage." In the sooty albatross the courtship display is simpler, whereas that of the mollymauks and the northern species is intermediate.

Immature birds evidently leave the islands of the southern ocean and move off into the direction from which the prevailing winds come. They may thus circle the earth in the zone of westerly winds. Banded wandering albatrosses from South Georgia, both breeding birds and immatures, have regularly been recovered off the coast of New South Wales (Australia). Similarly, banded royal albatrosses from Campbell

Fig. 6-3. Buller's albatross in flight intention posture.

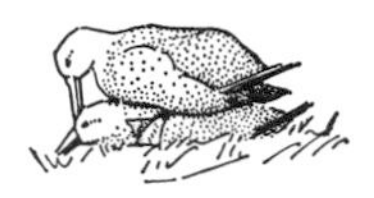

Fig. 6-4. Copulation in Buller's albatross.

Fig. 6-5. Royal albatross feeding its chick.

Island were found on the west coast of South America. There are about twenty-five records of albatrosses which strayed into the North Atlantic. Most records are of yellow-nosed albatrosses found off the coast of North America and of black-browed albatrosses in the eastern Atlantic and near Scandinavia. The latest find, on July 20, 1966, was of a black-browed albatross found with gannets *(Sula bassana)* near the Vestmann Eya Islands of Iceland.

The most common species is probably the black-browed albatross, which ranges over the seas right around the South Pole and breeds on many sub-Antarctic islands. Like the wandering albatross, it is known to many travellers through its habit of following ships. The royal albatross and the wandering albatross are the largest species, but exaggerated notions about their size were once held. In 1911 wandering albatrosses caught off the New South Wales coast, the wing span varied from 272 to 322 centimeters, with a mean of 300 centimeters.

The short-tailed albatross was much pursued by feather hunters in earlier times. On their only breeding place, the island of Torishima, they also had to endure a whole series of volcanic outbreaks, the last one in 1966. However, since these birds only become capable of reproduction after many years and before that only rarely visit the nesting place, such natural catastrophes destroy only that part of the population which is on the island at that time.

The Laysan and the black-footed albatrosses are still quite numerous. The United States Air Force undertook a large scale campagin to eradicate the black-footed albatross from Midway Island since the birds often collide with planes, causing many accidents. Nevertheless, their number is still estimated at 300,000.

Family: procellariids

The family Procellariidae is a group with many more species and is less uniform than other families. They are the typical tube-noses; what has been said above about the order as a whole applies particularly to them. This family includes all the tube-nose species except for the largest and the smallest. The L is 25–90 cm, and the weight is 100–4000 g. The egg weight is between fifteen and twenty percent of the female's body weight. There are two nose tubes which are side by side on the culmen of the beak and are sometimes fused to a single tube with a vertical separation. There are more than thirteen secondaries. Four subfamilies with fifty-five species are recognized (other scientists' estimates are fifty-one to sixty-two): 1. The FULMARS (Fulmarinae); 2. The PRIONS, known to sailors as WHALE-BIRDS or ICE-BIRDS (Pachyptilinae); 3. The HOOK-BILLED PETRELS (Pterodrominae); and 4. The SHEARWATERS (Puffininae).

No species of this group is such a pronounced glider as the large albatrosses, although the giant petrel *(Macronectes)* approaches the length, wingspan, weight, and flying ability of smaller albatrosses. Furthermore, none are as adapted to diving as the diving petrels

(Pelecanoididae), but many shearwaters *(Puffinus)* are skillful divers and swimmers, and members of several other genera show at least a moderate diving ability. The group shows a great variety of highly specialized adaptations for obtaining food, from the well-developed ability of prions to sift out food particles to the refuse consumption of fulmars. Several of the hook-billed petrel subfamily are said to live mainly on squids, although the special technique required to catch them is not known. Some species do almost all their feeding in the pack ice zone; snow petrels *(Pagodroma nivea)* are rarely seen more than a few hundred meters away from the Antarctic pack ice except in the breeding season, and Antarctic fulmars *(Fulmarus glacialoides)* often seek their food between floating ice floes.

Social behavior

Almost all tube-noses are sociable during feeding, and they wander in huge swarms that are more or less dense to the feeding grounds. Some of the most sociable of all birds are among the petrels. Many observers have reported swarms of prions, shearwaters, and hook-billed petrels numbering in the hundreds of thousands, or even millions, particularly in the southern oceans where plankton food is abundant. In the summer they are found in great concentrations in some areas. Winter flocks of shearwaters often spread out over larger areas. Many species avoid wintering in cool areas, presumably because the food is scarce and because the rough seas make feeding difficult. Wandering bands of sooty shearwaters and slender-billed shearwaters leave Australia and New Zealand every fall and move to the northern Pacific. They cross the equator and reach high northern latitudes before they return to breed in the southern spring. Other species, even though they remain in the Southern Hemisphere, seek out warmer latitudes in its northern part where they find food more abundant.

Breeding colonies

Social behavior is equally pronounced in tube-noses when they are breeding. Almost all of them breed in colonies; pairs of the larger species nest in sight of one another on open ground and the smaller species breed in colonies which often occupy hundreds of square meters either on the surface or in deep or shallow caves in the ground. Only albatrosses and the largest petrels breed on the open ground without cover; only they are large enough to ward off possible bird and introduced mammalian predators which might attack the nests. Smaller species, including nearly all the storm petrels, nest below the ground, in the cover of grass tussocks, in clefts between rocks, on rocky slopes, or on cliffs. These habitats protect them from attacks and favor the choice of a secure micro-environment with stable conditions for the eggs and young. The small tube-noses stay on land at night, returning to their nests shortly after sundown and feeding their young at night; before daybreak they leave again for the sea. They spend most of the day at sea, well out of reach of terrestrial enemies. On Antarctic and sub-Antarctic breeding places, skuas *(Stercorarius skua)* are their main

▷

Procellariids: 1. Hawaiian petrel *(Pterodroma phaeopygia).* 2. Manx shearwater *(Puffinus puffinus).* 3. Pintado petrel *(Daption capensis).* 4. White-chinned petrel *(Procellaria aequinoctialis).* 5. Arctic fulmar *(Fulmarus glacialis).*

1
4
3
2
5

1
2
3
4
5

◁
Storm petrels:
1. Magellan diving petrel *(Pelecanoides magellani).*
2. White-faced storm petrel *(Pelagodroma marina).*
3. Wilson's storm petrel *(Oceanites oceanicus).*
4. British storm petrel *(Hydrobates pelagicus).*
5. Fork-tailed storm petrel *(Oceanodroma furcata).*

predators, to which many fall victim, particularly late in the evening and in the early morning.

The nest density on oceanic islands and on breeding cliffs may be very high. The layers of peat on many islands in the temperate-cool zone are criss-crossed by a network of holes made by prions, shearwaters, and other small tube-noses. It is often difficult to walk in such places without breaking through the roof of a nest hole with every step. Many birds do not seem to get into the mood for breeding until they succeed in expelling others from their holes. Large aggregations of nesting sites are found, particularly on islands close to or in strong ocean currents. Rising currents, caused by elevations of the sea bottom, bring water which is rich in nutrients from deeper layers to the surface. This gives rise to an abundant growth of diatoms and algae and of plankton animals which live on these plants. Such areas supply food for great masses of birds which congregate there and which fully utilize the available space on the relatively small islands for breeding. On the other hand, many species of storm petrels feed far from their nest holes, and between two breeding periods they search for food over thousands of square kilometers. The young, particularly during their first year, are inclined to wander. Their migrations can be determined from the results of banding; the immatures of several species, when they are between fledging and the first breeding, wander further than the adults. The young of species which live long reduce competition for nest sites and food in the overcrowded breeding areas through their wandering, and are likely to reach areas where they can more easily find food.

In spite of the great dispersal, there are several indications that the young of many species return to the place of their birth, often to the same colony, when they are ready to breed. This accounts for the high incidence of subspecies formation among the tube-noses, since there is a tendency for breeding populations to remain isolated. How the birds find their way during their migrations has not yet been demonstrated experimentally. There are, however, clear indications of a remarkable homing ability in many species. The young probably learn their home area during the short time when they leave the nesting holes at night while they are half-fledged to exercise their wing muscles. The return may be facilitated by flock formation, since inexperienced younger birds fly with experienced older ones. Young shearwaters and probably those of many other species undertake at least the first part of their migration alone, since the parents leave the breeding grounds before the young are fledged.

Locomotion on land

The ability to walk is varied in procellariids. Most of them have a small pelvis which is presumably an adaptation related to their large eggs. Their legs are more suitable for movement in the water than on land. Many fly low over the water, placing the weight of the body on the wings, using the breast like the bow of a hydroplane and the small webbed feet as paddles. On land they walk in a similar way; they run for short stretches with beating wings, the body held horizontally or actually dragging on the ground. Few of them can stand upright like gulls or ducks. The legs have strong muscles and are used together with the beak in digging nest holes; the claws may be long and curved and are used mainly in fights with members of the same species.

The classification of this group into subfamilies, genera, and species, as well as their scientific names, is unfortunately not used uniformly by experts since the relationships of particular groups to one another are not yet clear. In the following, we use the classification suggested by W. B. Alexander and fourteen other ornithologists in 1965.

Subfamily: fulmars

The FULMARS (subfamily Fulmarinae) are large or medium-sized, heavily-built petrels; the L is 35–90 cm. The bill is high. The head, in the larger species, is large because of the bulky jaw muscles. The tail is short. They show great flying ability in a characteristic gliding and flapping flight, and they mainly eat various floating animal and vegetative matter. A few species have small lateral bulges in the upper mandible which aid the seizing, holding, and sifting of small plankton animals. Their habitat is the seas of the Southern Hemisphere; only birds of the genus *Fulmarus* are found in both hemispheres. There are five genera with seven species: 1. The FULMARS *(Fulmarus)*; 2. The GIANT FULMARS *(Macronectes)*; 3. The ANTARCTIC PETRELS *(Thalassoica)*; 4. The SNOW PETRELS *(Pagodroma)*; and 5. The PINTADO PETRELS *(Daption)*.

Fig. 6–6. Distribution of the Arctic fulmar *(Fulmarus glacialis)*

The ARCTIC FULMAR *(Fulmarus glacialis*; see Color plate, pp. 61/62 and p. 147) and the ANTARCTIC FULMAR *(Fulmarus glacialoides)* are closely related; some investigators place both into one species. They breed in high latitudes. The northern species is divided into an Atlantic and a Pacific subspecies which show slight differences in the shape of the beak, color, and size. The Antarctic fulmar shows no formation of subspecies. It breeds in an area around the South Pole on the Antarctic continent and adjacent islands. The nests are among snow and permanent ice, and it often seeks its food in icy water. It also follows cold currents toward the equator and may appear as a wanderer in the Northern Hemisphere. The northern fulmars, however, tend to remain in their home regions near their breeding places. The Atlantic Arctic fulmar *(Fulmarus glacialis glacialis)* has extended its breeding range remarkably fast in recent times, from Iceland as far as southern England, each year occupying an additional twenty kilometers of coastline. Presumably this is the result of the increase of whaling and commercial fishing which, with their refuse, opened new food sources for these birds. James Fisher has investigated this expansion into new breeding areas, which has been going on for over 200 years, in detail. Both species of this genus may have a light or a dark plumage; pale to dark smoky-gray and grayish-blue tints also occur. The proportion of dark birds increases towards the north; in Iceland only one percent of the

Fig. 6–7. Southern giant fulmar *(Macronectes giganteus)*. 2. Northern giant fulmar *(Macronectes halli)*. Breeding colonies: Falkland Island (F); South Georgia (Ge); South Orkney (O); Gough (Go); Marion (M); Crozet (C); Kerguelen (K); Macquarie (Mc); Chatham (Ch).

birds are dark, while in Spitzbergen this increases to ninety-five percent.

The GIANT FULMARS (genus *Macronectes;* the L is up to 75 cm, and the wingspan is almost 2 m) are by far the largest fulmars. They still breed on some islands of southern New Zealand; most of them, however, live in sub-Antarctic and Antarctic waters. In 1966 Bourne and Warham distinguished two species: the SOUTHERN GIANT FULMAR *(Macronectes giganteus)* occurs in several color phases with pale or dark plumage. It breeds in colonies, while the NORTHERN GIANT FULMAR *(Macronectes halli)* nests singly. In the northern species there are deviations in color; it has a well-marked white face. On Macquarie Island, both species live close together. The northern species lays eggs in August, the southern somewhat later.

The southern giant fulmars are great wanderers; particularly as immatures they appear at the coast of Chile, South Africa and Australia. Many thousands of nestlings have been banded; hopefully we will soon know more about their breeding cycle and the development of the population. Both species feed mainly on dead animals and on large plankton animals which they seize on the surface; they also catch the young of many seabirds on the water. On land they parasitize penguin colonies, which other tube-noses hardly do, and take eggs and young; they also feed on dead seals.

The ANTARCTIC PETREL (*Thulussoica antarctica*; the L is 45 cm) and the SNOW PETREL (*Pagodroma nivea*; the L is 35 cm) breed on the rocky outcroppings of the Antarctic coast. Snow petrels also breed further south in certain mountain regions up to almost 2000 meters above the sea and over 300 kilometers from the coast, and northward as far as southern Georgia and the Bouvet Islands. Both feed in ice-cold water. The presence of white snow petrels indicates nearby pack ice to seamen; the Antarctic petrels, however, extend far over the cold seas, from the edge of the pack ice as far as the northern edge of the cold water (the subpolar convergence). They catch their prey, small fish, crabs, and squids, by fluttering low over the water or by briefly alighting on it. Both species lay eggs in late November or early December, and the young leave the nest in March.

The PINTADO PETREL (*Daption capensis*; also called CAPE PIGEON; see Color plate, p. 147) has a similar breeding season but a much larger breeding area. It breeds on the Antarctic mainland and on all islands up to the northern edge of the cold water, and also in smaller numbers on many islands in the cold-temperate zone. Very large flocks inhabit the cold latitudes where they aggregate about large areas of plankton, swarms of young fish, whale corpses, and other areas which promise food. Small platelets located in the beak enable them to feed on small floating animals, but they also eat fairly large lumps of floating fat or meat, and they even tear vigorously on the corpses of large animals.

Subfamily: prions

The subfamily Pachyptilinae, with the single genus *Pachyptila*, is a divergent group of six small species. Their plumage is uniformly bluish-gray with darker wing feathers and blue or black beaks and feet. The L is 27–31 cm. The beak is more or less wide with well, or in the more primitive species, weakly developed sifting bulges which function much like the baleen of whales (hence the name whale birds). Their habitat is the southern oceans. It is possible that the blue petrel really belongs to this subfamily.

Prions are cave nesters whose breeding places are oceanic islands. When searching for food or when wandering, they gather into flocks often numbering millions, and they sometimes even mingle with other bird species. In flight the various species are difficult to distinguish; only the BROAD-BILLED PRION *(Pachyptila vittata)* can be relatively easily distinguished by its broad duck-like beak. The beak is more or less well developed into a straining device in all prions. They flutter low over the water with their beaks immersed, seeking food, and in this way they obtain small organisms swimming or floating on the surface or just beneath it.

In 1941 Fleming investigated the phylogenetic relationship of prions. They probably evolved from fulmar-like ancestors, to which the FAIRY PRION *(Pachyptila turtur)* and the THICK-BILLED PRION *(Pachyptila crassirostris)* are most closely related. Both have short, slightly flattened beaks like those of the fulmars, which have only little-developed lamellae for sifting small food particles from water. The other four species show a trend to increased size, an increasing flattening of the beak, better development of sifting platelets, and a more distinctive color pattern of the head plumage. The THIN-BILLED PRION *(Pachyptila belcheri)* of the Kerguelen and the Falkland Islands shows only slight deviations from the (presumed) ancestral form. The DOVE PRION *(Pachyptila desolata)* has a broader bill with small but distinct sifting platelets; SALVIN'S PRION *(Pachyptila salvini)* of Marion and Crozet Islands has a moderately wide beak with well-developed platelets. The final member of the group is the BROAD-BILLED PRION *(Pachyptila vittata)*, which has a broad triangular beak and highly effective lamellae. This species lives on small plankton animals, while the small-beaked species take predominantly larger prey, mostly crustacea and fish larvae. The breeding seasons are probably determined by the seasonal abundance of the food animals. Because of different breeding seasons, closely related forms in the same area, even on the same island, were able to develop into separate species. The broad-billed prions, for example, live on pinhead-sized plankton animals near their breeding islands off south New Zealand, and lay their eggs mainly during the first three weeks of September. Fairy prions, which nest close by, live on crabs which are fifteen millimeters long on the average, and they generally lay their eggs during the last three weeks of November.

The HOOK-BILLED PETRELS (subfamily Pterodrominae) are small to

Subfamily: hook-billed petrels

middle-sized petrels; the L is 25–45 cm, and the wingspan reaches 105 cm. Their bills are short and heavy, and have sharp cutting edges. The wings are relatively long, particularly in large species which are capable of gliding. There is often a checkered plumage pattern with several color tones in some species. There are twenty-five species in three genera: 1. The SOFT-NOSED PETRELS *(Bulweria)*; 2. The HOOK-BILLED PETRELS more specifically *(Pterodroma)*; and 3. The BLUE PETRELS *(Halobaena)*.

The larger species prefer to breed on the ground of temperate-zone or tropical islands, generally protected by rocks or bushes. Smaller species prefer islands in colder latitudes where they dig their nesting holes. None of them breed in the Antarctic, but both the smaller and larger species visit Antarctic and sub-Antarctic waters in search of food. Hook-billed petrels move far from their breeding places between the breeding seasons, and generally feed in large flocks far from land. Their food consists mainly of squids and small fish which they catch by dipping their bills just below the surface. Many species feed preferentially at night when the larger plankton animals are at the surface.

The smaller species are predominantly gray and white; they have relatively shorter wings than the larger forms and, very characteristically, flap and glide in flight. Of the twenty-one species of the genus *Pterodroma*, GOULD'S PETREL (*Pterodroma leucoptera;* the L is 30 cm) breeds on islands in the western and central Pacific. These species are all widely distributed over the Pacific. As a representative of the larger forms, the GREAT-WINGED PETREL (*Pterodroma macroptera;* the L is 40 cm) may be mentioned. It occurs on the southern oceans between the thirtieth and fiftieth degree latitudes, and breeds during the winter seasons on the islands of Tristan da Cunha, the Kerguel Islands, and in the Australo-Asiatic region. The HAWAIIAN PETREL (*Pterodroma phaeopygia;* the L is 43 cm; see Color plate, p. 147) and several other large species of this genus have a relatively restricted tropical and subtropical distribution. Three further species, including the great-winged petrel, live in cooler latitudes. Half of the southern species are winter breeders and half are summer breeders. The tropical species tend to have very long breeding seasons.

BULWER'S PETREL (*Bulweria bulwerii;* the L is 27 cm) and the two very similar but still almost unknown other members of this genus have a dark plumage and also differ in some skeletal features from the hook-billed petrels in a narrower sense. They inhabit the tropical and the adjacent northern latitudes of the Pacific, Indian, and Atlantic Oceans.

The blue-gray BLUE PETREL *(Halobaena caerulea)* has an L of 30 cm. With its white belly, forehead, and tail tip, and its partly blue bill and feet, it resembles the prions under which it has sometimes been classified. It inhabits southern oceans and is hardly ever seen north of forty-nine degrees southern latitude.

Subfamily: shearwaters

The Procellariinae are medium-sized birds. The L is 28–55 cm. The wings are long and narrow, and the beak is fairly long and finely sculptured. Their breeding area circles the earth, north as far as the Faroe Islands (more than sixty degrees north) and south to Macquarie (about fifty-five degrees south). There are seventeen species in three genera.

The WHITE-CHINNED PETREL *(Procellaria aequinoctialis)* is the largest of this subfamily, with an L of 55 cm. Its dark brown plumage is in contrast with the white bill and mostly white chin. It follows ships on the southern oceans, even during the day.

The BROWN PETREL *(Adamastor cinereus)*, with an L of 48 cm, is gray under the wings, but otherwise it has a white underside. It is placed within the genus *Procellaria* by many investigators. It can plunge into the water with opened wings from a height of up to seven meters and, as a wing-diver, remain under water for a long time. Like the previous species, it follows ships in latitudes between thirty and fifty-five degrees south.

Among the fourteen species of SHEARWATERS (genus *Puffinus;* the L is 28–55 cm), there are also many diving species which seize food from the surface, but they often dive for it for one-half to one minute, possibly longer. All live on the high seas, feeding mainly on fish, squids, and large crabs. They nest in holes beneath the grass or in mud on oceanic islands, but also between rubble and larger rocks. Several species nest on inland mountain slopes where they are out of the reach of terrestrial enemies. Australo-Asiatic shearwaters, called "mutton birds," were a valued food of the natives, and even today many hundreds of thousands of young, not yet feathered, are taken from nests in Tasmania and New Zealand.

Fig. 6-8. Distribution of the manx shearwater *(Puffinus puffinus)*.

Most shearwaters undertake a long winter migration. The MANX SHEARWATER *(Puffinus puffinus)*, with an L of 35 cm, migrates from its north Atlantic and Mediterranean breeding places as far as the east coast of South America. Other species of the Southern Hemisphere, like the Australian and New Zealand SLENDER-BILLED SHEARWATERS and the SOOTY SHEARWATERS *(Puffinus tenuirostris* and *Puffinus griseus)*, follow a circular path in a clockwise direction, moving from the south to the northern Pacific and even through the Bering straits. In the sooty shearwater the same holds for the Atlantic Ocean; besides, it wanders west and southward into the high latitudes of the Indian Ocean, although probably only the young and non-breeding adults move so far from the breeding areas at this time of the year. It should be mentioned that only the results of extensive banding experiments of Australian shearwaters by Serventy and Richdale have made the slender-billed shearwater, among others, one of the best known of the tube-noses. Much is now known of their breeding behavior, growth of the young, life-span, social structure of the colonies, and migration, thus furnishing the basis for the ecology of the whole order (Lack, 1966).

Fig. 6-9. Distribution of the Mediterranean shearwater *(Puffinus diomedea)*.

Fig. 6-10. Distribution of Leach's storm petrel *(Oceanodroma leucorhoa).*

The GREAT SHEARWATER *(Puffinus gravis),* with an L of 52 cm, reaches even more northern areas than does the sooty shearwater, for it wanders as far as Baffin Bay on its winter migration. Its black cap contrasts sharply with the white throat. It breeds only on the Tristan da Cunha group and on Gough Island, and hence is almost exclusively confined to the Atlantic Ocean. It rarely rises above the waves, which distinguishes it from the higher-flying MEDITERRANEAN or CORY'S SHEARWATER *(Puffinus diomedea)* with its albatross-like up-and-down flight in strong winds. This latter species has been classified in a separate genus *(Calonectris)* by some zoologists; it breeds on islands in the Mediterranean and the Atlantic.

Family: storm petrels by F. Goethe

The traveller at sea can hardly be more astonished than when he first sees storm petrels flying low over the high waves of the ocean. The unexperienced observer does not readily see that these swallow-like, delicate, and approximately swallow-sized birds are relatives of the generally strongly-built Procellariidae which seem so much better adapted to the rough conditions of the high seas. No wonder that, in the old tradition of seafarers, these fearless, delicately-built birds were called, in view of the dangerous ocean, "Mother Carey's chickens." Mother Carey was a corruption of "mater cara = dear mother," a name of the Blessed Virgin. They were also called "St. Peter's birds," which became contracted to "petrel." The last name referred to St. Peter's walking on the sea.

The STORM PETRELS (family Hydrobatidae) are the smallest of all sea birds; they are about the size of a swallow or starling. The L is 14–25.4 cm and the weight is 20–50 g. They have delicate beaks with a hook at the tip. Both nostrils are in a small common tube. The outermost primary is shorter than the following ones (in contrast to all larger tube-noses), and the hand and attached feathers are long; this makes their flight appear swallow-like. Unlike shearwaters, they can hover. The bones are delicate and contain little air. The plumage is dark brown or blackish on top, while the underside is a similar color or is partly whitish; some species have light to whitish wing coverts and upper tail coverts.

Fig. 6-11. Wilson's storm petrel *(Oceanites oceanicus):* lines along coastal waters = main areas of distribution
triangles = breeding areas
dots = areas of observation sites or sightings outside of breeding areas

There are nineteen species with many subspecies; today they are classified as a family and divided into two tribes with eight genera:

A. The LONG-LEGGED STORM PETRELS (tribe Oceanitini) have relatively long legs. The wings are not very pointed, and the tail is square. The more primitive forms are found predominantly in the Southern Hemisphere. There are five genera: 1. *Oceanites,* with two species, including WILSON'S STORM PETREL (*Oceanites oceanicus;* see Color plate, p. 148); 2. *Pelagodroma* with one species, the FRIGATE PETREL (*Pelagodroma marina;* see Color plate, p. 148). 3. *Fregetta* with two species, including the BLACK-BELLIED STORM PETREL *(Fregetta tropica);* and 4. and 5. the genera WHITE-THROATED STORM PETREL *(Nesofregetta)* and GRAY-BACKED STORM PETREL *(Garrodia),* each with one species.

The SHORT-LEGGED STORM PETRELS (tribe Hydrobatini) have short legs, very long and pointed wings, and usually a forked tail. They predominantly inhabit the Northern Hemisphere. There are three genera: 1. *Hydrobates* with one species, the BRITISH STORM PETREL (*Hydrobates pelagicus*; see Color plate, p. 148). 2. *Oceanodroma* with ten species, among them LEACH'S STORM PETREL (*Oceanodroma leucorhoa*), HORNBY'S PETREL (*Oceanodroma hornbyi*), the FORK-TAILED STORM PETREL (*Oceanodroma furcata*; see Color plate, p. 148), and the BLACK STORM PETREL (*Oceanodroma melania*, which is fork-tailed and long-legged, apparently a link between the two tribes); and 3. *Halocyptena*, with one species, the LEAST STORM PETREL (*Halocyptena microsoma*).

Fig. 6-12. Distribution of the British storm petrel (*Hydrobates pelagicus*)

It is characteristic for several species to fly just above the surface of the sea and dip one or both legs into the water with outstretched webs, thus giving rise to the impression that they are walking on the water. This foot immersion is believed to act as a brake, and is used as soon as the bird has spotted floating prey. Possibly it also causes an adhesion to the surface of the water, which might be necessary for these light birds. The British zoologist, Forbes, found interesting relationships between the structure and length of the legs and claws of the toes and this foot immersion during the famous "Challenger" expedition. The investigators Fisher and Lochley report about storm petrels as follows: "In rough as well as in calm weather, storm petrels follow the movements of waves and swells with astonishing grace, always keeping a few inches from the rising water surface. During storms, storm petrels are careful to avoid dangerous places such as wave crests where the wind is strong. They follow the troughs between waves and stay close to the windward slope of waves where the water surface is less disturbed and a good updraft carries them." Roberts, observing Wilson's storm petrels during feeding, found them gliding along the windward wave slopes. But if they rose only a few inches higher they were immediately carried away by the wind. This also happens when a strong wind suddenly turns in such a way that it stands parallel to the swell. During storms of changing directions, storm petrels are sometimes driven far inland.

A careful observer can soon notice differences in the manner of flight of the various species of storm petrels, even those of the same genus. The British storm petrel resembles, in its manner of flight, a bat rather than a swallow. The flight is a light wavering to and fro, with almost outstretched wings interrupted by only short glides. The flight of Leach's storm petrel resembles hopping and then floating; it often changes speed and direction and is sometimes reminiscent of a nighthawk (*Chordeiles minor*) or nightjar (*Caprimulgus europacus*), and at other times of a shearwater or a black tern (*Chlidonias niger*).

The moult of the primaries, as far as is known in three species, takes place from the end of the rearing period into the winter, and may last four months.

▷

Tropic-birds: 1. Red-tailed tropic-bird (*Phaëthon rubricauda*); 2. White-tailed tropic-bird (*Phaëthon lepturus*); 3. Red-billed tropic-bird (*Phaëthon aethereus*). Pelicans: 4. Brown pelicans (*Pelecanus occidentalis*).

1
4
2
3

1
2
3

Food of storm petrels

Storm petrels take surface plankton, i.e., passively drifting sea animals, from the surface of the sea; among them are cephalopods, particularly small squids, small crabs and the larvae of larger ones, small fish, coelonterates, and sea snails as well as feces and particularly the wastes of slaughtered whales. The storm petrels which follow ships, and not all species do this, eagerly take kitchen refuse thrown overboard. The American investigator of seabirds, R. C. Murphy, established the fact that storm petrels can sometimes dive during their search for food. He observed Wilson's storm petrels diving to a depth several times their body length. Yet on re-emerging they shot up into the air, apparently quite dry. In order to catch Leach's storm petrels for banding, the birds were lured by towing oily cod liver behind the boat. This suggests a good sense of smell in these birds.

Piping, squeaking, cooing, and chirping calls, often in long, rising and then declining sequences, are heard from gatherings of storm petrels at a rich food source or in their breeding colonies; the significance of these in the social life of the birds is hardly known as yet. However, their hearing must be good, for at the time of pair formation those sitting in underground holes can communicate with their partners in the air by calls.

Homing ability

The fact that storm petrels can find their way from their wintering areas, which are thousands of nautical miles away from their breeding places, across the high seas without landmarks to the very hole in which they nested the previous year, indicates a superior homing ability. The American investigator of bird navigation, Griffin, transported over two hundred Leach's storm petrels hundreds of miles away during the breeding season; they returned, evidently by the shortest possible route and without previous experience, to their nesting holes. Storm petrels are particularly active at night not only at their breeding places but also at sea, although they are sometimes found at sea in the daytime.

Reproduction

Storm petrels occupy nesting holes in their second year, but are not capable of reproduction until they are in their third year. They nest in colonies on islands, often near a mainland coats, where they have their nests in cracks in rocks or in holes which they dig in the ground. Only one species, Hornby's storm petrel, breeds on the mainland, in the saltpeter desert of Chile. Like their relatives, storm petrels lay only one egg which is incubated by both sexes, but possibly for a somewhat longer time by the male. The partners relieve one another, often at intervals of several days, and feed one another at the nest. The approach to or departure from the nest is made only at night. On land these weak-legged birds move on their bellies, using their wings for support. The young hatch after an incubation period of thirty-eight to forty-five days. They are rather helpless, and in some species they are still blind. They are fed with an oily food mush which the parents

◁

Pelicans: 1. Pink-backed pelican *(Pelecanus rufescens)*; 2. Australian pelican *(Pelecanus conspicillatus)*; 3. Dalmatian pelican *(Pelecanus crispus)*

regurgitate into the open beak of the chicks. After a period of fifty-two to sixty days in the nest, the young become capable of flight, although they have had no previous opportunity to practice this inborn ability during their development. As in most other tube-noses, the parents leave their young before they can fly. However, at that time the young are almost twice as heavy as adults and can draw on their fat reserves until they are ready to fly to sea.

Enemies of storm petrels

Certain birds of prey, including owls, are known to be enemies of storm petrels. It appears that the great black-backed gull *(Larus marinus)* has become dangerous to Leach's storm petrels in Nova Scotia since the recent increase of this gull in that area. Finally, rodents and above all, cats and dogs introduced by man on many islands, cause great losses to the petrels on their breeding grounds. Storms occasionally cause the death of many storm petrels which are blown off course. In the fall of 1952, over five thousand Leach's storm petrels perished in this manner in the area of Great Britain alone.

Storm petrels are distributed over all the oceans, but the Pacific is particularly plentiful in species, for its currents offer the best feeding possibilities. A few species remain within a shorter or longer distance of their breeding area, which is limited by climatic factors. Others, particularly those which breed in the higher latitudes of both hemispheres, replace one another as winter guests in the tropics and subtropics. In the course of this, they perform quite amazing migrations. It is interesting that the subspecies of the frigate petrel have quite different breeding seasons; those of the Northern Hemisphere breed from March to April, and those of the Southern, from October to December. Population figures are known for only a few species with some accuracy; there are several million Leach's storm petrels in North America, for example. In 1913 Murphy found Wilson's storm petrels in south Georgia in such great numbers they were almost beyond estimation.

Wilson's petrel *(Oceanites oceanicus)* of the south was thoroughly studied by Brian Roberts in 1940. It breeds on the Antarctic mainland and on the island groups of South Shetland, South Orkney, South Georgia, Falkland, Tierra del Fuego and Kerguelen in vast numbers. It carries out what is probably the longest migration in this group. Early in the southern fall the birds move far into the Atlantic to the north, and during the southern winter they are found in areas as far as Labrador and southern Greenland, even to the British Isles. The distance they cover twice each year must be at least 12,000 kilometers.

Family: diving petrels, by B. Stonehouse

The DIVING PETRELS (family Pelecanoididae) are peculiar tube-noses which are adapted to their distinctive way of life. They are short-winged, high sea divers; the L is 18–25 cm, and the weight is 120–220 g. The beak is small, and the nostril tubes are parallel at the base of the beak, short, and directed upward with a thin partition. The wings are short and wide. The flight is swift, fluttering, and whirring. In diving, the wings are used as flippers (wing divers); the primaries moult simultaneously, and the larger species limit their flying at this time. There is only one egg, weighing ten to fifteen percent of the bird's body weight; incubation is seven to eight weeks. The plumage is gray, blue-gray, or black and paler on the underside. There is only one genus *(Pelecanoides)* with five species:

1. The MAGELLAN DIVING PETREL (*Pelecanoides magellani;* see Color plate, p. 148); 2. The PERUVIAN DIVING PETREL *(Pelecanoides garnotii)* is somewhat larger; 3. The GEORGIAN DIVING PETREL *(Pelecanoides georgicus);* 4. The COMMON DIVING PETREL *(Pelecanoides urinatrix)* is the smallest species, with an L of 18 cm. It has several subspecies; 5. The KERGUELEN DIVING PETREL *(Pelecanoides exsul).* Most species range between the sixtieth and thirty-fifth degrees southern latitude in the belt of westerly winds; the Peruvian diving petrel ranges along the coasts of South America up to the tropics (to a latitude of six degrees south).

Diving petrels fly in a characteristic manner low over the water, diving into wave crests to snap up small crabs and fish. With their rapid wing beats and short-necked appearance, they resemble the auks of the Northern Hemisphere (family Alcidae; see Vol. VIII). They are the only wing divers among the tube-noses. They feed in flocks and breed colonially, nesting in burrows.

7 Pelecaniformes

Among the many bird groups whose members live on or at the water, the pelecaniformes are of especially striking appearance because of the peculiar structure of their feet. Their toes are joined by more or less well-developed webs, but the web in their case includes, in contrast to ducks and geese, the hind toe which is directed forward and to the inside. This "paddle foot" is found in all members of this order.

Order: Pelecaniformes

Birds of the order Pelecaniformes are medium sized to very large, and feed exclusively on animal food. Most species obtain their food from the sea. There are a total of six well differentiated families with seven genera.

1. The TROPIC-BIRDS (Phaëthontidae, genus *Phaëthon)* with three species; 2. The PELICANS (Pelecanidae, genus *Pelecanus*) with seven species; 3. The CORMORANTS (Phalacrocoracidae, genus *Phalacrocorax*) with twenty-eight species; 4. The ANHINGAS OR DARTERS (Anhingidae, genus *Anhinga*) with two species; 5. The GANNETS (Sulidae, genera *Morus* and *Sula*) with nine species between them; and 6. The FRIGATE-BIRDS (Fregatidae, genus *Fregata*), with five species. An additional family, the Cyphornithidae, lived in the Oligocene and Miocene, about thirty-five to fifteen million years ago.

Family: tropic-birds, by B. Stonehouse

The TROPIC-BIRDS (family Phaëthontidae with only one genus, *Phaëthon*) are birds of the high seas with a pigeon-like flight. The L is 80–100 cm, of which the body length is 30–45 cm. The wing span measures 92–109 cm and the weight is 300–750 g. They are predominantly white with a very long central tail feather; young birds lack these and have gray-white banded back and wing feathers instead. The short legs are far back on the body. Tropic-birds can hardly walk, but they can dig and scrape. There are three species: 1. The RED-BILLED TROPIC-BIRD (*Phaëthon aethereus*; see Color plate, p. 157) has an L of 100 cm. The plumage pattern of the immature birds persists on the back and wings in adults. 2. The WHITE-TAILED TROPIC-BIRD ((*Phaëthon lepturus*; see Color plate, p. 157) has an L of 80 cm. The white areas have a salmon-pink tinge. 3. The RED-TAILED TROPIC-BIRD (*Phaëthon rubricauda*; see Color plate, p. 157) is the largest species, with an L of 100 cm. The back is entirely white.

Tropic-birds usually nest on coastal cliffs, in cavities under rocks, or under vegetation which protects them from the sun and rain. On some Pacific Islands they also nest in trees. The nests are usually close together in colonies. Here the social courtship display of these beautiful birds can be seen. They fly excitedly in groups around their nesting places in undulating flight. In this display flight, the long tail feathers wave up and down.

Like many other sea birds, they lay only one egg. It is at first red or brown, but the pigment is water soluble and is lost during incubation due to moisture or rubbing. The chick hatches after an incubation period of forty-one to forty-five days. It is protected against the sun by a dense, silky, gray or yellow-brownish down plumage. It is fed by both parents from the third day on. It grows slowly and it takes eleven to fifteen weeks to fledge. Not infrequently, the young fall victim to the attacks of birds of the same or related species which are looking for nest sites. Among the red-billed and white-tailed tropic-birds of Ascension Island, these interactions within and between the two species have resulted in a low survival rate of the young and the evolution of a complex breeding "time-table." Red-billed tropic-birds breed every year, as was shown by banding experiments; however, the white-tailed tropic-birds breed every nine months, so this species is found breeding throughout the year, but as far as it is known, only on this group of islands.

Tropic-birds rarely hunt in flocks; they generally search for food singly or in pairs. In a nose-dive, they catch small flying fishes above the surface as well as other species of fish, squids, and perhaps crustacea from the surface or just below the water. They dive only slightly into the water. It seems that the average size of the prey is determined by the size of the beak; in areas where two species live, the sizes of prey overlap considerably, but the smaller species tends to catch smaller species of fish, as well as young and larval fish.

Family: pelicans, by J. Steinbacher

The largest birds of this order are generally known as the PELICANS (family Pelecanidae). The L is 170–180 cm and the wingspan is up to almost 300 cm in the Dalmatian pelican. The weight ranges from 7–14 kg; the maximum weight is reached before fledging. They appear clumsy, but because of the air in the bones and the skin, they are relatively light birds with large bodies, long broad wings, fairly long necks, and gigantic beaks. Between the branches of the lower mandible there is a distensible skin pouch; the upper mandible serves merely as a flat lid to cover it. The tongue is minute, the legs are short, the feet are large, and the four toes are connected by webs. There are seventeen cervical vertebrae; the preen gland has 6-9 slit-like openings. There are 20-24 tail feathers.

Pelicans float high on the water and they carry their wings slightly raised; since they have no wing pockets, the beak rests on the slightly curved neck. In flight the head is drawn back onto the shoulders. The flight is light and elegant; gliding often alternates with wing beats. One species makes plunging dives when feeding. The food consists exclusively of fish, which are scooped up in the bill; only the brown pelican is a plunge diver. They are sociable birds which fly in small groups or larger flocks, mostly in a diagonal line with respect to the direction which they are travelling. They search for food together, and often nest in very large colonies of up to several thousand birds, sometimes together with other water birds. They breed on a base of reeds and branches, and on shore often on just a few feathers. There are two to three plain bluish or yellowish eggs, often with a thick, chalky, calcareous layer which is at first white but soon discolors. The incubation period lasts thirty to forty-two days. The young are at first naked; the white or brownish-black down plumage appears only after eight to fourteen days. The fledging period is long, lasting twelve to fifteen weeks. The young become sexually mature in their third or fourth year. The sex differences are slight; the females are generally somewhat smaller, with shorter beaks. There are fossils from the Oligocene, about thirty million years ago. There is only one genus with seven species.

Fig. 7-1. Eastern white pelican in flight.

The two Eurasian pelican species, the EASTERN WHITE PELICAN (*Pelecanus onocrotalus*; see Color plate, p. 178) and the DALMATIAN PELICAN (*Pelecanus crispus*; see Color plate, p. 158) are of similar size (the weight is 9-14 kg), but they are distinguishable by plumage differences and behavior. The eastern white pelican has a white plumage tinged with rose color when freshly moulted, and in the breeding season there is an orange hump at the front edge of the forehead plumage. The Dalmatian pelican has a more silvery gray plumage. When the birds are seen from the front while at rest, one understands how the American zoologist Ruth Rose, who described William Beebe's Galapagos expedition, came to the somewhat anthropomorphic but apt characterization that "they look at us with that double chinned expression of disapproval unique to pelicans." Although the white pelican has a crest, it is much larger in the Dalmatian species which also has the long, soft, curled feathers of the head and neck. When seen from below in flight, the eastern white pelican is readily distinguishable by the dark end of the wing (the primaries are black and the secondaries are dark gray) from the Dalmatian pelican with its almost white wings (only the tips of the primaries are gray).

Fig. 7-2. Eastern white pelican (*Pelecanus onocrotalus*).

Fig. 7-3. Dalmatian pelican (*Pelecanus crispus*).

In Europe the white pelican prefers wide, low-lying swamps with rich vegetation and open, shallow water for feeding, as is found in the

Fig. 7-4. American white pelican *(Pelecanus erythrorhynchos):* 1. breeding area; 2. wintering area.

Fig. 7-5. Pink-backed pelican *(Pelecanus rufescens).*

Fig. 7-6. Australian pelican *(Pelecanus conspicillatus).*

Danube delta. The feeding places are almost always far from the breeding areas. Thus pelicans nesting in the northern part of the delta frequently fly in large flocks sixty kilometers to the coastal lakes in the south of the delta, or even one hundred kilometers and more up the Danube in order to fish. In Africa they are also found on barren offshore islands and rocks in the sea. Their need for company is strong, causing them to nest in colonies. The Dalmatian pelican, on the other hand, never occurs in such numbers. Its preferred breeding places are the smaller, protected lakes, but in the Danube delta it breeds with the white pelican in distinct groups within the colonies or at their edges. The nests are always in dense reeds where they are next to open water on a low platform of floating plant material. They are usually close together or in small groups. The Dalmatian pelican, in particular, sometimes breeds on high reed piles. The nests are so flat that the eggs often roll from one nest into another. Towards the end of the breeding season the nests flatten out, forming large platforms on which the young stand and sit close together. They crowd together into groups of ten or fifteen, or sometimes even more, birds. The young are of different ages and so of quite different sizes; hence the distinct breeding territories of individual pairs are no longer recognizable. The beaks of the young are still plump and their bodies are covered with woolly down; in the white pelican this is almost black, but it is white in the Dalmatian pelican. The parents feed their young predigested fish which the young take out of the throat pouches or the throats of the adults. During this feeding, the head and neck of the youngster might completely disappear into the parent's throat (see Color plate, p. 183).

Pelicans sit on their nests even before egg-laying, so data on the length of the incubation period are inaccurate. In captivity they incubate for hardly more than thirty days. At the beginning of the breeding season, pelicans are very shy and sensitive to every disturbance and hence they often abandon their nests. Floods and cold rainy weather cause great losses of eggs and young, so that more than one chick being raised in one nest is rare. At three to four weeks of age, the young can already escape into the reeds or water, and at ten weeks they leave the colony temporarily and begin to fly and to fish on their own. At eighteen weeks of age they are independent. The juvenile plumage of the white pelican is yellowish-brown on top and white underneath, while in the Dalmation pelican it is brownish-gray and white below; there is a short erect crest on the head.

Adult pelicans hardly use the few calls which they have available. They make hissing, blowing, groaning, or grunting calls. Occasionally one hears clattering sounds made with the two mandibles of the beak. In the breeding colonies the young are much noisier; they bleat like sheep, bark or squeak, and utter grunting contact calls. These are, however, only heard if one can remain in the colony unobserved or can approach it unseen. If one is seen by the birds, the young remain as silent as the adults.

Two- to three-year-old pelicans that are not yet sexually mature are regularly found far from the colonies; so are some older, mature birds. Evidently a certain proportion of the adult population does not breed every year, for reasons which are not yet known.

Numerous observers have described the peculiar manner of fishing used by pelicans. Its planned manner seems almost human and is always fascinating. H. A. Bernatzik, who observed it on Lake Malik in Albania, describes it as follows: "Since they are not divers, a number of them fish in the shallow waters by arranging themselves in a semicircle and chasing the fish towards the shore, as cormorants also do, or they gradually encircle them. They form chains of 'beaters,' scaring the fish by vigorous wing beats, and thus they block off large areas of the water surface. In narrow rivers they are said to occasionally divide into two parties in order to drive the prey towards one another. While doing this they may even swim in two or three rows, one behind the other." Thus they move slowly from deeper to shallower water until at last they merely have to scoop up their prey. Dalmatian pelicans, at this stage, often jump out of the water and with a splash submerge more deeply; this enables them to catch many fish, not only those of the shallow water as the white pelicans do.

Today both species are confined to the lower Danube in Europe, where about 3000 to 5000 pairs of white pelicans breed in the Danube delta. The Dalmatian pelican breeds here and in the Srebarna Sanctuary near Silistra, Bulgaria. The total estimated number is about 1000 pairs. Earlier breeding sites in the Balkans and Hungary have been abandoned, some in only the last few decades; they were, however, temporarily reoccupied, as was shown in the case of the Dalmatian pelican's breeding grounds in northern Greece in 1966. Records of vagrants were made, particularly in the eighteenth and nineteenth centuries, as far afield as northern Germany, Finland, France, and Spain. Thus a pelican was found at Königsberg in 1708, five were seen near Breslau in 1585, and 130 pelicans appeared on Lake Constance in 1763. Excavations have shown that Dalmatian pelicans nested on the lower Rhine in Roman times. Both species migrate as far as east and southeast Asia for the winter; those from the Danube delta go mainly to Egypt and beyond to east Africa, Mozambique, and Angola. However, hundreds of Dalmatian pelicans even today winter frequently in northern Greece.

Many tales and legends refer to pelicans and their strange appearance. They were known as domestic birds in ancient Egypt, as helpers with fishing in India, and as reputed helpers in the building of the Kaaba in Mecca by the Muslims. It was a symbol of maternal love in early Christianity, a bird which tears open its own breast to keep its

Fig. 7-7. Gray pelican *(Pelecanus philippensis).*

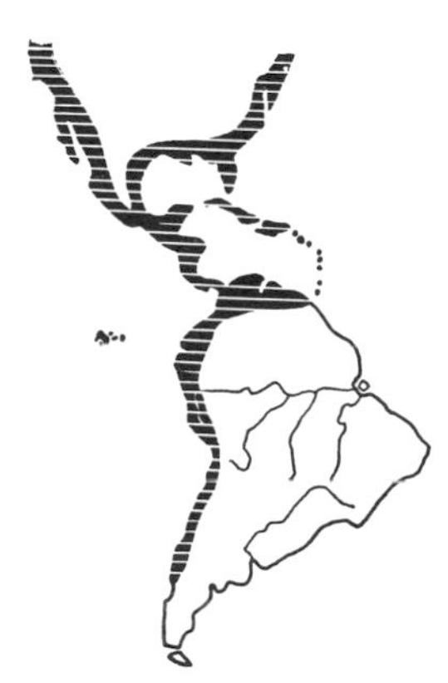

Fig. 7-8. Brown pelican *(Pelecanus occidentalis).*

▷

Cormorants: 1. Little pied cormorant *(Phalacrocorax melanoleucus)*; 2. Shag cormorant *(Phalacrocorax aristotelis)*; 3. Common cormorant *(Phalacrocorax carbo)*; 4. Spotted cormorant *(Phalacrocorax punctatus)*; 5. Pygmy cormorant *(Phalacrocorax pygmaeus)*; 6. Rough-faced cormorant *(Phalacrocorax carunculatus).*

1
2
3
4
5
6

1♂
2♀

◁
Anhingas: 1. Old World anhinga *(Anhinga rufa);* 2. American anhinga *(Anhinga anhinga).* The male is shown fishing underwater. Only very small fish are seized, as shown here; larger ones are stabbed with the beak.

young alive. This legend is probably based on the fact that in the Dalmatian pelican the reddish spot over the crop and the gular pouch, which appears during the breeding season, looks like a wound. The figure of the pelican as a martyr and a model of human mercy appears in the art of the Middle Ages in countless images, as well as on coats of arms. This belief has continued up to the present, symbolizing every form of mutual aid and Christian love of one's fellow man.

The life of most other pelicans, as far as details are known, resembles that of the two Eurasian species a great deal. Ursula Klös investigated the breeding biology of all pelican species at the Berlin Zoo. She found no differences in behavior. This might explain the frequent formation of mixed pairs in zoos. In Berlin, Dalmatian pelicans have paired with American white pelicans, pink-backs with Dalmatian penguins, and European white pelicans with pink-backs. Four hybrids were reared from these matings between 1964 and 1967.

Other pelican species

The PINK-BACKED PELICAN (*Pelecanus rufescens;* see Color plates, pp. 158, 183, and 221) is slightly smaller than the European white pelican and occupies its ecological niche in tropical and southern Africa, Madagascar, and southern Arabia. The AUSTRALIAN PELICAN (*Pelecanus conspicillatus;* see Color plate, p. 158) inhabits coastal lagoons and coastal islands as well as inland lakes in Australia and New Guinea. Its scientific name refers to the patch of bare skin around the eye. It is only slightly smaller than the European white pelican. The AMERICAN WHITE PELICAN *(Pelecanus erythrorhynchos)* is distinguished by a disc-like projection on the middle of the upper mandible of the beak, which only appears during the breeding season and then retrogresses. Its white plumage shows a yellow spot on the breast and wing coverts. The GRAY PELICAN *(Pelecanus philippensis)* is distributed from India to southern China, Java, and the Philippines. It is gray above, grayish-white below, and has dark brown feet. The BROWN PELICAN (*Pelecanus occidentalis;* see Color plate, p. 157) is, along with its southern subspecies, the CHILEAN PELICAN (*Pelecanus occidentalis thagus,* sometimes considered as a separate species), considered to be one of the most important guano birds of the rocky islands off the coast of Peru. It preys primarily on the sardine *Engraulis ringens,* which occurs there in great numbers. It has quite a different method of fishing compared to all other pelicans. It performs plunging dives to catch fish. The northern subspecies *(Pelecanus occidentalis occidentalis)* especially shows a perfection of its fishing technique which observers have described with enthusiasm again and again. From a height of a few up to twenty meters, the bird plunges down steeply or in a spiral, with its neck outstretched and its wings slightly closed, and disappears completely in the water. But, like a cork, it quickly bobs up out of the water again.

Fig. 7-9. In contrast to all other pelican species, the brown pelican is a plunge diver.

Family: cormorants, by G. F. van Tets

Like the pelicans, the CORMORANT family (Phalacrocoracidae) has only one genus, *Phalacrocorax.* Some ornithologists, however, distinguish the flightless cormorants *(Nannopterum)* and the pygmy cormorants *(Haliëtor)* as separate genera. The L is 48–92 cm, and the weight is 0.7–3.5 kg. They are equally well adapted for flying and for swimming. Relatively clumsy on land, they are still not as helpless as some other members of the order. They lie rather deeply in the water when swimming since the air spaces in their bones are very small. There are no external nares, and the edges of the beak are somewhat toothed. The head and neck have powerful muscles for closing the beak, which in part originate from special long, sesamoid bones behind the back of the head, and which are needed to maintain a grip on fish that have been caught. The plumage usually has crests in the breeding season. The sexes generally differ little in appearance, although sometimes they do in behavior. They are distributed worldwide, with twenty-nine species (according to some ornithologists, twenty-six to thirty-one), among them one which became extinct in 1852. There are more species than in all the other Pelecaniformes together.

Taxonomists and parasitologists are still not in agreement about the species and genus classifications of several cormorants of the Southern Hemisphere. According to their behavior and relationships with the environment, one can, however, classify all species except two into four groups: small or pygmy cormorants, large cormorants, shags, and guano cormorants. These groups are not systematically classified, and hence have no scientific name. Small and large cormorants can perch on trees and even on wires. In flight they hold the head above the long axis of the body shortly after take-off. Members of the two other groups never perch on trees or wires. Among them, the shags hold the head and neck straight forward in flight like geese and swans, while guano cormorants keep it lower, somewhat like loons. Small and large cormorants nest on trees or on the ground, using among other things sticks and twigs as nesting material. They live inland as well as on the coast, often feeding in lakes, ponds, and river estuaries. They cross islands and peninsulas in direct flight. Shags and guano cormorants nest on the ground, and only rarely use sticks and branches. They occur only in coastal areas. Shag species often seek food singly in rough water near rocky coastal stretches; guano cormorants, however, generally chase swarms of small fish in large flocks. Members of both groups rarely fly across islands or peninsulas, but prefer to follow the coast line. They are only occasionally found inland after storms or in dense fog.

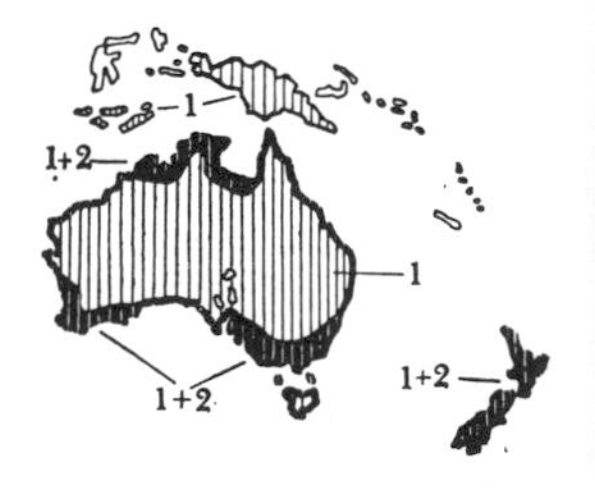

Fig. 7–10. 1. Little pied cormorant *(Phalacrocorax melanoleucus)*; 2. Pied cormorant *(Phalacrocorax varius)*.

The long-tailed, short-necked small cormorants occur in an area around the Indian ocean from South Africa to New Zealand. Their food consists of small fish which are evidently taken by mergansers (*Mergus,* Chap. 13) in other areas. There are four species, among them the LONG-TAILED SHAG *(Phalacrocorax africanus)* in Africa south of the Sahara, the LITTLE PIED CORMORANT (*Phalacrocorax melanoleucus;* see Color

Fig. 7-11. 1. Shag *(Phalacrocorax aristotelis)*; 2. Socotra cormorant *(Phalacrocorax nigrogularis)*; 3. Pygmy cormorant *(Phalacrocorax pygmaeus)*; 4. Long-tailed shag *(Phalacrocorax africanus)*.

plate, p. 167) from Java to the Solomons, New Zealand, and Australia, and the PYGMY CORMORANT (*Phalacrocorax pygmaeus*; see Color plate, p. 167; the L is 48 cm and the weight is 700 g), which lives as a very sociable bird in the Middle East and southern Europe, occasionally reaching Germany, presumably from southern Hungary.

The large cormorants are worldwide in distribution. Some species occupy entire continents. The COMMON CORMORANT (*Phalacrocorax carbo*; see Color plates, pp. 61/62, 167, and 255/256) has an L of 92 cm and a weight in the ♂ of 2.3 kg, and in the ♀ of 1.9 kg. It inhabits a particularly large region. The WHITE-BELLIED AFRICAN CORMORANT *(Phalacrocorax lucidus)*, which lives south of the Sahara, perhaps belongs to this species also. The only noteworthy population in Germany today lives in Pomerania. Three or four blue, chalky-covered eggs, each weighing 50 g, are in one clutch; they are incubated for 23-25 days. The young are naked and blackish in color at first, but their heads are pink. They can fly when they are 42-60 days old. These cormorants dive to a depth of one to three meters and remain underwater for about forty-five seconds when they are fishing. In eastern Asia they are trained to assist fishermen. Their call sounds like "krokrokro," and the display call is an "a-orr." The JAPANESE CORMORANT *(Phalacrocorax capillatus)* is a close relative. Less closely related to the common cormorant but similar to one another are the DOUBLE-CRESTED CORMORANT *(Phalacrocorax auritus)* of North America, the BIGUA CORMORANT *(Phalacrocorax olivaceus)* of South and Central America, the FLIGHTLESS CORMORANT (⊕ *Phalacrocorax harrisi)* of the Galapagos, and the LITTLE BLACK CORMORANT *(Phalacrocorax sulcirostris)* of Australia. The flightless cormorant is a seabird of very restricted distribution. Its closest relative is presumably the double-crested cormorant and not the shag and guanay cormorants which occur within its range. Although the flightless cormorant is as large as the common cormorant, its wings are only half the size. It looks peculiar when, after diving, it spreads its little wings for drying, as the cormorants do with their large wings.

The SHAG OR GREEN CORMORANT *(Phalacrocorax aristotelis*; see Color plate, p. 167), with an L of 75 cm and a weight of 1.8 kg, has a green iridescent sheen; it is the best known of the shag group, inhabiting rocky coasts in Europe and northwest Africa. The RED-FACED CORMORANT *(Phalacrocorax urile)*, the extinct SPECTACLED CORMORANT *(Phalacrocorax perspicillatus)*, which weighed 3.5 kg and inhabited Bering Island, and the PELAGIC CORMORANT *(Phalacrocorax pelagicus)* of the North Pacific resemble the shag. The RED-LEGGED CORMORANT *(Phalacrocorax gaimardi)*, with an L of 71 cm and a weight of 1.4 kg, inhabits the west coast of South America; along with the MAGELLAN CORMORANT *(Phalacrocorax magellanicus)* of Cape Horn, it is probably a close relative of the northern shags.

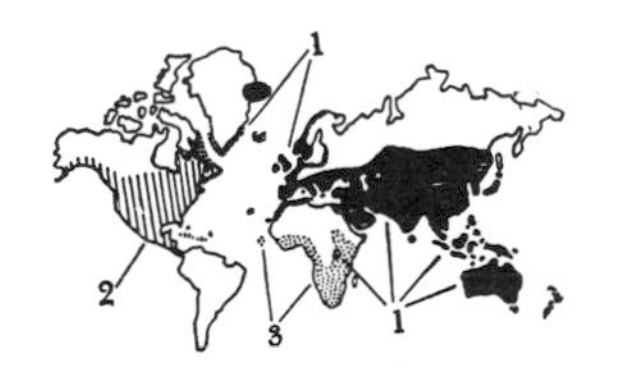

Fig. 7-12. 1. Common cormorant *(Phalacrocorax carbo)*; 2. Double-crested cormorant *(Phalacrocorax auritus)*; 3. African cormorant *(Phalacrocorax lucidus)*.

Members of the guano cormorant group live mainly in the colder areas of the Southern Hemisphere. The GUANAY CORMORANT *(Phalacrocorax bougainvillei)*, with an L of 68 cm, has a crest, a green ring of bare skin around the eye, and a white underside. This most important guano producer lives along the west coast of South America; the Peruvians and Chileans call it "guanay." To this group belong: the BRANDT'S CORMORANT *(Phalacrocorax penicillatus)*, with its black plumaged face, blue chin, and brown throat band, the WHITE-BREASTED CORMORANT *(Phalacrocorax fuscescens)*, with its black feet without a white stripe above the eye, and the ROUGH-FACED CORMORANT *(Phalacrocorax carunculatus*; see Color plate, p. 167). Two other species which are both crested and closely related to one another are included: the BLUE-EYED CORMORANT *(Phalacrocorax atriceps)*, with a white patch on the back, and the KING CORMORANT *(Phalacrocorax albiventer)*; both have crests.

Fig. 7-13. 1. Little black cormorant *(Phalacrocorax sulcirotris)*; 2. White-breasted cormorant *(Phalacrocorax fuscescens)*; 3. Spotted cormorant *(Phalacrocorax punctatus)*; 4. Rough-faced cormorant *(Phalacrocorax carunculatus)*.

Some species of cormorants are not sufficiently well-known to allow a determination of their relationships. One of the most interesting among them is the SPOTTED CORMORANT (*Phalacrocorax punctatus*; see Color plate, p. 167) of New Zealand. This species may be close to the ancestors of the cormorants, whose phylogeny began in the Eocene (about fifty million years ago) with a split of the ancestral group into cormorants and anhingas. The spotted cormorant uses sticks in its nest, which resembles those of the common and double crested cormorants. Yet it is not known to be a tree or an inland bird. The sticks in its nest suggest that at one time it was a tree nester. Presumably it was displaced from those nesting sites by the common cormorant, the little black and little pied cormorants, and an additional species, all tree nesters which also occur in New Zealand. Like the shags and the guano cormorants, spotted cormorants feed only at sea. Closely related to it is the CHATHAM CORMORANT *(Phalacrocorax punctatus featherstoni)* of Chatham Island.

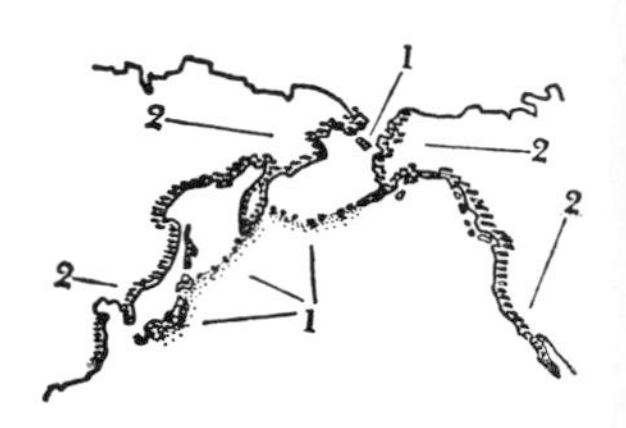

Fig. 7-14. 1. Red-faced cormorant *(Phalacrocorax urile)*; 2. Pelagic cormorant *(Phalacrocorax pelagicus)*.

Another interesting but little known species is the SOCOTRA CORMORANT *(Phalacrocorax nigrogularis)* of the coastline from Arabia to East Africa. It is the only bird of this order which builds its nests of stones, like some penguins do. Its eggs are spotted and brown.

As powerful fliers, cormorants move daily between feeding and resting places which may sometimes be one hundred kilometers apart. They find their prey at sea as well as inland, but they are rarely found out of sight of land when at sea. Some species breed beyond the Arctic Circle and winter in the tropics. Others breed on islands on lakes which are in tropical deserts and have water only during part of the year; they spend the dry season in coastal waters. When a number of cormorants fly in the same direction, they arrange themselves into a V formation like many other waterbirds. They alternate, rather like gannets, between short glides and sequences of wing beats; near cliffs and islands they soar occasionally in updrafts. Upon arrival at their nesting colonies, they often descend in a spiral with outstretched feet.

This rapid descent is accompanied by a loud, sharp whistle. It is thought that the whistle and the rapid descent scare off frigate-birds and jaegers and thus reduce persecution by these food parasites. Gannets descend in a similar manner.

Cormorants feed on a variety of fish, crabs, squids and amphibian larvae. They follow prey underwater and catch it with their hooked beak. Larger animals are brought up to the surface, flung into the air, caught, and swallowed head first. Cormorants are extraordinarily efficient hunters. They often gather in great numbers, form orderly lines on the water, and thus acting in concert catch fish, which at low tide are crowded together in river estuaries and bays. At sea and on large lakes they can often detect large swarms of small fish from afar by the aggregations of gulls. These abilities enable cormorants to satisfy their daily food requirements in a much shorter time than is the case with most birds; often half an hour is enough. Hence cormorants have much spare time during the remainder of the day which they spend resting, preening, displaying and playing.

Cormorants' bodies are particularly well adapted for swimming underwater. Their plumage is permeable to water. In a water bird this may seem paradoxical at first and it used to be regarded as a lack of adaptation. However, the plumage's water permeability is useful for a diving cormorant. It allows the air to escape from the plumage and so reduces buoyancy. There are even indications that cormorants fishing in salt water increase their specific gravity by swallowing stones. As in some other diving birds, transparent nictitating membranes serve as "diving masks". They move over the eye from the front to the back, enabling the bird to see underwater. The legs, as in other underwater swimmers, are far back on the body. Under water, cormorants push backward simultaneously with both feet. The jerky progression which results has given several observers the false impression that they use their wings under water. On the water surface they use alternating foot strokes; when taking off from the water they push back several times with both feet simultaneously, however.

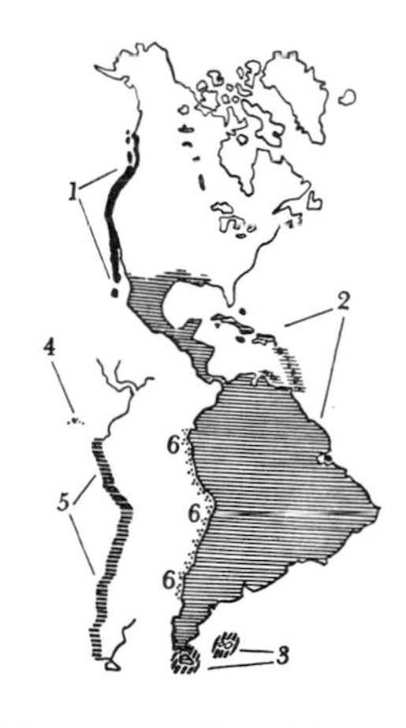

Fig. 7-15. 1. Brandt's cormorant *(Phalacrocorax penicillatus)*; 2. Bigua cormorant *(Phalacrocorax olivaceus)*; 3. Magellan cormorant *(Phalacrocorax magellanicus)*; 4. Flightless cormorant *(Phalacrocorax harrisi)*; 5. Red-legged cormorant *(Phalacrocorax gaimardi)*; 6. Guanay cormorant *(Phalacrocorax bougainvillei)*.

Once out of the water, cormorants have no difficulty in shaking water drops off their plumage, for the individual feathers have become water repellent through the application of preen gland oil. After fishing, cormorants briefly tread water and shake their wings back and forth. Then they seek out a nearby perch in order to digest their catch, which reduces their weight. Only then do they fly back to their roosting or breeding places. If a cormorant is disturbed during digestion, it often reduces its weight by regurgitating its stomach contents before flying off. When digestion has been completed, indigestible food components such as bones and fish scales are also regurgitated in a red mucous sac which is sloughed off from the stomach wall. When a cormorant is about to regurgitate such a mucous sac, gulls often fly by and catch it in the air. They are probably able to utilize nutrients which remain in the contents of these sacs.

Since cormorants live on fish, they are vigorously persecuted as competitors by man, being considered destructive to fisheries in some districts. There is generally no basis for this, other than that these "crows of the sea" are a convenient scapegoat for the problems of fishermen. This persecution seems particularly unjustified when carried out by so-called "sportsmen," for in sport fishing the fish are, after all, taken more for pleasure and amusement than to meet human food requirements.

Only when people purposefully crowd fish together in traps or where they are artificially kept in huge numbers in hatcheries are cormorants likely to do real damage. Apart from these unnatural and exceptional cases, it is likely that cormorants are more useful than harmful to fishery interests. After all, they feed on smaller and less palpable fish than people usually catch; like other carnivorous animals, they remove the less healthy members of the species they prey on and, unlike man, there is no danger that they will overfish particular waters, taking more fish than can be replaced by natural reproduction.

In many parts of Asia and Africa, cormorants are kept in cages and used to catch fish for people. Hans von Boetticher writes: "The tame birds perch on the sides of boats; they are fitted with a collar of hemp or soft leather, and released into the waves. As soon as they have seized a fish, they emerge and head for the boat. Since the collar prevents them from swallowing their prey, they are forced to deliver it to their master." In Japan cormorant fishing is generally carried out at night and has become a custom that is performed for the entertainment of spectators.

Like many other seabirds, cormorants belong to the "guano birds." Guano is the accumulation of feces of social birds or bats which forms in masses under their breeding colonies or resting places. It is rich in phosphorus compounds and other chemicals. Man removes these substances in large quantities from the soil each harvest, and hence they must be returned to it to maintain fertility. Guano is therefore a valuable phosphate manure and guano deposits are exploited systematically. The most productive guano deposits come from the vast cormorant colonies on the west coasts of South America and South Africa, where millions of these birds nest side by side. Rising currents on these coasts bring up phosphate compounds from the depths of the ocean. They fertilize the surface waters, which makes an unbelievably rich plankton life possible. This serves as food for larger animals and, through this food chain, the phosphorus finally enters the digestive tracts of the cormorants and becomes guano. An important link in this food chain in Peruvian coastal waters is the sardine, *Engraulis ringens.*

If, under rare conditions, the supply of phosphates from the sea currents is interrupted, a mass starvation of cormorants takes place. In order to encourage growth of the breeding colonies and to facilitate exploitation of the guano deposits, dog- and cat-proof walls have been built across peninsulas on the Peruvian coast and on the coast of southwest Africa, and large wooden platforms have been set up on posts in shallow water. In Florida as well, there are such artificial platforms as resting places for cormorants in suitable localities. Hence, what is a nuisance for some people provides a fortune for others.

Fig. 7-16. Common cormorant wing raising and lowering.

The adult plumages of the most important species of cormorants are shown on page 167, but the juvenile plumages differ more or less from these. Immature birds have irregular brown spots of varying tints in areas which in adults are black or white. Very few cormorants reach sexual maturity after only one year; usually they acquire the adult plumage when they are two years old or older, and then begin to breed. Within a particular species some individuals may breed earlier, but others do not until several years after acquiring the adult plumage. The plumage is similar in both sexes; after the end of the breeding season the black feathers become paler and eventually a dark brown. Before the onset of the breeding season, the birds acquire decorative feathers; in most species these are white. They vary in size according to the species as well as in distribution on the head, neck, rump, or thighs. The nuptial feathers on the head and neck are dropped soon after pair formation, but those on the rump and thighs are kept until the young are hatched. Bare areas of skin around the eyes and on the face, the throat, and the lining of the mouth are intensively colored or black in adults only during the breeding season, while during the rest of the year only the iris retains its bright color.

Fig. 7-17. "Invitation" to copulation.

Often several species of cormorants breed together in mixed colonies; on cliffs the larger species generally occupy the rocky projections, while the smaller ones are more often found in niches or recesses. Other seabirds frequently breed in cormorant colonies. The pelagic cormorant *(Phalacrocorax pelagicus)* of the North Pacific strikingly resembles the kittiwake (*Rissa tridactyla,* Vol. VIII) in its nesting behavior. Both species nest on particularly narrow rock ledges, building their nests of wet grass and hay which they pluck in rainy weather. In dry weather this cormorant gathers loose, dry grass, wets it in the sea, and builds it into the nest; it glues the outer rim of the nest to the rock with feces.

Fig. 7-18. Greeting its partner at the nest prior to changing places on the nest; "gargling" and throwing back of the head.

Even when the eggs are laid and the young have hatched, cormorants continue to add new nest material to the nest. This continues until the young are about to leave the nest. Nests on trees and cliffs are generally built more carefully than those on the ground. Cormorants have considerable difficulties in securing the first base of their nest platforms to certain species of trees. Tree nests are generally used for

Fig. 7-19. Threat posture on the nest.

several years. They finally fall down because the trees are weakened or killed by the bird's excretions and cannot carry the nests which become heavier and heavier. Derelict or unguarded nests are often taken apart by cormorants or other birds because nest material is scarce in many colonies. Cormorants may even have to travel a distance of several kilometers for it.

Fig. 7-20. Red-faced cormorant's posture which precedes taking wing.

The nesting places are occupied by males which proclaim themselves owners and try to attract females by a "masculine display." In most cormorants, this display consists of a repeated raising and lowering of both wings in which the closed primaries are kept behind the secondaries. The males of some species (among them the common cormorant) alternately show and then hide the white spots on their thighs as they raise and lower their wings in this display. The appearance of white against a dark background acts like a signal and may be visible at greater distances than the wing movements. The males of some other species, like the double-crested cormorant, which have no white thigh feathers replace this visual signal by repeated calls while waving the wings.

Fig. 7-21. Pelagic cormorant about to hop...

...and hopping.

The female cormorant selects a male and his nesting place. She lands beside her chosen male and seeks acceptance by a very complex behavior, both before and after landing. If the male is agreeable and prepared to accept the female as his mate, he stops the attraction display and begins a greeting display. If he does not accept her, he forces her out of the nest, flinging her off the cliff or the tree. If the nest is on the ground, the rejected female is forced to make a hasty withdrawal between the closely spaced nests and their hostile owners.

Pair formation is complete when the male leaves his female behind as a nest guard while he takes off to gather more nest material which she will build into the nest on his return. Later the two partners relieve one another in guarding the nest day and night until the young are old enough and strong enough to defend themselves. Occasionally the female brings in nest material, but this never occurs during the period of pair formation between the two birds. The birds drag sticks, branches, and feathers into the nest. In the beginning they use their feet while building the nest platform in order to hold the nest material temporarily in place.

At the nest site both sexes indicate their intention to take off before they leave the partner on the nest. Upon returning, both before and after landing, they perform certain ceremonies and are greeted by the partner who has remained at the nest. All these actions and the calls which accompany them are different in the various species of cormorants; it is also possible that certain individual differences facilitate the recognition of the partner. In courtship displays, males behave differently from females. Both sexes also behave quite differently on the nest than they do when away from it. On the nest the body is held

▷

Frigate-birds: 1. Great frigate-bird *(Fregata minor)*; 2. Magnificent frigate-bird *(Fregata magnificens)*. Boobies: 3. Brown booby *(Sula leucogaster)*; 4. Red-footed booby *(Sula sula)*; 5. Northern gannet *(Morus bassanus)*, a. juvenile plumage.

1
2
3
5
5 a
4

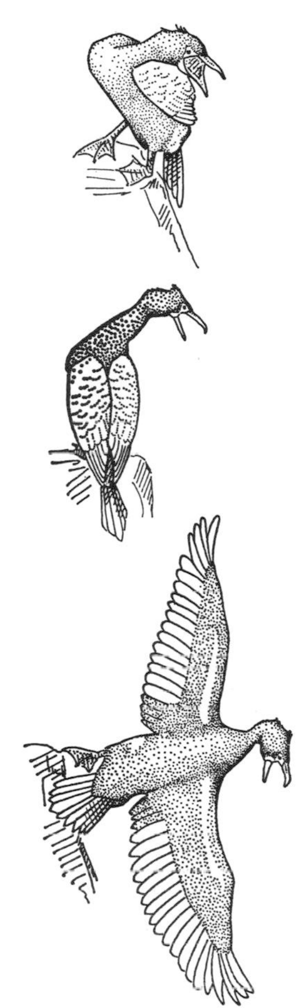

Fig. 7-22. Pre-flight postures in the take-off ritual of the pelagic cormorant.

horizontally, but away from it, vertically; on the nest the common cormorant *(Phalacrocorax carbo)* lowers its central head crest, while away from the nest, it is erected. It also lowers the crest when frightened while away from the nest.

The parents relieve one another during incubation, feeding, and brooding of the young. When it is cold they warm the eggs with the webs of their feet, and they also do this to the chicks while they are very young. When it is too hot they provide shade for the eggs or young and they cool them by placing wet hay or algae in the nest.

The young are naked during the first week after hatching. When begging for food, they raise their heads and call loudly with closed beaks. Older youngsters, when begging, aim at the throat of the parent and emphasize their urgency by beating their wings. They take food like other young Pelecaniformes, by inserting the head far into the parent's mouth and throat. In hot weather, young common cormorants, double-crested cormorants, and perhaps those of other species as well, beg for water by raising the wide open beak upward silently. The parent which is on the nest then flies to nearby water and swallows several beakfulls of fresh, brackish, or sea water. Then it returns to the nest and lets the water run into the open gape of the young, and occasionally also into that of the other adult which is on guard.

Finally the young begin to leave the nest temporarily. But they are still fed by the parents away from the nest and even on the water. They will then often fly behind their parents while begging for food. Soon after they can fly well they begin to catch prey by themselves.

Individual cormorant species differ, among other things, in the degree to which they accompany their expressive movements by calls. Many forms of behavior are carried out silently by some species, while others accompany them by specific calls. The calls may be the same in both sexes or they may be different. Before pair formation, the female of the common cormorant uses her own specific calls, but after pair formation her calls are just like those of the male. Cormorants have no alarm calls, but they do react to those of gulls and other birds. When disturbed, for example by the appearance of a sea eagle or a person, they assume an alarm posture: the head is raised up high and the plumage is sleeked down. If they are frightened they take off quickly and leave their nests without the usual preliminary flight-intention display. When cormorants are disturbed at the breeding colony or at their roosting places, they often form dense flocks which come down on the water at a safe distance.

The social life of cormorants and their social ceremonies are extraordinarily varied; much has been learned about the social behavior of all social animals, including man, from the general principles of behavior that were discovered by studying cormorants. The colonies can be observed from skillfully built blinds without disturbing the

◁
In display, male frigate-birds inflate their red throat sacs. This one is a great frigate-bird *(Fregata minor)*.

Pink-backed pelican *(Pelecanus onocrotalus)* while preening.

birds. Instead of blindly persecuting these highly interesting, beautiful, and predominantly useful birds, they should be protected in their natural environment.

Family: anhingas, by G. F. Tets

Two species of fresh water Pelicaniformes must be considered as a separate family because of a number of peculiarities of their body structure, although they are in many respects close to the cormorants. These are the ANHINGAS OR DARTERS (family Anhingidae, genus *Anhinga*). The L is 90 cm. The bill is straight and pointed, and is finely toothed on both cutting edges. The neck has twenty vertebrae and, when the bird is sitting or flying, it is bent into an "S" or even a "G" shape. The tail is long, stiff, and rounded at the end. They walk and swim underwater when diving. There are two species: 1. The AMERICAN ANHINGA (*Anhinga anhinga;* see Color plate, p. 168) has no subspecies. 2. The OLD WORLD ANHINGA (*Anhinga rufa;* see Color plate, p. 168) has three subspecies which some zoologists regard as distinct species, while others look on all anhingas as a single species.

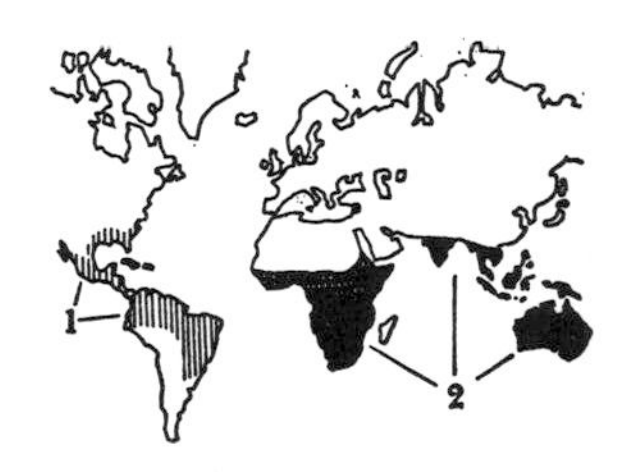

Fig. 7-23. 1. American anhinga *(Anhinga anhinga);* 2. Old World anhinga *(Anhinga rufa).*

Anhingas live on fresh water fish and other aquatic animals. Like herons, they have dagger-like beaks and long "G" shaped necks. But while herons wade in shallow water and stalk or wait for their prey above the water surface, anhingas hunt for their prey underwater. They swim slowly under water with partly opened wings, and attract the fish with their dark shadow. A special hinge and muscle arrangement on the eighth and ninth cervical vertebrae enables them to thrust the head rapidly forward so that the prey is stabbed and stunned. Fish taken from the stomachs of these birds show evident stab wounds on the side. Large fish are brought to the surface and flung into the air so that they can be swallowed head first.

Fig. 7-24. Australian gannet in a stretched-upward posture...

Like the plumage of the cormorants, the anhinga's is water permeable. This adaptation reduces buoyancy and enables anhingas to submerge silently without attracting the attention of prey animals or enemies. They often swim with only the head and the thin neck projecting above the surface, and when doing so they really resemble a swimming snake; hence they have another name, the snake-bird. After fishing, they shake the water out of their wings, as cormorants do, and go to a nearby resting place to dry out the plumage and digest their food. Thus they reduce their weight before flying to their roosts or nesting colonies.

In the air they soar and glide like pelicans, skillfully using thermal air currents. When they soar upwards in spirals without beating their wings and then glide downward into the next thermal air current, they look like flying crosses with their long necks and tails and wide wings. Anhingas are very well adapted to both air and water; they are gliders as well as spear fishers.

...greeting posture...

The distribution of black and brown plumage colors and of white patterns differs according to species, subspecies, and sex. Mature adult

...Bowing...

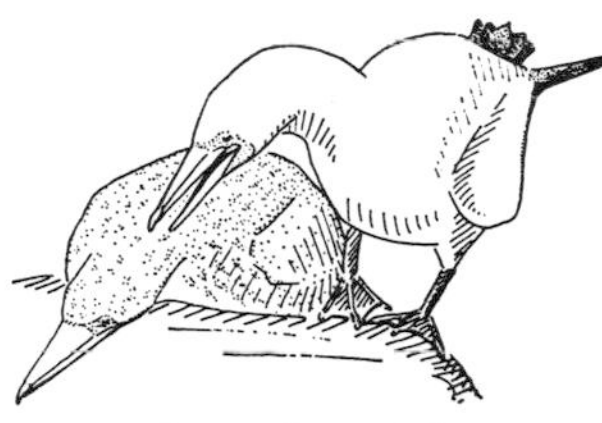

...and attacking. The behavior of gannets has many similarities to that of albatrosses (see Fig. 6-2); some investigators therefore believe that the two groups are related.

Fig. 7-25. 1. Northern gannet *(Morus bassanus)*; 2. Cape gannet *(Morus capensis)*; 3. The Australian gannet *(Morus serrator)* breeds on the coasts of Australia and New Zealand, the tropical boobies (genus *Sula*) on the coasts and islands of the tropical seas.

Family: gannets, by J. Warham

males show more black than females and immatures. In contrast to all other Pelecaniformes, anhingas become incapable of flight during the annual moult.

Their nests consist of sticks which are wedged together. Generally they are built on trees standing in the water or on branches projecting above the water. In some places they also nest on the ground in reed thickets. Often several pairs nest communally in the same group of trees, in which there may also be nests of cormorants, herons, or ibises. As in other Pelecaniformes, the males select the nest site and defend it, while the females look for a male with a nest site and try to get him to accept her. Once the pair has been formed, both partners take turns guarding the nest. The male gathers the nest material and carries it to the female, which builds the nest. Both sexes incubate alternately and both brood and feed the young together.

The behavior of the American anhinga during reproduction has been studied in detail in the southern United States. Courtship displays resemble those of cormorants, gannets, and herons. Thus the wing waving described earlier in the cormorant is found in a quite similar form in the anhinga as the male's attraction display. "Showing," in which head and neck are moved up and down with closed beak, serves to draw the female's attention to the nest site and to attract her. After pairing, the male emphasizes that he is the owner of the nest by a bow accompanied by snapping the bill. The bird raises both wings simultaneously, keeping the primaries folded behind the secondaries; at the same time it bends the neck and then snaps the head toward some nearby twig or a piece of nest material. The partners greet one another at the nest by flexing the neck at the throat. Sometimes they hold nest material in the beak while bending their necks in this way; otherwise the beak is held wide open during this action. The effect of these expressive movements is increased by the way the tailfeathers and the mane-like feathers on the neck, peculiar to the males, are carried. With these and other signals, anhingas can communicate and achieve the necessary cooperation for the successful rearing of their young.

The GANNETS OR BOOBIES are predominantly black and white seabirds (family Sulidae). The L is 70-100 cm, and the weight is 1.5-3.5 kg. The strong beak is pointed and conical, with fine serrations on the cutting edge near the tip. The bare parts of the face, throat, and feet are often colored. The strong feet have well-developed webs. External nares are absent. Gannets breathe through a specially constructed palate. There are two genera with nine species (or a single genus with seven species, according to other scientists):

A. The GANNETS (genus *Morus,* included by some investigators in the next genus) have three species which are sometimes treated as mere subspecies: 1. The NORTHERN GANNET (*Morus bassanus;* see Color plates, pp. 60/61 and 177) with a wing span of 180 cm and a weight of 3 kg; 2. The CAPE GANNET *(Morus capensis)*; and 3. The AUSTRALIAN GANNET *(Morus serrator)*. Both are very similar to the northern gannet.

B. TROPICAL GANNETS OF BOOBIES (genus *Sula*) are somewhat smaller, having six species: 1. The BLUE-FACED BOOBY *(Sula dactylatra)* is the largest booby species. It is powerful and colored like all gannets, but without the yellow markings on the head; 2. The RED-FOOTED BOOBY (*Sula sula*; see Color plate, p. 177) is the smallest species, showing a whole series of color phases from dark brown to white. The feet are red and the feathers are shiny; 3. ABBOT'S BOOBY (⊕ *Sula abbotti*) is rare; 4. The BROWN BOOBY *(Sula leucogaster)* is chocolate-brown above and predominantly white underneath. It is very widely distributed. 5. The PERUVIAN BOOBY (*Sula variegata*, Spanish "Piquero") is the most common species in the area of the Humboldt current, and is an important supplier of guano. 6. The BLUE-FOOTED BOOBY *(Sula nebouxii)* has a spotted plumage like young northern gannets, with bright blue feet.

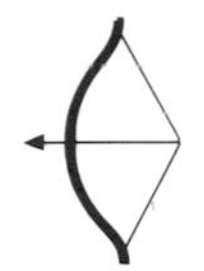

The gannets of the genus *Morus* are birds of the temperate seas, living mainly on fish. After the breeding season they perform long migrations, the young travelling furthest. Northern gannets migrate southward as far as West Africa, whereas cape gannets wander northward along both coasts of the African mainland. Australian gannets fly westward crossing the Tasman Sea. The tropical gannets wander much less. Many of them feed on flying fish and the red-footed booby also takes squids in the dark.

Fig. 7-26. Blue-footed boobies in courtship display.

All gannets obtain their food by diving. Where they are numerous, this method of feeding makes a fascinating spectacle, which Murphy describes in the Peruvian booby: "Once I saw them falling like lumps of lead everywhere, over the whole bay. They only closed their wings the instant before they dived into the water. They plunged down so fast that they were only visible as white streaks which were followed by a geyser of water."

Fig. 7-27. Lesser frigatebird in flight, seen from below.

Gannets have loud, raucous voices. In some of the tropical species, the calls of the two sexes are different and very characteristic. The females call sonorously in a low pitch, while the males only utter weak whistling sounds. These differences are due to differences in the vocal organs.

Red-footed and Abbot's boobies nest on trees. All species breed in colonies, almost always on islands. Generally, one to three eggs are laid once a year. But in some areas, the brown booby lays every eight months so that breeding pairs can be found at almost any time. Gannets also show pronounced expressive behaviors. This particularly applies to those of the genus *Morus*; their behavior was studied by Warham and Nelson. These gannets are very aggressive. The behavioral repertoire of the boobies is less varied, as was shown by the observations of Dorward and Verner.

▷
In this manner, a nearly full-grown pink-backed pelican *(Pelecanus rufescens)* takes food from the adult's gape.

▷▷
An African black-necked heron *(Ardea melanocephala)* above a papyrus swamp.

Although all Pelecaniformes, with the exception of the flightless cormorant, are excellent fliers, the FRIGATE-BIRDS (family Fregatidae) exceed the members of all the other families in aerial mastery. Almost half of their weight consists of the breast muscles and feathers; their load per unit of wing surface is extraordinarily small. They are therefore probably the most efficient soarers in rising air currents.

Family: frigate-birds, by B. Stonehouse

There is only one genus, *Fregata*, with an L of 75–112 cm, a wing span of 176–230 cm, and a weight of up to 1.5 kg. The beak is long and bent into a hook at the tip. The wings are narrow and the lower arm and hand bones are strongly elongated. The tail is deeply forked, often spread and then closed again in flight. The small feet are almost without webs. The ♀♀ are marked differently and are generally larger than the ♂♂. The young have a white head. They are restricted to tropical and subtropical seas; they live mainly where flying fish are common, in water of at least twenty-five degrees centigrade. There are five species: 1. The MAGNIFICENT FRIGATE-BIRD (*Fregata magnificens*; see Color plate, p. 177) is the largest species. The L is 103–112 cm, the wing span is 230 cm, the beak is 12 cm, the tarsus is 2.3 cm, and the weight is 1.4–1.5 kg. The ♀♀ have a white breast band and brownish upper lesser wing coverts. 2. The ASCENSION FRIGATE-BIRD (*Fregata aquila*) weighs 1.2 kg. The ♂♂ have a somewhat greenish gloss, while the ♀♀ are brownish on the upper breast, nape, and wing band. 3. In the CHRISTMAS FRIGATE-BIRD (*Fregata andrewsi*), both ♂♂ and ♀♀ have a white belly and a brown wing band; the ♀♀ have a black throat. 4. The LESSER FRIGATE-BIRD (*Fregata ariel*) has an L of 75 cm (♂♂) to 82 cm (♀♀). The ♂♂ are black to brown, with a conspicuous white spot on the side of the belly. 5. The GREAT FRIGATE-BIRD (*Fregata minor*; see Color plates, pp. 177/178) has an L of 95 cm. The ♂♂ have a broad, gray-brown wing band, and the ♀♀ have a white throat.

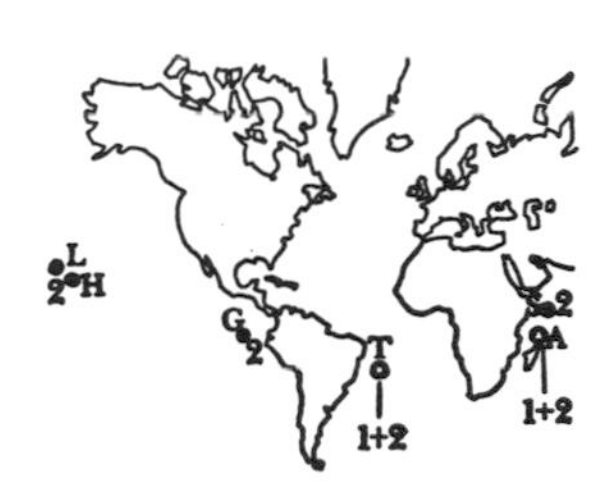

Fig. 7-28. 1. Lesser frigate-bird *(Fregata ariel)*, South Trinidad (T) and Aldabra (A), and also in the eastern and southern Pacific. 2. Great frigate-bird *(Fregata minor)*, Seychelles (S), Aldabra (A), South Trinidad (T), Galapagos (G), Hawaii (H), Layson (L), and also in the east and southwest Pacific.

Although frigate-birds sometimes fly far out over the sea, they tend to breed all year round and, as a rule, stay near their home islands. When a frigate-bird is seen at sea, land is generally not far away. Their ability to find their way is such that these birds are used to send messages in the South Sea islands, as are carrier pigeons elsewhere in the world.

Frigate-birds generally build their nests in low shrubs or trees, and only rarely on bare ground. The nesting colonies are generally close to those of other seabirds, particularly terns or gannets, which frigate-birds rob of their prey or young. In the courtship period, male frigate-birds develop a red, inflatable throat-sac, which is used in display. The males, with their black iridescent plumage, occupy nesting places in the colony, showing off with outspread wings and inflated throat sacs to the females which fly above them (see Color plate, p. 178). As soon as a possible partner approaches, they shake themselves and rattle their beaks and wings.

◁
A large white egret *(Casmerodius albus)* displays its nuptial plumes. These plumes formerly caused the death of many egrets, since they were worn on ladies' hats.

Frigate-birds lay only a single white egg which weighs about six percent of the mother's body weight. It is incubated for forty to fifty days. The young are naked when hatched, and stay in the nest for four to five months. They depend on their parents for another two to six months, but fly about the colony in groups and live on scraps. Playing high up in the air with feathers and bits of seaweed, they exercise their flight muscles and so acquire their magnificent mastery of flight.

Fig. 7-29. 1. Magnificent frigate-bird *(Fregata magnificens),* from the Galapagos Islands (G) to the west African coast, with subspecies in the eastern Pacific, the Caribbean, and the Atlantic. 2. Ascension Island frigate-bird *(Fregata aquila)* with two to three thousand birds found only on Ascention Island (A). 3. Christmas frigate-bird *(Fregata andrewsi),* Christmas Islands (Ch) and other islands in the East Indian Ocean.

Flying fish are the principal food of frigate-birds; they catch them in the air up to a few meters above the surface of the sea. They also chase gannets, pestering them so much that they regurgitate their food in order to escape more swiftly. The regurgitated food is caught by the frigate-birds in the air. They also take food scraps or young terns from the ground or water surface by diving down and grabbing them with their beaks while in full flight. They rarely swim, for their plumage is only lightly oiled and soon becomes wet. They can stay in the air during strong winds without effort, but upon landing their large wings are a hindrance. Therefore adults often have difficulties when landing on their nests, particularly when the trade winds blow strongly during the day.

8 The Herons

Order: Ciconiiformes

Most birds of the order Ciconiiformes differ conspicuously by their long legs from birds living on or near the water, like the loons and grebes, and the penguins, tubenoses, and Pelecaniformes. Unlike the ratites, the Ciconiiformes cannot however, use their legs for swift running; their gait is a much more measured walk.

Distinguishing characteristics

The L is 30–160 cm; in normal posture the height of the crown is 20–130 cm. The weight is 100–6000 g. They are almost always long-legged and long-necked. They have 16–20 cervical vertebrae, and the hind toe is well developed. All species live on animal food. The crop is absent but there is a well developed proventriculus and small caeca. The eggs are usually plain-colored (exceptions are the hermit ibis and the spoonbill). The young are altricial. Fossil remains of herons, storks, and ibises have been found dating from the transitional period from the Eocene and Oligocene, about forty million years ago.

There are five families: 1. The HERONS, EGRETS, and BITTERNS (Ardeidae); 2. The SHOEBILLS (Balaenicipitidae); 3. The HAMMERHEADS (Scopidae); 4. The STORKS (Ciconiidae); and 5. The IBISES (Threskiornithidae). Altogether there are fifty-nine genera and 115 species.

Family: herons, by H. Kramer

The HERON family (Ardeidae) is distributed over all parts of the world. The weight varies between that of the least bittern, which weighs just over 100 g, and the Goliath heron, which weighs 2600 g. There are twenty to twenty-one cervical vertebrae; the neck, which is almost immovable laterally, is S-shaped in flight (see Color plate, pp. 255/256) and is wedged in between the breast and the wings when the bird is at rest. The hind toe is long. There are twenty-four genera with sixty-three species, found mostly in the tropics and subtropics; they do not exist in the far north and in Antarctica.

Herons stalk their prey carefully or stand and wait for it, their long necks in the resting position, drawn back into an S-shape. The cervical spine is so constructed that a heron can thrust its head forward in a flash to stab its prey or to seize it with the beak. The neck is stretched maximally when the bird surveys the area. The changes of the position of head and body while the bird is stalking prey are balanced in such a way that the eyes are hardly moved; they remain focused on the prey. Koenig and Graefe's observations of European herons indicate a relationship between their structure and mode of life. The herons, which fish by standing in deep water, have longer legs than necks. The comparatively longest neck but shortest legs are found in the least bittern. When feeding, it holds onto reeds with them and must reach down far below its perch. In cattle egrets, which catch food in front of their feet, the neck and legs are of equal length. All herons have strong toes, which, when the bird climbs about in reeds, must firmly grasp the stalks; the toes are longest in the bitterns. The South American boat-billed heron *(Cochlearius cochlearius)* differs from all other herons by its wide beak and short neck.

Fig. 8-1. Common heron, greeting its arriving partner.

Herons do not oil their plumage with preen gland oil, for their preen gland is vestigial. Patches of powder downs on the breast, at the sides of the rump and in the groin area (not in all species) have taken over this function. Powder downs are feathers whose tips gradually crumble into a powder. They grow continually from the base and are not shed in the moult. Herons clean themselves by rubbing this powder into their plumage with their beaks and claws. This keeps the plumage water-repellent and removes fish remains and grease. The claw of the middle toe has a serrated edge which facilitates care of the plumage.

Fig. 8-2. Nuptial plume of a little egret.

The first downy plumage, which is generally gray or light brown, is immediately replaced by the first juvenile plumage. During the breeding season many species carry decorative plumes on the head, the base of the neck, and the back. As a rule the sexes are not distinguishable by the coloration of their plumage except for the smallest bitterns (genus *Ixobrychus*).

Herons are easily distinguishable from storks in flight by the way they draw their necks back. They fly well and with endurance in slow, quiet wing beats, but they do not soar like storks, even though some species know how to utilize thermals. Temperate-zone herons are generally migrants. But many species which breed in the tropics also migrate regularly according to the wet and dry seasons.

Most species of herons breed in colonies; many, however, are social even outside of the breeding season and spend the night in communal roosts. The nests are built in trees, in shrubs or reeds, or even on rocks. Usually both parents build the nest and relieve one another during incubation. The clutch generally consists of three to five eggs. Certain tropical herons lay only two eggs, but the smallest bitterns lay up to nine. The eggs are white, greenish, blue, or olive-brown, and a few species even have spotted eggs. The larger a species, the longer its incubation period. Little bitterns incubate about sixteen days, while the gigantic Goliath herons do so almost twice as long.

Fig. 8-3. 1. Common heron *(Ardea cinerea)*. 2. Great blue heron *(Ardea herodias)*. 3. Sokoi heron *(Ardea cocoi)*

Fig. 8-4. Once a male common heron has occupied a nest site, he attracts the attention of flying females by raising and lowering the head.

As a rule both parents tend the young. They bring food in their stomachs to the nest and then regurgitate it at first into the beaks of the young, and later onto the edge of the nest. Young of advanced age seize the beaks of their parents from the side and pull them down. The food is then regurgitated either directly into their beaks or onto the bottom of the nest.

The food of most species consists predominantly of fish, but also of frogs, salamander larvae, small mammals, and insects. Herons swallows their prey whole and digest fish almost completely. Indigestible remains, such as hairs and bones of small mammals and the chitin carapaces of insects, are regurgitated as pellets. The droppings of many herons kill plants near the nest trees and may also damage the trees themselves. However, plants with a particularly high nitrogen requirement, such as nettles, grow well beneath heron rookeries. Young herons leave the nest before they can fly properly. This occurs particularly early in species which breed in reeds or on the ground. After fledging, young herons often wander far in all directions, and they may then be seen far from home.

In spite of their large size, the eggs and young of herons are endangered by nest robbers. Heldt reports of the common herons of Schleswig-Holstein in northern Germany: "In 1952 in the colonies on the Staatshof and on Langenhemme, there were about twenty and fifty herons' nests, respectively, beside colonies of rooks. Since then the herons have had to give way to the rooks *(Corvus frugilegus)* because the latter have long specialized in plundering the nests of the herons when some disturbance causes them to fly off. Many herons' eggs which had been pecked indicated this as early as 1952, and this has since continued. The herons gave up both colonies." The adults have few enemies. Losses due to birds of prey are not very significant. The losses due to severe winters which occur in the Central European climate are quickly made up under normal circumstances. Man is the greatest enemy of herons when he pursues them or destroys their habitats.

The common heron

The most common heron species in Central Europe is the COMMON HERON *(Ardea cinerea;* see Color plates, p. 196 and pp. 255/256). In the northern parts of its range, it is a migrant which winters in southern Europe and Africa. In milder climates, such as in West Germany and the British Isles, it remains through the winter. It generally breeds in colonies which in Central Europe are, as a rule, located on trees. Colonies in reeds are known from the Netherlands, Denmark, and Lake Neusiedel in Austria. But even in reed nests, common herons show the behavior of tree nesters. For example, when in danger the young squat low in the nest, whereas those of the true reed herons escape from the nests.

After the return to the rookeries in the spring, the males first occupy the largest old nests. Later arrivals must be satisfied with small nests, or, if all are occupied, build a new one. When a male has selected a nest or nesting place, he calls incessantly from there for a female. By bending his neck, with his upright beak over his back, and then raising it up straight, he tries to attract the attention of females and lure them to the nest site. If this succeeds, the male begins to peck "symbolically" at twigs in the nest or on the nest site. The female soon pecks at the nest also. Mutual feather pecking and erection of the plumage lead to copulation, which takes place on the nest. Both partners take part in the building or repair of the nest. Until the young are fledged, the adults continually improve the nest.

Fig. 8-5. The bill-snapping movement of the common heron, which indicates readiness for mating.

The female generally lays three to five bluish-green eggs in two-day intervals into the shallow nest. In Germany, clutches are often complete as early as the end of March. Both parents incubate and usually start incubation as soon as the first egg is laid. The last youngster hatches after about two months. An adult which is about to alight on the nest "salutes" by raising its feather crest and greets its mate with special calls; its partner replies by stretching the neck up. However, even when the partner is absent, the returning bird salutes upon approaching the nest. When the young are larger, they too greet the parents.

Both parents feed the young, as was described earlier. At first the young chatter when any herons approach the nest. Later they learn to distinguish their parents from other herons, and they no longer chatter when strange adults approach. When the young are about three weeks old, they struggle so much for the best place that the young which were hatched last are pushed back and so much mistreated by their older and stronger siblings that they often perish. Because of this, only three young survive from large clutches.

When they are thirty days old, the young leave the nest for the first time. But only when they are about two months old do they fly well enough to finally depart from the colony. They are sexually mature at the end of their second year, but a number of females already reach this stage after the first year. Common herons can become quite old. One which was banded at the Rossiten Bird Observatory lived for over twenty-four years.

Other species of birds, such as cormorants, rooks, carrion crows, and raptors like the peregrine falcon *(Falco peregrinus)* or the black kite *(Milvus migrans),* often breed next to heron colonies.

Some heron rookeries are quite old. The oldest one in Germany is probably the one near the castle of Morstein on the Jagst River in Baden-Württemburg in southwest Germany. Old chronicles state that in 1586 it was already "many centuries old." There was even a "heron war" fought over this colony. The Baron of Crailsheim and the Margrave of Ansbach fought over the ownership of the herons which, in those days, were game reserved for the nobility. Hunting herons with falcons was particularly popular, and the heron population was therefore protected by strict laws. As late as the eighteenth century, the taking of young herons and destruction of the eggs was punished in the Palatinate by a fine of ten guilders.

The zoologist and nature writer Hermann Löns reports on the heron rookery of Wathlingen in Lower Saxony: "At one time it was cared for lovingly, protected from all disturbance, and a record of all occupied nests was kept. This was in the days when the cunning birds were hunted with falcons. Many a wild, colorful scene may often have taken place there, when a resplendent troop of riders galloped through the marsh, causing the water to spurt up high. When the falconer's practiced eye saw a heron taking off, he would remove the falcon's colorful hood and, accompanied by a sonorous hunting call, throw it into the air in the direction of the game. The hunt would then rush through swamps and sand after the heron."

These hunts for herons were not always so dramatic. The heron brought down by the falcon was not always killed. The Archbishop and Elector of Cologne, Clemens August, had a gyrfalcon which, in 1736, "had caught about thirty herons by itself." On June 19, 1738, this falcon (called the "Queen Gyrfalcon") brought down a heron which carried four bands, each with the name of the Elector and the dates 1725, 1728, 1734, and 1734. In the course of thirteen years, the Elector had therefore brought it down five times. Now he placed a band with the date 1738 on it and set it free, "but before this same heron was set free, it laid an egg in the falconer's garden." There is a picture of this heron with its five bands in the castle of Brühl. In the nineteenth century, the attitude towards herons changed. Löns reports: "For a while the secretive fishing birds were left unmolested out of habit, allowed to build their nests and rear their young, but fanatical materialists spoke too much against this 'pest of fisheries.' There were outings with guns of all sorts; there was good eating and better drinking, and the young herons were shot down from the old oaks by the hundreds and left as food for foxes and maggots."

Fig. 8-6. The small little egret and the large common heron prefer different depths of water and prey of different size in the same swamp areas. Thus they are not in competition.

In 1961, the Wathling colony mentioned by Löns still contained twenty-seven nests. A census in 1960/61 for Germany west of the Oder and Neisse Rivers shows a total of 6700 nests. Since then the number has probably decreased. Although hunting laws have at last been improved in favor of the heron, the increasing pollution of the waters creates new dangers for the herons. In the northern Rhine district, where there were 190 nests in 1961, there were less than fifty in 1966. It is, therefore, time for the common heron to be protected all year round.

In regard to the supposed damage done to fisheries by the common heron, Otto Koenig remarks: "The widespread notion of damage to fisheries cannot be maintained in the face of modern scientific data. An adult common heron consumes 330 grams of food per day. One third of this consists of fish, but two thirds consists of the larvae of diving beetles and dragonflies which are injurious to fish, and of many small rodents. These herons feed especially large numbers of field mice to their young. Dead ground squirrels *(Citellus)* have also been found in herons' nests.

Herons only catch animals swimming at the surface, for they do not like to immerse their beaks more than ten centimeters into the water. Thus the catch is mainly bleak *(Alburnus)* and small whitefish. The economically valuable carp is in little danger, since it is a bottom fish, and eels which are well hidden and largely nocturnal are not significant as prey. The motionless lurking pike is almost always overlooked by herons, which react to movement. In fish hatcheries, whitefish and other less valuable species are undesirable since they take too much food from the more valuable species. As destroyers of bleak and whitefish, herons are beneficial to fisheries. At one time bleak were caught by the ton for the manufacture of "pearl essence"; today, however, this species has no economic value. Furthermore, herons rarely catch fish longer than twenty centimeters. When larger fish are taken, these are sick, wounded, or were dead and floating on the surface.

Herons are not only beneficial to man by catching mice and countless water insects destructive of fish, but also because of their excrements. In breeding colonies, such as Lake Neusiedel, masses of fecal matter, rotting eggs, and dead young and adults fall into the water. This material is important for the increase of plankton, the basis of food of all young fish. Thus herons and spoonbills which breed in the same area become important resources for the fish industry."

On the Elbe River, common herons prey on a species of crab which fishermen regard as a nuisance. That herons do not avoid proximity to humans is shown by the rookeries established by wild herons in the Amsterdam Zoo and the Skanen Zoo near Stockholm.

▷

1. Goliath heron *(Ardea goliath)*; 2. Purple heron *(Ardea purpurea)*; 3. Great white egret *(Casmerodius albus)*.

▷▷

1. Mangrove heron *(Butorides striatus)*; 2. Common heron *(Ardea cinerea)*, a. adult; b. juvenile; 3. White-cheeked heron *(Ardea novaehollandiae)*. 4. Pied heron *(Notophoyx picata)*.

▷▷▷

1. Squacco heron *(Ardeola ralloides)*; 2. Cattle egret *(Ardeola ibis)*; 3. Whistling heron *(Syrigma sibilatrix)*; 4. Little blue heron *(Florida caerulea)*; a. adult; b. juvenile; 5. Little egret *(Egretta garzetta)*; 6. Chestnut-bellied heron *(Agamia agami)*.

The purple heron

In addition to the common heron, the genus *Ardea* is represented in Europe by the PURPLE HERON *(Ardea purpurea;* see Color plates, pp. 25 and 195). In Central Europe, it breeds in Czechoslovakia and Hungary and on Lake Neusiedel in Austria, where there were 273 nests in 1961. It also breeds regularly in Switzerland and the Netherlands. In recent years in Germany it has bred on Feder Lake in Swabia and on the Chiemsee, a lake southeast of Munich.

The purple heron prefers extensive reed beds for breeding and feeding. It has especially long toes with which it can readily grasp reed stalks. It wades only rarely with its relatively short legs. It prefers to fish in small ditches in reeds near the shore. Young purple herons leave the nest much earlier than young common herons, and when in danger

1
3
2

1
2a
3
2b
4

1
3
4 b
4 a
5
6
2

1
2
3
4
5

they escape into reeds. There they remain motionless until the disturbance has passed. Tree nests of this species are rare. Where small willow bushes are scattered through the reed beds, the nests are built on them. The colonies are generally smaller than those of the common heron. In the winter, purple herons move from their northern breeding areas to the steppes of Africa.

Relatives of the common heron

All other herons of the genus *Ardea* live outside Europe. The WHITE-CHEEKED HERON (*Ardea novaehollandiae;* see Color plate, p. 196, and Vol. X) occurs in Australia and New Zealand, and from there to New Caledonia and the Indonesian island of Lombok; the WHITE-NECKED HERON (*Ardea pacifica*) is found in Australia and Tasmania.

Fig. 8-7. Purple heron (*Ardea purpurea*)

The GREAT BLUE HERON (*Ardea herodias*) is closely related to the common heron and occupies its ecological niche in North America. It is somewhat larger than the latter, and has a black belly and reddish-brown thighs. It builds its nests on trees. The peculiar behavior of this heron just before the breeding season was described by the famous American bird painter and ornithologist, Audubon: "Just before the breeding season the male heron's behavior in looking for a mate is extraordinarily interesting. Before sunrise a number of them arrive on the edge of the wide sand banks or the savannah. For several hours they come from different directions, one after another, until forty or fifty are assembled. One can hardly believe there are so many in the area. In Florida I have seen hundreds which gathered like this in one morning. Their plumage is now at the peak of its beauty. Apparently there are no immature birds among them. The males step about with great "dignity" in postures which challenge their rivals. By croaking, the females try to attract the males' attention. They also utter "flattering" sounds and, as every male wishes to please them, he must expect the animosity of many opponents which, without much hesitation, open their mighty beaks, spread their wings, and rush at their enemy.

Every attack is carefully parried. Each stroke is met with a counterstroke. One would think that a single well-directed blow with the beak would be sufficient to kill, but the blows are fended off with the skill of experienced fencers. Although I have observed these birds for half an hour as they fought on the ground, I never saw one killed on these occasions. But I did see some fall down and get trampled several times, even after the breeding season had begun.

When these fights are over, males and females leave the "arena" in pairs. They are now paired for the coming breeding season; at least that is what I presume, since I never saw them twice at the same place. After pairing, they become fairly peaceful."

Later reports from the Florida coast of the Gulf of Mexico confirm this type of ground display. But it has not been observed in inland populations. Perhaps this is because the participants are mostly mi-

◁

1. European or black-crowned night heron (*Nycticorax nycticorax*); 2. Tiger bittern (*Gorsachius melanolophus*); 3. Boat-billed heron (*Cochlearius cochlearius*). 4. Little bittern (*Ixobrychus minutus*). 5. Eurasian bittern (*Boraurus stellaris*).

grants which arrive somewhat late in the breeding area, and therefore have a shortened courtship period. In South America the heron is represented by the lighter SOKOI HERON (*Ardea cocoi;* see Fig. 8-3).

The BLACK-NECKED HERON (*Ardea melanocephala;* see Color plate, pp. 184/185) is a close relative of the common heron in Africa, south of the Sahara Desert, and in Madagascar. Where both species occur, they sometimes breed in mixed colonies. However, they prefer different hunting areas. The black-necked heron looks for insects and rodents in dry grassland, whereas the common heron looks for fish at the edge of water. In the north of its African range, the black-necked heron is a migrant. During the rainy season, it breeds in the Sudan and migrates south during the dry season. The KING HERON *(Ardea humbloti)* breeds only in Madagascar.

The GOLIATH HERON (*Ardea goliath;* see Color plate, p. 195), with a length of 1.40 meters, is the giant among herons. It inhabits the swamps of tropical Africa and does not breed in colonies. It lives singly or in pairs outside the breeding season as well. The other two giant herons, the IMPERIAL HERON *(Ardea imperialis)* of Southeast Asia and the DUSKY GRAY HERON *(Ardea sumatrana),* which is distributed from Burma to northern Australia, are only slightly smaller.

The great white or common egret

The GREAT WHITE EGRET or COMMON EGRET (*Casmerodius albus;* see Color plates, pp. 186 and 195) of North America is one of the most magnificent of the white herons or egrets. It inhabits both the Old and New Worlds; its best known breeding area in Central Europe is at Lake Neusiedel in Austria. In 1961, 329 pairs bred there in eight colonies. There are other colonies in Hungary as well. In Europe this species breeds in reed beds, but in other areas they nest in trees.

In adult common egrets of the European subspecies, the beak is yellow in the winter and black in the breeding season. These egrets are well adapted to a sunny climate. While other herons at Lake Neusiedel seek out shade around noon, common egrets remain standing in the sun, showing no concern. Long loose decorative plumes are formed on the back as part of the breeding plumage. Around the turn of the century, these were in high demand as fashionable decorations for ladies' hats. Since these plumes are best developed in the breeding season, the adult birds were unfortunately killed at this time. The eggs were left to rot in the nests, and the orphaned young starved to death and entire colonies were destroyed. An American hunter shot forty-six egrets in a single day, and said, "We could have got more, if we had not become tired of shooting."

The feather trade involved millions of dollars. In 1903 a hunter could get $32 for an ounce of feathers; an egret plume was thus worth twice its weight in gold! In 1902, 1608 packets of egret plumes were sold in London, each weighing about thirty ounces; hence the total weight was about 48,240 ounces. To get one ounce of plumes, four egrets had

Fig. 8-8. Great white egret or common egret *(Casmerodius albus).*

to be killed. This meant that to supply the demand for plumes for one year, 192,960 egrets had to be killed, and about three times as many young or eggs perished. Fortunately, the plumes ceased to be fashionable, and the breeding populations were able to recover.

These egrets are now protected throughout the year at Lake Neusiedel, but the breeding colonies were endangered by systematic reed cutting. In order to determine the size and number of the colonies and their distribution over the area, Otto Koenig made helicopter flights over the reed beds. The egrets soon became used to these observation flights. The nests were photographed several times and the photographs were used for counts, as Bernhard Grzimek has also done in counting African flamingoes. Even the number of young in the nests could be determined in this way. Thanks to the efforts of Otto Koenig and the cooperation of the landowners, the first sanctuaries were established at Lake Neusiedel, in which the breeding colonies now remain undisturbed.

The yellow-billed egret

The YELLOW-BILLED EGRET *(Mesophoyx intermedia)* inhabits the hot areas of the Old World. In Japan, together with the little egret, it is a characteristic bird of the warm, rice-growing areas. In the huge heron rookery of Sagiyama near Tokyo, it is, next to the night heron, the most common of all species breeding there. The German ornithologist Jahn reports on this colony, which has been in the same location at least 170 years: "The local peasant families have always protected the birds. They were paid for this by the shoguns (the generals of the emperor) who passed through the village on their way to the temples of Nikko. Today also, the peasants anxiously protect the birds and tolerate no strangers there. The nests are like those of other species, found on every suitable site in trees of all kinds, at different heights. However, breeding in medium-sized giant bamboos is typical."

The reef heron and Chinese egret

The egrets of the genus *Egretta* are somewhat smaller than the large white egret. The EASTERN REEF HERON *(Egretta sacra),* distributed from Burma to Japan and New Zealand, is a coastal bird. Jahn was able to observe them in Japan, where they breed on inaccessible coastal cliffs. At low tide they search the exposed rocks for small marine animals. Reed herons occur in two color phases, a white one and a more common blue one. A close relative is the CHINESE EGRET (⏀ *Egretta eulophotes),* an endangered species of southeast Asia.

The SNOWY EGRET *(Egretta thula)* is an American representative of the genus *Egretta.* It is reported to attract and then seize small fish by standing in shallow water and making slow movements with its yellow toes. It can also take prey on the wing. Because of its beautiful plumes, it was formerly heavily hunted.

This was also true of one of its Old World relatives, the LITTLE EGRET (*Egretta garzetta;* see Color plate, p. 197). The Hungarian ornithologist, Sterbetz, writes: "The decorative plumes of the large white and little

The little egret

egrets served the Hungarian nobility and also the Turks, who had close economic and political ties with Central Europe in the Middle Ages, as expensive head decorations. High prices were paid for a bundle of plumes, for it was not easy to reach the breeding colonies in trackless swamps and obtain the plumes. Unfortunately, the colonies on the Danube and Theiss Rivers became involved in the international plume trade. In London, the center of the European plume market, egret plumes were valued very highly. Baron Kalbermatten, in particular, organized plume collecting expeditions which resulted in the complete destruction of egret colonies." All birds that were useful were ruthlessly exterminated, and the orphaned young died in masses.

Fig. 8-9. 1. Little egret *(Egretta garzetta)*, threatened or extinct in many parts of its breeding range. 2. Snowy egret *(Egretta thula)*.

When the destruction of the birds was stopped by strict laws in Hungary and the whole Carpathian area, the plume trade still flourished for a long time on the lower Danube. Almássy, in his description of the Dobruja, tells about the large plumage markets in Braila and Galata; egret feathers were exported from there to the west.

Drainage projects and the regulation of river beds further reduced the egret populations in eastern Central Europe. Since the beginning of this century, this species could not be considered a regular breeding bird of Hungary. But since 1947, it is once more at home there. The establishment of rice fields has supported its return, and the birds have spread very much. In the Danube delta as well, protective laws have ensured that the population is no longer endangered, and in southern France the little egret is a characteristic bird of the rice fields.

Well known colonies of little egrets in southwestern Europe are in the Camargue (southern France) and the Coto Doñana in southern Spain. In these colonies they generally breed together with night herons, purple herons, and cattle egrets. Such mixed colonies have also been observed elsewhere; evidently mixed breeding communities are frequent among herons.

While other herons catch their prey by standing still until it comes within reach, the little egret stalks its prey. It approaches it with cautious steps and then seizes it, or it scares up prey hidden in mud with a strong trembling movement of the foot. Long-legged herons hunt in deeper water, but the little egret prefers the shallows. As with other species of white egrets, dark-colored birds occasionally occur. Outside of Europe, particularly in eastern Africa, this color deviation is somewhat more common.

The two following species live in the Old World on the coasts of tropical seas. The SEA HERON bears the scientific name *Egretta dimorpha* ("two-shaped egret") because it occurs in both pale and dark slate-gray color phases. It breeds in Madagascar and Aldabra. The REEF HERON *(Egretta gularis)* breeds on the coasts of western and eastern Africa. Like the large white egret, it has especially long nape feathers. Its three different color phases are blue-gray with a white throat, spotted white,

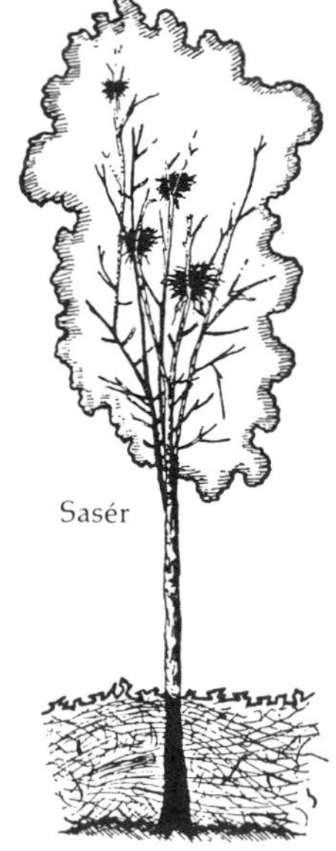

Fig. 8-10. Different nest sites of the little egret observed by Antal Festetics at Kisbalaton and Sasér in Hungary.

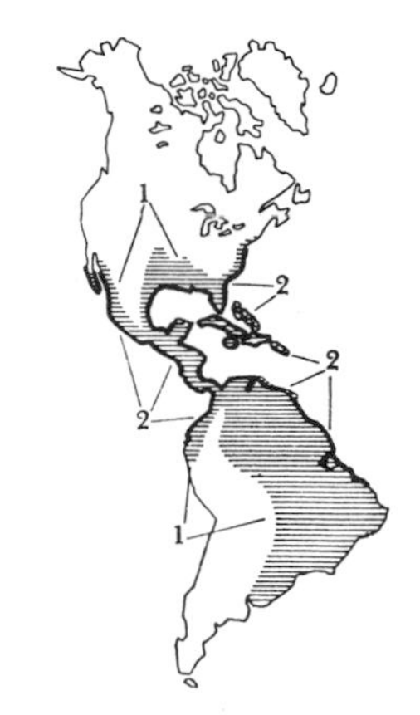

Fig. 8-11. 1. Little blue heron *(Florida caerulea)*; 2. Tricolored heron *(Hydranassa tricolor)*.

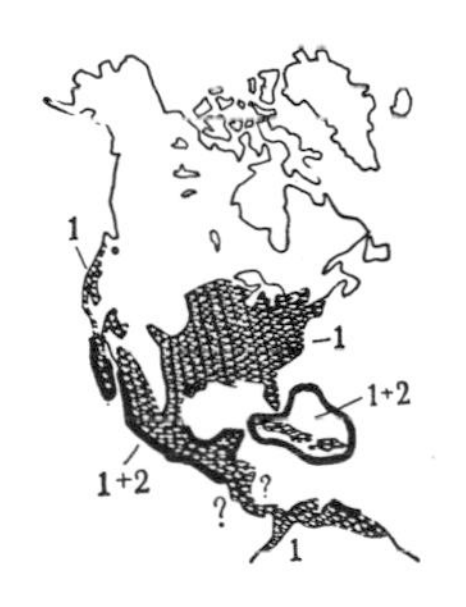

Fig. 8-12. 1. Green heron *(Butorides virescens)*. 2. Reddish egret *(Hydranassa rufescens)*.

Fig. 8-13. Squacco heron *(Ardeola ralloides)*.

and white. Birds of the dark phase are said to predominate in the northern part of the range.

Certain species are related both to the white egret of the genus *Egretta* and the green and crested herons (genera *Butorides* and *Ardeola*). To these belongs the BLACK HERON *(Melanophoyx ardesiaca)* of Africa and Madagascar. It has bright orange-colored toes. When hunting, it stands motionless in the water, the beak directed downward and the wings spread forward over the head, so that the tips of the feathers touch the water. It is assumed that fish seek out the shadow of this "umbrella" and are then seized by the heron. But it is also possible that the bird can see its prey better this way, since it would be undisturbed by reflections on the water surface. Its flight is quite fast. The breeding range of the PIED HERON (*Notophoyx picata*; see Color plate p. 196) extends from eastern Indonesia to northern Australia. The American LITTLE BLUE HERON (*Florida caerulea*; see Color plate p. 197) is, with the exception of the following species, the only inland heron which regularly has two color phases, a blue one and a pure white one. The young of this species are always white.

The TRI-COLORED HERON *(Hydranassa tricolor)* is very sociable; thousands may breed in a single colony. The REDDISH EGRET *(Hydranassa rufescens)* formerly occurred in Florida, where Audubon had observed the display of these beautiful birds. Since this species occurs in two color phases, and since groups of twenty or more gather for displays, they make a colorful picture for the observer. He sees "...how a purple-red male displays to a white female, while nearby a white male courts a purple female. And not far away, a white and a purple-red pair display." On the mainland the reddish phase definitely predominates, with less than ten percent of the birds being of the white phase. On the other hand, on the Bahamas there are populations of which almost ninety percent are of the white form.

When a reddish egret has spotted a swarm of fish, it runs with long, far-reaching strides through the shallow water. In so doing it tilts its body deeply forward and raises its head and neck. Abruptly it stops, makes half a turn, raises the wings, runs another few steps, hops into the air, and turns again. Suddenly it stabs at its prey. Each bird has a hunting territory which it defends against others of its species.

The colorful CHESTNUT-BELLIED HERON (*Agamia agami*; see Color plate p. 197) lives in tropical America from southern Mexico to Peru and Brazil. It is about as large as a night heron, having a very long neck and rather short legs. Its pointed beak is long and thin.

The herons of the genus *Butorides* are relatively inconspicuous. The AMERICAN GREEN HERON *(Butorides virescens)* breeds in the temperate zone in single pairs or small colonies, on trees or bushes near rivers and lakes. In the tropics and subtropics, it prefers coastal mangrove forests. When disturbed or excited, it flicks its tail. During the breeding season, its iris, which at other times is yellow, becomes orange-red, and the legs, which are yellow at other times, also turn this color. It has been observed that they save themselves from birds of prey by diving. Even young green herons can swim. As soon as they are fledged, they occasionally dive for prey. The GALAPAGOS HERON *(Butorides sundevalli)* is black, in contrast to other herons. It takes its name from its home, the Galapagos Islands in the Pacific.

Adjoining the range of the green heron to the south is that of the MANGROVE HERON (*Butorides striatus*; see Color plate p. 196). Its range reaches from Panama to Argentina and, in the Old World, from Africa through southern Asia to Australia and Polynesia in the south and Korea and Japan in the north. As the name indicates, it is found in mangrove swamps of tropical coasts. Inland, it breeds near waters shaded by trees or bushes. The Japanese ornithologist, Enomoto, found small and larger colonies of this crow-sized greenish heron on Gingko trees near the city of Osaka.

The RED-BELLIED HERON *(Erythrocnus rufiventris)* lives in east and southeast Africa. It inhabits reed beds. When disturbed, it only flies a very short distance and then alights again in the reeds. There it is said to remain motionless, like a bittern, but not in the bitterns' characteristic posture (see Fig. 8-19).

The squacco herons

The small crested SQUACCO HERONS of the genus *Ardeola* live in the Old World. Their necks, beaks, and legs are relatively short. They are sociable and show little fear of man. SQUACCO HERONS (*Ardeola ralloides*; see Color plate, p. 197) look inconspicuous when perching, but offer a striking sight when they spread their white wings. They are particularly numerous in the great heron rookeries of the Danube delta in southern Europe. As agile as rails, they can slip through reed beds. Their food consists mainly of insects. In the breeding season, the beak of this heron becomes a deep blue and the greenish-yellow legs become red.

The MADAGASCAR SQUACCO HERON *(Ardeola idae)* breeds from October to December, and then has a white breeding plumage. It leaves Madagascar after the breeding season and migrates to Africa. The INDIAN POND HERON or PADDY BIRD *(Ardeola grayii)* is distributed from the Persian Gulf to Malaya, while the CHINESE POND HERON *(Ardeola bacchus)* breeds in southeast Asia; the JAVANESE POND HERON *(Ardeola speciosa)* is restricted to the Indonesian islands.

Fig. 8-14. 1. Cattle egret *(Ardeola ibis)*; The arrows show recent immigrations.

Günter Niethammer, who has travelled all over Africa by jeep, writes about the CATTLE EGRET (*Ardeola ibis*; see Color plate, p. 197, and Vol. XIII): "It is found as far north as southern Spain, but much more commonly in Morocco, where it is seen in the fields everywhere in the winter, generally in sizeable groups. On cloudy days its shiny white plumage looks like snow patches. I saw this bird at least as often in the summer in Egypt, where it is particularly at home in Cairo and is strictly protected by the people. The cattle egret, even more than the scavenging black kite, is the characteristic bird of Africa's largest city.

They are found everywhere, not only in northern Africa but also south of the Sahara. On the Shari and Logone rivers, they are so common that the branches of the trees on which they roost bend under the load of the birds. In the northern Cameroon I once saw some mighty trees at a distance; they appeared to be covered with snow. About 5000 cattle egrets were sitting there beside one another. When I disturbed them, they took wing almost simultaneously and it was an incomparably impressive sight when, high above the trees, they caught the last rays of the sinking sun and reflected them many thousandfold from their bodies down to the ground, which was already shrouded in dusk.

The cattle egret is even found in the midst of the African jungle. One bird lived with Dr. Albert Schweitzer when my companion Michael Abs and I visited Lambarene on January 9, 1959. It became much more common south of the dense forest zone, in Angola right down to South Africa. This bird was our constant companion on the black continent during our long car journey from Morocco as far as Lüderitz bay in southwest Africa, and homeward via Pretoria, Rhodesia, Tanganyika, and Kenya."

It is doubtful whether the cattle egrets were already one of the most common African birds even a hundred years ago; in any case, they have increased considerably in recent years. This is because of their intimate relationship with domestic animals, from which they draw great benefit and which explains their common name. They show a preference for walking among grazing cattle, often sitting on their backs in order to catch insects and larvae. They even follow ploughs on cultivated land. It is understandable that this bird, which is beneficial to cattle and man, is not harmed but rather protected wherever possible. With the increase of cattle, the number of cattle egrets in the whole world rose from 695 million in 1939 to 800 million in 1953, and in Africa alone it increased from eighty to ninety-five million; hence the cattle egret had a good chance to further increase its population, to double and even multiply it several times over. The number of its descendants evidently increased faster than the number of cattle or its food supply, for the species was forced to enlarge its African range in all directions, so it is now only absent in true deserts and high mountains.

But even the gigantic African continent became too small for them. So one day, about thirty years ago, the birds crossed the Atlantic and discovered South America. There they found large herds of domestic animals as yet unaccompanied by birds of its type. The cattle egret thus found a vacant niche for itself and an abundant food supply.

Under these favorable conditions they were able to occupy extensive areas of northern South America, as well as Central and North America, in only three decades. Hence there was an astonishing increase in their numbers, for today there are many thousands, or even tens of thousands, of cattle egrets in the New World. Since 1948 this egret has also appeared in Australia, and in recent years it has moved from there to New Zealand. With such an ability to expand its range, it is surprising that the species has not advanced further into Europe. There has been no movement to the northeast from its traditional breeding places in southern Portugal and southern Spain, nor from those in northwest Africa and the Near East to the northwest. One could imagine that it would find a suitable new home in Central Europe. Single birds have wandered over the Balkan Peninsula and Hungary and north as far as Britain and Denmark, but they have disappeared again. The same thing happened to many cattle egrets introduced from India into England (also some into Ireland) since 1930. These attempted introductions have evidently been unsuccessful so far, although the bird seems adapted to living there and would help us to enliven a landscape which is becoming ever more lifeless. There have so far been only occasional breeding records in the Camargue in southern France. The more severe winters may possibly work against settling in Central Europe, for in the winter the birds would not find enough food. They would have to migrate to warmer countries in the winter, at least as far as southern Spain, and return in the spring to their new home in Central Europe. It is questionable whether a settlement of this beautiful bird could succeed in Central Europe for, unlike the Egyptian cattle egret, it would have to acquire the habits of a real migrant and, like these birds, find its way back as well. There are, however, other cattle egret populations in Africa which migrate regularly.

The cattle egret originally accompanied the herds of large wild mammals, as it still does in the wildlife sanctuaries of Africa. It feeds on the insects flushed up by their feet.

The well-known Viennese ethologist Otto Koenig has succeeded in breeding these egrets in captivity. In a large flight cage in the Biological Station Wilhelminenberg in the Vienna Woods in Austria, he was able to create an artificial breeding colony. The egrets found nesting material and food close together there. Thus they did not have to be away from the nesting place for two to four hours to seek food as in the wild. The young learned neither to wait for food nor to get their own. They begged continually from their parents and followed them to the food dish. The whole cage became, as it were, one big nest in which food was constantly passed from adults to young. With this state of affairs, the habitual greeting ceremonies of cattle egrets disappeared. Otto Koenig describes this degeneration of habits due to

Degeneration of behavior due to "prosperity" in cattle egrets

"prosperity" as follows: "The young remained with the parents for the whole year. In the following year, mating groups of three or four birds, consisting of adults and their young, were formed. Each male mated with practically every female within these groups. The young of the previous year did not seek food for their own young, but begged it from their parents. As all birds were continually in the colony, incubation relief was continuous. The eggs were constantly turned, and the egrets owning one nest sometimes incubated with two or three, one on top of another. The highest-ranking individual, the "grandmother," would actually sit on the nest. When a male tried to build a new nest away from the old one, his "sister" took all the nest material, carried it back to the parental nest, and tried to build it into the latter.

In later generations, with increased population density, the distinctions between families disappeared. One bird would fight with any other. A male would randomly copulate with any female. The number of hatched eggs declined drastically. Newly introduced, strange cattle egrets behaved normally during the first year, but then their behavior also degenerated." Koenig considers it possible that in the future the egrets in his flight cage will develop behavior patterns which are suitably adapted to their new conditions.

The boat-billed heron

The BOAT-BILLED HERON (*Cochlearius cochlearius*; see Color plate p. 198) lives in tropical America. Because of its peculiar wide, flat beak, it was formerly not grouped with the herons. Its beak, however, shows basically the same structure as that of night herons. It is, therefore, now included with the true herons. Like other nocturnal birds, it has particularly large eyes. It erects its long crest in courtship display, and males and females clatter vigorously with their beaks. The female crouches and stretches out her neck, while the male walks around her.

The whistling heron

The WHISTLING HERON (*Syrigma sibilatrix*; see Color plate p. 197) owes its name to the high-pitched whistle which it produces. It inhabits southern Brazil and Argentina. A group of medium-sized herons which are predominantly active at night and are therefore called night herons will be discussed next. Their heads, legs, and beaks are relatively short.

The night heron

The YELLOW-CROWNED NIGHT HERON *(Nyctanassa violacea)* does not hunt exclusively at night. Its face is black and white, while the rest of the plumage is bluish-gray. It has a preference for breeding in coastal areas and on islands. "Their food consists mainly of land crabs, which they catch skillfully, kill, and break into pieces," Maynard reports. "They eat all species, perhaps with the exception of the large white crab, a species which is often fourteen inches long and weighs about one pound. This animal is evidently too strong and too thick for the herons to overcome. However, they kill the black crab which, with its pincers, is almost a foot long. Their favorite food, however, is a smaller species of crab which resembles the black crab in shape. Because of the herons' predilection for it, the people of the Bahamas call it the "heron crab." It is very common. A group of crabs which are presumably safe from herons are the hermit crabs. They withdraw into their borrowed snail shell and guard the entrance with their pincers."

The CAPPED HERON *(Pilherodius pileatus)* breeds in eastern Panama and tropical South America. It is about the size of the European night heron, and is shiny white with black on top of its head. The NANKEEN NIGHT HERON *(Nycticorax caledonicus)* lives east of the range of the European species. It is somewhat larger than the latter, and has a chestnut brown back and reddish-yellow underside.

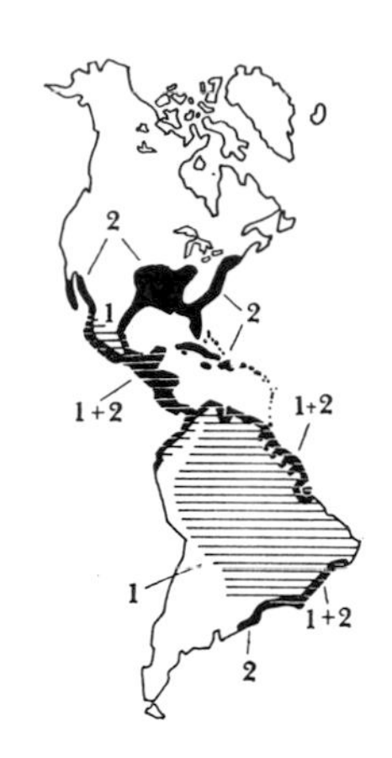

Fig. 8-15. 1. Boat-billed heron *(Cochlearius cochlearius)*; 2. Yellow-crowned night heron *(Nyctanassa violacea)*.

The EUROPEAN OR BLACK-CROWNED NIGHT HERON (*Nycticorax nycticorax*; see Color plate p. 198) inhabits a large part of the world. It does not acquire adult plumage until the third year. The white-spotted brown juvenile plumage is followed by a brownish intermediate plumage; the birds, however, sometimes breed at this stage.

As a rule, the night heron is active during the twilight hours and the night, but in the breeding season it also hunts food in the daytime. It prefers frogs, fish, insects, and tadpoles, as well as young birds, particularly those of other species of herons, and other warm-blooded animals.

Fig. 8-16. 1. European or black-crowned night heron *(Nycticorax nycticorax)*. 2. Nankeen night heron *(Nycticorax caledonicus)*.

Night herons breed in colonies of often no more than ten nests. The nests are on trees, bushes, or in reeds. The three to five blue-green eggs are incubated by both parents for almost a month.

In central Europe, night herons breed in Hungary and Austria, but within the last twenty years they have also bred in Bohemia and the Netherlands. They have been established in Bavaria in Germany since about 1950.

The interesting courtship behavior of the night heron was studied by Konrad Lorenz; Otto Koenig and his students studied its breeding behavior in particular. B. and L. Clormann report on their observations of a colony in lower Austria: "During the day the birds generally approach the colony flying at a considerable height. Just before or above the colony, almost every heron once again utters its flight call, glides down without wing beats to just above its breeding place, and then, if the young can already climb about and are perhaps out of the nest, circles the bushes looking for them. Finally the birds call and settle near their young. During dark nights the herons approach the colony flying low, only a few meters above the ground, but on moonlit nights they also fly high. In the dark the adults circle less around the colony, for the young are in the nests or at accustomed feeding places at night and need not be searched for.

The nests are scattered over about twenty approximately three-meter-high bushes. There are often loud squabbles within the colony both day and night, since night herons are very intolerant of their own kind outside the family units. Yet they do not like to fly away alone. Before a heron leaves the colony, it therefore often seeks a companion by flying from one bush to another, stimulating itself and others to fly away. These short preliminary flights are very characteristic of their colonial life. A bird rises up from its perch with the flight call, whereupon some others follow, also calling. The visual and auditory effect of the departing members of the species is sometimes so strong that others, which at first remained behind and then took off after all, make every effort to catch up with the already distant birds. It may be that the departing bird is followed by a string of ten other birds, each one eager to catch up with the one ahead, in which they generally succeed. This is particularly common in the early dawn, a time when the birds are in a strong mood to fly away and the young are already begging vigorously. The flight call of departing herons can even cause others, which have just come back from seeking food and are landing near their young, to join a departing group and undertake a long flight.

During the period of observation, ninety percent of the young acquired the ability to move, at least fluttering from one group of bushes to another. Gradually the peripheral nests were deserted and the young went to a narrow strip of bushes in the center of the colony. Here each youngster acquired a perch by fighting for it and was fed there by its parents. These movements of the young explain the adults' frequent circling over the colony. Apparently it was not always easy for them to find their young. Evidently the young were safer in the middle of the colony.

During the last week of observation the colony gradually broke up. The nesting area lost its attraction. The young left the colony, at first with their parents, and later alone, and then they searched for their own food, returning to the colony only at the approach of darkness and later not at all. This tendency was noticeable earlier in the adults. As soon as the young no longer needed constant protection, and particularly shade provided by the adults, most of them left the colony. While the young found cover in bushes, their parents sought protection from the noontime heat in the shady crowns of tall trees nearby. The increasing number of departures and the decrease of arrivals around noon clearly indicated the break up of the colony. By this time only a few small young that still needed constant care remained in only a few nests."

The night heron is a migrant over most of its breeding area. Birds of this species were seen in the Zambesi river in Africa, drifting downstream on floating islands of plant material.

Herons of the genus *Gorsachius* resemble the tiger herons, discussed below, in structure, calls, and mode of life. The MAGNIFICENT NIGHT HERON *(Gorsachius magnificus)* lives on the south Chinese island of Hainan and in Fukien, while the TIGER BITTERN (*Gorsachius melanolophus;*

see Color plate, p. 198) occurs in southern Asia, where it lives mainly on insects. The JAPANESE BITTERN *(Gorsachius goisagi)* breeds in eastern China, Formosa, and Japan. During daylight hours it sits quietly in the crowns of forest trees. Because of its nocturnal, secretive way of life, this solitary bird is rarely seen.

The tiger heron

The following herons are generally classified in the group of TIGER HERONS or TIGER BITTERNS. Like bitterns, they live solitarily, take up a concealment posture when in danger, and have a loud booming courtship call. Their plumage is also reminiscent of that of the true bitterns. However, in contrast to these, they have, like other herons, three patches of powder down; therefore, many taxonomists hesitate to place them with the bitterns and prefer to call them tiger herons. All live in the tropics.

The ZIGZAG HERON *(Zebrilus undulatus)* inhabits northern South America. It is only slightly larger than the little bittern, and has a short beak and a dark reddish-brown and black plumage. The BARE-THROATED TIGER HERON *(Tigrisoma mexicanum)* lives in Central America, and the FASCIATED TIGER HERON *(Tigrisoma fasciatum)* in South America. The LINED or BANDED TIGER HERON *(Tigrisoma lineatum)* lives in forested river shores in Central and South America. Heinrich Dathe, presently director of the Berlin Friedrichsfelde Zoo, was able to observe the behavior of a calling tiger heron in detail in the Dresden Zoo in 1941. The bird called for about nine weeks, from the end of March to the end of May, during the whole day and into the dusk; it was quiet at night, however. Dathe feels its call resembles the lowing of cattle even more than does the booming of the EUROPEAN BITTERN (*Botaurus stellaris;* see Color plate, p. 198). The WHITE-CRESTED TIGER HERON *(Tigriornis leucolophus)* of the west African rain forests and the NEW GUINEA TIGER HERON *(Zonerodius heliosylus)* belong to this group, as well as to the herons in the general sense.

Several species can be grouped as BITTERNS. All lead a concealed life in the shore vegetation of water, and, when endangered, assume the concealment posture described later. The smallest herons are in the genus of the LITTLE BITTERNS *(Ixobrychus).* In these, males and females have a differently colored plumage. Such sex differences do not occur in other members of the heron family.

The little bittern

The BLACK or MANGROVE BITTERN *(Ixobrychus flavicollis),* which occurs from India to North China and Australia, is a little larger than a turtle dove. There are several color variants in this species. Bushy shores of African waters are inhabited by the AFRICAN DWARF BITTERN *(Ixobrychus sturmii).* SCHRENK'S LITTLE BITTERN *(Ixobrychus eurhythmus),* which lives in Manchuria and Japan, prefers wet meadows and herb-covered shores. *Ixobrychus involucris* of South America is probably closely related to these two species.

Particularly closely related to one another are the three following species: The CHINESE LITTLE BITTERN *(Ixobrychus sinensis)*, a bird which breeds in east Asia where it inhabits the reed beds of shallow waters and flooded rice fields; the LEAST BITTERN *(Ixobrychus exilis)*, the world's smallest heron species, with two regularly observed color phases; and the OLD WORLD LITTLE BITTERN (*Ixobrychus minutus;* Color plate p. 198), which lives in a very secretive manner along still or very slow-flowing waters with shores overgrown by reeds, cattails, or willow bushes. The nests are built in reeds twenty to thirty centimeters above the water; sometimes they are in willow shrubs or even willow trees. On larger bodies of water, such as Lake Neusiedel, several pairs often breed only ten to twenty meters from one another, thus forming small colonies. However, this bittern's nests generally stand alone. Since they live very quietly, often on very small waters which are sometimes quite close to settlements, the species is more common than is often realized. The breeding population may, however, show marked fluctuations within short periods. The courtship call of the little bittern, a soft "wrroo," can only be heard at very close range, and the birds are rarely seen. They return from their winter quarters during the first half of May. The male begins nest building soon thereafter. If the female accepts the nest site, both continue to build, and both incubate. During the breeding season their beaks are a bright orange-red; nevertheless, adult males and females are readily distinguished since the female has a more strongly streaked camouflage plumage. The plumage of the young resembles that of the female. These bitterns move very slowly on the way to and from the nest. In the relieving process during incubation, the incubating bird first threatens the relieving partner and therefore must be pacified by the latter.

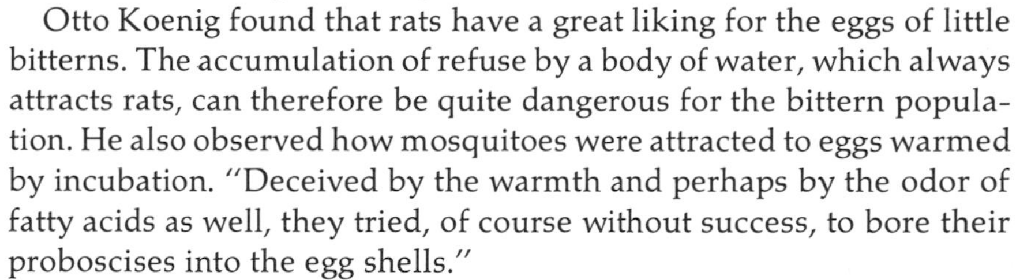

Otto Koenig found that rats have a great liking for the eggs of little bitterns. The accumulation of refuse by a body of water, which always attracts rats, can therefore be quite dangerous for the bittern population. He also observed how mosquitoes were attracted to eggs warmed by incubation. "Deceived by the warmth and perhaps by the odor of fatty acids as well, they tried, of course without success, to bore their proboscises into the egg shells."

Fig. 8-17. 1. Least bittern *(Ixobrychus exilis)*. 2. Little bittern *(Ixobrychus minutus.* 3. Chinese little bittern *(Ixobrychus sinensis)*.

The incubation period is at least sixteen days. The young leave the nest when only five to six days old, but at first they stay very close to it. At eight to ten days of age, they can climb about skillfully and, when one month old, they can fly.

The food of the little bittern consists largely of water insects; fish hardly make up a third.

When disturbed, they assume a concealment posture. Even small downy youngsters show this behavior. The neck and beak are stretched vertically upward, parallel to the reed stalks. This makes it difficult to detect the bird in the uniformity of a reed bed, particularly since it may remain immobile in this posture for a long time.

Kurt Gentz, who observed many little bitterns at pools in Saxony, believes this bittern takes on the concealment posture in order to look about and better see the cause of disturbance. Bitterns in this posture, however, have been observed bending slowly from side to side like the reeds when the wind blows them to and fro; the concealing significance of this posture should therefore not be underestimated. Apparently the readiness to fly is inhibited when this posture is assumed. When the birds incubate or brood, this inhibition is lacking; if they are surprised during these actions, they generally fly away.

The bitterns

The BITTERNS of the genus *Botaurus* are more than twice as large as the little bitterns. They also assume the concealment posture described above, and, moreover, they also have an impressive defensive threatening posture. They threaten with erected plumage, wings spread out and beak wide open, a posture similar to that of some owls, when they are approached too closely. All four species, the SOUTH AMERICAN BITTERN *(Botaurus pinnatus)*, the AMERICAN BITTERN *(Botaurus lentiginosus)*, the AUSTRALIAN BITTERN *(Botaurus poiciloptilus)*, and the EURASIAN BITTERN (*Botaurus stellaris;* see Color plate pp. 198 and 385/386) look much alike.

Fig. 8-18. 1. Eurasian bittern *(Botaurus stellaris)*. 2. Australian bittern *(Botaurus poiciloptilus)*. 3. American bittern *(Botaurus lentiginosus)*. 4. South American bittern *(Botaurus pinnatus)*.

The food of the Eurasian bittern consists mainly of fish and frogs. On Lake Neusiedel in Austria, it feeds almost exclusively on frogs, according to the observations of Otto Koenig. Apart from this, it also feeds on water insects, worms, and crustaceans.

The Eurasian bittern breeds in the lowlands on waters with extensive reed beds. The inconspicuous nest is built by the female alone, and projects only slightly above the water surface. Several nests are often found close together. These "colonies" are generally in the immediate vicintiy of a calling male. It has been possible to show that there is only one male for several breeding females. It was thus concluded that this indicates polygamy, but there is probably little relationship between a male and its females beyond the act of mating. Oskar Heinroth therefore interpreted the bittern's booming courtship call as a signal meaning, "Come here for mating." But booming presumably also indicates to other males that a territory is claimed.

Fig. 8-19. The posture assumed by the Eurasian bittern for camouflage in reeds...

...and its defensive threat posture.

Full clutches of five to six eggs are found from mid-April to mid-May. The young hatch after twenty-five to twenty-six days of incubation and leave the nest after about four weeks. But they are only completely capable of flight about a month later.

The bittern is mainly known for its loud booming calls. Many of its names, particularly folk names, are based on this: marsh ox, reed ox, moss cow, and others. The scientific name *Botaurus* is also derived from it, for "bos" means cow or ox, and "tauros" means bull. In the Walpurgis night of Goethe's "Faust," the "unisonous bitterns" are mentioned as well. Male bitterns call from mid-February to early July, both during the day and the night, but most frequently at dusk. At this time of the day, females fly around a great deal. Gauckler and Kraus, who studied the life of bitterns in Franconia, Germany, observed that males call more frequently as soon as a female flies over their booming places in the spring.

It was once believed that bitterns immerse their beaks into water when booming. A zoology professor on a field trip is said to have asked a fisherman about this. He was told: "Why don't you put your head in the water and then try to call out 'hurrah.'" This comparison is said to have pleased the students. Actually, a booming bittern stretches its neck forward and inflates its gullet with air. This produces a resonating chamber. The lower second part of the boom is produced during expiration, with the head and neck directed upward.

Today the Eurasian bittern is protected. In the past, it was a popular game bird, and many still-life pictures show a dead bittern among fruit, flowers, and dishes. There is also a self-portrait done by Rembrandt, showing himself as a bittern hunter.

▷
The hammerhead *(Scopus umbretta)* builds a huge nest with a small chamber

▷▷
Showbills *(Balaeniceps rex)* in a papyrus swamp.

9 Storks and Their Relatives

Family: shoebills, by C. W. Benson

The members of the remaining four families of the Ciconiformes are somewhat less adapted to life near the water than are the herons. Among them the SHOEBILL (*Balaeniceps rex;* see Color plate p. 216) differs so much from the usual type of heron and stork that it is regarded as the representative of a separate family, Balaenicipitidae.

Distinguishing characteristics

The standing height is about 115 cm, the L is 120 cm, and the wing length is 68 cm. The beak is extraordinarily high and wide; in connection with this, the skull is much enlarged and thus is pelican-like (convergent evolution). A slight crest is found at the back of the head. There is a powder down patch along the whole back, and the preen gland is very small.

Thirty years ago Bengt Berg wrote a widely read book on the "Abu Markub," meaning "father of the shoe," named after the shape of Arabian shoes. This book made this large stork-like bird with the enormous beak more generally known in Europe. The shoebill occurs only in the marshlands of tropical Africa and in general seems to be quite rare, but it is more common in Uganda and the Sudan. It is easily overlooked, since it stays along river shores in dense papyrus, but is often also seen on flat-flooded grassland with short grass. Its calm temperament makes it possible to approach it rather closely before it flies off. It flies well and sometimes soars in an updraft. It holds the beak pressed against the chest in flight and does this also on the ground. Generally it is solitary, but a few—up to seven—have been observed together. It feeds throughout the day, but preferably at night. It lives mainly on river fish, but also on frogs and snails. Usually it is silent, but occasionally a shrill laughter or a stork-like beak clattering is heard. The nest is a flat platform in reed beds or in papyrus. The eggs are bluish-white and covered with a calcareous layer; the clutch consists of two to three eggs. The young are covered with down and are altricial (i.e. remaining in the nest for some time after hatching).

▷▷▷
Marabous: on the left is the Lesser marabou *(Leptoptilos javanicus)* and on the right is the Greater marabou *(Leptoptilos dubius).*

▷▷▷▷
A pair of white storks *(Ciconia ciconia)* on the nest during defensive bill clattering. In the next moment the beaks will be laid, clattering, over the back. Wing pumping indicates that the birds are not greeting but are showing a hostile display towards a species member flying by. On the right, the male is recognizable by his lower wings while pumping.

▷▷▷▷▷
African marabous *(Leptoptilos crumeniferus)* and a pink-backed pelican *(Pelecanus rufescens).*

1
2
3
4
5

In Sudan, the breeding season begins in October as soon as the floodwaters recede. In the southeastern Congo and in Zambia, where the seasons are different from the Sudan, young have been found at the end of the dry season in October.

Family: hammerheads, by R. Liversidge

The relationships of the HAMMERHEADS (family Scopidae) are obscure, but they are generally placed with the Ciconiiformes. Some investigators place the single species of the genus *Scopus* with the stork, while others place it with the heron family. Like the storks, they have no powder down, and in flight they also stretch the neck forward. The comb on the outer edge of the claw of the middle toe, and the vocal organs are, however, heron-like. Strangely enough, their external parasites, the mallophaga, *Austromenopon* and *Quadraceps* belong to genera otherwise only found in plovers (Charadriiformes). No fossil forms are known, so that nothing can be said about their origin.

◁ Storks: 1. African wood ibis *(Ibis ibis)* is not considered a true ibis; 2. White stork *(Ciconia ciconia)*, bill clattering; 3. Black stork *(Ciconia nigra)*; 4. Jabiru *(Jabiru mycteria)*; 5. African saddlebill *(Ephippiorhynchus senegalensis)*.

There is only one species, the HAMMERHEAD (*Scopus umbretta*; see Color plate p. 215). The L is about 50 cm. The head, with its medium long, laterally compressed beak and its crest pointing to the rear, looks somewhat hammer-like. The sexes are of similar color; the young have orange-colored eyes and greenish-yellow legs. The hammerhead inhabits swamp and shallow water areas, generally looks for food in the water, and occasionally swirls up the mud with its feet to stir up prey. It is found singly or in pairs; family units of up to seven birds remain together for only a short time. Meinertzhagen describes them somewhat anthropomorphically: "Sometimes these comical-looking birds, particularly when two or three are together, are overtaken by an urge to play. They jump into the air and dance around one another. Thus, they look more foolish than nature probably intended and ridiculousness replaces every sign of dignity."

Fig. 9-1. 1. Shoebill *(Balaeniceps rex)*; 2. Hammerhead *(Scopus umbretta)*.

The flight is slow, as if swimming in the air as it were, and they frequently utter a shrill cry, particularly before it rains. They are active in the twilight and at night, and more rarely during the day. The nest of the hammerhead is an extraordinary, enormous structure which has impressed travellers in Africa and given rise to fairy tales among the natives. Its diameter is one and a half meters, and inside there is a small cavity only thirty centimeters across with an entrance that opens downward to one side. Three to six white eggs are incubated for about thirty days, and the young hammerheads leave the nest about fifty days after hatching.

Family: storks

The STORKS (family Ciconiidae) are birds of medium-size to very large. The beak, neck, and legs are long. The large and wide wings enable them to fly well and to soar, which saves much energy. When many birds fly together, there is no definite flight formation. They feed only on animals, and they build very large nests for their white eggs. There are species which have penetrated far into the temperate latitudes, and migrate far; others migrate over short distances. There are ten genera with eighteen species.

The white stork, by B. Grzimek and E. Schüz

The best known species is the WHITE STORK (*Ciconia ciconia;* see Color plates p. 222 and pp. 218/219). Males and females look alike. The young at first have blackish beaks and legs. The L is 110 cm; the wing length is 53-63 cm; the wing span reaches over 220 cm; and the weight is from 2.3-4.4 kg. There are three to five (rarely six) white eggs, with axes measuring about 710 x 510 mm and weighing 112 g.

The white stork, in many places, breeds in villages and even towns, and hence is well established in folk tradition. This is shown by its name in German folklore, Adebar, meaning bringer of good fortune. The name Klapperstorch (in English clattering stork) refers to the most striking form of behavior of the species. Apart from a short hissing sound which often precedes the bill clattering and the mewing of the young, storks have no other vocalizations. Beak clattering plays an important role as a greeting when a bird is about to alight on the nest, but also in warding off strange storks, even when they are still far away in the sky. When clattering, the stork bends the neck and head back until the crown of the head touches its back, so that the hyoid bone is displaced backward and a suitable resonance chamber is formed. When not used as a greeting but rather as a defensive signal, the bird drops its wings and then immediately raises them high above the back. This repeated up and down "pumping" is more prominent in males, which are mainly involved in guarding the nests, than in females. Soon after hatching, the young can bring the head over the back and make bill snapping movements. "One gets the impression that they would start beak clattering even in the egg if only there were enough room," so write O. and M. Heinroth when describing the handraising of young storks. The young were very well behaved towards Mrs. Heinroth, who fed them, but they attacked Mr. Heinroth with thrusts of the beak, although they submitted after being punished for this. Incidentally, a young stork once smashed a camera lens belonging to the Leipzig zoo director; it probably elicited the attack because it looked like an eye. Until just before fledging, young storks are not at all dangerous; they become immobile at the approach of an enemy. During frequent disturbances by strangers, this is more advantageous than a spirited but useless defense.

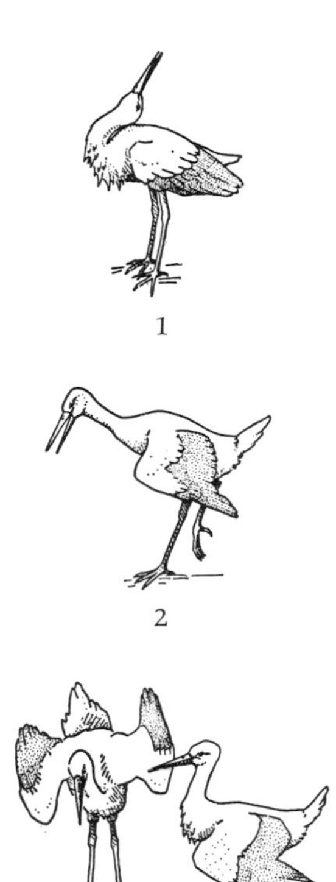

Fig. 9-2. 1. Greeting bill clattering of white stork; 2. Defensive posture (pumping); 3. Nest relief.

Food

Storks generally feed while walking; they wander over open shallow swamps, over meadows and fields searching for food. As a rule, prey is seized with a forward thrust of the beak, but animals may also be seized under water with sideways beak movements. The diet is quite varied. Earthworms play a great role in spring, and for feeding the young. In the summer they catch many insects, even crickets which are difficult to catch, but mainly they feed on the more easily caught grasshoppers. In some African languages the stork is often called a grasshopper bird or grasshopper eater. "In mid-December of 1942, great flocks of storks were busy eliminating grasshoppers," reports H.

Fig. 9-3. Mutual bill clattering.

Hamling from Rhodesia. "The area included a narrow, scantily wooded valley with scattered rocks and stones between grass and bushes on both slopes. Storks were distributed over the hills in small groups and mastered the difficult terrain with playful ease; they ran here and there and stabbed at their prey. Occasionally, when a grasshopper whirred over their heads, the birds would snap at it in the air. They were tireless, even excited, and made use of their wings again and again to move to another spot. Such a flock settled close to me and immediately began to move, probing every nook and cranny in the ground with their beaks. Often the storks at the rear of the group took wing, flew ahead over the others and came down again at the head of the flock. Then the birds in back "leapfrogged over them the same way." In Europe as well, insects play an important role in the stork's diet. In East Prussia, a stork's stomach, after a single feeding period, contained 76 cockchafers; another one had 730 leafwasp larvae; and a third one consumed 1315 grasshoppers, according to J. Steinbacher.

Vertebrates are of lesser importance as food. However, where dead fish drift ashore, or where fish are sick and generally less mobile and hence are easily caught, storks certainly eat fish. Lizards and snakes, even vipers are caught. Birds are occasionally taken. Among mammals, the mole and small rodents occasionally play an important role, as well as larger species such as rats and ground squirrels. When mice reproduce in masses, storks congregate to prey on them. The abundant rich food then results in the rearing of many healthy storks.

During the mowing of rape, Andresen one day observed a stoat (or short-tailed weasel) emerging from a mole hill. A stork, which was following the mowing machine, seized it at once. Whenever the weasel counter-attacked, the stork would fly up about a meter. After the bird had struck the stoat five or six times and thrown it into the air, it was dead. The stork then swallowed it with some difficulty. In exceptional cases, a stork which has become very tame and walks about a farmyard may form the habit of carrying chicks off for its young.

Soaring

Although well adapted in many respects, storks suffer greatly from cold and rainy weather. Not only young in the nest die of cold, in spite of being covered by the parents, but fully fledged storks as well are hindered in flight by wet plumage. The climate, therefore, determines the northern boundary of distribution. Storks also need warm temperatures, since they are dependent on thermal updrafts for soaring, as are many raptors. They can evidently reach high altitudes without effort in this manner. Soaring is also seen during migration and at the wintering grounds.

Turner, who met a flock of storks above the Serengeti while in a small plane, reports; "When meeting them in the air one must watch very carefully. At considerable altitude, sometimes up to 4500 meters, one may fly below a group of storks without suspecting it. The storks become disturbed and try to avoid the plane. Lowering their legs and closing their wings, they drop like stones. To be directly below a descending flock, plunging down behind the plane in all directions, is an unforgettable experience. However, the flying storks can generally be detected well in advance by the glint of the sun on their wings." High altitude flights during the breeding season are stimulated by suitable weather conditions.

The migration of storks

It has been known for a long time that the European white stork migrates far into Africa. In a book published in 1663, a Dutch ship's captain is mentioned, according to whom "the storks stay in Senegambia in great numbers in winter." In 1822 a stork was killed near Wismar, Germany, with an African arrow lodged lengthwise into its neck. The specimen, with the arrow that was eighty cm long, is still in the museum of the Rostock Zoological Institute. A total of eighteen cases of storks with arrows stuck in them have become known since then.

The migration of storks is a very impressive event. The flocks of the large, striking birds fly a straight course, flapping their wings only occasionally. The typical migration begins with slow circling and spiraling wanderings, whose direction is not easily determined. Storks seem to prefer to remain over land and avoid flights over open water as much as possible. Where this is necessary, as in crossing the Gulf of Suez at El Tor, the migration often attracts the attention of seafarers. Wellmann, an officer of the Hamburg-America line, described such an event on August 21, 1929, in the southern Gulf of Suez: "A column of storks moved from the northeast to the southwest. They flew quite low, one to three meters above the water in a very long band, about thirty meters wide. There were probably about 40,000 storks. They clearly avoided flying over the ship. As the bow of the ship, which was steering southeast, had passed the point where the two courses would have intersected, the endlessly long band made a marked bend toward the south. Only when the direction deviated too much from the southwest did the band break. Only a few storks flew over the ship, while the rest flew around it."

Their dependence upon upward air currents over land has led them to follow specific migration routes. The storks move on various routes from the breeding areas, which merge funnel-like into a narrow front, and these fronts follow geographic features in a certain way. A division of migration has developed in Europe along a line from Holland through the Kyffhauser hills in Thuringia, Germany, to the Lech River near the Austro-German border. Storks breeding southwest of the division migrate over western France to western Spain; from Spain, they and the Iberian storks cross the Straits of Gibraltar, wintering mainly in the Senegal and Niger regions. North and east of the division line live the so-called "eastern storks." A comparison of the number of birds using the two routes, based on an international census of 1958 by Schüz and Szijj, was made on the basis of the ratio of breeders for that year to the young of that year, and of non-breeders in recent years (all of which make up the departing population), which was 1:1:1.4. This showed that in Europe there were 425,000 eastern storks and 100,000 western storks. Corresponding to the marked decline, the central European portion of the western group numbers less than 2000; most European western storks now breed on the Iberian peninsula. They are joined in Africa by the storks of the Maghreb (northwest Africa), estimated by Bernis at 190,000. On the basis of counts at Istanbul, which is the narrowest point of the eastern route, Helps estimated that 207,000 storks had passed there during one entire fall migration. The European migrants on the southeastern route just mentioned, are joined by a considerable breeding population from Turkey.

What determines the choice of the different directions on either side of the migration division? To elucidate this, some nestling storks were taken from East Prussia, which has many storks, and transported to West Germany, where storks are scarce, and allowed to reach the fledging stage in cages. The young eastern storks were liberated only after the departure of the storks native to the west. They migrated not like storks that lived in this area to the southwest, but they flew off to the southeast, where they encountered the Alps and the plain of the Po River in northern Italy. The urge to migrate southeast was thus inborn. In other experiments, young East Prussian storks were raised outside of cages in the west. Now they migrate southwest. They had the opportunity to join western storks and so the social influence overrode the inborn migration direction. In spite of these results, and although other experiments show that birds can orient themselves according to the sun's position, the riddle of bird navigation is by no means solved. Since storks are social and live longer than most small birds, it is likely that, in the manner indicated by the second experiment, the contact with experienced members of the species is important in following the usual migration route.

Breeding area
Boundary of wintering area
Sightings of migrating birds
Migration division

Fig. 9-4. Breeding area, migration routes, and wintering areas of the western subspecies of the white stork (*Ciconia ciconia ciconia*).

Central European storks, on their southeast fall migration route converge west of the Black Sea, where the narrow front of migrating birds offers an impressive picture during the flight over Istanbul and Turkey. Here the numbers swell by the many storks which breed in Asia Minor. East of Iskenderun, along the north-south Amanus mountain range and Lake of Antiochia, almost all of these storks turn to the right. They then proceed southward along the valleys of the rivers Orontes, Leontes and Jordan. There is a deviation in this area on the spring return flight. It proceeds on a wider front than the fall migration. The birds fly further west and appear in great numbers on the coasts of Israel and Lebanon. The migration front is also widened to the east, since moisture opens up semidesert areas which are dry in late summer and autumn. Another peculiarity is the long duration of the homeward migration; there are still many migrants in the area of Syria and Israel as late as May. This is due not only to the availability of food, but also because there are many immature storks. The migratory urge of these birds weakens and so they remain in the eastern Mediterranean area. The punctual passage of breeding birds is thus marked by the large numbers of immatures. During the fall migration, the storks reach the Gulf of Aqaba, cross the Sinai peninsula in its southern part, the southern Gulf of Suez, and then cross over the bare mountains of the Arabian desert to reach the Nile River at Kena. The same is true of the spring passage in reverse. But there is always a smaller number of storks, particularly immatures, which do not leave the Nile at Kena, but pass into lower Egypt, a "dead end" for them. Furthermore, some storks evidently pass over a more western and northern part of the Sinai peninsula on the return migration, so that they reach the Mediterranean at El Arish. Hot desert winds (the Chamsin) may cause fatalities among storks crossing the Sinai desert and the Dshebel Hauran. Sometimes there are even mass deaths; Meinertzhagen saw hundreds of dead storks in this area.

The storks follow the Nile south of Kena. Whether they shorten their route by crossing the desert instead of following the loop of the Nile between Dongola and Berber is still an open question. There are few recoveries of banded birds between the tenth and fourth degree northern latitude. Storks may winter as far north as the Sudan or even upper Egypt; most, however, migrate further south even as far as Port Elizabeth.

Apart from the migratory drive, the availability of food plays a role. Large flocks may join together when there are invasions by migrating locusts *(Locusta migratoria* and *Schistocerca gregaria).* Since moisture dryness, and mass abundance of insects are variable with respect to place and time, the position of the wintering area varies considerably. The locust control officer, P. Lea of Pretoria, writes about 1959-60, that the estimates of the stork population "clearly show that in dry summers, like the present one, relatively few storks stay in the Veld in search of food—perhaps only a tenth of those present in the previous summer."

When they do not remain further north, the storks reach Zambia by about mid-November and stay at the southern point of their migration, east South Africa, during December. They begin the northward move in the same month or in January. Individual storks are, however, found at any time of the year along the whole migration route. These are sometimes sick birds, immature ones, or two-year-olds, as shown by band recoveries. Recently there have been a few cases in which storks nested near the southern coast of South Africa, east of the twenty-second and seventeenth degree longitude.

▷ Fig. 9-5. Mating takes place after complex courting behavior on the nest.

Although many young storks remain in Africa in their first summer, for they are sexually mature only when three or four years old, about eight percent settle later in the village where they were born, thirty-six percent settle within a radius of ten kilometers, and an additional twenty-two percent settle within ten to twenty-five kilometers of the place where they hatched. Of sixty banded storks, Hornberger found thirty-seven at the same nests a year later; after four years, eleven remained, and after six years there was only one. The maximal age of white storks is probably twenty years; in the Berlin Zoo, an Oriental white stork which is over thirty years old indicates that European white storks may also occasionally live this long.

Although white storks migrate in flocks, the returning birds arrive at their nests singly in March or April. The males usually arrive first and often occupy the old nest. The female usually arrives a few days later. If by accident a new female arrives first, and the female from the previous year arrives later, there may be vigorous fighting, and even blood may flow, but fatal injuries are rare. The victor is not necessarily the stronger bird; the rightful nest owner may show more vigor than a stronger rival which has a "less valid" claim on the nest. There may be attacks and fights during the breeding season as well. Two-year-old storks, which, as a rule, are not sexually mature until they are three and four years old, may become a nuisance through their attacks on breeding pairs. Such birds are not satisfied with empty nests; they try to conquer nests already occupied.

The nest of the white stork is quite an impressive structure. New nests are generally not completed in the first year. The nest is added to year by year until "castles," often weighing several hundred pounds, arise, but sooner or later they come down in a storm. Rags, items of clothing sometimes taken from clotheslines, and newspapers are carried into it; horse manure also seems to be a desirable lining. Storks originally nested on rocks and trees, but as storks have entered the vicinity of man, house roofs, chimneys, towers, and ruins have taken the place of rock. Power line poles may also substitute for trees, and so be used as nest sites. In eastern Central Europe, tree nests are not uncommon; as with nests on buildings, man plays a role in the nest site selection, when he provides platforms on trees which are accepted by storks. That iron supports are avoided is a superstition. House sparrows, starlings, and wagtails often breed as "subtenants" in the base of tall stork nests; in the near East, even the roller *(Coraxias garrulus)* does this.

The birds loosen up the nest during the first days after returning in the spring by stabbing and plucking beak movements. Since the nest mass has generally become a compact block, this "airing" is evidently important. The pair copulates a number of times. Before the male mounts the female, they circle one another. The male supports himself on his mate during mating by wing beats and makes excited beak movements down toward her; soft sounds can be heard. Soon the eggs are laid at two-day intervals. Since the white stork is altricial, they are not particularly large. The partners relieve one another from incubating at intervals of several hours. Storks can recognize their partners at a great distance.

At night the female is usually on the eggs. The young hatch after thirty-two days of incubation. Hatching is indicated when the parents stand a great deal and concern themselves with the floor of the nest; probably besides feeding, they loosen and arrange the nest floor. Both parents bring food, at first smaller prey, like earthworms. They lower the beak between the young and regurgitate the food. The young snap at it as it drops or take it up from the nest floor. Excess food is thrown into the air, caught, and swallowed by the adults. Larger young grab food from the parents' bills. The downy plumage grows progressively; after ten days a black edge appears along the wing as the first sign of the flight feathers. Finally, the young are white and black. At the approach of the parents, they sit up on their tarsi, beg, beat their wings vigorously, and sway to and fro on the nest. After delivering the food, when the parents get tired of the continuing begging of the young, they withdraw to a nearby chimney or other perch. At seven weeks of age, the young begin to spread their wings and make playful upward leaps if there is a suitable wind. The ability to fly matures without parental help; flying is therefore not learned, but merely practiced. At about two months of age the young are fledged, but at first they still stay on the nest for the night and part of the day. In areas of good food supply in good stork habitats, the young then gather in larger flocks, which occasionally perform group flights high in the air. Then about the twentieth of August, when there is good soaring weather, they leave their home region. As a rule, the parents stay longer in the breeding area. They continue to build on their nest as they did even when breeding. Quite often they copulate again, but they, too, leave their home during September.

Population figures

In Europe there is no bird that has been so well observed as the stork. The first counts were by Wüstnei and Cloduis in Mecklenburg in 1901, Braun in East Prussia in 1905, and K. Eckstein in Schleswig-Holstein in 1907. A continuous series of counts which are therefore particularly valuable, were made since 1928, by R. Tantzen in Oldenburg. Other counts were made by the Rossitten bird observatory and continued by Radolfzell Observatory at Lake Constance. In 1934 and 1958 there was an international census in areas extending beyond Germany. In 1934, East Prussia, with 16,600 pairs, had the largest population of storks in Germany, which then had a total of 30,730 pairs. Because of boundary changes, the results for Germany for 1958 were difficult to compare with the study just cited. But a careful comparison with the present German areas shows 9035 and 4800 pairs in the years of 1934 and 1958 respectively, a decrease of more than one half. The figures for these two years for the provinces are respectively: Schleswig-Holstein, 1776 and 953; Lower Saxony, 1925 and 998; Baden-Württemberg, 186 and 143 (in between a peak in 1948 with 253 breeding pairs); and Bavaria, 119, (149 in 1948), and 186. Toward the south the decline is thus less; in fact, in Austria and Hungary there was no decline. In Belgium, storks bred for the last time in 1895, in Switzerland in 1949, and in Sweden in 1954. In Holland the population declined in the twenty-four years from 1934 to 1958 from 273 to 56 pairs; in 1967 there were only 19 pairs left!

Attempted introductions

There have been attempts to supplement the decreasing stork population by translocations. Attempts by the Rossitten bird observatory in East Prussia showed that this was difficult. The most experience in this respect was gained in Switzerland. After the last breeding of wild birds there in 1949, Max Bloesch, in collaboration with the Sempach bird observatory, obtained a total of 500 storks from other European areas, especially from Algeria, between 1948 and 1967. They were reared in Altreu near Solothurn, and after the birds reached maturity there were several broods raised by these liberated birds, (in 1967, there were nine in Altreu, and one five kilometers away). There were even a few broods in the second generation. I. Waldvogel founded a society, "Stork friends of Alsace," and since 1958, A. Schierer has tried to reintroduce storks in Strassbourg, which once had great numbers of them. Where there is still a population of wild storks, care should be taken for the maintenance and building of new zest supports. Pamphlets on this are available from the Radolfzell and Ludwigsburg bird observatories in Germany. In endangered areas there is, however, a lack of returning storks rather than of nest sites.

Ecological requirements

White storks avoid dense forests; no doubt many parts of Europe became suitable for them only as a result of deforestation. B. Løppenthin believes that they did not settle in Denmark until the fifteenth century, since storks also need suitable nest sites. Swampy habitats are suitable feeding areas, but storks also feed on fields, as is evident at plowing time. Thus, fairly narrow river valleys with adjacent agricultural areas provide for its needs. Despite this adaptability, one can see that the nature of the soil and climate are essential for the production of animal food for storks. In East Prussia, the distribution of the stork coincides with that of agricultural productivity and the distribution of horses; poor sandy soils cannot support storks, while rich soils supply food for man and horse and the stork as well. Weather effects are evident as well. On the North Sea coast, a warm, sunny June is favorable, while in Hungary, with its continental climate, a moist June favors the food supply and this in turn, the number of young that can be raised from each egg laid, for the eggs hatch at daily intervals so that there are always one or two chicks smaller than the rest. These have a disadvantage when food is distributed, so that only under very favorable conditions are all the young raised. Breeding success or failure also depends on the maturity of the parent birds; young pairs breeding for the first time often fail as parents. They seem to regard a stunted youngster, generally the youngest, not as offspring, but as prey, and they try to swallow it. In reference to the Greek myth of Cronos, father of the gods, such behavior has been called "Cronism."

A still unsolved problem are the years in which the birds return late in spring, and, possibly due to some injury received in the wintering grounds or on migration, rear only few youngsters.

Dangers

Although at one time man opened up the country for storks and provided them with nest sites, and in Islamic areas, even honored them as pilgrims of Mecca, today technical developments are great dangers for them. Many storks, particularly young ones, perish on power lines; a few fall into open smoke stacks; and more collide with cars, railway engines, and air planes. The spread of firearms has lead to increased stork mortality, particularly in Syria and Lebanon, but also in Africa, where the natives formerly hunted storks only with arrows. The extent to which storks become victims of pesticides is still to be determined; there have been instances in the area of Hamburg and probably in Africa as well.

Fig. 9-6. 1. Distribution of the eastern subspecies of the white stork *(Ciconia ciconia asiatica)*, its migration routes and wintering areas; 2. Oriental white stork *(Ciconia boyciana)*, former breeding area, migration route, and wintering area.

In contrast to these dangers due to man, there are also losses due to natural events. Bad weather may force storks far off their routes, and for example in West Africa, blow them out into the Atlantic. Dry hot desert winds and hail cause deaths at high elevations in the southern half of Africa in particular; entire flocks of storks are occasionally killed by hailstorms.

There is a superstition that young storks feed and take care of their old parents. This was so firmly believed in ancient Greece that the law which required citizens to support their parents was called "Pelargonia," from the word "Pelargos," meaning stork. The best known stork fairy tale, which says that it brings babies, probably arose in north Germany. The *Atlas of German Ethnology* contains a map which shows that the belief that the stork brought babies was unequally distributed over Germany and Austria. In the Rhineland and other west German areas, babies were supposed to come from trees or wells or the midwife's bag. In the alpine area, where there are no storks, the stork legend was less widespread.

Other stork species, by E. Schüz

The ORIENTAL WHITE STORK (⊖*Ciconia boyciana*), with its black bill and red legs can also be regarded as a subspecies of the white stork, but it must have branched off early. It nests in trees of riverside woods or of open stands in the Amur and Ussuri area, reaching west as far as Blagoveshchensk (50° 20′ N and 127° 40′ E). Asiatic black bears sometimes rob its nests, and the natives like its flesh, while the Chinese make chop sticks from its bones. In Japan, where this stork was a

The Oriental white stork

symbol of long life, there were only thirteen in 1964, as compared with twenty-five in 1960. Since they do not reproduce in the wild any more, perhaps due to pesticide poisoning, huge flight cages have been built for the remaining pairs.

The black stork

The BLACK STORK (*Ciconia nigra;* see Color plate p. 222) is slightly smaller than the white stork; the L is 105 cm, the wing length is 52–60 cm, and the beak is 16–19 cm. The plumage of immatures is brownish-black. It is a forest dweller and is able to alight in the crowns of trees better than the white stork. It normally breeds on trees, but in Spain and in the East it breeds on rocks. There were about 140 pairs in East Prussia in 1935 and almost 200 in 1941; in the Silesian Bartsch lowlands there were still twenty pairs until recently. The last brood in Schleswig-Holstein occurred in 1938, and in Denmark, in 1951. Unfortunately the losses are in part due to hunting of the birds.

Fig. 9-7. 1. Abdim's stork; 2. Woolly-necked stork.

The black stork resembles the white stork in many respects, but is not much of a beak clatterer, although in excitement a few irregular clatterings may occur. In courtship and when excited, it spreads out its abundant white lower tail coverts and utters whistling, wheezing sound sequences with lowered extended neck. It may utter soft calls at the nest, and on rare occasions a pleasant-sounding flight call, "fujo" is heard. It migrates much like the white stork; almost nothing is known about its western route, since this has almost disappeared. The migration division evidently runs somewhat further east than that of the white stork. Somewhat more skillful and more inclined to ordinary flight than the white species, it does not strictly avoid the Mediterranean. Some of them winter in the Near East, and probably very few cross the equator during their migration. There are, however, a few breeding places south of the equator; in 1966, thirty-four breeding pairs were found in Rhodesia and on the Drakensberg.

Abdim's stork

ABDIM'S STORK *(Ciconia abdimii)* is a smaller species; the L is 75cm, and the wing length is 42–47 cm. It resembles the black stork in plumage, but has a white lower back. The white lower tail coverts are rigid and are as long as the tail. The call consists of a series of peeps.

Abdim's stork often forms large breeding colonies; Vierthaler once saw a huge monkey-bread tree with sixty-five nests in the Senaar. This species likes to nest near native settlements, and in the Sudan breeds on the roofs of huts. Since it arrives with spring rains, the natives welcome it as a "rain bringer." The flocks move toward South Africa in migration, and during the mid-November rainy season gather in Zambia by the thousands. They are twenty times as common there as the white stork, which arrives at the same time. When migrant locusts and other insects occur in masses, these storks join white storks to form great flocks. They consume "army worms," the numerous larvae of a moth. Abdim's storks also like to work over compost heaps for large larvae, such as those of the rhinoceros beetle.

The maguari stork

The MAGUARI STORK *(Euxenura maguari)* of South America (northeast Argentina to Guyana) resembles the white stork. The height is 110 cm and the wing length is 55 cm. The lower tail coverts are as stiff as the tail feathers and longer than these; the tail is forked as in the bishop stork. The shoulder feathers are black, the beak is bluish, and the legs are red. This stork breeds in swamps.

The openbill

The OPENBILL *(Anastomus lamelligerus)* is a small black stork; the L is 80 cm, and the wing length is 42 cm. It has a metallic iridescence, particularly on the tips of the feathers of the breast and underparts of the body, which are modified to black keratin platelets. The beak is open between the base and tip, and cutting edges have horny bristles.

The peculiar beak is not only useful for seizing frogs, but also serves particularly to open mussels and swamp snails (Ampullariidae). The shells of the latter are smashed before the snail is removed. This bird is mute but can clatter with the beak; it flies with slightly bent neck, yet still in a stork-like way. Occasionally, during aerial games they like to drop down like stones, with half open wings. The species breeds in trees, bushes and reed beds from Ethiopia to the Zambezi and Madagascar. It wanders northward after breeding and is then found, besides other areas, in the lakes and swamps of the Senegal.

The INDIAN OPENBILL *(Anastomus oscitans)* is closely related, with a white plumage with black only on wings and tail, and reddish-white legs. It is the most common stork species in India. Colonies consist of up to sixty nests. It can clatter its beak with retracted head, like the white storks. In sunny weather they soar upward in flocks. They also seem to migrate, for a large flock was seen in July at the northern tip of Ceylon.

The wood ibises

The WOOD IBISES of the genus *Ibis,* despite their down-curved beak (and the scientific name) are not true ibises but storks. The L is about 100 cm and the wing length is 45–50 cm. The primaries and tail are black, the beak is yellowish or reddish to blackish. There are three species: 1. The AFRICAN WOOD IBIS (*Ibis ibis;* see Color plate p. 222), is found in Africa and Madagascar. 2. The SOUTHERN PAINTED STORK *(Ibis cinereus)* is primarily white and so is sometimes called the milky stork. It occurs from Indo-China to Sumatra and Java. 3. The PAINTED STORK *(Ibis leucocephalus)* looks colorful, hence its English name. It also lacks a rosy hue; the upper head, cheeks, and throat are bare; it is a pale yellowish-red, the wings having a white band, and the breast having a black band. It is not rare in southwest China, Malaya, India, and Indo-China.

The painted stork, when seeking food, immerses its partly open beak into the water up to its eyes, and thus feels for its prey, consisting mainly of fish. In display it spreads out the lower tail coverts and clatters with its beak. The voice is a mere grunting. It breeds in colonies after the monsoon, from September to April, often with other colonial birds. There may be seventy to a hundred nests in a small space, sometimes spread over only about six trees. As Baker reports, there are, or used to be, colonies of four hundred to five hundred pairs.

The AMERICAN WOOD IBIS *(Mycteria americana)* forms a separate genus; the plumage, apart from flight feathers and tail, is white. The head and neck are bare, and like the legs are colored bluish-black. There is a horny plate on the forehead. It breeds in dense tree colonies from South Carolina to Buenos Aires.

In a Florida colony, 6000 pairs with 17,000 young were counted in a good year. The bird is also called the wood stork in America. American wood ibises live mainly on fish, which they often catch up to forty kilometers away from the nests. By gliding a great deal, they can easily cover such distances, since their flight requires very little energy. They can find fish by the sense of touch alone, without the aid of vision.

Egg laying begins in Florida at the beginning of the year, evidently stimulated by the beginning of the dry season, since they do not breed during strong winter rains. However, in the dry season when the fish concentrate in smaller areas, they can catch their prey more easily. The legs of these birds are often whitish from their own excrement. Kahl realized that this aids cooling by evaporation during great heat. The painted and white storks also have this habit.

Woolly-necked storks

The AFRICAN and INDIAN WOOLLY-NECKED, or BISHOP STORK *(Dissoura episcopus)* is striking because of its white downy plumage; it is white on the underside and lower tail coverts, while the rest of the plumage is black, and iridescent in parts. The L is 80 cm, and the wing length is 50 cm. There are two subspecies (occasionally rated as two full species): 1. the AFRICAN WOOLLY-NECKED STORK *(Dissoura episcopus microscelis)*, with feathered forehead, cheeks, and chin, is found from Ethiopia, the Sudan, and Senegal to South Africa. 2. INDIAN WOOLLY-NECKED STORK *(Dissoura episcopus episcopus)*, with bare forehead, cheeks, and chin, is found from India to Celebes; its population is being reduced by poachers, in part, as in Ceylon. It dances strangely at pair formation, breeding in Ceylon in February and March; it breeds as isolated pairs.

The BORNEO WOOLLY-NECKED STORK *(Dissoura stormi)* is somehwat smaller, with its beak slightly down-curved. The bare forehead is red, with an outgrowth; the sides of the neck are black. The population is much reduced, due to hunting.

The large storks

The AFRICAN SADDLEBILL (*Ephippiorhynchus senegalensis;* Color plate p. 222) is a particularly large representative of the Ciconiiformes; the L is 150 cm, the wing length is about 65 cm, the wing span is 240 cm, and the weight is 6 kg. The juvenile plumage is more brownish-gray and less contrasting than the adult's. Solitary pairs breed in trees or bushes. They are only seen as pairs or in small troops in swamps or along lakes, and are distributed from Ethiopia and Senegal to South Africa.

The BLACK-NECKED STORK *(Xenorhynchus asiaticus)* is almost as large as the African saddlebill and looks similar, but its black bill lacks the colorful saddle. It ranges in small numbers from India to north and east Australia, breeding on trees or rocks. In flight, it alternates wing beating with long periods of gliding. It mainly hunts fish. With zig zag jumps and wing beating, it runs through shallow water after its prey, and then seizes it.

The JABIRU (*Jabiru mycteria;* Color plate p. 222) is the third largest stork, with a wing length of 60 cm. It lives in America from Florida and southern Mexico to Argentina.

The marabous

The MARABOUS (genus *Leptoptilos*) has an L of 140 cm, a wing length of 70 cm, a wingspan of about 300 cm, and a weight of 5 kg. They are particularly large, carrion-eating storks. The beak is large and powerful; the head and neck are almost bare, only lightly covered with down. Two of the species have large throat pouches, containing only connective tissue (no crop) and presumably serve as a pad for the beak which often rests on it. There are three species: 1. The GREATER MARABOU (*Leptoptilos dubius;* Color plate p. 217), which is found from Central India to Borneo, formerly played a role as a street scavenger in towns like Calcutta. 2. The MARABOU (*Leptoptilos crumeniferus;* Color plate pp. 220/221), whose back, wings, and tail are slate gray with green iridescence, have large wings with white edges, and white underside. Many zoologists combine the greater marabou and the marabou into a single species. 3. The LESSER MARABOU (*Leptoptilos javanicus;* Color plate p. 217), has a wing span reaching 320 cm (probably the greatest wing span among land birds), no throat pouch, a horny plate on the head, and no white in the outer wing. It is found from Central India to Java and Borneo.

The huge wedge-shaped beak of these birds serves not so much to cut up meat, but mainly to cut open the abdominal wall of dead animals. The almost-bare, only sparsely down-covered head is also adapted for the insertion into large animal corpses, as in the bare head of vultures, with which the marabous often meet to feed on carrion. The appearance of these storks in the air is also vulture-like because of their huge dark wings and short tail, and because they draw the head back and glide majestically. The lower tail coverts consist of very fine, tattered feathers, which are in demand as trimming for ladies' clothes. This has led to a diminution of these important "sanitary workers" in some parts of Africa.

The marabou breeds in colonies on trees or rocks. M. P. Kahl examined a colony in Kenya in detail. Such colonies are found wherever there is a likelihood of carrion or refuse, such as areas rich in game, cattle rearing districts, and settlements. An adult marabou needs 720 grams of food per day. Its two or three eggs are laid in Africa at the end of the rainy season, so that the young are reared during the dry season. At that time there is more carrion, and aquatic animals are

◁

Ibises: 1. Hermit ibis (*Geronticus eremita*); 2. Hadada (*Hagedashia hagedash*); 3. Scarlet ibis (*Eudocimus ruber*), a. adult plumage, b. juvenile plumage; 4. Glossy ibis (*Plegadis falcinellus*); 5. Spoonbill (*Platalea leucorodia*); 6. Roseate spoonbill (*Ajaia ajaja*).

1
2
3
a
b
4
5
6

1
4
3
5
2
6

Fig. 9-8. Sacred ibis *(Threskiornis aethiopica)*

concentrated in the continually receding standing waters. The young fledge at the beginning of the next rainy season. Incubation takes 30 days; the young need 116 days until they can fly, and about 130 days until they finally leave the nest. Evidently only a fraction of the population breeds in any one year.

About eighty years ago the greater marabou was so numerous during the rainy season that "it was to be seen on the highest points of almost every house in Calcutta," as Baker reports. Today it is much rarer, and there are not any more "arrivals of large swarms" of this bird in Burma, which Oates once described. It breeds in India between October and December. The nests are in colonies on trees or on rocks.

Family: ibises, by H. Kumerloeve

Two groups of Ciconiiformes, both with peculiarly-shaped beaks, make up the family of the IBISES (Threskiornithidae). They are related to the storks, with the wood ibises forming a sort of link with the true ibises. Although the ibises, with their slender curved beaks, differ strikingly from the flat-billed spoonbills, they are nevertheless closely related. In zoos hybrids between the two have been reared; recently this occurred between a male Oriental ibis and a female spoonbill. This makes the usual division of these birds into two subfamilies appear to be doubtful, but this division is retained here for practical reasons.

◁ Flamingos: 1. James' flamingo *(Phoenicoparrus jamesi)*; 2. Andean flamingo *(Phoenicoparrus andinus)*; 3. American flamingo *(Phoenicopterus ruber ruber)*; 4. European flamingo *(Phoenicopterus ruber roseus)*; 5. Chilean flamingo *(Phoenicopterus chilensis)*; 6. Lesser flamingo *(Phoeniconaias minor)*. The Andean flamingo (2) is shown erroneously with a rear toe; however, the James and the Andean flamingo have no hind toe.

All members of the family Threskiornithidae are of medium size, with an L of 50–90 cm. The face and throat are more or less bare of feathers; the medium-length legs are sturdy. The vocal apparatus is only feebly developed; they only utter low sounds or are almost mute, although a few species (e.g. the hadada) utter far-reaching calls. Spoonbills can also clatter with the beak. The anterior toes are connected by webs. Both sexes are similar in color, the ♀♀ generally being somewhat smaller than the ♂♂. They are distributed over all warmer and tropical areas. Being very sociable they breed in large colonies and wander about or migrate in troops. In flight the neck is carried extended forward, as in storks. Two subfamilies are readily distinguishable by external characteristics: 1. The IBISES (Threskiornithinae), with their long, narrow, and markedly down-curved beak, probe for insects, mollusks, crustacea, and worms in mud and soil; occasionally they also catch larger prey. Wing beats alternate with spells of gliding; when in flocks all birds alternate from one form of flight to the other at more or less the same time. There are 17 genera with 20 species.

2. The SPOONBILLS (subfamily Plataleinae), with a beak which is flattened and widened at the tip, seize prey in side-to-side movements of the bill. They do not interrupt wing beats by gliding. There are two genera with six species.

The sacred ibis

The SACRED IBIS *(Threskiornis aethiopica)* has been a part of cultural history for 5000 years. In ancient Egypt it was revered as the embodiment of the god, Thoth, the god of wisdom, as well as the scribe of

Fig. 9-9. Sacred ibis.

the gods. This ibis-headed god is shown in many murals and sculptures, such as in the temple of Sethos, where it hands the hieroglyph of life to Osiris. It was used as a hieroglyphic symbol, and entire "cemeteries" of partially well preserved ibis mummies have been found at Sakkara near Cairo and at Hermepolis in middle Egypt. Apparently the bird was common in Egypt at the time of the Nile floods, but it has not bred there since the first half of the nineteenth century. It prefers lowlands near lakes and other waters, and also the vicinity of coasts. The breeding season depends on local environmental conditions; in Ethiopia it breeds between late March and May, in Uganda between February and July, in the Sudan July and August, and in Nigeria from May to July. Nests may be on the ground, often in papyrus thickets, or in bushes or trees. The clutch mostly consists of three to four eggs from which the young hatch after twenty-one days. Their downy plumage is dark. When feeding, the youngster grabs, once or several times, the beak of the parent, which then regurgitates the food, often with jerky, shaking movements into the beak of the young. The young are fledged in five to six weeks. Sacred ibises form lines of V-formations in flight.

Oriental and Indonesian white ibis

Closely related to the sacred ibis is the ORIENTAL IBIS *(Threskiornis melanocephala)* distributed throughout India, Pakistan, Ceylon, Burma, Thailand, and further east. Although it replaces the sacred ibis geographically, it must be regarded as a separate species. The INDONESIAN WHITE IBIS *(Threskiornis molucca)* is intermediate between the two. It is restricted to the Moluccas, New Guinea, Australia, and Tasmania.

The glossy ibis

In Europe there is still one species of ibis, the GLOSSY IBIS (*Plegadis falcinellus;* Color plate p. 237). It is the only species of ibis which has an almost world-wide distribution. However, it breeds widely dispersed and irregularly. In some areas it has enlarged its breeding range and appears to have crossed the ocean, for it has reached North America. It is replaced in the west, in Mexico, and along the gulf coast to about Louisiana by the slightly smaller WHITE-FACED GLOSSY IBIS *(Plegadis chihi).* In the Old World, the glossy ibis has had to give up former breeding areas, such as Sicily, Lake Fetzara in Algeria, and Ceylon. This is probably due to persecution, drainage operations, and other changes of the bird's environment. In central Europe the glossy ibis bred at Lake Neusiedel between Austria and Hungary until 1934; since then there have been only a few attempted nestings there. It still breeds on Lake Kisbalaton, in Hungary, where there were formerly up to 1000 pairs. There are other breeding places in the Balkans, particularly in the Danube delta where there are 4000–5000 breeding pairs. There are also local breeding areas in Asia Minor, and this species has bred in southern Spain and northern Italy.

Nests are among swamp plants or on trees. Both partners take a share in nest building, incubation, and rearing of the young. One clutch of three to four eggs per year is the rule, but replacement clutches and perhaps even second broods occur. Incubation is about twenty-one days, as in most ibises and spoonbills. The young are fed in the usual way, out of the parents beak into their throat. The half-grown young of several pairs aggregate when possible and are fed as a group. Similar behavior was observed in the hermit ibis and in spoonbills. The young are said to fledge when six to seven weeks old. As far as it is known, this ibis takes only animal food, mainly insects and their larvae, mollusks, crabs, and worms.

The hermit ibis

The HERMIT IBIS (*Geronticus eremita;* Color plate p. 237) is about the size of a goose; the females evidently are smaller. The lancet-like neck and nape feathers are often blown over the bare head by the wind; hence it received the German name, crested ibis. It breeds colonially. Courtship, mating and nest building take place on the Euphrates starting about the end of March; two to three or more eggs are laid beginning in early April. Incubation is twenty-seven to twenty-eight days, and the young are fledged after forty-six to fifty-one days, from about the first third of June. The food consists of insects, larvae, spiders, worms, and small reptiles and amphibia.

Fig. 9-10. 1. Glossy ibis *(Plegadis falcinellus);* 2. White-faced glossy ibis *(Plegadis chihi).*

To see this bird in the wild today, one must either travel to Morocco, where there are a few remote colonies on high cliffs, or to Birecik in Turkish Mesopotamia. There is a breeding colony in this town on the Euphrates, evidently the last one in western and central Asia. Until the seventeenth century this bird also bred in Europe. For breeding it needs steep cliffs, providing ledges or cracks, for the support of the large nests; hence it used to live in the alpine area. Former colonies were near Salzburg, Graz, Passau, and Kehlheim on the Danube, near Bad Pfäfers and Ragaz in Switzerland, and perhaps also in Italy. Excavations from the Middle Stone Age near Solothurn prove that the hermit ibis belonged to the Central European fauna at that time. The earliest written references to it, however, are from the early sixteenth century, mainly from Salzburg. It became more generally known through the description and illustrations first published in 1555 by Konrad Gesner, who called it *Corvus sylavticus,* forest raven. Evidently the species disappeared from the alpine area and thus presumably from Europe in the first half of the seventeenth century. At any rate, there are no more recent references to it. No doubt the disappearance of this large bird was due to persistent persecution. The young were taken from the nests and were regarded as a delicacy by the nobility. Protective laws, such as those passed by King Ferdinand in 1528 or the archbishops of Salzburg, were to no avail. Possibly the bird's extermination was also speeded up by gradual decrease of its food supply. Later this bird was forgotten and many naturalists regarded Gesner's forest raven as an invention.

In 1832 Hemprich and Ehrenberg discovered a new species, the hermit ibis, on the Red Sea. Later it was also found in north Africa and, in 1879, at Birecik on the upper Euphrates. Only in 1897 was it realized that the forest raven and the hermit ibis were the same species. Hermit ibises from Morocco have recently been reared in the Basel Zoo, and the offspring were given to other zoos to preserve the species. There are now ibises of this species in the zoos of Innsbruck, Heidelberg, and Berlin. At Birecik, 500 and 550 pairs nested in 1953, but today the number has decreased to about fifty. This is probably due to a reduced food supply. The use of pesticides has presumably also had a damaging effect. Disturbance due to the building of a new bridge over the Euphrates and the rapid increase of motor vehicle traffic may also have contributed to the decline of this colony.

Fig. 9-11. 1. Hermit ibis *(Geronticus eremita)*; 2. Bald ibis *(Geronticus calvus)*.

The Birecik ibises leave in July and presumably winter in Africa; they return from their winter quarters in mid-February. They are looked on as the harbingers of spring and the souls of the dead and therefore are highly appreciated. Their arrival used to be celebrated by a folk festival which was combined with a distribution of food to the poor. Unfortunately this custom is dying out.

The ibises busily bring in nest material from the Euphrates. The large nests are often only sixty to eighty centimeters apart; nevertheless there are no disputes. Adults feed not only their own young but also young of other pairs from beak to beak. All this takes place in the immediate vicinity of the cave dwellings and houses of the local population. No adult and no child would harm this harbinger of spring; at least this used to be the case.

The bald ibis

There is a related species, the BALD IBIS *(Geronticus calvus)* in eastern South Africa, particularly in the Drakensberg mountains. Its population is greater than that of the hermit ibis, but has also declined considerably in recent years.

The hadada

Two further ibises are characteristic of Africa, the HADADA (*Hagedashia hagedash;* Color plate p. 237) and the WATTLED IBIS. The English, the scientific, and the native name of the former species, Nanane, reproduce its loud call, which is uttered on disturbance or excitement. Nevertheless, at the nest the hadada is hardly less quiet than other birds. Pairs generally breed on their own in wooded ravines up to elevations of two thousand meters, but the birds descend to agricultural areas for feeding. The nests are used over and over; they are two to seven meters above the ground. In the Cape Providence, it incubates its three to four eggs in October and November. Both partners incubate and feed the young. The eggs hatch after twenty-six days; the young stay in the nest for about thirty-three days. Their mortality seems to be great. After the breeding season, the hadada are quite sociable, and up to thirty or more birds are often found together. At the approach of the next breeding season in August, they gradually disperse.

The WATTLED IBIS *(Bostrychia carunculata)* lives only in the highlands

▷
Lesser flamingos *(Phoeniconaias minor)* on an East African salt lake.

The wattled ibis

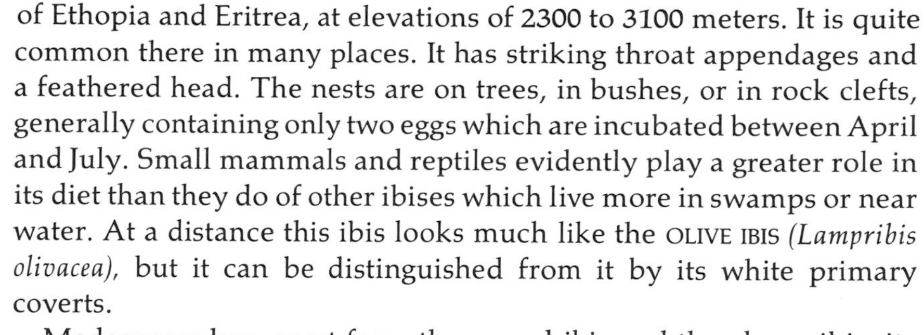

of Ethopia and Eritrea, at elevations of 2300 to 3100 meters. It is quite common there in many places. It has striking throat appendages and a feathered head. The nests are on trees, in bushes, or in rock clefts, generally containing only two eggs which are incubated between April and July. Small mammals and reptiles evidently play a greater role in its diet than they do of other ibises which live more in swamps or near water. At a distance this ibis looks much like the OLIVE IBIS *(Lampribis olivacea),* but it can be distinguished from it by its white primary coverts.

Other ibises

Madagascar has, apart from the sacred ibis and the glossy ibis, its own species, the CRESTED IBIS *(Lophotibis cristata).* It is characterized by longer feathers on the rear of the head, and much white on the wing. As the largest bird of the island's forests, it is distributed over the west and east of the island. Its nests are often on tamarind trees; three eggs seem to be the rule. The progressively drier climate and the consequently more frequent forest fires are reducing its habitat.

Of the ibises of south and southeast Asia, several other species are noteworthy besides the Oriental ibis mentioned above. The BLACK IBIS *(Pseudibis papillosa)* occurs in north India, Pakistan, Nepal, and Burma, and to the east of its range occurs the *Pseudibis davisoni,* which lives singly. In the same area the GIANT IBIS (⦶ *Thaumatibis gigantea*) occurs. The JAPANESE IBIS (⦶ *Nipponia nippon*) is of particular interest. Up to 1900 it was widely distributed in Japan and up to about 1920 in north China, Manchuria, and the Ussari area. In Korea it was still not uncommon in 1936. Today, however, this beautiful white or gray ibis with its red face, black beak and dark red legs has almost disappeared. According to the latest information, only nine remain in Japan, and it is not known whether any survive on the mainland. Their favorite habitat was groups of trees surrounded by swamp land. Drainage and other changes of the landscape, persecution, and disturbances during the war have so reduced its numbers that there is little hope of survival of this species.

The STRAW-NECKED IBIS *(Carphibis spinicollis;* Color plate, Vol. X*),* is characteristic of Australia and also occurs in Tasmania. It has pointed feathers on the front of the neck. South America also has several species of ibis, among them the PLUMBEOUS IBIS *(Harpiprion caerulescens),* the BUFF-NECKED IBIS *(Theristicus caudatus),* the SHARP-TAILED IBIS *(Cercibis oxycerca),* and the related GREEN OR CAYENNE IBIS *(Mesembrinis cayennensis)* and the BARE-FACED IBIS *(Phimosus infuscatus).*

◁ Mute swans *(Cygnus olor)* are not only found as ornamental birds on park ponds, but they also breed on many central European waters.

White and scarlet ibis

Better known than all these are the WHITE IBIS *(Eudocimus albus)* and, even more so, the beautiful SCARLET IBIS (*Eudocimus ruber;* Color plate p. 237). The two replace one another geographically. The white ibis is mainly Central American and has a tendency to spread northward; the scarlet ibis is confined to South America. Mixed pairs of the two are occasionally found, and so some ornithologists look at them as one species. In behavior, particularly in breeding, habitats, and food, the two are essentially similar. The nests stand in colonies of variable size, often on mangrove islands, but also on willows, opuntia, and other plants. Pair formation takes place in the small nest territory which is defended by both partners. Nest material is brought mainly by the male, and is used for building by the female. Eggs are laid between the end of March and the middle of May. The white ibis lays three to four eggs; the scarlet ibis, which in general is a little smaller, often lays only two. The young hatch after twenty-one to twenty-three days, are tended by both parents, and begin to leave the nest about three weeks later. At two years of age they are sexually mature. The scarlet ibis was formerly pursued ruthlessly because it makes good eating. Since 1953 it has been protected, for example at the Caroni Sanctuary in Trinidad and also in Venezuela. In the Caroni Sanctuary there are about 3000 breeding pairs. P. Allen, writing in 1961, described how countless brown, whitish, or blue and pink young ibises crowd together in the mangrove thicket with a "canopy" of adult birds fluttering above them; he adds enthusiastically, "The living beauty of the scarlet ibis is an enrichment of every country it inhabits. Its continuing protection is an enrichment for those who protect it."

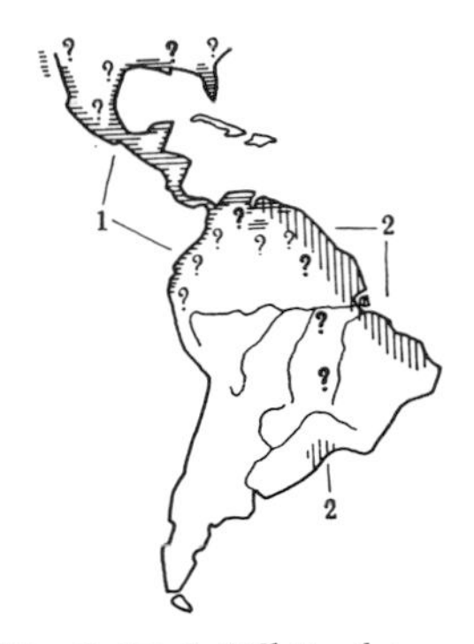
Fig. 9-12. 1. White ibis *(Eudocimus albus)*; 2. Scarlet ibis *(Eudocimus ruber)*.

The spoonbills

Compared to the ibises, there are far fewer species of spoonbills. Only two species deserve particular emphasis, the SPOONBILL (*Platalea leucorodia*; Color plate p. 237) and the ROSEATE SPOONBILL *(Ajaia ajaja)*. The remaining species, the AFRICAN SPOONBILL *(Platalea alba)*, the far eastern LESSER SPOONBILL *(Platalea minor)*, the AUSTRALIAN YELLOW-BILLED SPOONBILL *(Platalea flavipes)*, and the ROYAL SPOONBILL *(Platalea regia)* of Australia and parts of Indonesia, are regarded by some investigators as being subspecies of the spoonbill. The spoonbill breeds, like most birds of the ibis family, in colonies of varying size. In central Europe it breeds only in the Netherlands, Austria (at Lake Neusiedel), and Hungary, although exceptionally this occurs in Czechoslovakia and on the East Frisian island of Memmert. There are other important breeding places in the Danube delta, in southern Yugoslavia and Albania, in southern Spain, and in northwestern Asia Minor. Spoonbills return in late March from their African or southeast Asiatic winter quarters. Soon both partners are involved in nest building. They like to build on broken reeds or other marsh plants, as well as on willows, alders, or on Salicornia, often near the ground or over water. High tree nests are unusual in Europe, but common in Asia, for example, at Lake Manya in northwest Anatolia. The three to five eggs, rarely more, are laid in April or May. The young hatch after twenty-one days and are cared for by both parents. Four weeks later they begin to leave the nest, and after another four weeks, they are able to fly. Even the newly-hatched young show the characteristic spoon-like bill. The adults take insects and larvae, crustacea, mollusks, and other water animals as well as spawn.

Fig. 9-13. 1. Spoonbill *(Platalea leucoradia)*; 2. African spoonbill *(Platalea alba)*; 3. Yellow-billed spoonbill *(Platalea flavipes)*.

Fig. 9-14. Spoonbills at rest (left), and sleeping (right).

Fig. 9-15. Spoonbill feeding youngster.

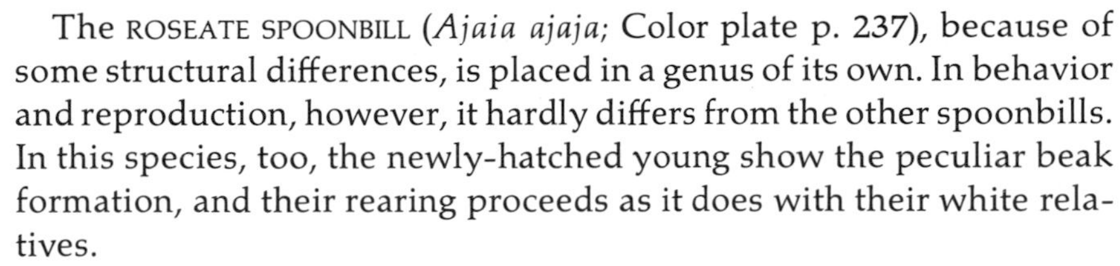

The ROSEATE SPOONBILL (*Ajaia ajaja;* Color plate p. 237), because of some structural differences, is placed in a genus of its own. In behavior and reproduction, however, it hardly differs from the other spoonbills. In this species, too, the newly-hatched young show the peculiar beak formation, and their rearing proceeds as it does with their white relatives.

As in the scarlet ibis and in flamingos, the beautiful roseate color disappears in zoos, even on what was formerly the usual diet. If the birds are given carotene-rich food, the whitish feathers are again replaced by a colorful roseate plumage after the next moult.

The roseate spoonbill has been able to maintain its numbers only in very remote areas of South and Central America. Although according to information from America it is rarely hunted for its feathers, its numbers showed a steady decline until the 1930's. A considerable recovery since the 1940's is due to the efforts of the Audubon society and other bird protection associations. Unfortunately this success has more recently been questioned again, for the opening up of land has displaced these birds from certain colonies and, moreover, they are also endangered by pesticides used in mosquito control. Great efforts are being made in America to ensure their preservation. Fortunately, the outlook is considerably better than it is in the case of the hermit ibis or the Japanese ibis. The American ornithologist, Allen, has written about the roseate spoonbill: "Where they occur, they deserve to be guarded like crown jewels."

Fig. 9-16. 1. Lesser spoonbill *(Platalea minor);*
2. Royal spoonbill *(Platalea regia).*

Fig. 9-17. Roseate spoonbill *(Ajaia ajaja).*

10 Flamingos

Family: flamingos, by A. Studer-Thiersch

The FLAMINGOS (family Phoenicopteridae) are long-legged water birds highly adapted to taking small water animals. Because of their overly long legs, they were formerly grouped with the Ciconiiformes, but some zoologists considered them to be related to the Anatidae. It is, therefore, appropriate to set up this family as a separate order, Phoenicopteri.

Distinguishing characteristics

The L is 80–130 cm from the tip of the beak to the tip of the tail, but it may reach 190 cm if measured to the tips of the toes of the extended legs; the weight is 2500–3500 g. The long neck is curved, with 19 cervical vertebrae. The legs are very long, ratio of thigh to leg to tarsus as 1:4:4 in the greater flamingo. The feathers have an after-shaft; there are 12 primaries, and 12 to 16 rectrices. The skeleton, muscles, and air sacs are formed as in storks. There is a very strong ventriculus and a well developed caeca. The feathered preen gland has eight openings. The copulatory organ is retrogressed. The voice is goose-like. There are three genera with five species, all rather similar to one another. 1. The GREATER FLAMINGO *(Phoenicopterus ruber)* has four toes; subspecies include the AMERICAN FLAMINGO (*Phoenicopterus ruber ruber*; Color plate p. 238), with an L of 110 cm, and the EUROPEAN FLAMINGO (*Phoenicopterus ruber roseus*; Color plate p. 238), with an L of 130 cm. 2. The CHILEAN FLAMINGO (*Phoenicopterus chilensis*; Color plate p. 238) has four toes and an L of 105 cm. 3. The LESSER FLAMINGO (*Phoeniconaias minor*; Color plate p. 238) has four toes and an L of 80 cm. 4. The ANDEAN FLAMINGO (*Phoenicoparrus andinus*; Color plate p. 238) has three toes and an L of 120 cm. 5. JAMES' FLAMINGO (*Phoenicoparrus jamesi*; Color plate p. 238) has three toes and an L of 90 cm.

The food of all flamingos consists of small swimming crustacea, algae, and unicellular organisms which they sift out of the water with the beak, which has been transformed into a filtration apparatus. The lower mandible is large, and at its cutting edge, looks as if it is inflated; the upper mandible, on the other hand, is small and fits on the lower

The filter beak

one like a lid. Broad areas inside the beak have lammellae. A sharp angle about the middle of the beak ensures that, when sifting for food, the upper mandible faces downward, and also results in that the cleft between the two mandibles remains small along its entire length when the beak is open. Moreover, the cleft is further narrowed by horny outer lammellae which project laterally from both mandibles.

Thus only particles up to a particular maximal size can be sucked into the beak with the water. Suction results when the bird retracts its thick fleshy tongue with the beak slightly opened; this reduces the pressure in the beak, and so water enters. Then the bird closes the beak and moves the tongue forward, expelling the water while the food particles are caught on the lamellae. At the next retraction of the tongue, food particles are carried into the oral cavity by the bristle-like projections of the tongue, and simultaneously water once more enters the beak.

The efficiency of this filtration arrangement varies in the different flamingo species, depending on the size and shape of the upper mandible and the extent of the surfaces which bear inner lammellae. Thus the lesser, Andean, and James' flamingos eat mainly minute blue algae and others, while the larger species can take considerably larger food particles like small crustacea and mollusks, but are less efficient in retaining smaller organisms. The differences in regard to the major foods are shown in the fact that the species with the fine mesh filtration mechanism sifts just beneath the water surface, while the species with the coarser weave largely work in the mud beneath the water. They also often stir up the mud by tramping with their feet. Flamingos are not, however, capable of a directed selection of food particles; rather, the uptake of food proceeds mechanically. Only the numerous tactile organs on the tongue can check the food taken in.

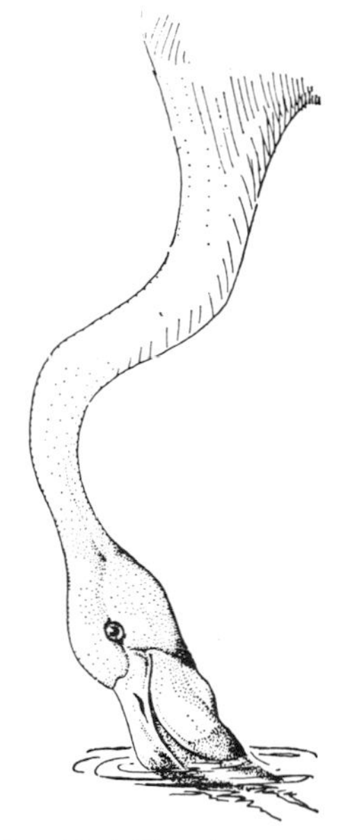

Fig. 10-1. Flamingos, when feeding, hold the beak "reversed," with the upper mandible facing down.

The differences in the filtration apparatus makes it possible for several flamingo species to live in the same area and even to feed in the same lake without disturbing one another. Thus the ranges of the lesser and European flamingos, as well as that of the Chilean, Andean and James' flamingos, coincide in part, while species with similar filtration apparatus, and hence similar food, always have separate areas of distribution.

The type of food and the manner of feeding of flamingos assume an abundance of prey of fairly uniform size, particularly since the birds live in large flocks of up to several hundred thousand individuals. Such conditions are found in salt lakes and brackish coastal lagoons of warm areas. In such waters there is, under suitable conditions, often a mass multiplication of a single animal species, such as the saline shrimp *(Artemia salina).* In these inhospitable areas hardly any other birds compete with flamingos for food, for only they can filter the floating organisms out almost dry, and thus avoid too great of a salt intake. For drinking, flamingos prefer less saline or salt-free water, which they find in springs, or during rain, on the ground; they even drink rainwater running down over their plumage.

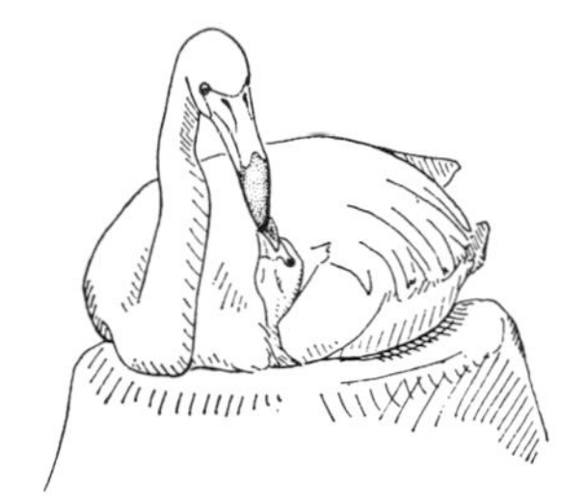

Fig. 10-2. Flamingos feed their young with a bright red nutritional liquid secreted from the esophagus.

Their manner of feeding readily differentiates flamingos from the related orders Ciconiiformes and Anseriformes, to which they do, however, show numerous similarities. Skeletal characteristics and the structure of other organs indicate a connection with the Ciconiiformes; egg protein characteristics resemble those of herons, while the mallophage, voice, webbed feet, and the tongue suggest a relationship with the Anseriformes. Fossils have not clarified the question of the systematic position of the flamingos. But they indicate that the group is very old, and existed as early as the Upper Oligocene, about thirty million years ago, before many other avian orders had yet evolved. The flamingos illustrate a phenomenon often observed in the animal kingdom, namely that old forms are often displaced by younger ones and can only continue to exist if they can adapt to certain extraordinary conditions.

Flamingos of the genus *Phoenicopterus* are often seen in zoos, and when one speaks of flamingos in general, it is usually these which are understood. In all three species, the sexes are hardly distinguishable by external features, and only size can indicate the sex of individual birds.

Although flamingos live in areas that are difficult to reach and are very sensitive to disturbance, particularly early in the breeding season, a certain amount is known about their behavior. They swim readily, sometimes feeding with submerged head while swimming. They fly with an extended neck, and they occasionally soar. Whether flamingos which breed at high altitudes (in Afghanistan at 3000, in the Andes up to 4000 meters) perform regular migrations is not yet known. They breed in huge colonies very close to or right in the water. Their colonies are not always permanent, and this is probably connected with changes in the water levels of their breeding areas. The factors which initiate breeding are still largely unknown, but favorable conditions for nest building are probably essential. Yet breeding fails to occur in some years with apparently favorable conditions, while on the other hand, flamingos sometimes breed even on rocky ground where they can hardly build nests. In the period before breeding, and to a lesser extent, at any time of the year, a number of males and females standing together in a loose group perform ritualized stretching and preening movements, always in the same sequence. This ceremony has developed from signals indicating to change location, and its function is to strengthen cohesion of the group.

Behavior, reproduction

In the time before breeding this behavior helps to gather, out of the gigantic flock, a group of birds equally ready for reproduction, which will later breed together in one part of the colony. This group behavior

Fig. 10-3. At these localities flamingos have bred irregularly in recent years.

1. American flamingo (*Phoenicopterus ruber ruber*);
2. Chilean flamingo (*Phoenicopterus chilensis*);
3. Andean flamingo (*Phoenicoparrus andinus*);
4. James' flamingo (*Phoenicoparrus jamesi*).

1. European flamingo (*Phoenicopterus ruber roseus*);
2. Lesser flamingo (*Phoeniconaias minor*).

is not connected with mating, but it does facilitate pair formation. Nest sites are selected by the females a few days before egg-laying. Actual nest construction is also started by them just before this point. The nest is made of heaped up mud in the shape of the base of a cone, with a shallow depression for the eggs. After egg-laying, both males and females continue busily to build on the nests for several days. They use only such materials as mud, stones, mussels shells, grass, feathers, etc., which they can reach with the beak while standing or lying on the nest site. They heap these materials up around themselves. The height of the mud cone varies according to the nature of the ground; it may be forty centimeters high or be altogether absent on rocky ground. The full clutch consists of a single, elongated chalky white egg averaging 9 x 5.5 cm in size. It is incubated for twenty-seven to thirty-one days by both partners. The newly hatched chick has a white downy plumage, a straight red beak, and thick, swollen, red legs. This leg swelling is hardly noticeable on the second day, and the red color of the legs and beak is replaced by a deep black after seven to ten days. If there is no disturbance, the chick, on its own initiative, leaves the nest when it is four to seven days old. It is then constantly accompanied by the parents and defended against other birds which come too close. During this period both the parent birds and the youngster often climb into empty nests. Soon the parents leave the youngster alone even more frequently, and the youngster joins others to form loose groups with a few adults around. When two to three weeks old, the young grow a second gray downy plumage and their beaks begin to bend. At about four weeks of age, the first contour feathers appear on the shoulders. The beak filtration mechanism develops slowly; in seventy-day-old young which can already fly, the lamellae are not yet fully functional. Up to this age the young can hardly take up the food characteristic for the adults. They therefore depend largely on a highly nutritious liquid secreted by the parents in the region of the esophagus and proventriculus. This secretion has a nutritional value comparable to that of milk; its content of carotenoids and blood give it a bright red color. In the first few days the chick generally receives this liquid while it is being covered by an adult on the nest, its head showing beside the bend of the parent's wing. Later the youngster stands in front of and below the parent when being fed, facing in the same direction. Parents know their own young by their voice and will feed no others, even when the young are already gathered in groups. The first juvenile plumage of flamingos is predominantly a grayish-brown. When three quarters to one and a half years old, they moult into a pale plumage, somewhat like that of the adults, in which only a few feathers still have gray-brown spots near their tips. Adult plumage is assumed at the age of three to four years. Young flamingos continue to grow after they are able to fly. Males only reach full size when one and a half to two years old. They breed for the first time when about six years old. Corresponding to the long growth period and the late onset of breeding, flamingos have a long life span. Some flamingos breeding in the Basel zoo have been in captivity for more than thirty years.

Zoo flamingos lose the red color of their plumage in a few years, but if the birds are given red carotinoids or other suitable food additives, the actual plumage colors are retained.

The behavior of the lesser flamingo is much like that of the two *Phoenicopterus* species. The most marked difference known so far is in the ceremony which, in the latter, leads to group formation. At the corresponding period lesser flamingos walk in large crowded groups, in which the birds touch one another with upright necks, taking rapid steps to and fro. Changes of direction are made by the whole group at the same time. The ritualized stretching and preening movements characteristic of the greater flamingo group are rare in the lesser flamingo group. In areas where both the European and the lesser flamingo occur, both species sometimes breed in the same colony.

The birds of the genus *Phoenicoparrus* have a beak very similar in structure to that of the lesser flamingo. The color of the beak differs and there is a complete absence of the fourth toe, which is present, although very small, in birds of other two genera. Hardly anything is known so far about the Andean and James' flamingos; the latter was even believed to be extinct until it was rediscovered in 1957. In the Andes it lives with Chilean flamingos in the same flock, and they are said to breed together. It may therefore be assumed that these two species differ not only in the structure of their beak, but also in their behavior.

11 Anseriformes: Screamers, Ducks and Geese

Order: Anseriformes

Distinguishing characteristics

A large number of goose-like or fowl-like birds which live near water and at least temporarily go into water are grouped in the order Anseriformes. They are ground and waterbirds; the L is 28–170 cm, and the weight is 200–13,500 g. The nostrils connect with one another, and the lower mandible has a long process at the angle. The sternum has two indentations or two foramina at the rear; these are absent in the fossil giant duck *(Cnemiornis)* of the glacial period. Two pairs of muscles are located between the sternum and the trachea. The neck is extended in flight. There are ten to eleven primaries, the fifth secondary is absent (diastataxic wing), and there are twelve to twenty-four tail feathers. Many down feathers are found in the fully developed plumage. The unspotted eggs are light in color. The young are nidifugous, have a dense downy plumage, and are tended for a long time by one or both parents, except in one brood parasitic species. They are distributed over all continents except Antarctica.

The flamingos and the Ciconiiformes are the nearest relatives of the Anseriformes. Before the Eocene, which was more than fifty-five million years ago, the primitive Anseriformes, as birds of shallow water, had separated from the long-legged, shore, and very shallow water-inhabiting flamingos. The primitive Anseriforme-flamingos and primitive Ciconiiformes arose as parallel groups from a primitive Anseriforme-ciconiiform stock. The living Anseriformes are divided into two families distinguished by the absence or presence of horny lamellae in the beak: 1. SCREAMERS (Anhimidae); and 2. DUCKS and GEESE (Anatidae).

Family: screamers, by R. Kinzelbach

The SCREAMERS (family Anhimidae) are almost goose-sized birds of fowl-like appearance, with fairly thick, long legs and feet without webs. The weight is 2–3 kg. Similarities to the Anatidae include the hexagonal plates on the leg (not on all), the double retractor muscles of the trachea, and the two indentations in the rear of the sternum. Two sharp, spur-like outgrowths of the fused carpal bones on the wing,

F. Reimann

Bird migration on Lake Constance in the late fall:

Like the sea coast, large inland lakes, among them Lake Constance, offer many opportunities to see passing migrants. ☐ Pelecaniformes: 1. Cormorant *(Phalacrocorax carbo).* ☐ Ciconiiformes: 2. Common heron *(Ardea cinerea),* adult. ☐ Ducks and geese: 3. Mute swan *(Cygnus olor),* adults and (a) juveniles; 4. Mallard *(Anas platyrhynchos);* 5. Pintail *(Anas acuta);* 6. Shoveller *(Anas clypeata);* 7. Gadwell *(Anas strepera);* 8. Red-crested pochard *(Netta rufina);* 9. Common golden-eye *(Bucephala clangula);* 10. Tufted duck *(Aythya fuligula);* 11. Common pochard *(Aythya ferina).* ☐ Grebes: 12. Great crested grebe *(Podiceps cristatus),* in winter plumage; 13. Black-necked grebe *(Podiceps nigricollis),* in winter plumage. ☐ Rails (see Vol. VIII): 14. Little drake *(Porzana parva).* ☐ Shorebirds and gulls (see Vol. VIII): 15. Black-headed gull *(Larus ridibundus),* in winter plumage (Breeding plumage with chocolate-brown head); 16. Common tern *(Sterna hirundo);* 17. Common curlew *(Numenius arquata);* 18. Dunlin *(Calidris alpina);* 19. Common snipe *(Gallinago gallinago);* 20. Ruff *(Philomachus pugnax).* ☐ Passerine birds (see Vol. IX): 21. Swallow *(Hirundo rustica);* 22. Sand martin *(Riparia riparia).*

which are covered externally by keratin, are characteristic of screamers. The interior of the whole skeleton is permeated by air sacs to a great degree, and these continue into an air-containing tissue which takes up most of the dermis. But in contrast to the gannets, the feather shafts do not reach this air tissue in the screamers. The ribs lack uncinate processes. The plumage does not show the usual extension of contour feathers and down into pterylae. There is no copulatory organ.

Though the screamers have been known in Europe for a long time, their internal structure was only examined in the last third of the nineteenth century. This clarified their position in the zoological system. Before that they were grouped with the rails or the Ciconiiformes. Because some of their characters are primitive, they were even classified as the closest living relatives of the extinct archaeopteryx. Today it is generally believed that the screamers and the ducks and geese have a common ancestor. This view is strengthened by the fact that the Australian magpie goose shows certain similarities to the screamers in its manner of moulting, the absence of webs on the feet, the plates on the tarsus, the formation of the sternum, and the tendency to perch in trees.

There are two genera with marked differences in their internal structure, the HORNED SCREAMERS *(Anhima)*, with fourteen tail feathers, and the CRESTED SCREAMERS *(Chauna)* with twelve tail feathers; altogether there are three species with no subspecies. 1. The HORNED SCREAMER (*Anhima cornuta;* Color plate p. 267) has an L of 80 cm and inhabits the flood forests of the Amazon delta. 2. The CRESTED SCREAMERS (*Chauna torquata;* Color plate p. 267), with an L of 90 cm, are found in swampy pampas areas of the La Plata States; and 3. The BLACK-NECKED SCREAMER *(Chauna chavaria)*, with an L of 70 cm, is found on forest rivers of Colombia and Venezuela.

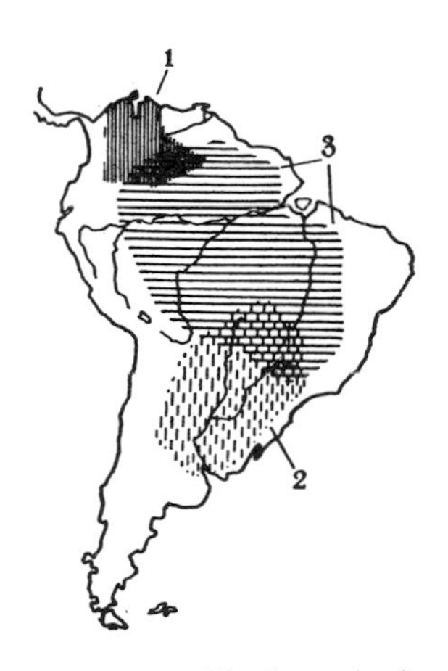

Fig. 11-1. 1. Black-necked screamer *(Chauna chavaria)*; 2. Crested screamer *(Chauna torquata)*; 3. Horned screamer *(Anhima cornuta)*.

Outside the breeding season, the horned screamers live in troops of five to ten birds. The crested screamers are, however, found in larger flocks which circle above the waters in their habitats in the evenings, calling melodiously. In contrast to the Anatidae, they can glide well. Apart from goose-like calls, they occasionally utter a throaty sound resembling drumming. The horned screamer is less vocal, uttering a call which can be described as "ee-oo." In spite of their unwebbed feet, screamers swim very well; the crested screamer will even climb onto the leaves of floating plants from the shore. These birds calmly walk about the shore or in shallow water; their food is entirely vegetarian and they obtain some of their food while swimming. They readily perch on the branches of trees and, when disturbed or pursued, generally take to trees.

The sharp spurs on the wings are sometimes used as weapons in fights connected with pair formation. They are used even more against enemies; with the spurs, screamers can fend off hunting dogs and other opponents successfully. There are no external differences between the two sexes. The "horn" of the horned screamer is formed from an unbranched modified feather shaft and can be up to fifteen centimeters long; it is present in both sexes.

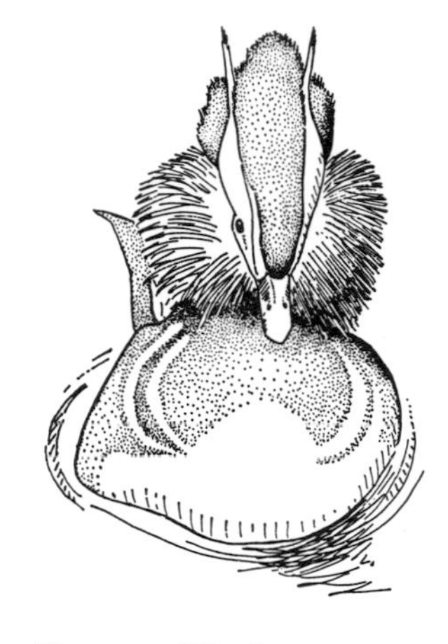

Fig. 11-2. Displaying mandarin duck.

The large nest is either near or actually in shallow water and is built of plant materials. In the case of the crested screamers, it contains five to six eggs; in the case of horned screamers there are only two eggs which have smooth yellowish-white shells and are incubated by both parents. The young are precocial, and with their yellow down they resemble goslings. They remain in the nest for only a short time, and then wander about led by their parents. Young screamers are often caught and tamed in their home countries. They readily take to captivity and can be kept with chickens in farmyards; they are even said to defend their companions against birds of prey and other enemies. Occasionally one finds them walking about at liberty in South American parks and zoos. The crested screamer in particular is often seen in European zoos and occasionally it has reared young there.

Family: ducks and geese, by U. and H. G. Klös

All other members of this order are included in the family of ducks and geese (Anatidae). They have horny lamellae on the interior of the beak near the cutting edge. There are webs between the anterior toes. The upper mandible has a particularly hard process, the "nail", at its tip. There are large nasal cavities; because of this, geese breathe faster when reacting to olfactory stimuli. In the interior of the beak there are special tactile receptors, the Grandry corpuscles in addition to the corpuscles of Herbst which other birds also have. There are sixteen to twenty-five cervical vertebrae. The preen gland is particularly large. They cannot soar or glide to any extent; a few species are flightless, while others have a rapid flight. The nest down, which the female plucks from her own body, is placed under the eggs and pulled over them when the female leaves the nest. There is a long penis which, when evaginated, turns to the left. Copulation almost always occurs on water. The downy young have a long tail, and take to water soon after hatching in the company of one of the parents (except in the one brood parasitic species). The down feathers of the young are on the tips of the juvenile feathers, which emerge later. There are three subfamilies: 1. the MAGPIE GOOSE (Anseranatinae) with only minute webs; 2. the GEESE and relatives (Anserinae) with larger webs and small scales on the tarsus and toes; and 3. the DUCKS and relatives (Anatinae) which also have large webs; there are almost always large scales or plates on the front of the tarsus and on the toes.

Distinguishing characteristics

The various forms of the ducks and geese

The Anatidae are made up of many species that are colorful and of many shapes. There is a wide range of changing forms and ways of life, from the minute AFRICAN PIGMY GEESE to the TRUMPETER SWAN, which weighs thirteen and one half kilograms, and from the inconspicuous GREYLAG GOOSE to the colorful plumage of the KING EIDER. Nevertheless, the different species have common characteristics which justify their grouping in one family. Thus all are water birds, even species which live mainly on land like the CAPE BARREN GOOSE and the HAWAIIAN GOOSE. Since, as swimmers, their plumage must always be greased, they have a particularly large preen gland. Males of all ducks, geese, and swans have a copulatory organ which is evaginated from the cloaca during copulation. Such an erectile penis is found in only a few orders of birds. The lamellae, placed in a row in the interior of the beak, have become a sifting apparatus, which is more or less specialized in various species. They provided the basis for an earlier name of this group, Lamellirostres.

Plumage colors

In the true geese and swans the sexes are similar in plumage and there is a single moult each year; in the others, however, with few exceptions, the two sexes differ in color and the contour feathers are moulted twice a year. In both groups there are, however, some species with a less extensive second moult. In the oldsquaw and the female pochard, there are even three plumage changes in a year. As a rule, female ducks have an inconspicuous plumage all year round, in which dull browns and grays predominate. Males of most species, however, change before the period of pair formation in the fall, from an eclipse plumage which resembles that of the female to the nuptial plumage.

The color on one or more parts of the body may be as bright as a torch; the metallic green on the head of the male mallard, shoveller, chestnut teal, and some mergansers, the bright orange on the head of the red-crested pochard or the deep brown of the head of the pochard, illustrate these bright nuptial plumages. Colors are, however, not the only decoration in nuptial plumages. There are modifications of the shape of single feathers or of whole regions of the plumage. Thus there are curled hoods, as in the muscovy duck, or crests as in the patagonian crested duck, the marbled and falcated teal, the tufted mandarin, the wood duck, and certain mergansers. There are also manes on the neck, which are found in the maned wood duck and the Orinoco goose. The attractive central tail feathers in the mallard, with their upward curl, are well-known; even more striking are the sickel feathers of the falcated teal and the elongated lancet-like flank feathers of the plumed tree duck. Especially peculiar are the wide, upward-bent inner vanes of the shoulder feathers of the mandarin duck, which in the swimming bird look like orange sails. In the pintail, the much prolonged tail feathers characterize the nuptial plumage, but the male oldsquaw has even longer tail feathers.

Fig. 11-3. Hooded merganser with depressed and erected hood. The striking colors of many ducks are often signals which are made more conspicuous by certain movements or by spreading the feathers.

In some Anatidae, parts other than the plumage contribute to the decorative effect of the nuptial plumage. A highly colored fleshy hump at the base of the beak is present in the breeding season in the mute and black-necked swan, the shelduck, the king eider, the rosy-billed pochard, and the scoters. In the mute swan, the rosy-billed pochard, and the common and velvet scoters this hump sits on a bony projection of the skull. The color of the beak also becomes more intense in many species as the nuptial plumage is being assumed; in the shoveller and the gadwall it becomes deep black, in the rosy-billed pochard and freckled duck it becomes bright red, and in most stiff-tailed duck species it becomes a brilliant blue.

The protective coloration of the females

The inconspicuous plumage of most female ducks can be explained by the fact that, while incubating, they must remain unseen as much as possible and must blend with their environment. This does not apply to the males, which take no part in incubation. On the contrary, their conspicuous colors effectively support the species-specific movements involved in pair formation and the maintenance of the pair bond. But there are exceptions to this rule. In many, sometimes closely related species, the males of one species have bright nuptial plumage, such as in the mallard, while in others, for example the Florida duck, and the spot-billed and gray ducks, both sexes have a similar dull plumage. The reverse, a bright plumage in both sexes, is seen in the Chiloë wigeon and the shelducks.

In all ducks and geese, the flight feathers are moulted only once a year and they are dropped simultaneously, so that the bird is incapable of flight for a short period. Only the Australian magpie goose is an exception to this; it sheds its primaries one after the other and retains the ability to fly throughout. Oskar Heinroth writes: "The time of shedding of the flight feathers is related, at least in many species, to reproduction. In ducks, where only the female incubates and tends the young, males shed the flight feathers earlier than females. The latter shed these feathers only when the young are half grown. In the swans, the sequence is reversed; females become flightless soon after breeding and the males later, at a time when females can already make use of their wings to some extent. During the slow development of the young, the parents thus moult one after the other so that one of them is always ready to guide and particularly to defend the youngsters. The ability to fly is reached by the young and regained by the adult male at about the same time, in late summer, and the family is then ready to begin the fall migration." Incidentally, there are 25,000 contour feathers on the body of a whistling swan, and about eighty percent of them are on the long neck.

The so-called nest-down grows among the ordinary down on the underparts, before the start of the breeding season, in birds which will later incubate, which in most cases are the females. The nest-down feathers are longer and often of a different color than the ordinary down feathers. Before the clutch is complete, the incubating bird plucks out the nest-down and pads the nest cavity and the rim of the nest with it. When the parents leave, the eggs are often first covered by this down. It is interesting that the nest-down is dark in species which nest in the open and whose eggs need a protectively colored cover, but white in hole-nesting species. Thus the American merganser, which nests in holes, has white down, but the red-breasted merganser, which nests in the open, has dark nest-down.

Many ducks, including the shelducks, have a conspicuous speculum on the wing. In the simplest cases the speculum is white, as in most diving ducks and the whistling or tree ducks. It may be well demarcated by a black edge as in the gadwall, or it may be an area with a metallic glint as in most of the other species of feeding ducks. These conspicuous specula serve as signals; for example, in birds which fly in a flock, they aid in the cohesion of its members. The same function, to maintain contact, is served by white tail feathers, the bright tail, or the wing bands.

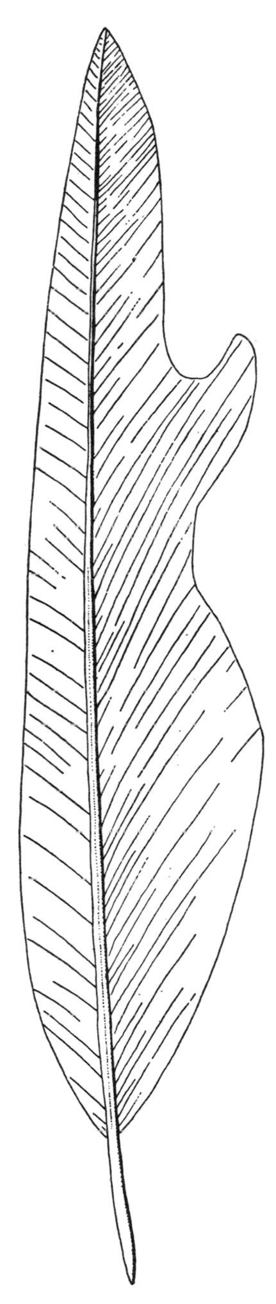

Fig. 11-4. The sound-producing wing feathers of the lesser wandering tree duck. The peculiar shape of this primary causes a characteristic sound during flight.

Instead of visual contact signals, there may also be acoustic ones. Thus, a few species of ducks have special flight feathers which, when strongly agitated, produce a sound. In the Indian whistling duck these feathers produce a whistling sound, and in the golden-eye and the common scoter they produce a ringing sound. The rushing wing sound produced by flying mute swans is well known; it is often audible for more than one hundred meters and helps the birds maintain contact. The whooper swan, on the other hand, makes no sound in flight, but the structure of its trachea is such that it can utter loud trumpet-like calls so that the members of a migrating flock can communicate.

A few further peculiarities about the color of nuptial plumages are worthy of mention. With very few exceptions, green colors in birds are not due to pigments but are brought about by a particular construction of the feathers. As von Boetticher puts it, the birds are therefore "not green, they merely look green." An exception to this rule is the male common eider; the soft green on its plumage is due to a particular pigment, a green lipochrome.

The pink-headed duck owes the pink color of its head and neck to a pigment which is unique in birds. The salmon-pink hue on the breast and belly plumage of some species of mergansers, on the breast of male common eiders, and on the flank of the red-crested pochard, is due to a light-sensitive pigment.

All Anatidae have webbed feet. The webs extend from the second to the third and from the third to the fourth toe. The webs have retrogressed only in the magpie goose, which has already been mentioned several times as an exception, and the Hawaiian goose. In the latter, the retrogression of the webs is an adaptation to its environment on the lava slabs of the Hawaiian archipelago; it makes the toes more mobile and facilitates climbing on the rock slabs. In contrast to these very few species which live predominantly on land, others, particularly the scoters, rarely go ashore except during reproduction. Anatidae do not have to make any particular effort to remain on the surface of the water or to swim. Their buoyancy is due mainly to the air held in the plumage. Heinroth demonstrated this in an interesting experiment on a freshly killed drake mallard weighing 1337 grams. "The wings were held in the natural position by a thread passed around the bird's body and it was so immersed in the water in a vessel. It displaced 2060 milliliters of water. Then we plucked it and immersed it again in the same vessel. It now weighed 1270 grams and displaced 1390 milliliters of water. Therefore, in the plumage which weighed 67 grams, 650 milliliters of air had been held. The specific gravity of the bird with its plumage was about 0.6, while that of the plucked bird was 0.91, so that when floating it only barely projected above the water."

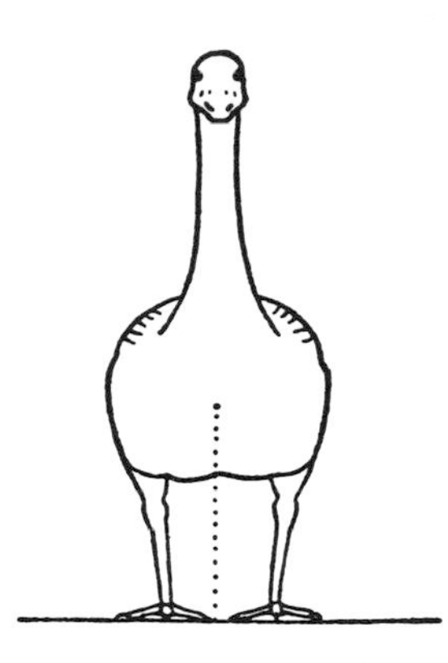

Fig. 11-5. The wider the body the more the bird waddles (walking Greylag).

Birds of this family take great care not to get water under their plumage. The closed wing is covered by feathers projecting up from the side of the breast so that, of the wing feathers, only the scapulars and the primaries are generally exposed. The body plumage is continually and carefully covered with oil from the preen gland and so forms a layer impermeable to water. As Heinroth expressed it: "The birds float as in a trough, in which not only the body remains dry but the wings are also protected from the water."

A particular adaptation to the requirements of swimming is the broad cross section of the body of most Anatidae. This broad, barge-like body maintains its balance despite wind or waves. Another adaptation is the shortening of the thigh and tarsus. The tarsus functions like the arm of a lever in swimming, and in slow swimming is almost the only part of the leg that moves. Only in faster swimming is the thigh involved as well; it is drawn back with partially extended knee, the lower leg serving merely to transmit power.

To reduce resistance to the water when the tarsus moves forward, the webs and toes are folded together and the toes are bent. In pushing back, the toes and webs are fully extended and form an effective oar-like surface. As in walking, so in swimming the legs move alternately. Only the mute swan in its agressive display swims with jerky movements, pushing back with both legs at the same time; however, it swims no faster by this method.

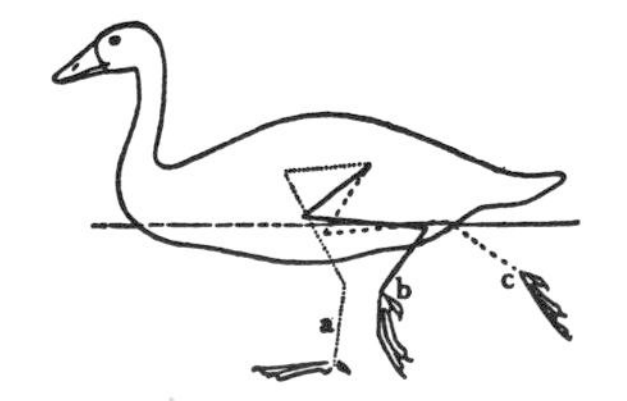

Fig. 11-6. Position of the leg of a goose when standing (a), and swimming: forward movement of foot with folded web (b), backward movement of foot with outspread toes and web (c).

Zoo visitors often express surprise at the swimming ability of little ducklings. In view of what was said above, it is evident that their swimming is a purely "passive" skill. One is not surprised by the fact that a cork or piece of wood floats. It is truly surprising, however, that the little feather balls, often only a few hours old, are already able to dive, for diving requires effort. Actually all Anseriformes can dive, and this skill is practiced, probably as a protective measure, particularly by the young. The diving ducks, mergansers, and scoters have a pronounced diving skill. Because of their somewhat higher specific gravity, they float deeper in the water than the species which rarely dive, and the rear of their bodies lies deeper in the water than the breast. This body posture would result in an upward movement in diving, but this is counteracted by pushing the feet backward beyond the body's center of gravity. Thus the feet serve not only as oars but also determine the depth of underwater swimming; according to whether the diving bird moves its feet behind, at, or in front of the center of gravity, it will swim downward, horizontally, or upwards when under water. In diving, in contrast to swimming, both legs are pushed backward at the same time. The pocket formed by the breast and the flank feathers covering the closed wing is particularly tight in diving ducks, preventing the wing from getting wet. The duration of dives is generally sixty to seventy seconds, although the velvet scoter dives voluntarily for as long as two minutes. The depth of dives is said to reach up to nineteen meters in eiders.

All species of the family use flapping flight, but they are unable to glide to any extent. Among them are some which fly fast and with great endurance. The greatest altitude of flight was established in wild geese at 10,000 meters above sea level. On land these birds move with a peculiarly waddling gait. To maintain their balance, the center of gravity must be shifted above the supporting leg. Because of the wide body and the relatively short legs, the displacement of the center of gravity can only be effected by a rotation and a shift of the body to the side. This motion causes the characteristic waddling of ducks and geese.

The most obvious common characteristic of geese and ducks is the beak. A similar beak structure is found only in the flamingos and shearwaters. At the edge of the beak there is a row of lamellae which, together with the tongue, form a fine shifting apparatus. "Except in the mergansers, the tongue is fleshy, sensitive, and usually toothed at the edges, and has a certain similarity to that of flamingos," described Heinroth. "When geese chew, it serves to push bits of food between the teeth of the beak, while in dabbling ducks and swans it functions as a suction piston during the so-called chattering. Chattering, in this connection, does not mean a call; it is the removal of food particles from the water, accompanied by a characteristic sound. There is quite a peculiar sound when a large number of ducks work over the shallows of a pond at night. In chattering the beak is continually opened a little and closed again; the tongue sucks the muddy water in through the tip of the beak and on closing the beak squeezes it out again at the base of the beak between the two mandibles. Thus, small food particles are detected by the fine nerve endings and held back by the lamellae. No other birds (except the flamingos) seek and find food in this manner."

The sifting apparatus is particularly pronounced in shovellers and the pink-eared duck, which feed on the smallest floating organisms.

The lamellae are coarser in the geese and serve to tear off grass by means of a lateral movement of the head. In the mergansers they have become small teeth, which enable them to grasp slippery fish, their prey. Lüttschwager has counted the lamellae in many species of ducks and found that in general their number is twice as high in surface-feeding ducks as in diving ducks.

A duck's beak has great tactile sensitivity, for it contains many of Herbst's and Grandy's corpuscles. These lie mainly on the inner surfaces of the tips of both mandibles. The following comparison gives an indication of the marked sensitivity of duck and goose beaks: in 1893, Geberg discovered that there were on the average twenty-seven touch receptors per square millimeter near the edge of the upper mandible in the mallard, while on the tip of the human index finger there were only twenty-three. The outer surface of the beak is largely covered by a soft membrane; only the tips of both mandibles carry a hard horny "nail." An exception are the mergansers; their narrow beaks are covered throughout by a hard horny surface (rhamphotheca).

According to their manner of obtaining food, the Anatidae can be divided into several groups. Swans, shelducks, and surface feeding ducks "up end"; that is, they immerse the head and neck with the rear of the body projecting almost upright above the water. In this way they can feel over the bottom of shallow waters with the beak, and obtain food by straining it out of the water. Diving ducks also generally get their food from the bottom, but they reach greater depths and dive completely below the surface. Lastly, the mergansers chase fish beneath the surface.

The true geese, swans, and whistling ducks are entirely vegetarian. Mergansers, scoters, eider ducks, and the South American torrent ducks take only animal food. The rest take both plant and animal food. The amounts of food taken are at times considerable. In the digestive tract of an eider duck, 114 mussels were found, some of which were already partially digested within their shells; the gullet and stomach of a velvet scoter contained forty-five oysters.

The vocal sounds of the Anatidae are as manifold as the family itself. In geese, swans, and whistling ducks, the voices of the two sexes are very similar; they are, however, different in male and female ducks. Males of the Anatidae have a sac-like dilatation at the bifurcation of the trachea. This bulla is completely or, as in the diving ducks, only partly ossified. The Cape Barren goose is, among others, an exception lacking a bulla. The whistling ducks, in contrast to the others, have a symmetrically placed bulla in both sexes. In the magpie goose the trachea forms a large loop lying between the skin and the breast muscles, while in the whooper and whistling groups of swans there is a similar loop within the hollowed-out sternum. Because of the resulting resonating chamber, the voice of these birds is particularly

Fig. 11-7. In this posture the drake mallard utters the attraction and warning call "rab."

The "rab rab" chatter of a mallard pair calms both members.

With this gesture a female mallard rejects strange males.

loud and far reaching. The mute swan lacks such a resonating chamber; its trachea is straight and hence its voice is insignificant, which, with some exaggeration, has given it its English name. The hissing unique to geese is produced by a rapid expulsion of air through the narrow glottis.

Calls enable the birds to communicate with one another, and are part of their modes of expression. Besides Oskar Heinroth, one of the founders of ethology, Konrad Lorenz and his co-workers have become experts in the "language" of geese and ducks as a result of intensive studies of a number of geese and duck species in particular. Sounds are generally coupled with quite definite and specific expressive movements. A comparison of these movements and calls of particular species permits conclusions about the degree of their phylogenetic relationship. Thus behavior patterns can be clues, which are as valid as structual characteristics, for taxonomy.

Calls which are not part of pair formation include the contact call, the warning call, and the call to attract others. The realm of calls related to pairing is more exclusive. The males of some species (the common teal, the garganey, the red-crested, rosy-billed, and common pochards, and the golden-eye) are, according to Heinroth, altogether mute except in connection with mating. In the group of birds related to the geese, in which both sexes incubate, the calls of the two sexes are almost identical, but they differ fundamentally in the ducks. The "language" of the mallard has been known for the longest time and has been the most thoroughly studied as a result of the work of Konrad Lorenz and W. von de Wall. The most significant mallard calls include the call of attraction and warning, a monosyllabic "raab," which is similar in both sexes. A repetition of this call indicates the intention to move to another place. The members of a pair or of a group of siblings maintain contact with the double syllabic contact call "rab-rab," with emphasis on the second syllable. The movement of rejection of females, an unmistakable expression, is accompanied by a sequence of sharp "quak" calls, and is used by a female to reject a pursuing male. The "decrescendo" call is generally a six syllable "quagagagagagag," descending in pitch, which is most often uttered by unpaired ducks, and by paired ones only when their partner has flown away. This call is released by the sight of ducks of the same species in flight.

According to von de Wall, the courtship display, with its typical movements and calls, is divided into two separate segments in the surface feeding ducks, the Anatini: the social display, as Heinroth calls it, and the directed display. While the social display has the purpose of attracting as many ducks of the same species as possible and thus to make pair formation possible, the function of the directed display is to facilitate the cohesion of the individual pair and to finally lead to copulation.

One of the expressive movements of the social display is the nodding of female ducks. Konrad Lorenz writes about it: "The female uses this movement only when several drakes have gathered and have indicated their mood to display by the erection of the head feathers and by shaking themselves. Then the duck suddenly dashes among the drakes in a peculiar flat body posture and with strongly emphasized nodding movements of the head, swimming around as many of them as possible in short arcs. In nodding, the head is held so low over the water that the chin touches the surface." Characteristic of the drakes are the "introductory shaking," the "grunt whistle," "head-up, tail-up," and an "up-and-down" movement. In the introductory shaking, the drake swims imposingly high on the water with retracted head, strongly erected head plumage, and tightly compressed body plumage.

Fig. 11-8. Many members of the species join in the social display. This facilitates pair formation. The nod-swimming of the female belongs to the mallard's social display...

...as well as the head-shaking of the drake...

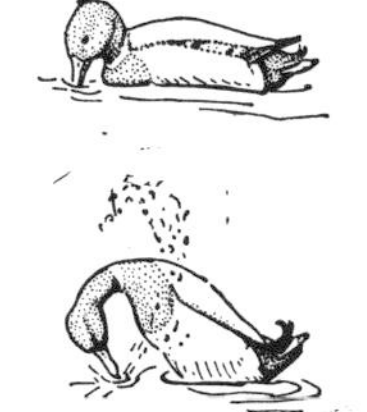

...the drake's grunt whistle in which he holds the neck strangely bent,

and the head-up, tail-up of the drake.

"This posture is only distinguished," writes Konrad Lorenz, "from the posture which precedes normal shaking in the mallard in that it is held for a longer time, often for several minutes. In normal shaking, the head retraction and erection of the head plumage last only a few seconds and increase in intensity. The expert can predict exactly when the actual shaking will occur from this increase in intensity, just as one can predict the moment of sneezing from the preceding facial distortion in man. In the social display of male mallards, not only is the expected shaking delayed much longer, but when at last it happens it does not bring with it a release of the preceding posture. Rather, the first shaking, in which the head is tossed up in a peculiar, inhibited, cautious, and yet hasty fashion, is followed in a few seconds by a second and a third one. The intensity of the movement increases every time until the shaking movement seems to pull the bird high out of the water, as if it were a seizure. When this stage has been reached, one of the three display movements now to be described follows almost regularly (grunt whistle, up and down, and head-up, tail-up), in which all drakes take part. The whole gathering then, for the time being, shows a release of tension and stops displaying, or it begins again after a short pause with an introductory shaking of lesser intensity."

The grunt whistle requires a peculiar bent posture of the drake, which probably stretches the trachea. At the climax of this movement there follows, according to Lorenz, "a loud, sharp whistle, followed by a deep grunting sound, while the head is raised once more and the body sinks back onto the water surface. The grunting sounds as if the air compressed during the whistle were escaping." The head-up, tail-up is one of the most conspicuous movements of the drakes. "The drake first pulls the head back and upwards with retracted chin, while uttering a loud whistle, and simultaneously raises the tail with its strongly erected rump feathers, so that the whole bird becomes pecu-

▷

Screamers (see Chapter eleven): 1. Crested screamer *(Chauna torquata)*; 2. Horned screamer *(Anhima cornuta)*. Anatidae: 3. Magpie goose *(Anseranas semipalmata)*

1
2
3

1
2
4
3
5
6

Fig. 11-9. The inciting of the female belongs to the mallard's oriented display.

Sham-preening is part of the drake's oriented display.

The pumping of a mallard pair is a prelude to copulation.

After copulation pull-up of the male's head is followed...

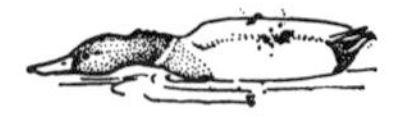

...by nod-swimming.

◁

Whistling and tree ducks: 1. Black-bellied tree duck *(Dendrocygna autumnalis);* 2. White-faced whistling duck *(Dendrocygna viduata);* 3. Fulvous tree duck *(Dendrocygna bicolor);* 4. Plumed whistling duck *(Dendrocygna eytoni);* 5. Wandering whistling duck *(Dendrocygna arcuata);* 6. Spotted whistling duck *(Dendrocygna guttata).*

liarly short and high. The elbows are raised at the same time, so that the projecting curled tail feathers remain visible from the side. This phase lasts for one twentieth of a second, and then the body resumes its normal posture. Only the head remains raised for a moment, and the beak is then directed at a particular duck present during the social display of the drakes; in paired drakes this is always the mate." Nodding and a change of the head plumage into a narrow, high, and very shiny disc, with rotation of the back of the head towards the female, also belong to these movements of expression. The up and down movement is characterized by a rapid dipping of the beak into the water, followed by a quick raising of the head, during which the breast remains deep in the water. At the highest position of the head a whistle is heard which is followed by a rapid "rab rab" call.

Inciting by the female is part of the oriented display. "The duck," says Lorenz, "turns towards her partner or the drake she has selected as her partner, swims after him, and at the same time threatens over her shoulder towards the other males of the species." The call which accompanies this is a bleating "queggeggeggeggeggeggegg" which is generally emphasized on the third syllable. Also part of the oriented display is drinking toward one another, the "meaning" of which was recognized by Heinroth: "If two ducks meet on the water and they drink facing one another, there will be peace. The development of this symbolic action may perhap be explained by the idea that animals which eat and drink together have no bad intentions towards one another." As a form of male display, drinking to one another is often combined with sham preening.

The prelude to copulation is "pumping." The duck and the drake perform up and down pumping movements of the head, in which the head is raised slowly and lowered in jerks. The reverse movement, down slowly and upwards in jerks, is a flight intention signal. In the post-coital display, the drake performs a movement called "jerking up"; immediately after copulating it throws the head and neck backwards over the back, sometimes still holding the duck's nape feathers in the beak, so that her head is jerked up too. Then follows swimming.

Like most species of birds, ducks and geese live in monogamy. Only wild muscovy ducks and the comb duck are exceptions to this. The duration of the pair bond is, however, variable. In most species of ducks it ends with the laying of the first egg, while in others, such as the blue-winged teal and the shoveller, the males remain near the nest until the young hatch, and only then do they gather into continually growing flocks of males. In the true geese, the males are always by the nest, defending the female and the clutch and, after the young hatch, taking a great part in leading them. Since young geese are led for nearly a year until the start of the next breeding season, the parents stay together through the winter and mate again with one another in the spring. Thus a permanent pair bond results in geese. It is often so strong that a widowed bird will not pair up again for the rest of its life.

"Engagement" and the mating season

In many Anatidae, mating is preceded by a so called "engagement period." The birds meet in the fall for social displays and to form pairs. At this time the gonads are still completely in the quiescent state, and if there is a copulation it is without the normal consequences. Only in the following spring are the gonads fully developed, and so the mating season proper begins.

Fig. 11-10. Courtship feeding is part of the red-crested duck's display. The drake brings the duck a piece of a plant.

In the great majority of species, copulation takes place on the water. Whistling ducks, South American sheldgeese, and coscoroba swans stand in the water while mating. Magpie geese, however, according to Delacour, copulate, as a rule, on their nest on land. The penis is normally hidden in the cloaca and is evaginated and erected for copulation. Release of the ovum from the follicle in the female and the release of sperm in the male depend not only on excitement, but also on external factors. Thus every breeder of waterfowl knows that when there is bad weather in the mating period, clutches are smaller and the eggs are often infertile. The spermatozoa live long. In the mallard, Elder and Weller found that spermatozoa retained life and fertilizing capacity in the oviduct for twelve to fifteen days.

In spring, when the gonads are at the peak of their development, there are attempts to "rape" strange females in the mallard and pintail and a few other species. Heinroth describes these raping flights: "In a shallow bay of the pond the female of the mallard, pair A, seeks food by up-ending; her mate is close by keeping watch. A hundred feet away a second pair B, comes down on the water. Male A quickly swims towards the strange female B, finally flying towards her, but she takes flight at the last moment and a wild aerial chase begins. The pursued duck rises higher and higher, the strange drake A behind her; she tries to escape by swerving and suddenly flying slowly, and both are followed by male B since he does not know where his mate will end up. Thus one sees two drakes following a duck, and this is generally interpreted as if both were chasing the duck; in reality, however, a strange male is chasing the female of a pair which belong together. Gradually male A gives up and returns directly to his mate."

Nest building

Most of the species build their nests on the ground. Some of the genus *Tadorna,* like the common shelduck, prefer burrows in the ground as nest sites, although some other species of ducks, among them the mallard, also nest on trees. The Orinoco goose, many of the Cairini like the maned wood duck, the mandarin, and the wood duck, the Brazilian teal, the pygmy geese, the comb duck and its relatives, the goldeneyes, and some mergansers breed preferably in tree cavities. Nests of the magpie goose, the coscoroba swan, and many diving ducks and stiff-tailed ducks are often found in dense swamp vegetation on the water. The yellow-billed pintail, according to Wetmore, sometimes breeds in deserted parrot nests.

In general, Anatidae except for the snow goose, Brant goose, barnacle goose, red-breasted goose, and Canada goose are not colonial nesters. But there may be great accumulations of nests, for example on islands. Thus Salomonsen reported that 1000 Eider duck nests were found on an island the size of about ten hectars; Hammond and Mann reported 2000 Gadwall nests on an area of 4000 square meters. It is understandable that waterbirds, like ducks and geese, build their nests as close as possible to water. In many surface-feeding ducks, nests may be up to one and one-half kilometers away from water; most nests, however, are no more than forty-five meters from it.

The nest construction is simple. As far as it is possible, the birds make a hollow in the ground and pull in stems and leaves from around the nest site, as far as they can reach with their outstretched necks. Probably all species cover incomplete clutches with plant material when leaving the nest. Shortly before the last egg is laid, the females pluck the nest-down and line the nest with it.

The females usually incubate

In general only the females incubate. Only in the magpie goose, the black swan, and the whistling ducks do the male and female relieve one another for incubation. Occasionally a female will lay its eggs in strange nests before it has built its own nest. Such a trend toward brood parasitism is particularly common among species which nest in holes in trees. In this way, almost unbelievably large clutches may arise. Delacour reports eighty-seven eggs in a red-head nest. Highly developed brood parasitism occurs in the black-headed duck. It always lays its eggs in strange nests and does not concern itself about its offspring. It prefers to lay into the nests of the equally black-headed rosy-billed pochard, but often lays in the nests of quite unrelated birds, even in those of birds of prey.

Clutch size varies considerably. In most species the female lays one egg per day until the clutch is complete; in others, as in the mute swan, there is an interval of two to three days between each egg. This is a considerable accomplishment. Consider the mandarin duck as an example: the female, weighing only 500 grams, lays thirteen eggs, each weighing 50 grams, within only thirteen days. It thus produces altogether 130 per cent of its body weight. Incubation begins with the laying of the last egg and lasts twenty-one to forty-three days. The young hatch during the season which provides the best weather and the best feeding conditions. In the northern temperate zone this is in the months of April to June; south of the Equator this occurs from September to December. In the tropics there could be breeding throughout the year, and yet the individual species have also evolved preferred breeding seasons there. As has been established in zoos, most species adapt their breeding season to the new conditions when transported to Europe. Thus South African species breed in spring instead of in the fall. Only Australian black swans and the cereopsis retain their old breeding seasons and still lay eggs in the middle of the European winter.

▷ Swans: 1. Coscoroba swan *(Coscoroba coscoroba)*; 2. Black swan *(Cygnus atratus)*; 3. Black-necked swan *(Cygnus melanocoryphus)*.

The young of the Anatidae are among the best developed precocial birds. They are covered with a dense nest-down which gets greased by mother's abdominal plumage and is water repellent. Only a few hours after the hatching of the last chick, the flock with the mother leaves the nest and heads for water. Leaving the nest site is difficult for the young of tree hole nesters because of the drop involved. Some mallards breed every year in the Berlin Zoo on the nine-meter-high rock of the lion enclosure. Every year it is a problem when the mother, with quick wing beats, flies down from the cliff, and the young, like little balls of wool, tumble after her. In spite of the height the falls never injure them; they are light, and their bones are soft and pliable so that they can act as springs. Ducklings feed independently from the first day on; the adults, or in most ducks the female alone, merely warm and protect their offspring. In some swans, the care of the young includes carrying tired youngsters on the parent's back. Goose families often return to their nests in the evening during the first few days. The female duck leads the young until they are able to fly. In the shoveller and blue-winged teal, this takes thirty-nine to forty days, in the mallard fifty-six to sixty, and in the red-head and canvasback fifty-four to seventy days. Geese, as already mentioned, lead their young until the next breeding season.

Fig. 11-11. Swan parents sometimes carry tired chicks on their backs (Black-necked swan).

The young of most duck species are sexually mature when nine to eleven months old; thus they breed in the second summer of their life. Whistling ducks in general breed when one year old, the true geese when three years old, and the swans at four to five years of age.

Some species of ducks and geese prefer quite definite types of waters. Thus sea geese, as the name suggests, live on coasts and islands. Most true geese inhabit shallow pools and flooded areas. Diving ducks are found in great numbers in swampy waters, while the Egyptian and Orinoco geese prefer rivers, the golden-eyes prefer deep lakes, and the peculiar torrent ducks prefer swift-flowing mountain streams. The food supply also plays an important role in the selection of habitats. Thus grazing geese always live near tundra or grasslands. Some species have such limited adaptability that they are tied to a certain type of habitat, as for example, the wood duck. It not only uses certain types of waters in which it finds food, but also wooded areas with nest hole cavities. The mallard, on the other hand, accepts almost any type of water, is not choosy about nest sites, and therefore is one of the most widely distributed species.

Displacement by danger from man

As man needs more land for agriculture, the more the Anatidae, like many other wild animals, are displaced from their ancestral habitats.

1
2
3

1
2
3
4
5

◁ Swans: 1. Whistling swan *(Cygnus columbianus columbianus)*; 2. Bewick's swan *(Cygnus columbianus bewickii)*; 3. Trumpeter swan *(Cygnus cygnus buccinator)*; 4. Whooper swan *(Cygnus cygnus cygnus)*; 5. Mute swan *(Cygnus olor)*.

The less adaptable species suffer the most, and therefore, many of them are already threatened with extinction. Modern methods of watershed management are particularly harmful for birds so dependent on water; straightening of a river course, dams, and hydroelectric power plants deprive many species of their nest sites embedded among reeds or curved shore lines. Marshes are drained, and so countless young die in the drainage ditches since they cannot climb the steep sides. Entire large bays of the sea, like the Zuider Zee in Holland, have been taken from the sea and thus from water birds; harbors extend where formerly there was one nest next to another. In Sweden, according to Wolf and Errington, a single river system lost eighty-eight percent of its water surface as a result of drainage. As counter-measure to these changes of the habitat, one can only try to create as many waterfowl sanctuaries as possible.

Sanctuaries

Milne reports on the sanctuaries of the northwest American coast: "On the migration route followed by waterfowl, which is parallel to the Pacific Coast of America, the most magnificent wildlife sanctuaries are those in the basin of the Klamath River, where it crosses the border between southern Oregon and northern California. Four sanctuaries form one group, each a well utilized area with emphasis on ducks and geese. Together these sanctuaries extend over an area of 580 square kilometers, mostly marsh and shallow lakes. Their shores and islands are bordered by a sedge *(Scirpus)* known in that area as tule grass. Large swamps supply important food for the waterfowl. For many years this area was heavily hunted on a commercial basis. Today it is a breeding area which produces about 100,000 ducks each year. In addition, 375,000 or more geese winter there, which is perhaps half of the migratory geese of the western flyway. Most of the ducks of the whole of northwest America use these sanctuaries about halfway between the Klamath falls in Oregon and the city of Sacramento in California in order to rest for the cold months."

In Germany these bird sanctuaries are naturally much smaller, yet far more numerous than is generally known. There are bird sanctuaries on almost all East and North Frisian Islands in the North Sea in which, beside countless other water birds, shelducks and eider ducks can breed protected. But outside these sanctuaries, German waterfowl are threatened, recently as a result of the increasing masses of people which require more and more space for their leisure activities. There are weekend houses, the noise of bathers and power boats on many quiet lakes, where formerly the bittern boomed and ducks nested in the reeds. Dümmer Lake in Saxony and Lake Neusiedel in Austria are shocking examples of this.

Oil pollution

Oil pollution, which has arisen in the age of technical developments, is another threat to sea birds. With the increased consumption of mineral oils and the increased number and size of oil tankers associated with this, the surface of the seas is becoming more and more polluted. In 1954, thirty-two seafaring nations signed an agreement in London, according to which it is illegal for ships to pump oil-containing bilge water into the sea or to dump oil cargo except in emergencies. However, accidents involving tankers do occur, and their huge oil slicks spread death and destruction among seabirds. Experts estimate that still about 1,000,000 tons of oil is spilled on seas annually. A sad example of oil pollution was the giant tanker "Torry Canyon," which was stranded and broached off the British coast in 1967, and from which a huge oil slick moved to the English and French coasts. Birds which come into contact with such oil die either through the gumming up of their external or internal nostrils, or as a result of the enforced uptake of oil into the digestive tract. The plumage gummed up with oil fails to provide enough protection against cold. This weakens the birds; they are unable to take food, lose weight, and eventually die of general exhaustion.

When the Danish tanker Gerd Maersk had an accident in January, 1955, in the mouth of the Elbe, its cargo of 8000 tons of oil was pumped into the sea. Herbert Ecke describes the losses of water birds which resulted from this catastrophe: "On the island of Sylt alone, 650 dead or dying birds were found, among them 600 black scoters, ten long-tailed ducks, besides about twenty red-breasted mergansers and ten red-throated loons. The total is given as 2272, of which 2132 were black scoters and 140 other birds of nineteen species." Altogether it is estimated that about half a million seabirds perished as a result of the Gerd Maersk disaster.

According to Weller and the "Red Data Book" of the International Union of Nature Conservation, IUCN, the following Anatidae are today in an endangered state: the three flightless subspecies of the teal (⊕ *Anas aucklandica*), the Madagascar teal (⊕ *Anas bernieri*), the Mexican duck (⊕ *Anas platyrhynchos diazi*), the Laysan teal (⊕ *Anas platyrhynchos laysanensis*), the Hawaiian duck (⊕ *Anas platyrhynchos wyvilliana*), the freckled duck (⊕ *Stictonetta naevosa*), the Brazilian merganser (⊕ *Mergus octosetaceus*), the Tule goose (⊕ *Anser albifrons gambeli*), the Aleutian Canada goose (⊕ *Branta canadensis leucopareia*), the giant Canada goose (⊕ *Branta canadensis maxima*), the Hawaiian goose (⊕ *Branta sandvicensis*), the cereopsis (⊕ *Cereopsis novaehollandiae*), and the trumpeter swan (⊕ *Cygnus cygnus buccinator*). Already extinct are the crested shelduck (*Tadorna cristata*), the pink-headed duck (*Rhodonessa caryophyllacea*), the Labrador duck (*Camptorhynchus labradorius*) and the Auckland Island merganser (*Mergus australis*).

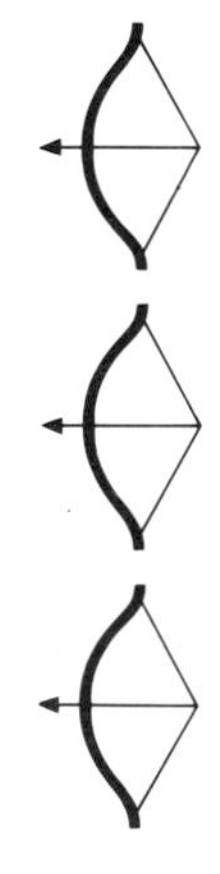

12 Magpie Geese, Geese and Their Relatives

Subfamily: magpie geese, by H. G. and U. Klös

The first subfamily of the Anatidae, the MAGPIE GEESE (Anseranatinae) consists of only one genus and species, the MAGPIE GOOSE (*Anseranas semipalmata*; Color plate p. 267). The Latin genus name means "duck goose." The magpie goose is, however, neither a duck nor a goose. It is not certain whether this bird, which in many particulars is exceptional, represents an old and therefore primitive form of the geese or whether it should be looked upon as a link between the geese and the screamers. The species name *semipalmata* (half-handed) indicates one of the peculiarities: the webs are very small and extend only between the most basal parts of the toes. Another peculiarity is the very long hind toe which is attached low on the tarsus. The plumage is similar in both sexes and the striking distribution of black and white has suggested its English name. The magpie goose is the only species of the Anatidae which does not moult all its flight feathers at once and is therefore always able to fly. The high bony ridge on its head is also unique. Its trachea is much prolonged and folded into a loop as in most swans, but does not, as in the latter, lie deep in the sternum; rather, it lies between the skin and the muscles on the left side of the breast. Heinroth has aptly called this trachea an "outboard trumpet." This explains the purpose of the loop; it amplifies the vocalizations of the magpie goose into a shrill trumpeting quality. Adult males have tracheal loops which are 150 cm long, while in females it is somewhat shorter (see Fig. 1-16).

Fig. 12-1. Magpie goose (*Anseranas semipalmata*)

The home of the magpie goose is in the swamp areas in South New Guinea, North, East and South Australia, and Tasmania (see map Fig. 12-1). It lives socially in groups, and prefers to perch on trees, showing great skill in maintaining its balance on the thinnest branches, due to its long hind toes. It feeds on grasses, the fruits of land and water plants, and occasionally on plant bulbs. Males and females incubate in alternation for thirty-five days on a clutch of five to fourteen eggs. The nest is built on the muddy ground. Klaus Immelmann reports on these peculiar birds in Australia: "The magpie goose is the characteristic bird of the wide coastal plains. Last year, a million of these peculiar goose-like birds were counted. They breed at the end of the rainy season in wild rice, but when the cultivation of true rice began, they preferred this to their natural habitat. This soon made them to be the greatest problem of the rice farmers, for they not only graze on the rice, but also trample it down. For a long time, shooting these birds was regarded as the only solution. But this was protested by zoologists and conservationists, for the magpie goose is not only a peculiar and beautiful bird which should be preserved, but is, because of its peculiar and primitive position among the Anseriformes, of particular interest to zoologists. Australian ornithologists worked on this particular problem for several years. At last they came up with a solution. Magpie geese breed at the end of the rainy season in the most densely overgrown areas of swamps and lagoons. In nest building they first trample down the plant stalks to form a platform for the nest. Then they tear off nearby stalks and lay them on this base. To construct these nests, they require a quite definite depth of the water, and the water level may be too high or too low. Since this became known, it has been possible to reduce the water levels in the rice fields below the minimal level required by the geese during the breeding season. Being unable to build nests in the rice fields, they are forced to return to the lagoons in order to breed. Today a peaceful and close coexistence of geese and rice is established."

Subfamily: Anserinae, by H. G. and U. Klös

The second subfamily of the ducks and geese are the GOOSE RELATED (Anserinae). Its members, like the magpie goose, are similarly colored in both sexes. Many species, such as the whistling ducks and the red-breasted goose, have a colorful and conspicuous plumage. They moult once a year, and at the most only small areas of the plumage are moulted twice yearly. Their plumage shows no iridescence, and their tarsi and toes carry small hexagonal scales or plates. Males and females have almost identical calls and both sexes take part in rearing the young. In the black swan and some species of the whistling ducks, the male even takes part in incubation. There are two tribes: 1. The WHISTLING DUCKS (Dendrocygnini) with one genus and eight species; and 2. The GEESE, Anserini, with four genera and twenty species.

Whistling ducks

The scientific name of the genus, whistling ducks (or tree ducks), *Dendrocygna,* meaning female tree swan, is as misleading as the alternative name, tree ducks. These long-necked and long-legged, graceful birds are neither swans nor ducks, nor do they show a predilection for trees. They, like swans, belong to those birds related to the geese. All have high-pitched whistling calls which sound like "teereeree" or "see ree ree," which are produced with the aid of the tracheal bulla present in both sexes. The whistling flight sound of the lesser or Indian whistling duck *(Dendrocygna javanica)* arises from a special modified flight feather (see Fig. 11-4). Almost all species show an emphasis of the color or pattern of the flanks, created by spots in an otherwise inconspicuous plumage, such as in the SPOTTED WHISTLING DUCK (*Dendrocygna guttata*; Color plate p. 268) and the BLACK-BILLED or CUBAN TREE DUCK ( *Dendrocygna arborea*), or through a white color as found in the BLACK-BELLIED TREE DUCK (*Dendrocygna autumnalis;* Color plate p. 268) and the LESSER or INDIAN WHISTLING DUCK (*Dendrocygna javanica*). The flank feathers are slightly prolonged in the WANDERING WHISTLING DUCK (*Dendrocygna arcuata;* Color plate p. 268) and the FULVOUS TREE DUCK (*Dendrocygna bicolor;* Color plate p. 268), and very markedly in the PLUMED WHISTLING DUCK (*Dendrocygna eytoni;* Color plate p. 268), in which they may reach a length of fourteen centimeters. The WHITE-FACED WHISTLING DUCK (*Dendrocygna viduata;* Color plate p. 268) is particularly attractively marked. Whistling ducks live mainly on plants, which they feed on preferably at night. During the day they rest, often in large flocks on the shores of waters.

Fig. 12-2. Post copulatory display of the fulvous tree duck.

Members of pairs appear to be joined for life, showing their solidarity by their mutual preening. The nests are generally built on the ground. Both partners alternate in the incubation of the strikingly short and rounded eggs for twenty-seven to thirty days, and then also lead the young together. The young of all eight species have a very similar downy plumage. Its most striking feature is a light stripe running from one side of the base of the beak horizontally around the back of the head to the other side.

Goose tribe

A similar downy plumage in the Anseriformes is only found in the young of the COSCORBA SWAN (*Coscoroba coscoroba*; Color plate p. 273), which is an intermediate between the whistling ducks and the true swans. Much in its external appearance is reminiscent of the whistling ducks, but other aspects of its appearance and behavior show a relationship to the swans. This white bird, with the striking coral-red legs and beak, is a native of southernmost South America.

Swans

The tribe GEESE (Anserini) begins with the coscoroba swan. This tribe includes, as the largest and most imposing of the ducks and geese (Anatidae), the SWANS (genus *Cygnus*), with five species. In his comedy "The Swan," Molnar says: "Swans should never go ashore but should always sail about majestically on the water. When they go ashore, they look like geese." Zoologically, the swans are geese, although they are geese with a strikingly elongated neck which enables them to feed in deeper water. The lore (the area between the base of the beak and the eye) is bare of feathers. There are five species in two well-distinguished groups.

Fig. 12-3. In its display swimming, the mute swan pushes with both feet at the same time and darts forward creating a foaming bow wave.

In Europe, the MUTE SWAN (*Cygnus olor;* Color plates pp. 244, 255/256 and 274) which is a form that has become almost domesticated, has increased extraordinarily on many waters in Germany, such as on the Havel and Alster rivers. The threat and display posture of the mute swan is generally known. In this posture, the neck is curved into an S and the head is drawn back; the elbows are raised so that the wing feathers project from the sides like white sails above the back. With a foaming bow-wave, it swims towards its opponent in this posture. It has become a symbol of striking beauty among birds; in fairy tales it is even shown with the crown of a bewitched prince. Like all members of this tribe, mute swans pair for life. They mate after a very simply display on the water and nest preferably in dense reed beds.

Bengt Berg tells about the mute swans on Lake Tokern in Sweden: "When they have chosen a nest site, they begin to bite off the reeds, stalk by stalk around the chosen spot, and from these pieces the nest is built. The stalks are never braided together or laid down in a particular pattern. The birds simply throw down one stalk after another, until after a few days there is a mound which seems still large enough even after it has become pressed down at its base by its own weight and become solid. When the nest is large and rigid enough to carry the female, she scoops out a shallow depression in the center and lays her eggs; often when the reeds are still wet she lays her first grayish-green egg, which she hides under reed stalks when she leaves the nest. Then she lays the remaining five to seven eggs at a little more than twenty-four hour intervals."

After thirty-five days, during which the male has defended the breeding territory against all intruders, the gray chicks hatch. They are then often seen swimming around with their mother in the lead and the father bringing up the rear. Among the almost domesticated mute swans, there is a variant in which the adults have flesh-colored instead of black legs, and in which the chicks are white instead of gray.

The closest relative of the mute swan is the AUSTRALIAN BLACK SWAN (*Cygnus atratus;* Color plate p. 273). It also often carries the neck in an S-shaped curve and arches up its wings. This distinguishes it from the whooper swan group. With its very dark plumage, it deviates from the usual snow-white swan type. But it makes an impressive picture with its bright white flight feathers, its deep red beak with the white band across the tip, and its white "nail."

The BLACK-NECKED SWAN (*Cygnus melanocoryphus;* Color plate p. 273) is smaller, short-necked, and has short wings and legs. It inhabits about the same area as the coscoroba swan and, in winter, gathers in great flocks in the Fjords of Tierra del Fuego and the south Chilean Islands. Its white, downy youngsters spend their first days almost entirely on the backs of the parents, hidden under their protecting wing feathers. According to our observations in the Berlin Zoo, the dark neck only begins to show when the young are three months old.

The WHOOPER SWAN (*Cygnus cygnus cygnus;* Color plates p. 274 and pp. 385/386), the TRUMPETER SWAN (♆*Cygnus cygnus buccinator;* Color plate p. 274), the WHISTLING SWAN (*Cygnus columbianus columbianus;* Color plate

▷

Field geese: 1. White-fronted goose *(Anser albifrons);* 2. Lesser white-fronted goose *(Anser erythropus);* 3. Pink-footed goose *(Anser fabalis brachyrhynchus);* 4. Western bean goose *(Anser fabalis fabalis);* 5. Greylag goose *(Anser anser).*

▷▷

1. Swan goose *(Anser cygnoides);* 2. Domestic Chinese goose (Domestic form of *Anser cygnoides*); 3. Bar-headed goose *(Anser indicus);* 4. Toulouse goose (a heavy domestic breed of *Anser anser*).

▷▷▷

1. Ross' goose *(Anser rossii);* 2. Emperor goose *(Anser canagicus);* 3. Snow goose *(Anser caerulescens,* blue phase).

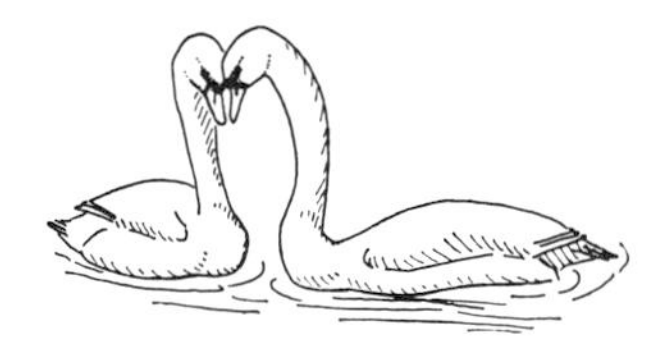

Fig. 12-4. Posture expressing "tenderness" in a pair of mute swans.

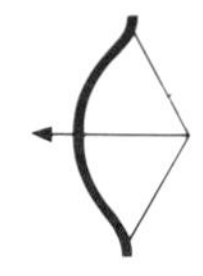

1
2
3
5
4

1
2
3
4

1
2
3

1
2
3
4
5

p. 274), BEWICK'S SWAN (*Cygnus columbianus bewickii;* Color plate p. 274 and pp. 385/386), and JANKOWSKI'S SWAN *(Cygnus columbianus jankowskii)* form a closely related group of swans which are very similar in structure and behavior. Their beak has no hump and is yellow and black. The five subspecies belonging to two species differ in the distribution of these colors. They carry the neck vertically and the beak horizontally. All have a tracheal loop, which is particularly long in the trumpeter swan, giving rise to a strikingly deep trumpeting call. The whooper swan, with its high-pitched call, has the shortest trachea. Both species inhabit the northern temperate hemisphere. The whooper swan, as a shy bird, prefers remote hidden breeding places, which the male defends vigorously. Its breeding range is joined in the north by Bewick's swan and the northeast Siberian Jankowski's swans, which nest in the wide tundras of Eurasia. Both species migrate southwards in the fall and winter in great numbers to areas on the German Baltic and the North Sea coasts.

Whooper swans migrate in smaller groups, Bewick's swans in larger ones; in flight they form a slanted line. Hilprecht writes about them: "The whooper swan flies without making any particular sounds, but it calls a great deal in flight and on the water. Its calls are long drawn-out, trombone-like calls, which, beginning nasally, sound like "ang." They sound lyrically soft and are also spoken of as being like silver bells. Bewick's swans also call a great deal. Their calls are monosyllabic, clear and strong, sounding like "goohk" or "koohk"; in excitement it turns into sequences somewhat like "goo-a-loohk." The calls of large flocks of Bewick's swans may be heard over several kilometers."

The whistling swan which breeds in North America beyond the Arctic Circle belongs to the same species *(Cygnus columbianus)* as the Eurasian Bewick's swan. The whooper and trumpeter swans are also only subspecies of the same species, *Cygnus cygnus.* The trumpeter swan has an almost completely black beak and, with a weight of about thirteen and a half kilograms, is the heaviest of the Anatidae, apart from the feral mute swan. It was formerly widely distributed in North America, but with the increasing settlements of white immigrants on the continent, its numbers steadily decreased. It was hunted not only for its meat but also for its feathers. By the beginning of the twentieth century the decline of this subspecies had become alarming. Two large sanctuaries were therefore set up in the United States, in which the trumpeter swan found suitable conditions; furthermore, it was protected by laws against hunting. As a result, the population of this swan slowly increased again and it is now estimated at 2100. This subspecies could therefore be removed from the list of birds threatened with extinction. The sale of these birds is illegal; they may only, in special cases, be donated by the Fish and Wildlife Service. This beautiful swan is therefore very rarely seen in Europe, where it is kept

Fig. 12-5. Whistling and tree ducks (genus *Dendrocygna*).

Fig. 12-6. Swans (genera *Coscoroba* and *Cygnus*).

Fig. 12-7. *Cygnus columbianus* (whistling swan, Bewick's swan, and Jankowski's swan).

Fig. 12-8. 1. Whooper swan *(Cygnus cygnus cygnus)*; 2. Mute swan *(Cygnus olor)*.

◁

Brents: 1. Hawaiian goose *(Branta sandvicensis)*; 2. Brent goose *(Branta bernicla)*; 3. Red-breasted goose *(Branta ruficollis)*; 4. Canada goose *(Branta canadensis)*; 5. Barnacle goose *(Branta leucopsis)*.

only in the Berlin Zoo and in the Wildfowl Trust of Slimbridge in England. In 1965 it reproduced for the first time at Slimbridge.

Field geese

The true geese may be divided into FIELD GEESE (genus *Anser*) and BRENTS or SEA GEESE (genus *Branta*). In general they can walk well and with endurance, and they are excellent swimmers. Some species even rest at night on the water. Their wing beats produce a swishing sound. Field geese fly in slanted lines or V-formations, while the sea geese fly in groups with no particular formation. Field geese are also distinguished by peculiar grooves in the neck plumage; they can be divided into several groups.

Fig. 12-9. Threat posture of the whooper swan.

The "wild goose" of Central Europe is the GREYLAG GOOSE (*Anser anser;* Color plate p. 281 and Vol. XI), which occurs in two subspecies and now breeds in Germany only east of the Elbe River. Its behavior has been studied and described in detail by Konrad Lorenz. He has contributed the following account of his first incubator-hatched greylag goslings:

Behavior of the greylag geese, by K. Lorenz

"I waited until the chick, beneath the electrical heating pad which replaced its mother's warm abdomen, was strong enough to hold its head up and to take a few steps. Holding its head to the side, it looked up at me with a large dark eye. It only used one eye, for like most birds, the greylag goose uses one eye when it wants to see something distinctly. It looked at me for a long, long time. When I moved and said a short word, its intense attention was released and the little gosling gave a greeting. With outstretched neck, it uttered rapidly and in several syllables the greylag contact call, which in a young bird sounds like a fine, eager whispering. It greeted exactly like an adult greylag and in a manner which it would repeat thousands of times during the rest of its life. The greatest expert would not have been able to tell that it had performed this ritual for the first time in its life. I did not yet know what a heavy responsibility I had taken by withstanding the examination of the little dark eye and by releasing its contact call by a small word. I wanted to keep this gosling, together with others which were hatched out by my turkey hen, with a fat white domestic goose which was incubating the remaining ten greylag eggs. When my chick had gathered its strength, three others had just hatched under the domestic goose. I carried my little one into the garden. The fat white goose sat in the dog house from which it had ruthlessly expelled the rightful owner, my dog Wolfi the First. I pushed my gosling far under the warm belly of the old goose, convinced I'd carried out my part. But I still had much to learn.

Artificial rearing

"There were a few minutes during which I remained seated in front of the goose in happy meditation, when from beneath the white one there came a soft, questioning whisper, 'wee wee wee wee?' The old goose calmly answered with the same contact call, only at her pitch, 'gang gang gang gang.' But instead of being comforted by this, as any

Fig. 12-10. Field geese (genus *Anser*)

Fig. 12-11. 1. Greylag goose *(Anser anser)*; 2. Lesser white-faced goose *(Anser erythropus)*.

sensible goose child should have been, mine came out from among the warm feathers, looked up with one eye at its foster mother's face and, 'weeping' loudly, ran away from her, calling 'pfeep, pfeep, pfeep.' This is approximately the sound of the 'piping of being left alone' of the greylag gosling, which in one form or another is found in the young of all nidifugous birds.

"Fully erect, continually piping, the poor gosling stood halfway between the goose and me. Then I made some slight movement and at once the weeping ceased and the gosling came toward me with outstretched neck, eagerly greeting 'weeweeweewee.' It was touching, but I had no intention of playing the role of mother goose. I seized the little one and once more pushed it deep under the belly of the old goose and took off. I hadn't taken ten steps when behind my back I heard 'pfeep pfeep pfeep' and the poor goose child came running in desperation. It could not yet stand but could only sit on its heels, and in walking slowly it was still quite insecure and wobbly. But in its great distress it was able to manage a fast run. In many gallinaceous birds this strange, but apt, sequence of the maturation of different movements is even more pronounced. Young partridges and pheasants, for example, can run before they can walk slowly or stand still.

"The way in which the poor gosling came after me, 'weeping.' stumbling, and falling over itself, but running surprisingly fast, would have brought tears from a stone. This determination conveyed an unmistakable meaning, that I, not the white domestic goose, was its mother. Sighing, I shouldered my little burden and carried it back to the house. Although it only weighed a hundred grams then, I knew quite well how heavy it would become, and how much hard work and time it would take to carry out my obligation properly. I pretended to have adopted the gosling, rather than having been adopted by it, and so I ceremoniously christened it Martina.

"The rest of my day passed just as it passes for a goose mother. We went to a meadow to graze on tender grass and I was able to convince my child that chopped egg with nettles was good to eat. The goose child, in its turn, succeeded in convincing me that, at least for the time being, I could not leave it alone for even a minute. If I did, it became so desperately fearful and cried so piteously that I gave up after a few tries and fixed up a little basket in which I could carry it around all the time. At least I could move about freely as long as the gosling was asleep. But it never slept for a long period. During the first day this did not mean much to me, but it did at night. For the night I had fixed up a wonderful, electrically heated cradle for my goosechild, which had already replaced the mother's warming plumage for several young birds. When, fairly late in the evening, I placed little Martina under the heating pad, it answered right away contentedly with the rapid whisper which, in goslings, indicates readiness for sleep and sounds like 'wirrr.' I put the box with the heated cradle in a corner of my room and crept into bed. I was about to fall asleep when Martina softly and sleepily said 'wirr' once more. I kept still. Then came, a bit louder and more questioning, the contact call 'weeweeweewee?' Selma Lagerlöf has genially translated its significance as, 'Here I am, where are you?' I still did not answer, dug deeper into my pillow, and fervently hoped the gosling would fall asleep again, but no; still I heard the 'wee wee wee' contact call, but with a threatening approach to the piping of being left alone. And then it came, loud and penetrating, 'peee...peep.' I had to get out of bed and go over to the box. Martina received me happily, greeting 'wee wee wee.' There seemed to be no end to it, since she was so relieved to no longer be alone in the night. I gently pushed her back under the electric blanket and with a 'wirr', we both fell asleep at once. But after less than an hour, about half-past ten, I heard another questioning 'wee wee wee' and the events described above were repeated. This happened again at a quarter to twelve and again at one o'clock. Then, at a quarter to three, I got up to make a rearrangement of the experimental conditions. I placed the cradle within reach by the head of the bed. When, predictably, at half past three the questioning 'Here I am, where are you?' came again, I answered in broken greylag language, 'gangangang,' and tapped on the electric blanket. 'Wirrr,' said Martina. 'I'm falling asleep again; good night.' I learned to say 'gangangang' without waking up. I believe even today if I were asleep and someone softly said 'weeweewee,' I would answer in the same way.

To be left alone means death

"Martina was a wonderfully well-behaved child. It was not obstinacy that it did not like to be left alone for even a moment. It must be realized that, for such a young bird in the wild, it would be fatal to lose its mother and its siblings. It is biologically meaningful that such a little "lost sheep" cannot think of eating or drinking or sleeping, but uses every spark of energy, up to total exhaustion, in crying for help which may enable the mother to find her young again.

"If one keeps several young wild geese which have some contact with one another, it is possible, with a bit of severity, to gradually train them to remain unattended. But a single one would literally cry itself to death. This deep instinctive aversion against being alone bound Martina firmly to me. She followed me everywhere and was quite content when she could lie down under my chair when I would work at my desk. She was not troublesome; it was enough if I replied with an inarticulated grunt, whenever she asked with her contact call, whether I was still there and alive. In the daytime she asked every few minutes, and at night about once an hour.

"Since for Martina's sake I had to act as a goose-mother anyway, I did not, as originally planned, put the other nine goslings which hatched out under the turkey hen into the care of the domestic goose.

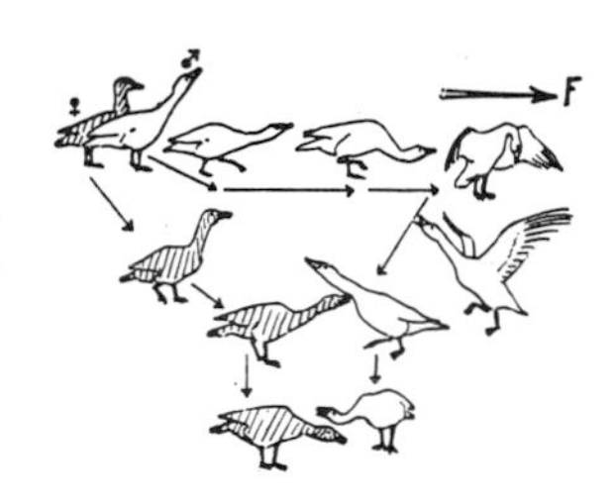

Fig. 12-12. Triumph-calling of greylag geese is a mutual ceremony which strengthens pair or family bonds. The gander (white) walks toward an enemy (E) and chases it away. It turns around and returns to the goose (shaded) in the threat posture, uttering a "rolling" call containing many high tones. Both birds utter this call and the ceremony ends with mutual cackling.

Ten little goslings take less rather than more of their guardian's time than a single one, because for them it is less critical if they are left alone. Strangely enough, Martina did not establish a sisterly relationship with the nine, although she often met them in the course of the day, particularly during walks made in common. Although after some initial squabbles she was accepted as a sister by the others, she herself cared little for them, certainly did not miss them when they were absent, and was always ready to walk away from them with me. Although the nine, like Martina, looked on me as their mother, they kept to each other as much as they followed me. That is to say, they were only happy and quiet when, first, they were together, and, second, when they were with me. At first I tried to take only two or three of them along with Martina on my walks. Since, to cover longer distances, I simply put the goslings into a basket and carried them with me, and since for my observations three or four birds were quite enough, I would gladly have left most of the goslings at home. But this was not possible since such a minority, separated from their siblings, were, in spite of my presence, always restless and anxious, inclined to utter the piping call of being left alone, to stand still, and were reluctant to follow. This reaction to the absence of the sibling flock was less individual, but more with respect to the numbers. If I took the majority along with me, leaving only two or three at home, they followed readily and were quiet. But those left at home almost cried themselves to death. On my outings I therefore had to take along either Martina alone or all ten goslings. Two years later when I again adopted some goslings, having learned my lesson I restricted myself to four.

"I spent an unbelievable amount of time with my ten youngsters in that first "summer of the geese," and I learned an unbelievable amount from them. It is a fortunate science in which an essential part of the research consists of crawling about the shores of the Danube and swimming with a flock of geese.

Animals cannot be hurried

"Animals cannot be hurried; if one wants to study them one has to live with them, and if one wants to live with them, one has to adapt to their tempo of life. A person who is not naturally lazy just cannot do this. A constitutionally active, energetic person would go mad if he had to spend a summer acting like a goose among geese, as (with interruptions) I have done. Wild geese spend at least half of the day lying still and digesting. They spend at least three quarters of the other half grazing. The times interspersed between the periods of grazing and digesting are the ones during which observation is essential. Altogether they amount to only an eighth of the daytime which they spend awake. If what they do during this eighth of the day were not so interesting, wild geese would be extremely boring animals.

Moods and calls

"The calls by which greylag geese indicate their mood, their intention to walk away, to swim, or to fly away, are of particular interest.

Even the smallest goslings react in an inborn manner to the finest nuances of this quite complicated vocabulary. The normal contact call, the well-known soft and rapid goose chattering, is made from time to time, even when the birds are at rest, when they graze, or when they walk slowly. Because of its strong overtones, this call sounds peculiarly broken and has six to ten syllables. The number of syllables and the intensity of the high-pitched overtones are directly related in the usual contact call, but are inversely related to the loudness of the call. The more syllables there are, the higher pitched and softer the chatter is. If these three characteristics are pronounced, it indicates the utmost comfort, and so the birds have no tendency to move in the near future. Chattering with many syllables, high-pitched and soft, therefore means in human terms, 'It's good here, let's stay,' with the secondary significance of maintaining contact. To the degree that the tendency to change the place becomes noticeable in the goose, the contact call changes in a way that the number of syllables decreases, the high-pitched overtones disappear, and the chattering becomes louder. With a decrease in the number of syllables, the tendency of the goose to walk forward increases. With this, the individual syllables of a group of calls are spread out, while at the same time the groups follow one another more quickly. In this way there finally arises a series of slow staccato calls, no longer divided into groups. When this degree of excitement has been reached, one can predict with certainty that the goose will fly up in the next moment.

"When there is no mood to fly off, but the goose is about to make the intended move by walking or swimming, it uses a special call which indicates just that. Chattering with between three and four syllables, the goose utters a loud, well-demarcated, metallic-sounding, three syllable call, with the strongly pronounced middle syllable being about six tones higher than the other two, which is somewhat like "gangingang." Parents leading chicks which are not yet able to fly often get into the mood of making a move, with emphasis on the intention not to fly. One hears this call particularly often in domestic geese leading young. This always strikes the expert as somewhat comical, since these fat fellows can hardly fly anyway, so that their continuous "assurances" that they will move on foot and not on the wing are quite superfluous. Since all these mood-indicating sounds are inborn, the birds themselves naturally are not aware of this.

"Equally inborn, as already mentioned, is the innate understanding of this whole vocabulary expressing mood by every young greylag goose. One or two day old chicks already respond to all the fine variations described. If one utters the contact call sharply, with fewer syllables, the young stop feeding, raise their little heads, and slowly the whole group gets into the mood of moving off, and they begin to walk forward.

"The response of goslings to the "gangingang" is particularly pretty

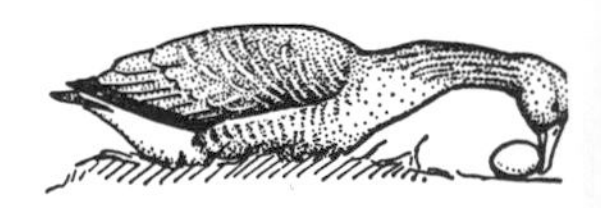

Fig. 12–13. Like other ground-nesting birds, greylag geese try to return eggs which have rolled out to the nest. This inborn reaction is also performed with a substitute for the egg, in this case a rubber ball.

and, if used sparingly, can always be well-demonstrated. It is interesting that little goslings in particular regard this utterance of the parents as "meant for them," when lured by some particularly tasty plant, they have remained behind during the march. In such cases the "gangin-gang" strikes them like a whip, and they come racing along with their little wings extended, behind the parents or the human parent substitute. This reaction of little Martina gave me an opportunity for a bit of deception.

"Although her name was not derived from the attraction and contact call of the species, we had hit, with Martina, the best call name that had been given to any bird with us at Altenberg. For if one pronounced her name in the tone and the absolute pitch of the greylag "gangin-gang" and with strong pronounciation of the "i," it released with certainty the response described above. Martina would come racing along like a well-spurred horse. Hunters and others knowledgeable about dogs were surprised by the response to her name, which I had "taught" the little gosling in hardly a week. Only I had to make sure there were no other little "untrained" geese within hearing range, or they, too, would have come rushing along as if pulled on a string.

"Just like the meaningful responses to all variants of the contact call is inborn in goslings, so is the response to the warning calls of the adults. This call consists of a single, generally soft, nasal "gang," with a bit of an "r" sound, so that the call is perhaps best rendered in letters as "ran." One can imitate this hoarse call most effectively by pronouncing the syllable while inhaling. In response to this call, all goslings raise their heads to look around, and their otherwise almost continuous contact calls cease at once. If one pronounces the call more loudly, adult geese get into the mood for taking wing, and they first seek to find a place from which they have a clear view all around, and from where they can fly up readily. Little goslings, on the other hand, run quickly to their mother or the human parent-substitute and bunch up close to them for protection. The fearful mood of the young lasts until they are "reassured" by the parents. Thus, the goose-parents do not need to warn a second time to keep their young quiet; hence, they can concentrate on the danger. When this danger passes, the young are released by a soft contact call gaggling, to which the young respond by a greeting ceremony with extended necks.

"Just as quickly as spring passes into summer, so the lovable down balls become beautiful gray birds with silvery flight feathers. The transition from one to the other is charming. The unharmonious intermediate stages between the chick and the full-grown young are touching; their feet are too large, their joints are thick, and their movements are clumsy, but this stage is concentrated into a few weeks in the greylag goose. It is a wonderful moment when the new harmony of the full-grown bird is reached, when the wings are strong and open for the first flight."

The domestic goose, by M. Lühmann

Descendants of greylags have been made into domestic birds for a long time by man. Domestic geese were already kept all over Europe in ancient times; the Romans found large flocks of white geese in Germany. Raising geese remained unchanged in Europe for a long time. In West Germany there has been a decline in the numbers of domestic geese only since 1956. With changes in agriculture management the number of geese decreased from 1.99 million to 830,000 in 1966.

▷

Hen mallard *(Anas platyrhynchos)* with newly hatched ducklings.

▷▷

1. Pink-headed duck *(Rhodonessa caryophyllacea)*; 2. Pink-eared duck *(Malacorhynchos membranaceus)*; 3. Freckled duck *(Stictonetta naevosa)*; 4. Torrent duck *(Merganetta armata)*; 5. Flying steamer duck *(Tachyeres patachonicus)*.

▷▷▷

Shelducks: 1. Cereopsis *(Cereopsis movaehollandiae)*; 2. Andean goose *(Chloephaga melanoptera)*; 3. Ashy-headed goose *(Chloephaga poliocephala)*; 4. Ruddy-headed goose *(Chloephaga rubidiceps)*; 5. Kelp goose *(Chloephaga hybrida)*; 6. Magellan goose *(Chloephaga picta)*; 7. Abyssinian blue-winged goose *(Cyanochen cyanopterus)*; 8. Orinoco goose *(Neochen jubatus)*; 9. Paradise shelduck *(Tadorna variegata)*; 10. South African shelduck *(Tadorna cana)*; 11. Australian shelduck *(Tadorna tadornoides)*; 12. Radjah shelduck *(Tadorna radjah)*; 13. Ruddy shelduck *(Tadorna ferruginea)*; 14. Egyptian goose *(Alopochen aegyptiacus)*.

Today most of these geese are still strong, well-established domestic geese. Such self-incubating geese were formerly found on most farmsteads, and on particular goose-grazing areas which cannot be utilized otherwise; the number of birds in the flocks, then and now, sometimes reaches several thousand. Meanwhile the so-called high-egg-yield geese have taken on an increasing significance. Most of them are the high-yield geese, (Rheinische Vielleger), of which individual establishments keep a breeding stock of 400 or more to produce young for sale. Besides this, there are also some highly selected goose strains.

The white Emden goose weighs ten to twelve kilograms; the Toulouse goose (Color plate p. 282), distinguished by a lappet under the chin, is somewhat lighter. Both breeds lay thirty to fifty eggs, but hardly incubate them. The large spotted or white Pomeranian geese incubate, as do the white Diepholz geese, which weigh only five to seven kilograms in which, by selective breeding, two to three broods per year have been established. The medium weight curly goose is distinguished by extraordinarily prolonged, curly feathers, particularly on the shoulders and the upper back; apart from that it has very shortened flight and tail feathers.

Crests of varying size are found frequently among our domestic geese, but no particular breed has arisen in relation to this. The many colors of land geese also have not so far led to the formation of breeds of particular colors. The sexes differ in plumage color in the American Pilgrim goose; the males are white, but adult females are pale dove-gray with white head and neck.

Other field geese, by H. G. and U. Klös

A second group of field geese is formed by the two species with a white forehead, the LESSER WHITE-FRONTED GOOSE (*Anser erythropus*; Color plate p. 281) and the WHITE-FRONTED GOOSE (*Anser albifrons*; Color plates p. 281 and pp. 385/386) which occurs in several subspecies. The BEAN GOOSE (*Anser fabalis*; Color plate p. 281 and pp. 385/386) is darker and browner, but otherwise much like the greylag goose; among its subspecies is the PINK-FOOTED GOOSE (*Anser fabalis brachyrhynchos*; Color plate p. 281) which has a shorter beak, is paler on the back, and has a pink beak and legs. All these species breed in the far north of Eurasia and migrate a greater or lesser distance, since the severe winters of their breeding areas provide no food at that season. They therefore migrate southward in flocks before the onset of winter and gather along open sea shores, bays, and river estuaries.

1
2
3
5
4

1
2
3
4
5♂
5♀
6♀
6♂
♀ 10 ♂
7
8
9♀
♂
13♂
11♂
14♂
12

Fig. 12-14. White-fronted goose *(Anser albifrons).*

Favorite goose-wintering areas in Europe are the Severn estuary in England, the swamps of the Guadalquivir in Spain, Dutch Friesland, Lake Müritz (Color plate pp. 385/386), the waters of East Friesland (Color plate pp. 61/62), and northern Oldenburg in Germany. During the winter months, one can see thousands of white-fronted and pink-footed geese on the frozen seashores, or among dry grasses and on half-frozen cultivated fields in East Friesland. In the lake country of Mecklenburg, Germany, in October/November of 1967, about 20,000 wild geese spent the night in several roosts. The fall and the spring migration of the northern geese depends, according to Uspenski, on the accumulation or melting of the snow cover in their breeding areas. In the high Arctic, where the land is only free of snow for two months, only the brant goose and the snow goose breed. They lay eggs immediately after their arrival, and the growth of their young falls into a period which is very short for this group of birds.

Domestic Chinese goose, by M. Lühmann

The SWAN GOOSE (*Anser cygnoides*; Color plate p. 282) is distinguished by an almost swan-like long beak. It is the ancestor of the DOMESTIC CHINESE GOOSE (Color plate p. 282). The latter is distinguished by a large hump on the forehead, which is formed by the frontal processes of the nasal bones and only develops at the onset of sexual maturity, reaching a considerable size in a few weeks. Through selective breeding, the posture of the domestic form has become peculiarly exaggerated. In Germany, a gray variant with pale underparts and an almost white neck, with a dark brown stripe on the nape, is widely distributed, but of no particular economic importance. In America, on the other hand, a white Chinese goose is of economic significance; flocks of these geese are driven into strawberry and asparagus beds to keep down grasses. Chinese geese are fertile to only a limited degree when crossbred with European domestic geese. But the Steinbacher fighting geese are derived from a cross between these two species. In Russia, too, there are large goose populations derived from a cross between European and Chinese domestic geese and which therefore, like the Steinbacher geese, originate from two different wild species.

Fig. 12-15. Bean goose *(Anser fabalis).*

The remaining geese, by H. G. and U. Klös

The BAR-HEADED GOOSE (*Anser indicus*; Color plate p. 282) is medium-sized and slender, and has two characteristic black bars across the back of the head. It breeds around lakes at high elevations in Central Asia, and migrates southward to India for the winter. One of the most beautiful field geese species is the EMPEROR GOOSE (*Anser canagicus*; Color plate p. 283) of Alaska and the Aleutians. It is rare in European zoos and waterfowl collections, and is bred even more rarely. In Germany, breeding in captivity was not successful until 1960 and then occurred simultaneously in the Berlin Zoo, the Berlin-Friedrichsfelde, and the Hagenbeck animal park. The emperor goose prefers a cold climate, and a sudden heat wave may kill a flock of goslings all at once.

◁ Shelducks: Common shelduck *(Tadorna tadorna).*

The ARCTIC SNOW GOOSE (*Anser caerulescens*; Color plate p. 283 and Vol. XI) has a pure white plumage except for the black primaries. Yet its scientific name *caerulescens* means bluish. This is because the smaller subspecies (LESSER SNOW GOOSE, *Anser caerulescens caerulescens*) contains in addition to white birds, birds of dark slate-gray plumage, and so this blue phase justifies the name. The large subspecies, the GREATER SNOW GOOSE *(Anser caerulescens atlanticus)* does not have a blue color variant. ROSS'GOOSE (*Anser rossii*; Color plate p. 283) is considerably smaller and is further distinguished by a warty growth at the root of the beak during the breeding season. Some years ago, ornithologists were worried about its status, since only two thousand were found in its wintering area in California. Since then it has been realized that it associates not only in wintering areas, but even on its breeding places with lesser snow geese, from which it is difficult to distinguish at a distance. A closer look at the situation revealed that there was, fortunately, no reason for concern. In February, 1966, the population of Ross' goose was about 30,000.

Brents or Sea geese

The BRENTS OR SEA GEESE (genus *Branta*) are distinguished from the field geese by a delicate, entirely black beak; no lamellae are visible externally on the upper mandible. Among them is the CANADA GOOSE (*Branta canadensis;* Color plate p. 284 and Vol. XI). It is the most common goose of North America, with many subspecies which show astonishing differences in size. The largest forms live in southern North America. Among them is the GIANT CANADA GOOSE (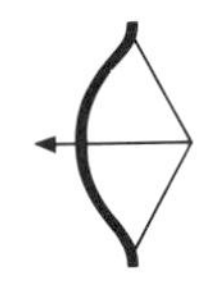 *Branta canadensis maxima),* weighing up to eight kilograms, with the same body size as the whooper swan. It used to breed in the prairies of central North America, but was believed to have become extinct about twenty years ago, until in 1962/63 it was rediscovered by American biologists in the area of Rochester, Minnesota. Its population is now estimated at 10,000 birds. Toward the north, the subspecies become smaller and are more delicate in body structure. Smallest of all is the CACKLING CANADA GOOSE *(Branta canadensis minima);* weighing hardly one and one-half kilograms, it is only as large as a domestic duck.

Fig. 12-16. Brents (genus *Branta*).

Canada geese have been introduced in various parts of the world, including England and Sweden. In 1905, forty-eight were brought to New Zealand, where the species remained confined to a small area until 1930. Then it began to increase explosively and spread over the entire south island. Local concentrations of geese led to the complete grazing off of green grasses, and hence control measures against the geese had to be instituted. In spite of this, there were still 20,000 Canada geese in New Zealand in 1964. Domestic forms of Canada geese are being bred as is done from the greylag and swan geese.

Fig. 12-17. Brent goose *(Branta bernicla).*

The small, dark BRENT GOOSE (*Branta bernicla;* Color plates pp. 61/62 and p. 284), of which there is a far northern form with a light underside, among other subspecies, used to winter in great numbers on the European coasts. According to Naumann, "their flocks used to darken the sky." Unfortunately this is not the case any more. While up to 1931 about 10,000 brent geese used to winter in Holland, there were only

Fig. 12-18. Barnacle goose *(Branta leucopsis).*

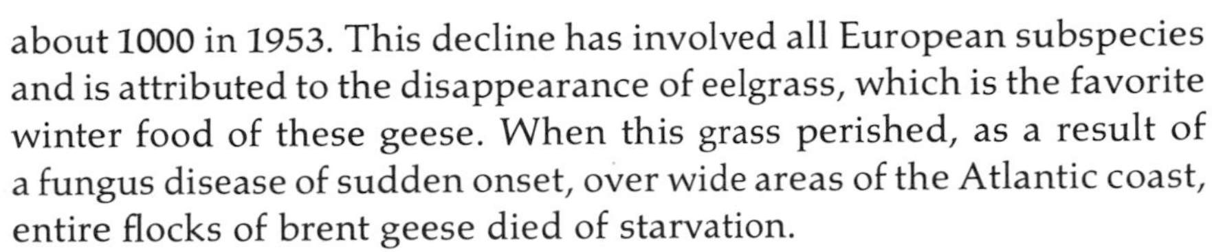

about 1000 in 1953. This decline has involved all European subspecies and is attributed to the disappearance of eelgrass, which is the favorite winter food of these geese. When this grass perished, as a result of a fungus disease of sudden onset, over wide areas of the Atlantic coast, entire flocks of brent geese died of starvation.

The BARNACLE GOOSE (*Branta leucopsis;* Color plates p. 284 and pp. 385/386) is as much dependent on the sea as the brent goose. Its habitat is therefore always very close to the seashores. This nicely marked bird is a frequent guest of the north shores of Germany during the winter. The most colorful species of the brent geese is the small, thick-necked RED-BREASTED GOOSE (*Branta ruficollis;* Color plate p. 284). It breeds in the wooded tundra of northern central Siberia.

The most interesting, and for conservationists probably the most expensive history among all the sea geese, is that of the insignificant-looking HAWAIIAN GOOSE or NENE (*Branta sandvicensis;* Color plate p. 284). Its home is the Island of Hawaii. There it inhabits the mountain slopes between the volcanoes Mauna Loa and Mauna Kea, at an elevation of 800 meters. This is an inhospitable area with fog and frequent rain showers. Instead of green meadows, there are only fields of lava, the latest of which are still bare and rough; only the older ones are eroded. In these eroded lava fields there are small grass flats with bushes and some shallow pools scattered about like oases. These "Kipukas" are the refuges of the Hawaiian geese. The Hawaiian goose has become so adapted to these conditions that, in the course of evolution, it has become a land goose, with unusually long legs, long, strong toes, and much reduced webs. In the eighteenth century, about 25,000 of these birds lived there. However, the population decreased alarmingly up to World War II as a result of biologically damaging hunting regulations, the activities of feral pigs and dogs, and the introduced banded mongoose. When the Hawaiian goose became protected during World War II, it was almost too late. At that time, there were only about thirty Hawaiian geese in the wild, and thirteen others were living on farms.

The fact that this species was saved from complete extermination is the achievement of two men; the farmer, Herbert Shipman in Hawaii, and the well-known bird artist and breeder of waterfowl, Peter Scott in England. With much idealism and at great expense, Scott began a breeding program at Slimbridge (England) with a few Hawaiian geese sent to him by Shipman. The first young hatched in 1952. Since then more than a hundred have been reared at Slimbridge, and it was possible to release in Hawaii a considerable number of those born in Europe. In 1966 there were once more 285 Hawaiian geese living in the wild, and the world population was about 500 birds. The steps taken to save the Hawaiian goose were supported by the U.S. alone, at a cost of $15,000 a year.

13 Ducks and Related Forms

Subfamily: Anatinae, by H. G. and U. Klös

All the remaining ducks and geese are grouped as ducks and related forms in the subfamily Anatinae. The first tribe of these to be considered here are the SHELDUCKS (Tadornini). Apart from many duck-like characters, they also have, at least externally, goose-like features. Thus, as in the geese, the males and females of most species are similarly colored; their English names designate some species as geese, and others as ducks.

Tribe: shelducks

The striking, colorful COMMON SHELDUCK (*Tadorna tadorna;* Color plate p. 296) lives as a breeding bird on the German coasts, among other areas. Among the coastal people of Germany, it is also often known as the "cave goose," since, like other related species, it incubates its large egg clutches in burrows in the ground. Shelducks have become widely known because of their "moult migration" to the Great Knechtsand, a North Sea island, which, since it was used as a bombing target after the Second World War, worried biologists and conservationists. Friedrich Goethe writes about this moult migration:

Fig. 13-1. Shelducks (tribe Tadornini).

"The gathering of flocks of moulting shelducks in the coastal shallow between the Weser and Eider rivers, particularly on the Great Knechtsand where masses of them moult, is one of the greatest phenomena in the bird life of the Heligoland Bight." He continues: "The origin of the tens of thousands of moulting shelducks which gather here every year was a great riddle. Evidently they were not merely the breeding birds of the German coasts, for these were estimated by the Heligoland bird observatory to be about 700 pairs at the most." Recoveries of banded birds showed that most shelducks of Central and Northern Europe gather there. Why they do so is still unknown.

Fig. 13-2. Common shelduck *(Tadorna tadorna)*

Just as colorful as the common shelduck is the AUSTRALIAN RADJAH SHELDUCK (*Tadorna radjah;* Color plate p. 295) which occurs in two subspecies. Its relatives include the RUDDY SHELDUCK (*Tadorna ferruginea;* Color plate p. 295), which lives in North Africa and northern Eurasia, the extinct CRESTED SHELDUCK (*Tadorna cristata;* Color plate p. 307), the

Fig. 13-3. Ruddy shelduck *(Tadorna ferruginea).*

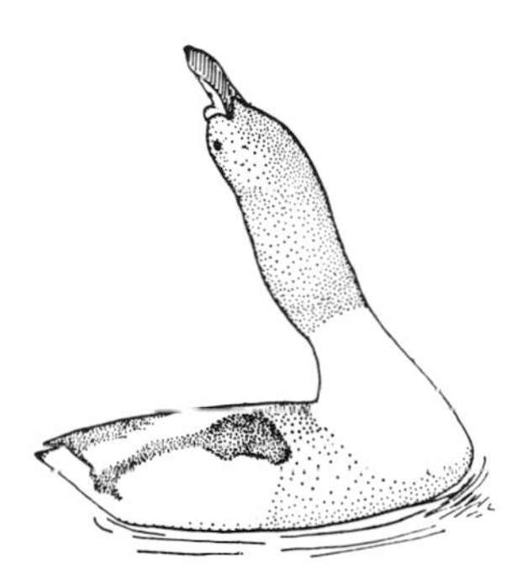

Fig. 13-4. The common shelduck calls in this posture.

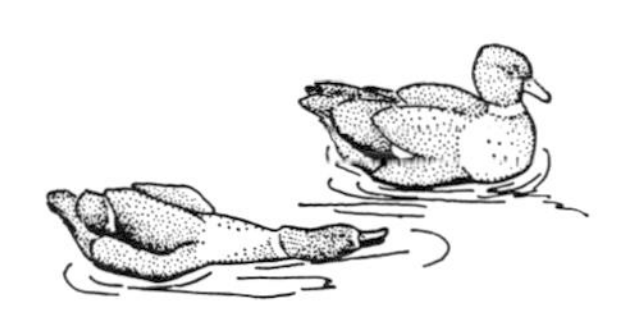

Fig. 13-5. Casarca females "incite" more often and more intensely than other Anatinae (Australian shelduck).

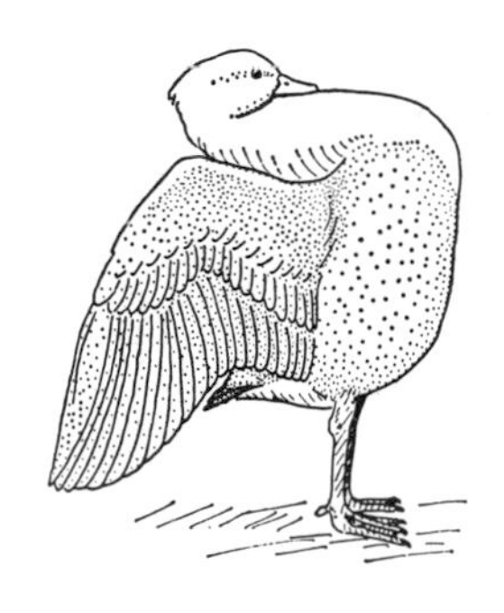

Fig. 13-6. Orinoco goose in display posture.

CAPE SHELDUCK (*Tadorna cana;* Color plate p. 295), the AUSTRALIAN SHELDUCK (*Tadorna tadornoides;* Color plate p. 295), and the NEW ZEALAND or PARADISE SHELDUCK (*Tadorna variegata;* Color plate p. 295), which together form the group CASARCAS. In all of these, the plumage is predominantly brown, although females have a tendency to have white facial marks. It is particularly striking that in the Paradise shelduck, the females have a white head, while the young look like the black-headed and otherwise dark males. A particularly striking characteristic of the casarca shelducks is the continuous inciting of the females. Oskar Heinroth writes about this:

"As soon as a member of the species or genus in the wider sense approaches a pair, the female makes a sort of sham attack on the stranger by either running toward it, or at least showing the intention of doing so. In so doing, a female holds the neck outstretched like a goose, with the head and beak just above the ground, continually uttering its anger call." If the intruder does not respond to this, the female's behavior is intensified. "Furiously, the female rushes around her male pointing her head with lowered beak and excited calling, again and again pointing her head at the intruder, until at last the male attacks the stranger and, if possible, chases it away. It is only necessary to remove the females from the otherwise hostile pairs of casarcas for complete peace to reign once again."

In the same manner, females which are not yet paired play their suitors off against one another, preferring the strongest one. Heinroth reports, "Usually it is not the male but the female which selects a partner and incites him to attack other members of the species. If the chosen mate is repeatedly the loser in these encounters, she will join a stronger male and repeat this behavior toward it."

The EGYPTIAN GOOSE (*Alopochen aegyptiacus;* Color plate p. 295 and Vols. XII and XIII) inhabits tropical Africa. Like the other species of shelducks, with the exception of the ORINOCO GOOSE (*Neochen jubatus;* Color plate p. 295) and the ABYSSINIAN BLUE-WINGED GOOSE (*Cyanochen cyanopterus;* Color plate p. 295), it is marked by striking white wing coverts and a black and bronze-green speculum.

The genus SHELDGEESE *(Chloephaga)* includes a number of South American shelducks. Some of these inhabit, or used to inhabit, the grasslands of the south of the continent in very large flocks. The ANDEAN GOOSE (*Chloephaga melanoptera;* Color plate p. 295) is a strong, somewhat plump-looking bird, with a violet-reddish, shiny speculum. Males and females have the same plumage, as is also the case in the very graceful ASHY-HEADED GOOSE (*Chloephaga poliocephala;* Color plate p. 295) of South Chile, Argentina and Tierra del Fuego; and in the RUDDY-HEADED GOOSE (*Chloephaga rubidiceps;* Color plate p. 295) of Tierra del Fuego and the Falkland Islands. The best known species, the MAGELLAN GOOSE *(Chloephaga picta)*, is distinguished by a marked sex dif-

Explanation for the following color plates. The subfamily Anatinae includes altogether 119 species; 104 of these are shown in color on pages 293 to 296, pages 317 to 320, and on the following pages.

Surface feeding ducks

Plate I

1. Silver teal *(Anas versicolor)*
2. Bronze-winged duck *(Anas specularis)*
3. Bahama pintail *(Anas bahamensis)*
4. Chiloë wigeon *(Anas sibilatrix)*
5. European wigeon *(Anas penelope)*

Plate II

1. Marbled teal *(Anas angustirostris)*
2. Cinnamon teal *(Anas cyanoptera)*
3. Blue-winged teal *(Anas discors)*
4. Falcated teal *(Anas falcata)*
5. Pintail *(Anas acuta)*

Plate III

1. Gadwall *(Anas strepera)*
2. Garganey *(Anas querquedula)*
3. Common teal *(Anas crecca)*

Plate IV

1. Mallard *(Anas platyrhynchos)*
   a) drake in nuptial plumage
   b) in eclipse plumage
2. Domestic runner duck (a domestic form of the mallard)
3. Baikal teal *(Anas formosa)*
4. South American green-winged teal *(Anas flavirostris)*
5. Spotbill *(Anas poecilorhyncha)*

Plate V

Whistling ducks

1. Black-billed whistling duck *(Dendrocygna arborea)*

Shelducks

2. Crested shelduck *(Tadorna cristata)*, extinct
3. Crested duck *(Lophonetta specularioides)*

Surface feeding ducks

4. American wigeon *(Anas americana)*
5. Black duck *(Anas rubripes)*
6. Australian black duck *(Anas superciliosa rogersi)*
7. Yellow-billed duck *(Anas undulata)*
8. African black duck *(Anas sparsa)*
9. Chestnut-breasted teal *(Anas castanea)*
10. Gray teal *(Anas gibberifrons)*
11. Madagascar teal *(Anas bernieri)*, threatened
12. Yellow-billed pintail *(Anas georgica)*
13. Red-billed pintail *(Anas erythrorhyncha)*
14. Hottentot teal *(Anas punctata)*
15. Cape teal *(Anas capensis)*
16. Salvadori's duck *(Anas waigiuensis)*
17. Ringed teal *(Calonetta leucophrys)*
18. Red shoveller *(Anas platalea)*
19. New Zealand shoveller *(Anas rhynchotis variegata)*
20. Blue or mountain duck *(Hymenolaimus malacorhynchos)*

Eider ducks

21. Spectacled eider *(Somateria fischeri)*

Diving ducks

22. South American and African pochard *(Netta erythrophtalma)*
23. Red-head *(Aythya americana)*
24. Ferruginous white-eye *(Aythya nyroca)*
25. Baer's pochard *(Aythya baeri)*
26. Australian white-eye *(Aythya australis)*
27. Ring-necked duck *(Aythya collaris)*
28. New Zealand scaup *(Aythya novaeseelandiae)*

Perching ducks and geese

29. Indian pigmy goose *(Nettapus coromandelianus)*
30. Eastern Hartlaub's duck *(Cairina hartlaubi albifrons)*

Sea ducks and merganser

31. Barrow's goldeneye *(Bucephala islandica)*
32. Labrador duck *(Camptorhyncus labradorius)*, extinct
33. Chinese merganser *(Mergus squamatus)*
34. Brazilian merganser *(Mergus octosetaceus)*

Stiff-tailed ducks

35. Black-headed duck *(Heteronetta atricapilla)*

Plate VI

Diving ducks

1. Common pochard *(Aythya ferina)*, the canvas-back *(Aythya valisneria)* of North America is similar.
2. Tufted duck *(Aythya fuligula)*
3. Greater scaup *(Aythya marila)*
4. Red-crested pochard *(Netta rufina)*
5. Rosy-billed pochard *(Netta peposaca)*

Sea ducks and merganser

6. Common goldeneye *(Bucephala clangula)*
7. Smew *(Mergus albellus)*
8. Common merganser *(Mergus merganser)*

Plate VII

Eider ducks

1. Steller's eider *(Polysticta stelleri)*
2. King eider *(Somateria spectabilis)*
3. Common eider *(Somateria mollisima)*

Sea ducks and merganser

4. Hooded merganser *(Mergus cucullatus)*
5. Red-breasted merganser *(Mergus serrator)*
6. Old squaw (Long-tailed duck) *(Clangula hyemalis)*
7. Harlequin *(Histrionicus histrionicus)*

1♂
2♂
3♂
4♂
5♀
5♂

1♂
2♂
4♂
3♂
5♂

♂
1
♀
♂
2
♀
♂
3
♀

2
1♀
1b ♂
1a ♂
3♂
4
5

1
2♂
3
4♂
5
6
7♂
8♂
9♂
10
11
12
13
14
15
16
17
18♂
19♂
20
21♂
22♂
23♂
24♂
25♂
26♂
27♂
29♂
31♂
30♂
32♂
28♂
35♂
33♂
33♀
34

1♂
2♂
3♂
4♂
6♂
7♂
5♂
8♂

1♂
4♂
3♂
2♂
3♀
7♂
6♂
5♂

♂
1 ♀
2♂
2♀
♀
3
♂

ference in the plumage. The head, neck, and the breast are white in males, but reddish-brown in females; furthermore, the female has fine cross bars on the breast. In the smaller subspecies (*Chloephaga picta picta;* Color plate p. 295) males occasionally have a barred breast. The sex differences are even more obvious in the KELP GOOSE (*Chloephaga hybrida;* Color plate p. 295), which also inhabits southern South America. Here the males are altogether white and the females are grayish-brown. Kelp geese are great rarities in zoos since they feed mainly on seaweed and kelp.

The ash gray CEREOPSIS or CAPE BARREN GOOSE (⦶ *Cereopsis novaehollandiae;* Color plate p. 295), with its peculiar short, very high, and apple green beak, pink legs, black toes, and webs, deviates in many ways from the other shelducks. Nevertheless, systematists include it in this tribe. The breeding area of this bird is on small islands off the west coast of Australia, as far as the Bass Strait. Peter Scott observed them on a journey and writes about them: "We saw our first Cape Barren geese at the westernmost end of Big Dog island. Three pairs were scattered far apart on the rocky ridges of the island. The members of each pair stood close together, often on a projection of the rocks. They looked powerful and majestic, and somehow appropriate to their environment. On the next island there were another three pairs." Since these peculiar shelducks were intensely hunted for tasty meat, their population has greatly declined. In 1962 there were no more than 2000. As a result of strict protective measures, the population is again recovering.

In southern South America, there are three peculiar shelducks, which, externally, resemble massive domestic ducks. These are the STEAMER DUCKS (genus *Tachyeres*). While the FLYING STEAMER DUCK (*Tachyeres patachonicus;* Color plate p. 294) is a plump but nevertheless good flyer, the other two species, the MAGELLANIC FLIGHTLESS STEAMER DUCK (*Tachyeres pteneres*) and the FALKLAND FLIGHTLESS STEAMER DUCK (*Tachyeres brachypterus*) have lost the ability to fly, since their wings are shortened and cannot lift the heavy body.

Raethel writes: "Steamer ducks, in order to move faster when escaping or attacking, use not only their large paddle feet, but also the small narrow wings which they use alternately, not simultaneously, like a person in a kayak uses his paddle. This causes much splashing and foaming, a scene reminiscent of a stern wheeler of the nineteenth century traveling at full speed." The steamer ducks owe their name to this manner of locomotion. Hans Krieg describes another peculiarity of these birds. "When we approached them rapidly and directly, they dived. We noted that in addition to diving in the usual manner (head first) of all diving birds, they have a second way which is comparable to the gradual submersion of a submarine. The heavy body, which in any case lay deep in the water, sank deeper and deeper, and disap-

◁
Scoters and sea ducks: 1. Velvet scoter (white-winged scoter, *Melanitta fusca*); 2. Black scoter (*Melanitta nigra*); 3. Surf scoter (*Melanitta perspicillata*); 4. Bufflehead (*Bucephala albeola*).

peared except for the head, which projected above the water like a periscope."

The last species of the tribe of shelducks, the CRESTED DUCK (*Lophonetta specularioides*; Color plate p. 307), seems to be a transition to the tribe of the surface feeding ducks. With two subspecies, it inhabits waters and uplands in southern South America. It is a slim, graceful bird with a long tail and a striking crest.

Surface feeding ducks

Among the remaining members of the ducks and related forms, the large tribe of the SURFACE FEEDING DUCKS (Anatini) is contrasted with the DIVING DUCKS (Aythyini). This does not mean, however, that surface feeding ducks cannot dive, but rather that they do not dive as much, and they only dive to a depth of about one meter. They get most of their food, which consists predominantly of plants, by dabbling, immersing the head, neck, and front of the body under water, and holding the tail up in the air while they graze on the bottom. After dabbling, they have to flap their wings vigorously a few times to shake out the water which has entered the wing pockets. In swimming, the tail is carried horizontally and does not touch the surface of the water.

Since their legs are inserted far back on the body, these ducks are poor walkers. All of them lack a lobe on the rear toe. The wings are fairly long and pointed, and because of this, surface feeding ducks can readily take wing from the water without a preliminary run. In about half of the species, the plumage of both sexes is similar in color and generally inconspicuous as in the gray ducks; sometimes, however, it is colorful in both sexes, as in the Chiloë wigeon. In the other half of the species, which mainly inhabit the temperate zone of the northern hemisphere, the male has a colorful nuptial plumage. The females of the various species, with their inconspicuous gray-brown plumage, are often very difficult to distinguish. The downy young are also similar to one another in color and pattern. A shiny, metallic-looking speculum on the wing is lacking only in the gadwalls, the marbled teal, the yellow-billed pintail, the Madagascar teal, and four other aberrant species. Surface feeding ducks live on fresh water and adapt only temporarily to sea water conditions when migrating. They generally have large clutches of six to sixteen relatively small eggs, which they incubate for twenty-one to twenty-five days.

Most species of surface feeding ducks belong to the genus *Anas*. They are all listed in the systematic section of this book and are shown on the color plates pages 303 through 306.

The AFRICAN BLACK DUCK (*Anas sparsa*; Color plate, p. 307) seems to be closely related to the shelducks in many respects. With three subspecies, it is native to equatorial and South Africa. In these black ducks both sexes are dark colored, with white spots on the wings and flanks. Being shy, unsociable birds, they live in pairs on the waters of forested areas, and so are difficult to observe. In contrast to other surface feed-

Fig. 13-7. Genus *Anas*.

ing ducks, they dive regularly and also take animal food. The BRONZE-WINGED DUCK (*Anas specularis*; Color plate p. 303) lives on mountain slopes in Chile and the Argentine.

The mallard

Of the six subspecies of the MALLARD (*Anas platyrhynchos*), the European MALLARD (*Anas platyrhynchos platyrhynchos*; Color plate p. 306 and pp. 255/256) is by far the best known. Since the mallard is so common, the beauty of the male is easily, though unjustly, overlooked. Mallards are adaptable in their requirements. For this reason, they became the sole ancestral form of domestic ducks. The various subspecies inhabit Europe, Asia, and North and Central America. While the mainland subspecies are still numerous everywhere, the population of two small insular forms is now very small; these are the LAYSAN TEAL (*Anas platyrhynchos laysanensis*) and the HAWAIIAN DUCK (*Anas platyrhynchos wyvilliana*).

In Central Europe, the mallard incubates ten to twelve greenish-gray eggs for twenty-eight days in early spring. The downy young are blackish-brown and are marked by a yellowish face and underparts, and a black line through the eye. Many forms of behavior of the mallard have already been discussed in chapter eleven, such as the different calls of the two sexes, pair formation in the fall, and courtship display, which has given ethologists a basis for comparisons with other surface feeding ducks. In the course of his research, Konrad Lorenz asked himself, "Why are newly incubator-hatched mallard ducklings shy and intractable, in contrast to incubator-hatched greylag geese? Young wild geese regard the first object they meet, e.g. a man, as their mother, and follow him faithfully. Young mallards, however, just wanted to get away from me." Lorenz assumed that young ducklings have an inborn response to the mother's call of attraction, but not to her appearance. He writes about this: "When a brood of young wild ducks were due to hatch, I placed the eggs in the incubator and, when the young hatched, took them into my care as soon as they had become dry; I then quacked at them in my best imitation of the mallard contact call. This was successful. The ducklings looked up at me with confidence, and were evidently not afraid of me. When, still quacking, I slowly moved away, they followed me in a tight little bunch like ducklings follow their mother." (Color plate p. 293).

Fig. 13-8. Mallard (*Anas platyrhynchos*).

Fig. 13-9. Mallard drakes.

Domestic ducks by M. Lühmann

Mallards have been domesticated since ancient times, but the process of domestication still continues in our day. European domestic ducks are mostly Peking ducks of American origin. These white ducks weigh up to three kilograms and are reared by the million in special duck hatcheries, where they are killed at about eight weeks of age.

The other breeds of domestic ducks are only of importance today with a view toward producing new breeds. The German Peking duck holds itself more upright than the American strain; its white feathers have a strong yellow tint. The Vierländer ducks are heavy white birds of the "farm duck" type. The dark mallard-like Rouen ducks are marked by a strong dewlap of the keel of the sternum; they carry their body horizontally as does the white Aylesbury duck, which has a pink beak and feet. Pomeranian or Swedish ducks have a blue or black plumage with a white bib, while Cayuga ducks are black with a green sheen. Saxon ducks and the somewhat smaller Streicher ducks are marked by differences in size and color. Incidentally, the blue-gray color of domestic ducks is not inherited as a genetically pure characteristic; the offspring of birds of this color are black, blue-gray, or pale gray in the proportions of about 1:2:1, in accordance with Mendelian laws, which is an indication that "blue" ducks are the result of a cross between black and pale gray parents.

Fig. 13-10. Drake common teal (left), Drake garganey (right).

Drake gadwalls.

Man has been able to increase the egg production of ducks to over 250 eggs a year much earlier than that of chickens; nevertheless, the high yield breeds have not been maintained in Europe in great numbers, since duck eggs occasionally transmit typhoid and paratyphoid, and are therefore seldom bought. An example of a high yield duck formerly was the very mobile RUNNER DUCK (Color plate p. 306). Today these ducks, with their very upright posture, their slim body, and their scanty plumage, are mostly kept for show. Crosses with other domestic ducks produced the liver-colored Orpington duck and the brown Khaki-Campbell duck. The latter was formerly of considerable significance in Holland as an egg producer.

Drake European wigeon.

The small domestic ducks, weighing only one to one and a half kilograms, include the high yield duck, (Hochbrutenten), the emerald duck, and the dwarf duck. The first is widely distributed in the Frisian coastal area and is used as a decoy by duck hunters. The emerald duck has a black plumage with green gloss. In dwarf ducks, a breeder expects a short head and beak, a compact body, and pure color phases. High yield flying ducks, however, are only occasionally bred in pure color strains; their color is therefore often quite variable.

Drake pintails.

There are hardly any particular deviations of the body shape among domestic ducks. Only crested ducks with a head crest are still bred as a medium weight breed. Occasional crested birds also occur among farm ducks and high yield flying ducks. How such crests are inherited has not been finally clarified; occasionally there are a few birds with crests among the offspring of smooth-headed ducks. The behavior of domestic ducks resembles that of the mallard to a great extent, although most breeds have lost the drive to incubate.

Shovellers.

The remaining surface feeding ducks by H. G. and U. Klös

The COMMON TEAL OR GREEN-WINGED TEAL, (*Anas crecca*, Color plate p. 305) is a common inhabitant of small lakes and pools overgrown with reeds. It is unusually shy; often only the loud, monotonous "kreck, kreck, kree-ee-eck-kreek" call of the drake indicates its presence. It is the smallest of the Central European ducks. The drake has a beautiful nuptial plumage from October to June, the most prominent characteristic of which is the green stripe on the chestnut-red head. There

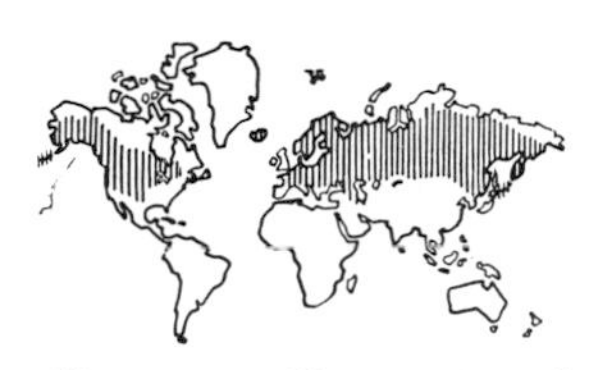

Fig. 13-11. Common teal *(Anas crecca).*

Fig. 13-12. Gadwall *(Anas strepera).*

Fig. 13-13. European wigeon *(Anas penelope).*

Fig. 13-14. Pintail *(Anas acuta).*

are one Eurasian and two North American subspecies.

Common teals are migrants which move south in large flocks in winter. "When a flock of teal descends onto a small lake in the winter migration, thirty or a hundred bright green specula momentarily shine as they turn over the water before the whole flock comes down and becomes invisible in the reeds." In the words of Kaltenhäuser; "Every observer remembers the beauty of these most skillful fliers among ducks and geese."

Closely related to the common teal is the CHILEAN TEAL (*Anas flavirostris*; Color plate p. 306); males and females are similar in color in all four subspecies. An isolated form is the BAIKAL TEAL (*Anas formosa*; Color plate p. 306) of eastern Siberia. The name *formosa* has nothing to do with the island of Formosa, but means beautiful; it refers to the drake's nuptial plumage with its white, sand-colored, and green head. The FALCATED TEAL (*Anas falcata*; Color plate p. 306) breeds in north Asia; the drake has shoulder feathers prolonged into a sickle shape in its nuptial plumage, thus showing some affinity to the gadwalls.

The GADWALL (*Anas strepera*; Color plate p. 305 and pp. 225/226) inhabits a large area of the Norhern Hemisphere (see Fig. 13-12); a smaller species which is probably extinct, COUE'S GADWALL (*Anas strepera couesi*), used to occur on the Fanning islands south of Hawaii. The absence of a metallic sheen on the speculum differentiates the gadwall from most other surface feeding ducks. Its beak has a system of especially fine lamellae; it lives even more exclusively on plants than its relatives, and is closely related to the wigeons.

The EUROPEAN WIGEON (*Anas penelope*; Color plate p. 303 and pp. 385/386) breeds in northern Eurasia and migrates in winter as far as Africa, India, China, and Japan. The drake's nuptial plumage is marked by a light-colored crown patch on its reddish-brown head. The AMERICAN WIGEON (*Anas americana*) of North America looks rather similar. In the CHILOË WIGEON (*Anas sibilatrix*; Color plate p. 303), both sexes have a colorful plumage.

The PINTAIL (*Anas acuta*; Color plate p. 304 and pp. 255/256) derives its name from the prolonged tail feathers of the drake's nuptial plumage. The slim neck is also striking. Pintails breed in temperate Europe, Asia, and North America, and migrate far south in the winter. During migration they are excellent fliers; the Asiatic subspecies even crosses the Himalayas. Hundreds of pintails were seen on the moor land lakes of the Brahmaputra valley in Assam. As Makatsch discovered in Hungary, pintails there like to build their nests in the immediate vicinity of lapwings: "The incubating pintail duck evidently relies more on the attentiveness of the lapwings rather than on her drake, which during the long laying period always remains near the nest." Two dwarfed insular forms, in which both sexes are equally inconspicuously colored, occur on the Kerguelen and Crozet island.

The seven species of the Garganey group are very similar in their behavior and in some features of their coloration. They all have blue-gray middle and lesser wing coverts. The GARGANEY (*Anas querquedula*; Color plate, p. 305 and pp. 385/386) is native in northern Eurasia, including Germany. It has only one call, which resembles the noise made by running a finger over the teeth of a comb. It prefers standing or slow-flowing waters with rich vegetation as a habitat. It lays its eight to eleven eggs in May to June, and incubates them twenty-three days. Its closest relative is the BLUE-WINGED TEAL (*Anas discors*; Color plate p. 304), whose male has a blue-gray head. With two subspecies it inhabits central North America as far as the Atlantic coast. North and South America are the homes of the CINNAMON TEAL (*Anas cyanoptera*; Color plate p. 304). The drakes of all five subspecies are marked by a bright chestnut-brown nuptial plumage. The enlarged bill of the cinnamon teal forms a link with the shovellers.

Of all the shovellers, the COMMON SHOVELLER (*Anas clypeata*; Color plate pp. 255/256) of Eurasia and North America is the most colorful. All shovellers carry a particularly fine and dense system of lamellae in their broadened beaks, which enables them to sift out the smallest food particles from the waters. A peculiarity is their habit of swimming in keel formation, which is often observed.

The PINK-HEADED DUCK (*Rhodonessa caryophyllacea*; Color plate p. 294) is an isolated type. It differs so much from other surface feeding ducks that it has been placed in a separate genus. Its peculiar rosy-pink head plumage was noted in chapter eleven. Its home is in India, and it is probably extinct.

Four other species of surface feeding ducks each represent a separate genus. There is the BLUE OR MOUNTAIN DUCK (*Hymenolaimus malacorhynchos*; Color plate p. 307) which inhabits the mountain streams of New Zealand; its beak has a soft fringe of skin on each side of the tip. In the PINK-EARED DUCK (*Malacorhynchos membranaceus*; Color plate p. 294) of Australia and Tasmania, this fringe of the beak forms a flap which hangs down from the tip of the spatula-like beak. An inconspicuous red "ear patch" is responsible for its English name. The rarest of the Australian ducks is the FRECKLED DUCK (⍿ *Stictonetta naevosa*; Color plate p. 294). It lacks a wing speculum, and the male has no tracheal bulla. According to some sources, the male's beak becomes red in the breeding season.

Deviating most from the other surface feeding ducks is the TORRENT DUCK (*Merganetta armata*; Color plate p. 294). This slim, long-tailed species has a bony spur on the wing butt in both sexes, for which it has also been called the armed or spurred duck. The several subspecies of torrent ducks live in the rushing mountain streams of the Andes, up to elevations of 3600 meters. The females of all subspecies are similar. Their underside is chestnut-brown, the top of the head, the neck, and the back are gray, which makes these ducks, which lie very

Cairini

▷
1. Mandarin duck (*Aix galericulata*); 2. Carolina wood duck (*Aix sponsa*).

▷▷
1. Comb duck (*Sarcidiornis melanotus*); 2. White-winged wood duck (*Cairina scutulata*); 3. Muscovy duck (*Cairina moschata*), wild form. a) white domestic form; 4. Spur-winged goose (*Plectropterus gambiensis*).

▷▷▷
1. Brazilian teal (*Amazonetta brasiliensis*); 2. Maned goose (*Chenonetta jubata*); 3. African pigmy goose (*Nettapus auritus*).

Fig. 13-15. 1. Garganey (*Anas querquedula*); 2. Marbled teal (*Anas angustirostris*).

Fig. 13-16. Shoveller (*Anas clypeata*).

1 ♂
1 ♀
2 ♂

2
1 ♀
1 ♂
3 a ♂
♀ 3
♂
♀
4 ♂

1
2 ♀
2 ♂
3 ♂

♂ 1 ♀
3
2 ♂
4 ♂

◁
Stiff-tailed ducks:
1. Ruddy duck *(Oxyura jamaicensis)*; 2. White-headed stiff-tail *(Oxyura leucocephala)*; 3. Musk duck *(Biziura lobata)*; 4. White-backed duck *(Thalassornis leuconotus)*.

low in the water when swimming, almost invisible. The males, however, differ in the degree of lightness in the plumage. However, all of them have a white head with a characteristic black stripe running from the eye back and down the neck. Both sexes have the cherry-red beak.

Torrent ducks are surprisingly good climbers; they use their stiffened tail feathers as supports, moving with almost unbelievable skill through the foam and eddies of waterfalls. Paul A. Johnsgard writes: "By far the most impressive and unforgettable spectacle in the behavior of torrent ducks is their unbelievable ability to master the most impossible rapids, fighting upriver against an overpowering current, or turning, rolling over bottoms up, and shooting down the rushing rapids, unconcerned about rocks, and almost disappearing among spray and foam. I have observed adult torrent ducks falling down falls several meters, mainly when they tried to escape from danger." After detailed observation of their behavior, Johnsgard classified the torrent ducks as relatives of the tribe of the perching ducks and geese (Cairinini).

Eider, diving, and perching ducks by H. G. and U. Klös

Tribe: eider ducks

The EIDER DUCKS form a separate tribe (Somateriini). The smallest species, STELLER'S EIDER (*Polysticta stelleri*; Color plate p. 309), takes up a somewhat isolated position within this tribe. The female plumage of all eiders resembles that of the female mallard; drakes, however, are conspicuously black and white colored ducks. In some species, there is a shimmer of green and rusty-red on the breast and on the sides of the head; it is due to a pigment peculiar to these species. Eider ducks are good swimmers and excellent divers; they can even withstand rough seas. Their food consists mainly of marine mollusks, crustacea, sea urchins, coelonterata, fish, and fish spawn. They breed in the Arctic of the Old and New World. They migrate southward to ice-free waters in winter. Thus Steller's eider or the beautifully colored KING EIDER (*Somateria spectabilis*; Color plate p. 309) of the far north are occasionally found on the German coasts.

The COMMON EIDER (*Somateria mollissima*; Color plate p. 309) with its very characteristic head profile, is not just a winter visitor to Germany, but a few pairs also breed on the islands of Sylt, Amrum, and Juist. Eiders nest on the ground, often in colonies, and always lay their four to six olive-green eggs near the shore. They surround the nest with a particularly fine down, for which they have been intensly persecuted. In Norway and Iceland, some of these breeding colonies are fenced in today, and the eiders breeding in them are utilized. The first two clutches and their down are collected, while the third clutch is left to the birds. Surprisingly enough, eiders seem to appreciate these places where they are protected from predators, for they regularly return to them. The period of incubation is twenty-eight days. Right after hatching, the young rush to the water and immediately show themselves to be excellent swimmers and divers. The adult plumage is only

Fig. 13-17. Eider (Tribe Somateriini)

attained in the third year. The SPECTACLED EIDER *(Somateria fischeri)* of the north coast of Siberia is distinguished by the peculiar spectacle pattern on its predominantly green head.

Tribe: diving ducks

The DIVING DUCKS (tribe Aythyini) differ from surface feeding ducks by their short, compact, almost drop-shaped bodies, and their large feet which are inserted far to the rear, carrying an appendage on the hind toe. The wings are short and pointed, and so these quite heavy birds have to run, fluttering over the water surface for a short distance, before they can become air borne. Then, however, they show themselves to be skillful fliers. Their head is generally quite large, and the beak is flatter than in surface feeding ducks. The tail is carried low and in swimming is partly immersed in the water. Compared to the surface feeding ducks, there are also marked differences in the structure of the tracheal bulla, which comes to a point and has openings closed by membranes. The specula of diving ducks have no colors with metallic iridescence, but are mostly white, or at least light in color.

Fig. 13-18. Diving ducks (Tribe Aythini)

There is a marked sex difference in almost all species. In temperate climates, the males have a nuptial plumage which often differs considerably from that of the female, but the females never have the spotted plumage which is so common in surface feeding ducks, like the female mallard. Diving ducks acquire the adult plumage in the first year and the females can breed at this age. The nests are truncated cones with a depression in the narrower upper end; they are built on land, hidden among reeds or grass. The flight feathers of the chicks appear quite late. Evidently it is not necessary for them to quickly acquire the ability to fly, since they can escape all dangers by skillful diving. With the exception of a few subtropical and tropical species, diving ducks are migrants. They live on animal and plant food.

Fig. 13-19. Drake common pochard

A well demarcated group of birds which externally still look rather like the surface feeding ducks are the colorful SOUTH AMERICAN ROSYBILL (*Netta peposaca;* Color plate p. 308), the dark brown SOUTH AMERICAN POCHARD (*Netta erythrophthalma;* Color plate p. 307), which inhabits Africa and South America with its two subspecies, and the RED-CRESTED POCHARD (*Netta rufina;* Color plate p. 309 and pp. 255/256), which breeds in Eurasia, and also in southern Germany on Lake Constance and in a few other places in Germany. The male red-crested pochard is distinguished in its nuptial plumage by a bright orange to light chestnut-colored head plumage and a red beak. Like all diving ducks, these pochards like to be out in the middle of their home waters, and are therefore easy to observe. During courtship there are tumultuous "rapes" of the females in all of the diving duck groups. In courtship the male red-crested pochard utters a strange sneezing sound while it moves the beak sideways. As in other diving ducks, the female's call is a croaking. The six to fourteen stone-gray eggs are incubated for twenty-six to twenty-eight days.

Fig. 13-20. Red-crested pochard *(Netta rufina).*

Fig. 13-21. Common pochard *(Aythya ferina).*

Fig. 13-22. Greater scaup *(Aythya marila).*

The EUROPEAN POCHARD (*Aythya ferina;* Color plates p. 308 and pp. 255/256 and pp. 385/386) owes its name, "table duck", to its tasty meat, and is the most striking diving duck of Central European waters. Except for the breeding season in April and May, these pochards live sociably in fairly large groups. The CANVASBACK *(Aythya valisneria)* of North America is very similar, but is larger and has a heavier beak. It feeds largely on wild celery *(Valisneria)* from which it derives its scientific name. The delicate FERRUGINOUS DUCK (*Aythya nyroca;* Color plate p. 307) breeds in southern Eurasia. It is a shy bird which tends to hide in the dense reed beds of waters. The TUFTED DUCK (*Aythya fuligula;* Color plate p. 308, pp. 61/62 and pp. 255/256, native to Germany, is marked by a decorative crest. It is not infrequently seen on park waters or rivers in the midst of our cities in winter. Further species of the genus *Aythya* are listed in the systematic section of this book, and are shown on color plates, pages 307 and 308.

Tribe: perching ducks and geese

The tribe Cairinini, wood ducks or perching ducks and geese, includes species of quite different shapes and sizes. According to the latest research, the inconspicuous BRAZILIAN TEAL (*Amazonetta brasiliensis;* Color plate p. 319), with two recognized subspecies, belongs in this group, as well as the RINGED TEAL *(Calonetta leucophrys),* which was formerly placed with the surface feeding ducks. The pigmy geese are the smallest species of this tribe; among them is the AFRICAN PIGMY GOOSE (*Nettapus auritus;* Color plate p. 319), the most colorful. It is replaced in Ceram, Buru, New Guinea, and northern Australia by the GREEN PIGMY GOOSE *(Nettapus pulchellus),* which has fine black and white wavy marks on the breast and on the flanks; in southern Asia and northern Australia it is replaced by the INDIAN PIGMY GOOSE *(Nettapus coromandelianus),* which occurs there in two subspecies. The largest bird of this group is the SPUR-WINGED GOOSE (*Plectropterus gambiensis;* Color plate p. 318) of Africa, which is armed with a strong wing spur.

Fig. 13-23. Ferruginous duck *(Aythya nyroca).*

All birds of this group show a preference for wooded habitats. They have strong, pointed claws, and therefore can perch readily in trees. Almost all nest in cavities in trees. The chicks are marked by little pointed claws and long stiffened tail feathers. The gait of these ducks is peculiar, almost giving the effect of limping, since the birds only nod their heads at every second step. Almost all have a strong metallic green color. This is most intensive in the WHITE-WINGED WOOD DUCK (*Cairina scutulata;* Color plate p. 318) and the MUSCOVY DUCK (*Cairina moschata;* Color plate p. 318).

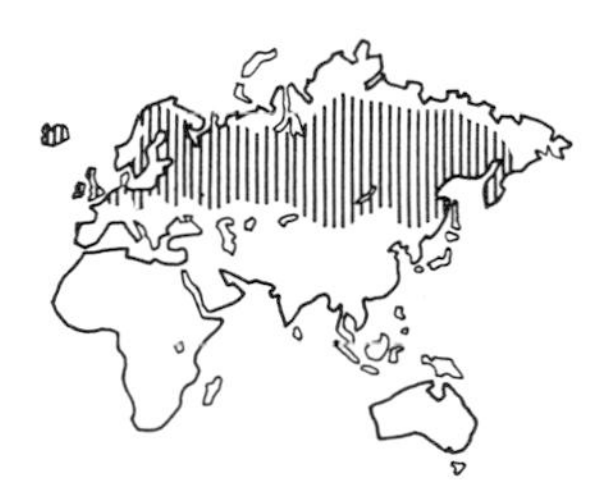

Fig. 13-24. Tufted duck *(Aythya fuligula).*

The domestic muscovy duck, by M. Lühmann

The muscovy duck is the ancestral form of the domestic muscovy duck (Color plate p. 318). South American Indians had domesticated this bird long before the discovery of America. It was brought to Europe by the Spaniards. Since then, it has displaced European domestic ducks to a considerable degree on German poultry farms. The sexes differ remarkably in size: males weigh about three and one half to four kilograms, while females weigh only half as much. Apart from an increased number of broods and some differences in color, domestication of the muscovy duck has not resulted in marked changes. In addition to almost black birds, there are white, pale ash gray, brown, and barred ones. When pale ash gray birds are mated with ducks of the wild plumage type or black ones, the color of the offspring is blue; in later generations of hybrids the colors segregate as they do in domestic ducks. In a cross of solid-colored and white muscovies, blotched offspring are produced; their offspring in turn segregate into white, blotched, and uniformly colored birds.

Fig. 13-25. Pigmy geese (Tribe Cairinini).

In the last decade, muscovy ducks have spread more and more on poultry farms, particularly in non-European countries. The reasons for this are their frugal requirements, the certainty that they will breed several times a year, and their great care of the young. Therefore, they offer advantages as birds for fattening in the course of changes in agricultural practices.

Fig. 13-26. Scoters, sea ducks, and mergansers, (Tribe Mergini)

Hybrids between muscovies and ordinary domestic ducks are sterile. The offspring from the crosses of a domestic duck with a male muscovy have about the same weight in both sexes, but they do not lay eggs at all. Offspring between a muscovy hen and a domestic drake are much rarer; this is due to the different mating behavior of the two species. The sexes of these hybrids show considerable differences in weight; they lay small eggs weighing about forty grams, whose embryos do not develop. In some parts of Europe, hybrids between male muscovies and domestic ducks are reared extensively for fattening.

Fig. 13-27. Black scoter (*Melanitta nigra*).

Other perching ducks, by H. G. and U. Klös

The reddish-brown HARTLAUB'S DUCK (*Cairina hartlaubi;* Color plate p. 307) of West and Central Africa has a particularly attractive blue-gray wing speculum. It is very closely related to the white-winged wood duck and the muscovy ducks. While other ducks of this tribe nest predominantly in tree holes, nests of the African species have been found in termite nests or in the huge nests of the HAMMERHEAD (*Scopus umbretta*). The spur-winged goose, however, seems to prefer to nest on the ground. Very large egg clutches are characteristic of these birds. Thus fifty-four partly broken eggs were found in a single nest of the COMB DUCK (*Sarcidiornis melanotus,* Color plate p. 318). There is no satisfactory explanation for this. Some ornithologists attribute the large clutches to a lack of nest sites and assume that several females lay their eggs into the same nest. Others, like Pitman, believe that the birds are polygamous. There are two subspecies of the clumsy-looking comb duck: the South American form has dark sides, while the African and Indian subspecies have white flanks.

Fig. 13-28. Velvet scoter (white-winged scoter, *Melanitta fusca*).

There were originally no ducks of this tribe in Europe. Two particularly pretty and dainty species, the WOOD DUCK (*Aix sponsa;* Color plate p. 317) of the United States and southern Canada, and the MANDARIN

Fig. 13-29. Harlequin (*Histrionicus histrionicus*).

Fig. 13-30. Old squaw *(Clangula hyemalis).*

DUCK (*Aix galericulata;* Color plate p. 317) of east Asia, have been kept as free-flying birds by breeders of waterfowl for a long time. They are therefore encountered in European parks as wild birds. Attempts to introduce the wood duck in Europe have been made, particularly by Heinroth, who described them in full detail. In the drake mandarin duck, the sail-like shoulder feather is particularly striking; if it is lost, it can be replaced three times within one moulting period. It disappears among the shoulder plumage in flight, so that the bird looks quite different then. Also included in this tribe is the MANED GOOSE (*Chenonetta jubata;* Color plate p. 319) of Australia and Tasmania. It lives in pairs or troops along the wooded upper course of rivers, and grazes like a true goose.

Tribe: sea ducks and mergansers by H. G. and U. Klös

The SCOTERS, other SEA DUCKS, and the MERGANSERS form yet another tribe, Mergini. They are closely related to the Cairinini and could be called birds of that tribe which hunt their food (mollusks, crustacea, insects, and fish) under water. Being excellent divers, they move clumsily on land. Their wings are short and their flight is rapid. In most species, the males are magnificently colored, but they have no iridescence. Apart from the AUCKLAND ISLAND MERGANSER *(Mergus australis),* which has not been seen alive since 1905 and is probably extinct, and the BRAZILIAN MERGANSER *(Mergus octosetaceus)* of South America, all birds of the Mergini tribe inhabit the cold or temperate zones.

Fig. 13-31. Barrow's goldeneye *(Bucephala islandica).*

The scoters form a well defined group within this tribe. They include the BLACK SCOTER (*Melanitta nigra;* Color plate p. 310), which inhabits northern Eurasia and the outermost northwest of North America, with two subspecies, the SURF SCOTER (*Melanitta perspicillata;* Color plate p. 310), of North America, and the VELVET OR WHITE-WINGED SCOTER (*Melanitta fusca;* Color plate p. 310), which inhabits northern Eurasia and northwest North America, with four subspecies. They nest on the ground near fresh water lakes, and leave the fresh water with their fully fledged young to spend the rest of the year at sea. The velvet and black scoters are winter visitors of the German North Sea and the Baltic coast.

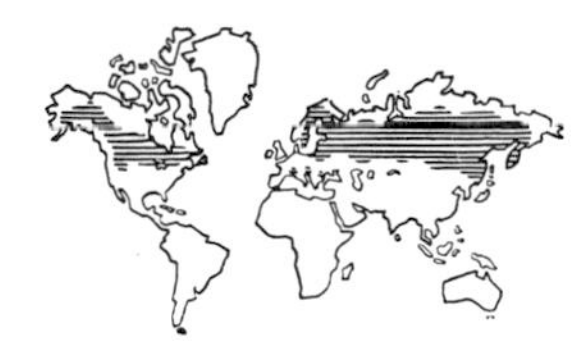

Fig. 13-32. Common goldeneye *(Bucephala clangula).*

The colorful HARLEQUIN DUCK (*Histrionicus histrionicus;* Color plate p. 309) is a link between the scoters and the old squaws and goldeneyes. In the breeding season, harlequins live on rushing mountain streams and, in this, resemble the torrent ducks and the mountain ducks. The striking plumage of the male is not easily recognized in the spray and foam of mountain streams. After the end of the breeding season, Harlequins migrate to the seacoast, the females with the young, somewhat later than the males. Here too they seem to be happiest in the strongest surf.

Fig. 13-33. Common merganser *(Mergus merganser).*

The OLD SQUAW OR LONG-TAILED DUCK (*Clangula hyemalis;* Color plate p. 309) breeds on the north coasts of the Old and New Worlds. It is marked by the long tail of the drake and the fact that males and females have two entirely different plumages in summer and winter. To this must be added the eclipse plumage of the drake, so that males moult three times and even four times in parts of the plumage. Two other species, BARROW'S GOLDENEYE (*Bucephala islandica;* Color plate p. 307) and the COMMON GOLDENEYE (*Bucephala clangula;* Color plates p. 308, pp. 255/256, and pp. 385/386), which also live in the north of Eurasia and North America, are very similar to one another; the females in particular are difficult to distinguish. The white facial spot is crescentic in the drake Barrow's goldeneye, while in the common goldeneye it is round. The most southern breeding area of the common goldeneyes in Germany is the pond areas of Lusatia. The BUFFLEHEAD (*Bucephala albeola;* Color plate p. 310) of the north of western and central North America shows some relationship to the mergansers.

The genus MERGANSERS *(Mergus)* is characterized by its streamlined body shape. All have a thin beak with saw-like edges and a sharp, bent-down nail. Such a beak is very suitable for catching and holding slippery fish, which these birds hunt while swimming underwater. Only in the Brazilian and the Auckland Island merganser are the plumages of males and females the same; in all other species the sexes differ in color, and the males develop a nuptial plumage. All mergansers have a more or less developed crest in both sexes. This is most pronounced in the HOODED MERGANSER (*Mergus cucullatus;* Color plate p. 309) of North America. In excitement it erects the hood, its white patch expanding to form a bright signal.

Three merganser species occur in Europe: the very small SMEW (*Mergus albellus;* Color plates p. 308 and pp. 61/62), the RED-BREASTED MERGANSER (*Mergus serrator;* Color plate p. 309), with a subspecies in Greenland, and the COMMON MERGANSER or GOOSANDER (*Mergus merganser;* Color plates p. 308 and pp. 61/62). The males of the three subspecies of the common merganser are generally paler than those of the red-breasted merganser. Females of these two species are, however, difficult to distinguish. The breeding behavior of the two is strikingly different: the common merganser nests in cavities and lines its nest with pale down, while the red-breasted merganser breeds on the ground, more or less in the open, and uses dark nest down. The eggs are incubated for thirty-one to thirty-two days by the red-breasted merganser, while the common merganser takes thirty-four to thirty-five days. The CHINESE MERGANSER (*Mergus squamatus;* Color plate p. 307) of southeast Siberia and Manchuria is rather similar to the red-breasted merganser. Breeding places are so far known only from mountain streams in Ussuria. It owes its alternative name, the scaly-sided merganser, to its flank feathers with their black edges, giving the effect of scales.

The last tribe of the Anatinae, the STIFF-TAILED DUCKS (Oxyurini), are in an isolated position in the taxonomic system. Among other features, they are characterized by flat, wide beaks and stiff, quite long tails,

▷

Above: Golden eagle *(Aquila chrysaetos)* in North America. Below: Philippine monkey-eating eagles *(Pithecophaga jefferyi)* in the Frankfurt Zoo. These were bought before the director's decision mentioned in Chapter 15.

▷▷

Golden eagle nest in the Austrian alps.

▷▷▷

African white-backed vulture *(Pseudogyps africanus)*, coming down by a dead animal.

▷▷▷▷

An Egyptian vulture *(Neophron percnopterus)* uses stones to open an ostrich egg. It flies to the ostrich nest with a stone in its beak and beats or throws the stone onto the egg until the shell breaks. Baron van Lawick and Jane van Lawick-Goodall discovered this use of a tool and photographed the process.

▷▷▷▷▷

Above: Trained goshawk *(Accipiter gentilis)* on the falconer's fist and on a pheasant. Below (from left to right): Sparrowhawk *(Accipiter nisus);* a common buzzard *(Buteo buteo)* has found a dead rabbit and is eating it; African white-backed vultures *(Pseudogyps africanus)* at the remains of a dead zebra.

Tribe: stiff-tailed ducks (Oxyurini), by H. G. and U. Klos

1
2
3
4
5

Fig. 13-34. Red-breasted merganser *(Mergus serrator).*

Fig. 13-35. Smew *(Mergus albellus).*

Fig. 13-36. Stiff-tailed ducks (Tribe Oxyurini).

◁ New World vultures: 1. Andean condor *(Vultur gryphus);* 2. California condor *(Gymnogyps californianus);* 3. King vulture *(Sarcoramphus papa);* 4. Black vulture *(Coragyps atratus);* 5. Turkey vulture *(Cathartes aura).*

which in swimming either dip below the water or are carried rigidly erect. In most species, males and females have a different plumage. An exception to this is the WHITE-BACKED DUCK (*Thalassornis leuconotus;* Color plate p. 320), which is native to Africa and Madagascar and has two subspecies. Particularly deviant species include the BLACK-HEADED DUCK (*Heteronetta atricapilla;* Color plate p. 307), which as a brood parasite lays its eggs in the nests of other birds, and in many respects, the MUSK DUCK (*Biziura lobata;* Color plate p. 320 and Vol. X) of southern Australia and Tasmania, whose male has a conspicuous appendage under the beak.

The main group of stiff-tails includes the MASKED DUCK *(Oxyura dominica)* of the West Indies and the WHITE-HEADED STIFF-TAIL (*Oxyura leucocephala;* Color plate p. 320) of Eurasia. In both species the male has a deep chestnut-colored nuptial plumage and, as in all ducks of this genus, a bright blue bill. Females have a dull, dark brown plumage. The RUDDY DUCK (*Oxyura jamaicensis;* Color plate p. 320) inhabits North and South America with three subspecies. In the area of Valparaiso (Chile) it lives next to the ARGENTINE RUDDY DUCK *(Oxyura vittata)* with which it does not interbreed. It is represented in Australia by the AUSTRALIAN BLUE-BILLED DUCK *(Oxyura australis)* and in Africa by the AFRICAN MACCOA DUCK *(Oxyura maccoa).* The courtship display of the stiff-tails is particularly interesting. The male erects the tail rigidly and offers the female a view of the bright white lower side of his tail. Frequently the male not only erects his tail but also pumps his neck full of air. When he beats his beak against his neck, he squeezes air out of the plumage; the air rises in bubbles to the water surface, where they burst. The eggs of stiff-tails are relatively large; the males take part in rearing the young.

Christopher Savage observed white-headed stiff-tails in West Pakistan and writes about their way of life: "The waters preferred by the stiff-tails are the brackish lakes Khabbaki, Kallar Kahar, and Nammal in the salt area, with a salt concentration of 1760, 8060, and 3180 parts per million respectively. They are, on the average, one to one and a half meters deep with only little surface vegetation, but they are rich in underwater plants like *Ruppia maritima, Melilotus indica, Hydrilla verticillata, Potamogeton nodosus,* and green algae." He says further, "The stiff-tails in general kept to themselves. They were busy seeking food through most of the day, while most of the other ducks which were present rested in the daytime. One could easily approach them in a boat up to thirty to thirty-five meters before they escaped by diving. They would only occasionally fly up, and only because other birds nearby did so. Their start and flight resemble that of grebes; after a long run, they clumsily flew with fast wing beats, but rarely higher than one and a half meters. In the water they held their tails at a forty-five degree angle, except in moments when they felt disturbed. With fluffed-up plumage, they often looked like African white-backed ducks."

14 Raptors

Order: raptors

Among birds, the ability to hunt living prey is most highly developed in the RAPTORS (order Falconiformes). These birds were also called birds of prey, which, however, is not a specific name since many birds of other orders also hunt for living prey. They are particularly distinguished by the "weapons" they use to overcome their prey, namely the short, hooked beak, with the upper mandible strongly curved, and in particular the strong feet with long toes, and the highly developed sharp claws which are an excellent tool for grasping. By far the majority of raptors are adapted to capturing prey, which may be insects, amphibia, reptiles, small birds, or small mammals. A few species prefer a vegetarian diet; some, for example the vultures, utilize every type of dead animal. But, also buteos and even eagles do not avoid carrion, nor do kites, which, moreover, like to beg for prey from other raptors. Only a few species, including certain goshawks, true falcons, and above all some eagles, like the golden and sea eagles, are able to overwhelm larger prey animals. Many species of raptors are adapted to take specific animals as prey.

Distinguishing characteristics

Raptors are easily distinguished from the members of other orders of birds. The body is strong, compact, and wide-breasted; the large head is generally rounded, and only rarely elongated. The neck is normally short and strong, and only seldom is it long. The rump is short, the breast and limb muscles are strong, and the short, hooked beak is laterally compressed. The cutting edges of the upper mandible project like scissors over those of the lower mandible. The beak is covered by a soft cere at the root of the upper mandible. The foot is short, strong, with long toes, and an outer toe which in some species can be rotated (to either the front or the rear). The claws are more or less bent, and when strongly bent they are pointed and form grasping tools for seizing prey. The flight and tail feathers are large. There are ten primaries, twelve to sixteen secondaries, and twelve to fourteen rectrices. The plumage often extends over the whole tarsus. The large eyes have highly developed vision. The hearing is good, but olfactory ability is found only in some New World vultures. The ♀♀ are larger than the ♂♂, or may be of equal size (an exception are some vultures). The young develop slowly and often have several plumage changes before the adult plumage. The distribution is worldwide. There are four families: 1. The NEW WORLD VULTURES (Cathartidae); 2. the SECRETARY BIRDS (Sagittariidae); 3. GOSHAWKS and related forms (Accipitridae); and 4. FALCONS and related forms (Falconidae), with a total of 291 species.

The life of raptors, by H. Brüll

In view of the extensive changes which the natural environment has undergone at the hands of man, we have today a quite different understanding of the role of predators in their various habitats. From their relationship with other animals, and in some cases, plants, it can be seen that they are not just "robbers," as used to be assumed, but they play rather a definite role in the balance of nature. They remove animal corpses and feed to a great extent on sick or weak animals, thus ensuring that the population of their prey animals remains healthy and able to compete.

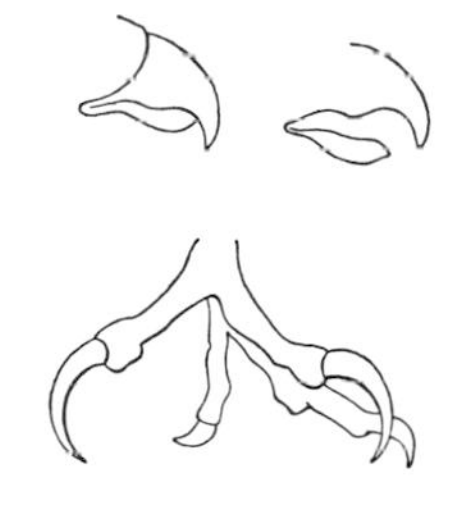

Fig. 14-1. Beak shape for tearing and hooking, with a smooth cutting edge (above), and foot with curved claws used in killing prey, which is characteristic for this group (below).

This applies particularly to the members of the two large families of the goshawks (Accipitridae) and the falcons (Falconidae). According to their abilities when seizing and killing prey, they may be divided into two functional groups: 1. the grasping killers, having a beak for tearing, hooking and cutting; and 2. the grasping holders, having a beak for tearing, hooking, and biting. The first group includes most goshawks, the second, the pigmy and true falcons. The two groups are distinguishable not only on the basis of their beak structure, but also according to the structure and use of their feet, which have become converted into grasping tools.

The feet of grasping killers, as a rule, have short toes; only bird and bat hunters are an exception. The hind toe and the inner front toe have stronger or particularly strongly developed claws. The center claws and outer toes are clearly weaker, even among those birds that prey on small animals. These features are particularly evident on the feet of eagles and the goshawk. The feet of grasping holders, on the other hand, have claws which show no such differences in length. Only the claw of the hind toe is a little longer in these birds. Owls, which are not related to the raptors but belong to the functional type of grasping holders, also have this type of foot. The beaks of the grasping killers have sharp cutting edges on the sides of the upper mandible, as well as a tearing hook. With such beaks they can cut into the tough skin even of larger animals. Notable exceptions to this are those species which make special use of the beak to crack snails or animals with strong chitin carapaces. Thus the round-winged DOUBLE-TOOTHED KITE *(Harpagus bidentatus)* has two points (or "teeth") behind the hook of the beak on either side of the upper mandible. The beaks of grasping

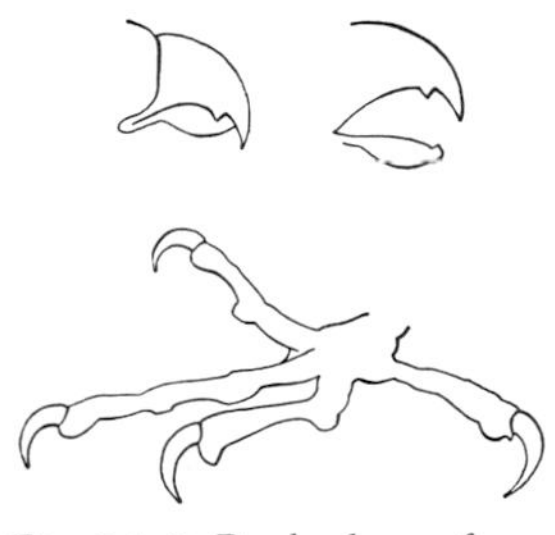

Fig. 14-2. Beak shape for tearing and biting (cutting edge with notch or tooth) (above), and foot suited for holding prey (below). This group of birds holds its prey with the beak.

holders, on the other hand, serve to split open the back of the prey's head, whereby the animal is held with the foot. In the falcons, the upper mandible has a "tooth" behind the hook which fits into an indentation of the lower mandible.

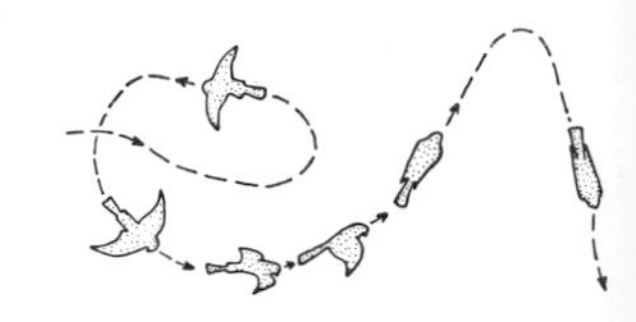

Fig. 14-3. The common buzzard performs a display flight in its territory when other members of the species fly through.

Fig. 14-4. Display flight of the goshawk.

When prey has been seized and killed by the foot or the bite of the beak into the back of the head, it is plucked more or less carefully if it is a bird or mammal; from insects the coarse pieces of chitin, such as the wing covers of beetles are removed. Nevertheless, much indigestible material, such as feathers, hairs, or pieces of chitin, are consumed with the morsels of food. The food is predigested in the crop by gastric juice which is "pumped up," and is then squeezed into the stomach, which dissolves everything that is digestible. Feathers and hair are gathered into clumps and are regurgiated through the beak as "pellets," generally after sixteen to eighteen hours.

Raptor pellets, in contrast to those of owls, do not generally contain bones. Only while feeding young, when the adults have to kill prey for the young several times a day, are there bones in raptor pellets. This is because they remain in the stomach for a shorter period of time. In addition, during the time when they are feeding young, the adults take only coarse pieces of the prey with many bones for themselves, feeding the softer parts to the young. The size of the pellets of the various species corresponds to the capacities of their gizzards.

The wings and flight muscles of the goshawks and falcons are adapted to their particular functions in catching prey. Those which hunt insects and small animals have, as a rule, a large wing surface, and they move the wings more slowly; they seize prey by an aerial form of stalking and gliding. Goshawks and sparrowhawks, on the other hand, have wings with a smaller surface which are moved by muscles having larger cross sections, and therefore move with greater efficiency; they seize prey by "stalking" and at the start of flight. With a high initial speed in their very maneuverable flight, they surprise their prey and seize it after only a short distance. Because of this, goshawks are popular with falconers (Color plate p. 332/333).

They share this popularity with the true falcons, particularly the peregrine falcon and the gyr falcon, which have a strong breast musculature together with a minimal wing surface. They attack after watching from afar, and then from high up they make astonishingly long-distance swooping attacks. It is understandable that, in comparison with these efficient species, the slowly gliding buteos which are adapted for small mammal prey or the vultures which seek dead animals could never play a role as hunting companions for man.

All raptors of the functional group of grasping killers, with tearing, hooking, and cutting beaks, build their own nests. Nests may be on the ground, but are usually found on trees. They may be newly built, or the birds may utilize old nests of other raptors or even of crows and

Fig. 14-5. Mating "invitation posture" of the female goshawk. The white lower tail coverts are displayed.

ravens as a base. In contrast, the grasping holders with teari
ing, and biting beaks use ready-made platforms, namely ledg
or old nests in trees.

Nest-building is initiated by selecting a nest platform at the time of courtship. In courtship display, the goshawks and falcons show off in aerial display flights, and they also attract the attention of their partners by courtship calls. Often a pair, when once formed, e.g. in the goshawk, which is a non-migrating resident, remains together for many years. If a strange, sexually-excited male trespasses the territory of the pair, it is often killed, plucked, and eaten, probably by the female of the pair. Thus Kramer and I found six killed and plucked male goshawks, all in the spring, in the breeding territories of resident pairs.

Fig. 14-6. Contrasting spots (here the black eyes and beak in a white downy plumage) are releasers for the feeding in the goshawk.

Males and females attract the attention of their partner by certain postures, such as special plumage postures, which may be enhanced by characteristic calls. Beyond this, postures play a role in the interactions with other members of the species and toward other inhabitants of their habitat. During the courtship period, the "display" flights of the conspicuously colored adult Montagu's, hen, and marsh harriers are particularly striking. The display flights of kites, honey buzzards, goshawks, and falcons take place high in the air accompanied by excited calls. In displaying adult goshawks, the loosened under tail coverts, which form a bright white ball, are particularly conspicuous. This display is used very impressively by the female to indicate readiness for mating, and also plays a role in all meetings of the mates at the nest.

The contrasting markings on the downy young play an important role as releasers of feeding actions of the parents. The eyes of the nestlings are dark, shiny black spots which stand out on the white or grayish-white downy heads, even in species in which the iris becomes yellow or orange with increasing age. In the goshawks there is also a black beak, in addition to the dark eyes. The effectiveness of these is enhanced by waving movements ("begging movements") of the head and species-specific "begging calls." Begging movements and calls have to be performed at an intensity above a certain threshold in order to release parental feeding.

The adult holds up a bright red food morsel and does not offer it directly to a particular nestling. Thus the strongest and quickest one, the one which reacts first to the food releaser, the red color, gets the food. The significance of the contrasting markings is readily recognized when one observes nestlings before and after they are satiated. When they have enough food they turn the head, with its tightly-filled crop, away from the adult. As long as they show their markings the adult tries to feed them. When a single sparrowhawk nestling, a few days old, fell on its back during feeding and hence could not turn away its markings, the adult continued to offer food morsels, and layed some of them on the head of the youngster between its eyes and beak.

Fig. 14-7. "Covering" posture of the goshawk.

If a nestling does not keep up with its siblings in its development, it becomes weaker and weaker. Generally this is the last youngster to be hatched. Naturally the intensity of its begging behavior decreases and reaches subthreshold levels for the adult; hence the exhausted youngster is eventually not recognized any more as a nestling. It may then become prey for the stronger nestlings. Such behavior has nothing to do with cannibalism; such a biological process is part of the selection process by which the strongest survive. In raptors this can take place while they are still in the nest.

Threat posture of the goshawk.

In the course of the nestlings' development, there is a progressive change in the role of the adults. Originally they are food dispensers, from which the young beg, and then they divide the prey among the nestlings, and gradually they become mere food suppliers. The juvenile plumage displaces the down more and more; it remains for a very long time on the head. Here, feathers appear first in the eye ring, which increases the conspicuousness of the contrasting markings. The feet of the young become more and more active. They strike playfully into the nest cavity. With the beginning of the coordination between beak and feet, change in the attitude of the young is reached. The nestlings still beg from the adult flying to the nest with prey, but the thin voices of the initial begging calls are now replaced by a strong, excited call. In response to this, the adult lays the food in one piece on the nest rim. The prey generally is minus the head, so that the bloody wound forms a red mark. The adult no longer tears off pieces and offers them to the young. As soon as a youngster, now largely covered by the juvenile feathers, gets hold of the prey, it defends it at once against the adult, as well as its siblings, by the characteristic covering posture, (see Fig. 14-7).

Attack posture of the goshawk.

With this posture, trained goshawks and falcons also defend their prey against the falconer. A raptor also shows it when another resident of its territory wants its share of the prey. An actual attack may result from the covering of the prey, in which the plumage is erected, the wings are opened wide, and the tail is spread. The covering posture is, however, not restricted to the defense of prey. Females at the nest, which take the main role in tending the offspring, use it to protect downy young from sunshine or heavy rainfalls. This behavior can also be regarded as a form of defense.

The territories occupied by individual pairs of raptors during the breeding season are divided into the hunting area and the nest area. This division, as well as the size of the hunting and nest areas, offer a good insight into the function of the two. The hunting area of hunters of very small animals is small, becoming larger with the size of the prey of the raptor species. Some, like the migrants among the raptors, hunt over several continents. Others, like the goshawk and sparrowhawk, do not hunt near their nests during the breeding season, and only actions directly related to the nest and the young are performed in it, such as the preparation and distribution of the prey. Hunting is clearly separated from care of the eggs and young. In species which often breed socially, hunting and nesting areas cannot be separated.

The moult is adapted to the activities of the two sexes at the nest. In most goshawks and falcons it begins and is completed during the breeding season, which in turn is determined by the period of reproduction of the particular prey animals, because young animals are more easily hunted, young mammals are often not yet fully mobile, and young birds cannot yet fly properly. An adaptation to these circumstances, as well as to the different activities of the two sexes, can be recognized in the moult of the primaries in the goshawk and sparrowhawk. Females of both species first shed the innermost primary when laying the second egg, and the next three primaries are shed within fourteen to sixteen days. Thus a gap arises in the wing, which strongly hinders hunting. However, this restriction is compensated for by the ability of the male to supply the female with enough food during incubation and the feeding period so that the young can be reared without the female's participation in hunting.

Two to three weeks after the onset of the moult in the female goshawk, the male begins to moult by shedding the innermost primary. But in contrast to the female, the primaries of males are shed at such long intervals that the first feather shed is almost totally replaced by a new one before the next one is shed. Thus no gap which would impair hunting flights occurs in the wing of male goshawks and sparrowhawks. The different sequence of the moult is thus exactly adapted to the roles of the partners within the context of the family. The female only takes part in hunting flights when the fledged young are perched on branches and their begging flights have begun. These relationships were established for these two species in the wild. Among the falcons, corresponding observations have been made on trained birds and on peregrine falcons in the wild. They begin the moult with the shedding of the seventh primary, counting from the outside to the inside (i.e. with the fourth if one counts from the body toward the wing tip). The peregrine falcon, which lives largely in the air, does not show such an evident wing gap as the goshawk, although in its case there are also different moult patterns between the two sexes. The different moult pattern in the two sexes is particularly impressive in the raptors which prey on larger animals. In general, the female of the raptors is mainly, or even exclusively, concerned with incubation and feeding of the nestlings, while the male brings in prey and hands it over to the female with species-specific ceremonies within the nest area.

In view of the ordered relationship of all living things in their environments, we must reconsider our judgement in line with biological facts. Thinking in terms of species should be replaced by thinking in terms of habitats. Only thus can we replace anthropomorphic concepts of "birds of prey inspired by lust to kill and intoxicated with blood" by a better understanding of the raptors, and thus overcome the disastrous classification of organisms as useful and harmful. Only when we finally give up classifying the relationships among living creatures on this basis, shall we be able to at least preserve the last remains of the original order of life. The role of predators in the population of other inhabitants of their environments has nothing at all to do with our concepts of usefulness or harmfulness; rather the predator has an extraordinary significance with the maintenance of the balance of nature.

Like all predators, raptors irresistably respond to animals which seem impaired in their movements. These they recognize as prey. These impaired animals are always hunted preferentially and are readily seized. Hence raptors, like other predators, select among other inhabitants of their environment, a function which should not be underestimated. On the other hand, many ground-nesting birds (nightjars, partridges, female black grouse, and ducks) lure the enemy away from the nest or the young through simulated disability. Man too is generally deceived by this behavior. Since ancient times, hunters of raptors used impaired movements of the prey, such as a caged pigeon, as a stimulus to attract goshawks. Falconry also utilizes these emphatic prey signals. In addition there are secondary prey releasing signals which include shapes, colors, and movement patterns of prey species. If a falconer trains his bird to hunt only one species, this leads to a far-reaching narrowing in the choice of prey. Such narrow specializations are rare in nature.

The prey species compensate for losses to predators by their annual reproduction, as long as natural conditions prevail. Disturbing human influences, however, cause special conditions, although remnants of the natural order can persist if man would only interfere with the necessary common sense and respect for life. Therefore, in our time, in which we so extensively alter landscapes and nature, we have more and more come to recognize the importance of worldwide protection of raptors. This must, however, be supported by thorough studies on the role of raptors in particular environments.

15 New World Vultures, Secretary Birds, Hawks, Old World Vultures, and Harriers

Family: Cathartides
by K. E. Stager

Formerly, the NEW WORLD VULTURES (Cathartidae) were classified with those of the Old World on the basis of superficial similarities. The differences between the New World vultures and other raptors, however, including the Old World vultures, are fundamental not only in the skeleton, but also in the muscles and other soft parts. The similarities between the two groups of vultures are only external; for example, both have bare necks and heads. Today we know that such similarities are due to convergence, for both groups fulfill the same function by removing carrion and refuse within their respective environments.

The New World vultures are an old family; they have no close relatives among the raptors. Hence they are often classified as a separate suborder, Cathartae. As the name indicates, they inhabit only the New World. Some of them belong to the largest flying birds, such as the condor of the South American Andes and the almost extinct California condor. Bones of a New World vulture, which must have had a wingspan of about five meters, have been found in glacial deposits in Nevada; it is called *Teratornis incredibilis,* meaning "unbelievable bird monster." Today the family consists of five genera with seven species, of which five will be mentioned.

Fig. 15-1. 1. Andean condor *(Vultur gryphus);* 2. California condor *(Gymnogyps californianus)* former and present distribution.

A. CONDORS (genera *Vultur* and *Gymnogyps*): 1. The ANDEAN CONDOR (*Vultur gryphus;* Color plate, p. 334) has an L of 132 cm, a wingspan of 290 cm, and a weight of 11.35 kg; 2. The CALIFORNIA CONDOR (⦶ *Gymnogyps californianus;* Color plate, p. 334) has a wingspan of about 290 cm. B. The KING VULTURE (*Sarcoramphus papa;* Color plate, p. 334) has an L of 79 cm. C. Genera *Coragyps* and *Cathartes:* 1. The BLACK VULTURE (*Coragyps atratus;* Color plate, p. 334) has an L of 64 cm; and 2. The TURKEY VULTURE (*Cathartes aura;* Color plate, p. 334) has an L of 74 cm.

All New World vultures feed mainly on fresh or decomposing animal corpses; some species, however, also catch living mammals and birds. None build nests. They simply lay their eggs on the ground, in tree cavities, between rocks, or in cavities in cliffs along the sea.

The habitat of the Andean condor is not limited to the highest peaks of the Andes; it often visits the shores of the Peruvian coasts where it consumes dead fish, whales, and seals. In many seabird colonies on the islands off the Peruvian coast, it robs Guanay cormorants of their eggs and seizes shearwaters as they emerge from their breeding burrows. In the Andes it lives on the corpses of large mammals. It is said to attack and to be able to kill animals up to the size of a calf.

Fig. 15-2. King vulture *(Sarcoramphus papa).*

The CALIFORNIAN CONDOR was formerly distributed over the entire area of North America from British Columbia to Florida. Today it is almost extinct, since its population is estimated at only fifty to sixty birds in the mountains of southern California. This majestic bird has become a victim of the progress of civilization. It had no enemies before man came into its habitats. After the settlement of western North America by the white man, many condors were poisoned by strychnine, which was put into dead cattle by cattle breeders to destroy wolves and coyotes. The natural food supplies of the condors decreased more and more, and many were also shot by thoughtless hunters. Unfortunately, reproduction of the condor is very slow. Since a young condor needs two years of parental care, females only lay one egg every two years. The marked losses cannot, therefore, be replaced by the offspring, and this is the cause of the decline in recent years.

Fig. 15-3. Black vulture *(Coragyps atratus).*

California condors prefer the meat of large mammal carrion in the following order: first cattle, followed by sheep, deer and horses. When seeking food, the gigantic birds range up to eighty kilometers per day from their resting or breeding places, and so they are exposed to further dangers. Those still surviving are carefully protected, and yet the survival of this magnificent species is in doubt.

The most attractive of all New World vultures is the king vulture. It inhabits dense tropical forests. Little is known about the habits of this colorful bird; some aspects of its anatomical structure and its behavior suggest that it is among the few birds which find food by their sense of smell. At any rate, king vultures appear at a dead animal at once, even when it lies on the ground in the thickly overgrown primeval forest. They find their food in places where vision is much restricted, and where without a good sense of smell they would hardly be able to find decomposing meat.

The black vulture is among the most common in the New World vulture family. It is probably the most social bird of the family and gathers in great flocks of hundreds and sometimes thousands, even in inhabited places at the edges of large cities. Black vultures are residential; they sit on roof tops and trees until they find an opportunity to seize a piece of carrion or human refuse. Their diet is the least selective of all vultures; they accept just about any kind of refuse.

▷ The secretary bird *(Sagittarius serpentarius)* prefers to hunt snakes; it is protected against their bites by its long, scale-covered legs and it kills with its feet.

The black vulture finds food exclusively by vision, and is also guided to food by observing other members of the species. Although

1
2
3
6
4
9
10
7
5
11
8
12
F. Reimann

many people find it abhorrent that this bird swallows feces and digs in garbage heaps, it is still appreciated as a "street cleaner" in most areas. Many small towns of tropical America and the slums of many cities could hardly rid themselves of refuse without the cleaning activities of this bird.

Fig. 15-4. Turkey vulture *(Cathartes aura).*

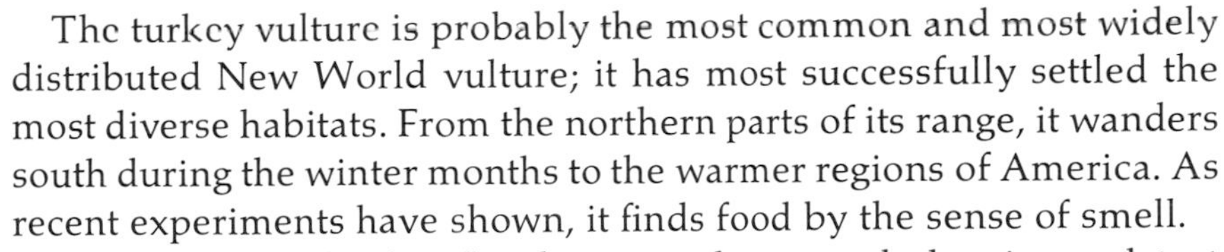

The turkey vulture is probably the most common and most widely distributed New World vulture; it has most successfully settled the most diverse habitats. From the northern parts of its range, it wanders south during the winter months to the warmer regions of America. As recent experiments have shown, it finds food by the sense of smell.

When seeking food, it flies low over the ground; thus it can detect olfactory stimuli which cannot be perceived at higher altitudes. It can detect even small amounts of food by its odors which other members of the family are unable to detect.

Family: secretary birds, by R. Liversidge

The second family of the raptors, the SECRETARY BIRDS (Sagittariidae) are thought by E. Stresemann, the eminent German ornithologist, to be a separate suborder (Sagittarii); others group it with the hawks and falcons in the suborder Falcones. There is only one species, the SECRETARY BIRD (*Sagittarius serpentarius;* Color plate, p. 345).

The height is about 100 cm, the wingspan is 200 cm, and the face is bare. The head has a crest. The central tail feathers project beyond the others by one half; the outer tail feathers are the shortest. The long legs are feathered down to the tibio-tarsal joint.

The secretary bird is such an extraordinary-looking raptor that at one time it was even classified with the bustards. This family probably developed in Africa, south of the Sahara, where today it is generally distributed in all suitable areas. However, in the Upper Eocene and Lower Oligocene, about twenty-five million years ago, as well as in the Miocene, about twenty million years ago, two related species also lived in southern France.

◁
Kites: 1. Red kite *(Milvus milvus);* 2. Black kite *(Milvus migrans);* 3. Square-tailed kite *(Lophoictinia isura);* 4. Snail kite *(Rostrhamus sociabilis);* 5. Plumbeous kite *(Ictinia plumbea).* Honey buzzards: 6. African cuckoo falcon *(Aviceda cuculoides);* 7. Cayenne kite *(Leptodon cayanensis);* 8. Honey buzzard *(Pernis apivorus).* White-tailed kites: 9. Black-shouldered kite *(Elanus caeruleus);* 10. African swallow-tailed kite *(Chelictinia riocourii);* 11. Swallow-tailed kite *(Elanoides forficatus);* 12. Pearl kite *(Gampsonyx swainsonii).*

The bird owes its name to the long crest feathers which project from the back of the head, like the feather pen behind the ear of an office worker of the last century. When hunting, the secretary bird spreads these feathers like a fan. It strides "majestically" about over the steppe, either alone or with its mate. When pursued, it relies on the speed of its legs, although it partially spreads the wings when running. It only flies when hard pressed. It seeks its food with the beak, but also stamps on grass tussocks with the feet to scare up lizards, grasshoppers, or other small animals. Its kick is strong and it uses it mainly to kill snakes, which are its preferred prey. It spreads its wings which seems to distract snakes when dealing with them (Color plate, p. 345).

The nest is a large flat structure of branches tied together with grasses. Generally it is located three to ten meters up in the top of a thorn bush or thorny tree. Although the nest has a diameter of almost two meters, it is difficult to detect in the dense foliage. The female lays two to three white eggs which average 77 x 56.3 mm in size. The clutch is primarily incubated by the female. The young hatch after forty-five days and remain in the nest for eighty days, although on occasion they may remain in the nest for up to ninety-eight days.

Family: hawks and related forms

The different raptors, like the goshawks, buteos, kites, harriers, Old World vultures, and eagles, are all placed in the family Accipitridae, in the modern zoological system. There are nine subfamilies: 1. WHITE-TAILED KITES (Elaninae); 2. HONEY BUZZARDS (Perninae); 3. TRUE KITES (Milvinae); 4. GOSHAWKS (Accipitrinae); 5. BUTEOS (Buteoninae); 6. OLD WORLD VULTURES (Aegypiinae); 7. HARRIERS (Circinae); 8. SNAKE-EAGLES (Circaetinae); and 9. OSPREYS (Pandioninae). Altogether there are 198 species. Their distribution is worldwide, and they are predominantly diurnal.

Subfamily: white-tailed kites, by H. Brüll

The KITES (subfamily Elaninae) are a group of raptors which prey mainly on small animals; their toes correspondingly have only weak claws. Eight species in five genera are recognized, of which the following are mentioned here:

A. KITES, more specifically (genus *Elanus*): 1. The BLACK-SHOULDERED KITE (*Elanus caeruleus;* Color plate, p. 346) has an L of 35 cm. It is often found in Africa, India, Malaya, parts of Indonesia, and as a very rare vagrant in Germany; 2. The AUSTRALIAN BLACK-SHOULDERED KITE (*Elanus notatus*) has an L of 34 cm, and is found in Central Australia; 3. The LETTER-WINGED KITE (*Elanus scriptus*), with an L of 34 cm, is found in Central Australia; 4. The WHITE-TAILED KITE (*Elanus leucurus*), with an L of 40 cm, is found in parts of North and South America.

B. The AFRICAN SWALLOW-TAILED KITE (*Chelictinia riocourii;* Color plate, p. 346), with an L of 35 cm, is found from Senegal to Ethiopia.

Fig. 15-5. Secretary bird (*Sagittarius serpentarius*).

C. The BAT HAWK (*Machaerhamphus alcinus*), with an L of 45 cm, is dark brown to black. It has a white spot on the throat, the breast, and around the eye, and there are pale bands on the wings and tail. The beak is deeply cleft. The upper mandible has a keel. The large yellow eyes have thickly feathered lores; hence the face appears nightjar- or owl-like. The wings are long and pointed, the tail is short, and the white-blue feet have long, slender toes, which are falcon-like, being suitable for seizing flying prey, mainly bats.

D. The PEARL KITE (*Gampsonyx swainsonii;* Color plate, p. 346) has an L of 18–22 cm.

E. The SWALLOW-TAILED KITE (*Elanoides forficatus;* Color plate, p. 346) has an L of 60 cm.

Fig. 15-6. 1. Black-shouldered kite (*Elanus caeruleus*); 2. White-tailed kite (*Elanus leucurus*); 3. Australian black-shouldered kite (*Elanus notatus*); 4. Letter-winged kite (*Elanus scriptus*).

Rats of several species are the main prey of the black-shouldered kite; they constitute about eighty per cent of its prey. It also hunts mice, lizards, small snakes, frogs, insects, and crabs, and only rarely does it catch birds. It is particularly active in the early morning hours and shortly before dusk. Its attack on prey is preceded by hovering, like that of the kestrel, although it has slower wing beats. It is most agile

Fig. 15-7. Bat hawk *(Machaerhamphus alcinus).*

in hunting. Sometimes it hides itself in the foliage of isolated trees, suddenly rushing out to seize its prey. The three other species of the genus *Elanus* have short toes and weak claws, and so correspondingly prey on smaller animals.

The African swallow-tailed kite lives mainly on insects, primarily Orthoptera and Hymenoptera, as well as on lizards and mice. It glides above its prey on the ground and then pounces on it, but it seizes insects in the air. Its swallow-like wings and deeply forked tail are well adapted for this activity. Since there is plenty of prey for these birds in bushy landscapes and grass steppes, they associate not only in the breeding season but also at other times. A dozen or more will often hunt together. Larger groups of these kites gather about the nests of stronger raptors. Thus six were seen on the edge of the nest of a short-toed eagle, and on another occasion forty were seen at the nest of a griffon vulture with one youngster. This suggests that these small kites wait for a share of the prey of larger raptors.

Fig. 15-8. Pearly kite *(Gampsonyx swainsonii).*

The bat hawk is quite a peculiar bird. The conspicuous white marks on its plumage are probably important in the twilight as recognization marks, since these kites only leave their forest perches to hunt early in the morning and late in the evening. They usually move about near limestone cliffs, house roofs, or other places inhabited by bats. Their flight is falcon-like, swift and strong. They seize mainly bats with their long claws, but they will also take large insects, as well as weaver finches and other small birds. They move the prey up to the beak with the foot and swallow it whole in flight. Nothing is known as yet about the size of their hunting areas. Their density of population does not, however, seem to be great. As far as it is known, only two pale blue-green eggs are laid in their nests, which are lined with green leaves. Nests are built in forks of high trees, and are sometimes used for several years.

Since they are predators of small animals, white-tailed kites do not generally require large hunting territories. Thus the Australian black-shouldered kite *(Elanus notatus)* and the white-tailed kite of the Americas *(Elanus leucurus)* tend to breed in colonies. In western Queensland, Australia, groups of twelve to eighteen pairs have been found five to eight kilometers apart from one another. Similarly, white-tailed kites in southern California were found only about 200 meters apart, although the individual pairs confined themselves strictly to their nest territories.

Fig. 15-9. Swallow-tailed kite *(Elanoides forficatus).*

Nests built by the adult kites themselves are on trees at different heights, according to the habitat. The nest cavities are sometimes lined with green grass or leaves. Old nests built by other birds may be used as the base for kite nests. Both adults build the nest and relieve one another during incubation and rearing of the young. The division of labor of the sexes, customary in other Accipitridae (see Chapter 14) during the breeding season, is also observed in white-tailed kites. The female does most of the incubating and later the feeding of the young, while the male catches prey, brings it to the nest, and passes it to the female with species-specific rites. The clutch consists of two to six eggs, generally three to four, which are incubated for about thirty days. The adults feed the young until they are seven to ten weeks old.

Fig. 15-10. Swallow-tailed kite with a lizard.

The population of black-shouldered kites on the small island of Masira in the Arabian Sea shows how dependent reproduction is upon the food supply. They are the only raptors on this barren island. There are no rodents on the sandy shores and the jagged lava cliffs on the island's interior. There are no trees suitable for nesting. Yet a small population of kites lives here, supporting itself, as do the gulls, on refuse of the Arabian fishing villages. The nests are on lava cliffs. Prey consists of fish refuse and a few lizards; the adults therefore only rear one young per pair.

The eggs of the North American population of the white-tailed kite and the swallow-tailed kite of South America have attracted egg collectors, who have gone to great trouble to obtain them. In this way, both species, as is true of so many other species, were largely exterminated. In addition, swallow-tailed kites defend their nests very bravely, and are therefore only too easily shot.

The small pearl kite was, for some time, classified with the falcons because of its shape and manner of flight. However, since it moults the primaries from the innermost (first) toward the outermost (tenth) as do all Accipitridae at least during the first moult, it must be classified with them. It is therefore placed in this subfamily with the kites.

Little can be said about its life as yet. It is found at forest edges and in the lightly wooded savannahs of West Nicaragua and South America. There may be several nests in one tall tree. How they are built is not known. The clutch consists of three elliptical, white and brown spotted eggs. Its prey, mainly insects and lizards, is seized both in the air and on the ground.

The swallow-tailed kite leads us to the groups of honey buzzards and true kites. Audubon reports that it was widespread in Louisiana and the Mississippi area in 1840. Today there are breeding populations only in some parts of southern Florida and Mexico, and Central and South America. Its strong decline is mainly attributable to supposedly "scientific" egg collectors and shooting of the birds at the nest.

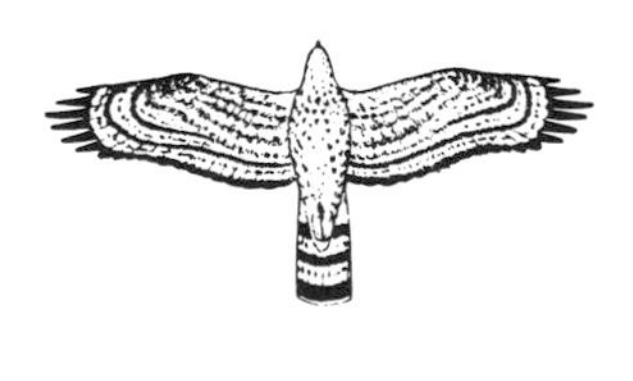

Fig. 15-11. Flight silhouette and foot of the honey buzzard *(Pernis apivorus).*

During the breeding season, the kites may circle a great deal, often very high. When hunting, they search the ground and then swoop with great speed down onto various small animals, which they seize and eat while airborne. It is reported that in the southeastern United States they seize water snakes which lie sun bathing on the shores of swamps and rivers. They also prey on worms, larvae, dragonflies, wasps, cicadas, beetles, grasshoppers, tree toads, frogs, and lizards. The fairly large swallow-tailed kite feeds, therefore, primarily on insects and small animals from which a high population density might be expected.

Formerly, North American swallow-tailed kites migrated in large flocks as far as Ecuador. But tropical swallow-tails do not migrate; rather, outside the breeding season, they range in large flocks over areas which offer much prey. About fifty were seen over the lowland forests of Caicara in Venezuela. The nest is usually on a Cuban pine near cypress swamps. It is built of small twigs, bark, and Spanish moss. Old nests are not deserted, but repaired. Both sexes build the nest, incubate two to four eggs, and rear the young.

Subfamily: honey buzzards, by H. Brüll

The HONEY BUZZARDS (Perninae) are the second Accipitridae subfamily; there are five genera with a total of twelve species, of which the following will be mentioned:

A. The CRESTED BUZZARDS (genus *Aviceda*) have an L of 35–45 cm: 1. The AFRICAN CUCKOO FALCON (*Aviceda cuculoides*; Color plate p. 346); 2. The MADAGASCAR CUCKOO FALCON (*Aviceda madagascariensis*); 3. The CRESTED BAZA (*Aviceda subcristata*); 4. The JERDON'S BAZA (*Aviceda jerdoni*); 5. The BLACK BAZA (*Aviceda leuphotes*).

Fig. 15-12. 1. African cuckoo falcon (*Aviceda cuculoides*); 2. Madagascar cuckoo falcon (*Aviceda madagascariensis*); 3. Crested baza (*Aviceda subcristata*); 4. Jerdon's baza (*Aviceda jerdoni*); 5. Black baza (*Aviceda leuphotes*).

B. The LONG-TAILED HONEY BUZZARD (*Henicopernis longicauda*) has an L of 55 cm.

C. The HONEY BUZZARDS more specifically (genus *Pernis*): 1. The HONEY BUZZARD (*Pernis apivorus*; Color plate, p. 346) has an L of 51–58 cm; 2. The MALAYAN HONEY BUZZARD (*Pernis ptilorhynchus*).

D. The CAYENNE KITE (*Leptodon cayanensis*; Color plate p. 346).

E. The HOOK-BILLED KITE (*Chondrohierax uncinatus*).

The members of this subfamily are also predators of small animals and insects. The crested buzzards hunt at forest edges, over clearings, and on the shores of rivers and lakes. Their upper mandible has a "tooth" on each side, behind the hook. The usefulness of this striking structure becomes clear when one thinks of the large insects with strong chitin carapaces, such as beetles, and particularly those of the tropics. With this peculiar tooth the carapaces are easily cracked.

Although the crested buzzards particularly like to take insects, they occasionally show preference for frogs, crabs, bats, lizards, mice, and shrews, and more rarely for small birds and their eggs. The Madagascar cuckoo falcon appreciates chameleons in so far as they are common in its habitats, even when insects or grasshoppers are abundant.

Fig. 15-13. 1. Honey buzzard (*Pernis apivorus*); 2. Malayan honey buzzard (*Pernis ptilorhynchus*); 3. Long-tailed honey buzzard (*Henicopernis longicauda*).

Crested buzzards are twilight hunters, being particularly active at sunrise and in the twilight. One of these small raptors may suddenly emerge from the foliage of a tall tree to seize a flying insect or to take a caterpillar from a nearby tree or bush. In the savannahs of East and Central Africa, crested buzzards also take their prey from the ground, much as Central European kestrels do. Their nests are on tall trees; the nest cavity is lined with grasses and roots and is continually lined with fresh green leaves as well. Thus they bring green leaves to their nests like the honey buzzard, common buzzard, and goshawk. The clutch of one to normally three eggs is incubated by the female for thirty-two days, while the male supplies the food.

Fig. 15-14. 1. Cayenne kite *(Leptodon cayanensis)*; 2. Hook-billed kite *(Chondrohierax uncinatus)*.

The more insect eaters prefer wasps, bees, bumble bees and other stinging insects, the more the skin in front of their eyes is covered with scales. This is very evident in the honey buzzard, which is encountered in Germany from May to September or October. Moreover, its upper mandible lacks the bulge of the cutting edge which, as a rule, is found in the tearing hook and cutting beaks. It also deviates from other raptors in the use of its feet. It uses them mainly for scratching, to dig out wasps' nests in the ground. The birds often work themselves so far into the ground that they do not notice an approaching person. Since such a bird is surrounded by excited wasps while working, it fails to hear even the loud cracking of branches among the buzzing of so many wasps. One can get so close to a digging honey buzzard that one could pick it up if it were not for the stinging wasps.

Wasps and bees minus their stingers have often been recovered from the stomachs of honey buzzards. It is therefore assumed that the bird skillfully bites off the terminal segment from the insects. On the other hand, Bartels reports that the Malayan honey buzzard swallows wasps and bees with their stingers, although both are often found without their stingers in its digestive tract. Apart from bees and wasps, this bird takes worms, larvae, insects, spiders, frogs, lizards, and certain small snakes, as well as the young of ground nesting birds and small mammals, or, in other words, those animals which it can seize while walking on the ground. Furthermore, it does not avoid sweet fruit like blueberries and other woodland berries.

According to A. E. Brehm, the honey buzzard may act as a food parasite and, for example, beg food from a goshawk. But there has been no proof of this assertion as yet. A honey buzzard which was shot on a dead hare did not have hare flesh in its crop, but rather, fly maggots which it had taken from the corpse. Two other honey buzzards shot on the meadow of a pheasant hatchery were not after the pheasant chicks; they were hunting grasshoppers, which were scampering about among the pheasants. Unfortunately, the appearance of the honey buzzard in flight is easily mistaken for that of the goshawk, and this is often disastrous. Although the honey buzzard is adapted to small prey, with an evident preference for stinging insects, it inhabits quite a large territory with a diameter of about three and one half kilometers. It requires both coniferous and deciduous woods with clearings in its habitat. There are many wasps where forests border on meadows; therefore the honey buzzard prefers such areas.

Honey buzzards return from their African winter quarters in May and are once more found in their breeding territories of Central

Europe. Then they show attractive display flights by circling high in the air, which are only rarely accompanied by calls. The male rises above the female, then plunges down at her with folded wings, swerves off skillfully, and once more rises upward. Both partners build the nest; it is either a new one or they use old nests of the common buzzard or goshawk as a base. The cavity is lined with leafy twigs wherever possible; these are constantly replaced and loosened with the beak until the period of feeding of the young. Both partners also build nests which are not used for egg-laying. They sit on the nest in the incubation posture, even when there is as yet no egg. About mid-June the female lays from one to three, normally two, eggs and incubates them for thirty to thirty-five days, during which she is relieved by the male either for twelve or for five hours, although in many cases she is not relieved at all.

During the feeding of the young, there is no such strict division of labor in the honey buzzard as there is in other hawks. Both parents hunt. Caterpillars or grasshoppers which they have seized are carried to the young in the crop and then regurgitated. The parents pick out the larvae from the combs of wasps' nests they have brought in, and offer them in the beak to the young. In some cases the hen takes prey from the male and distributes it to the young. After the hatching of the young, the parents begin to moult and only complete the process in the wintering grounds.

The young remain in the nest for forty-two to forty-four days. In contrast to other raptors, they do not squirt their droppings far from the nest but deposit them on the rim of the nest. Hence one does not find the characteristic deposits beneath the nest of a honey buzzard with young. After fledging, the young return to the nest platform again and again, because the adults deposit prey there. Often only females do so, since the males frequently have already left. With time, many wasp combs accumulate on the nest, for in contrast to most other hawk-like raptors, the adults do not remove food remains. The pellets, which are indigestible food remains regurgitated out through the mouth, consist mostly of chitin particles and wasp comb fragments, and only rarely of feathers or hair of birds or small mammals. On migration, honey buzzards gather in September and October in numbers of up to several hundred birds. Occasionally, honey buzzards fall prey to goshawks (four proven cases) or the eagle owl (two proven instances).

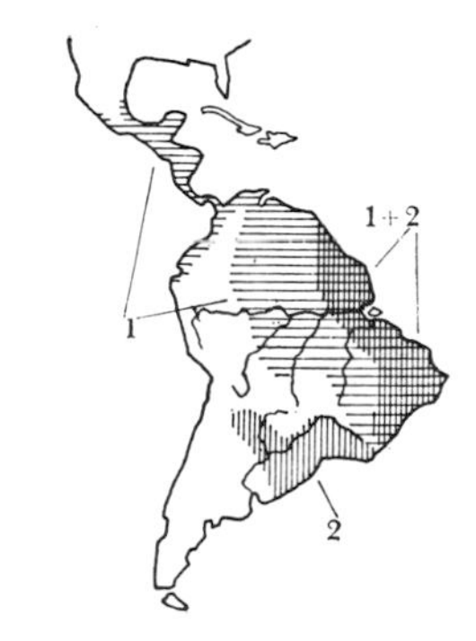

Fig. 15-15. 1. Double-toothed kite *(Harpagus bidentatus)*; 2. Rufous-thighed kite *(Harpagus diodon)*.

Little is known of the life of the closely related Cayenne kite. It inhabits largely unexplored forests along rivers, where it is difficult to observe. It is said to feed on wasps and birds; the tooth on the lower mandible behind the hook indicates that it is also able to crack strong chitin and other carapaces. This also applies to the hook-billed kite, which inhabits the same areas. It is reported to prefer the forest interior and to feed on insects, reptiles, and birds. In Cuba, where it has become rare, it particularly takes tree snails of the genus *Polymita.*

The last two species are close to the subfamily of the TRUE KITES (Milvinae), one of which, the genus *Harpagus,* is a link to the honey buzzards. The eight true kite genera are made up of ten species of which only the following will be discussed:

A. Genus *Harpagus:* 1. The DOUBLE-TOOTHED KITE *(Harpagus bidentatus)* has an L of 30-35 cm. The wings are rounded, and the upper mandible has two "teeth" on each side behind the tearing hook. 2. The RUFOUS-THIGHED KITE *(Harpagus diodon)* is very similar, with red feathers.

B. Genus *Ictinia:* 1. The MISSISSIPPI KITE (*Ictinia misisippiensis;*) has an L of 32.5-37.5 cm. 2. The PLUMBEOUS KITE (*Ictinia plumbea;* Color plate, p. 346) is very similar.

C. Genus *Rostrhamus* and *Helicolestes:* 1. The SNAIL OR EVERGLADE KITE (*Rostrhamus sociabilis;* Color plate p. 346). 2. The SLENDER-BILLED KITE *(Helicolestes hamatus)* eats snails.

D. Genus *Haliastur:* 1. The BRAHMINY KITE *(Haliastur indus)* and 2. The WHISTLING HAWK *(Haliastur sphenurus).*

E. TRUE KITES more specifically (genus *Milvus*): 1. The RED KITE (*Milvus milvus;* Color plate p. 346) has an L of 61 cm. 2. The BLACK KITE (*Milvus migrans;* Color plate p. 346 and pp. 385/386), with an L of 56 cm, has several subspecies; among them is the EGYPTIAN SCAVENGER KITE *(Milvus migrans aegyptius)* and the SIBERIAN BLACK KITE *(Milvus migrans lineatus).*

F. Genera *Lophoictinia* and *Hamirostra:* 1. The SQUARE-TAILED KITE (*Lophoictinia isura;* Color plate p. 346) with an L of 50 cm, is found in pairs or in large flocks, often near narrow water courses. It breeds in the accompanying forest belts. Its food consists of caterpillars, reptiles, and young birds. 2. BLACK-BREASTED BUZZARD KITE *(Hamirostra melanosterna)* is larger, with an L of 60 cm. Its food consists mainly of lizards and rabbits, as well as bird eggs, including those of bustards and of the emu. It is said (?) to open emu eggs with stones, which it drops on them from above, or by beating on the egg with a stone in its beak.

The double-toothed kite evidently lives on insects with strong chitin carapaces which it cracks with the "teeth" on the edges of its beak. It also catches reptiles. It has been observed to fly from its perch, gliding down onto a lizard which it has spotted in a nearby tree. It then hops with outspread wings on the shaking twig to keep its balance after catching its prey. The nest is a flat bowl of branches in a fork of a large deciduous tree. The clutch consists of three to four white eggs.

The Mississippi and the plumbeous kites are recognizable by an elegant, strong gliding flight which often takes them to considerable heights. They sometimes fly in company with swallow-tailed kites and New World vultures. In sunny weather, they seize and devour insects while soaring on updrafts. In unfavorable weather they seek a branch

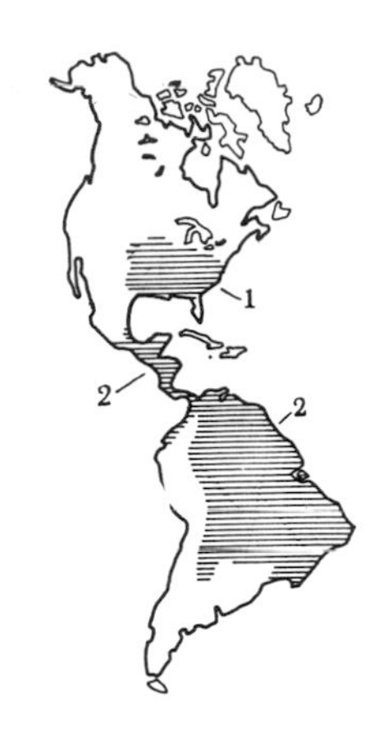

Fig. 15-16. 1. Mississippi kite *(Ictinia misisippiensis);* 2. Plumbeous kite *(Ictinia plumbea).*

Fig. 15-17. 1. Snail kite *(Rostrhamus sociabilis);* 2. 2. Slender-billed kite *(Helicolestes hamatus).*

▷ Goshawks: 1. White goshawk *(Accipiter novaehollandiae),* a) dark, b) light phase; 2. European goshawk *(Accipiter gentilis gallinarum),* a) juvenile, b) adult plumage; 3. European sparrow hawk *(Accipiter nisus);* 4. African little sparrow hawk *(Accipiter minullus);* 5. African long-tailed hawk *(Urotriorchis macrourus);* 6. Pale chanting goshawk *(Melierax musicus);* 7. Savannah hawk *(Heterospiza meridionalis).*

a
1
b
2
a
b
3
6
4
5
7
F. Reimann

1
a
b
2
3
4
5
6
7
F. Reimann

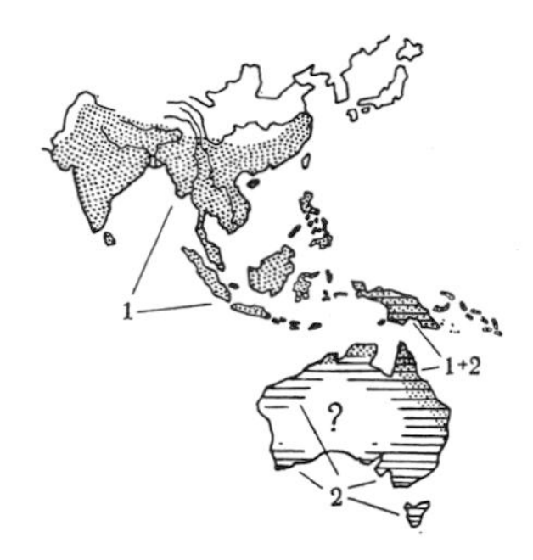

Fig. 15-18. 1. Brahminy kite *(Haliastur indus)*; 2. Whistling hawk *(Haliastur sphenurus)*.

Fig. 15-19. 1. Red kite *(Milvus milvus)*.

Fig. 15-20. Black kite *(Milvus migrans)*.

Buzzards: 1. Common buzzard *(Buteo buteo)*, a) dark, b) pale phase; 2. Rough-legged buzzard *(Buteo lagopus)*; 3. African red-tailed buzzard *(Buteo auguralis)*; 4. Red-tailed hawk *(Buteo jamaicensis)*; 5. Slate-colored hawk *(Leucopternis schistacea)*; 6. Bay-winged or Harris's hawk *(Parabuteo unicinctus)*; Blue buzzards: 7. Gray eagle buzzard *(Geranoaëtus melanoleucus)*.

as a perch, from which they seize prey and to which they return again and again. Apart from insects, they take frogs, lizards, and occasionally small snakes. Their nests, which are large flat bowls, stand either in high trees in parks or plantations, or low in mesquite oaks as in Texas or Oklahoma. The clutch of one to three white eggs is incubated by both sexes for twenty-nine to thirty-one days. Both share in the rearing of the young. They attack approaching birds or mammals on the wing in defense of the nest.

The beaks of the snail and slender-billed kites are adapted to a very specific function. The narrow, sickle-shaped bills of both species resemble a pair of curved forceps, and they enable the birds to get snails out of the shells. Both almost exclusively eat fresh water snails of the genera *Pomacea* and *Ampullaria*. During the hot part of the day these snails stay in the water, but early in the morning and in late afternoon they crawl about among the shore vegetation. At these times, both species of kites work over the territories in low level flight with slowly beating wings, like the marsh harrier, and after a brief stay over a certain spot, they swoop down to seize a snail. These birds use particular places where they open and eat their prey. Such spots are marked by the accumulation of snail shells; 200 to 300 shells may be found there. When the snail is seized, it withdraws into its shell. To pull it out, the kite's narrow beak is passed between the lid and the shell into the interior to loosen the snail's body; sometimes the bird waits quietly until the snail emerges from the shell and then pulls it out completely.

The drainage of the Everglades in southern Florida has threatened to bring about the extinction of the snail kite. In 1900 it was still widespread there. Today, the snails on which the kites feed occur in drainage ditches; more and more of them have, moreover, become intermediate hosts for a lung parasite which is dangerous to the kites. The snail kite is more numerous in suitable areas in South America today.

In display flights, the birds circle about at considerable heights, chasing and swooping at one another. These display flights look quite attractive. The birds breed in colonies in Argentina; for example, twenty to a hundred nests have been found only a few meters apart. The male builds the nest and interrupts this activity only to catch snails for himself and the female. The nest is a platform of branches on a clump of sedges or the stump of a dead tree. Two to four white eggs are laid and incubated by both sexes. Both feed the young, which fledge less than a month after hatching. Snail kites from southern South America migrate northward at the onset of the cool season, while those from North America migrate southward.

The brahminy kite has a much more varied diet. Although it can sieze small animals such as insects, crabs, frogs, lizards, small snails, mice, and shrews, it mainly feeds on refuse. Its preferred habitats are harbors, cliffs and sandbanks along rivers, mangrove swamps, and swampy rice fields in India and Indonesia. Here the birds may be found singly or in pairs or in larger flocks, or sometimes even in hundreds, particularly where many intestines and other refuse are lying about. Thus brahminy kites, for example, fight over such food with crows in the harbors of Ceylon. They readily rake floating dead fish, and young and wounded birds, but also live fish which are either taken out of fishermen's nets or seized from the air in shallow water. When seeking food a bird soars about at a certain altitude with constant head movements, being on the lookout for some tidbit. When it has spotted something of interest, it circles and then swoops down. During the monsoon, land crabs as well as winged termites, which emerge from the rain-saturated ground, become its main diet. It is reported that in western Australia its relative, the whistling hawk, keeps the rabbit population down.

Fig. 15-21. 1. Square-tailed kite *(Lophoictinia isura)*; 2. Black-breasted buzzard kite *(Hamirostra melanosterna)*.

The brahminy kite breeds twice a year, in June and again in December. The nests are loose structures of twigs and branches high up in a tree fork. Rags, paper, leaves, and dry mud are also used in nest construction. At times the nest tree stands in water and then becomes a sort of fortress against enemies, which above all include monkeys, next to man. The clutch consists of two, or possibly three eggs which are incubated by both parents for four weeks. Both adults, but mainly the female, tend the young.

Fig. 15-22. Flight silhouette of the red kite *(Milvus milvus)*.

Like the brahminy kite, the true kites feed in part on dead animals, and occasionally on human corpses as well. Above all, the Siberian black kite, which occurs between the Urals and Japan and Formosa, feeds on human corpses, since there it is customary to lay them naked out on the steppes. The Egyptian scavenging kite lives preferably in human settlements, in which all refuse is thrown onto the streets. Therefore, kites, like some Old World and New World vultures, provide a minimal degree of cleanliness on the streets by their feeding activities.

Fig. 15-23. Flight silhouette and foot of the black kite *(Milvus migrans)*; the spread of the toes measures 5.5 cm; in the red kite 5.8 cm.

The black and red kites also breed in Central Europe (see Fig. 15-19 and 15-20). Both species migrate within their ranges; they return to their northern breeding places during March. The black kite prefers districts rich in water, since, like the brahminy kite, it feeds on dead fish and other animals from the water. It also takes nesting materials out of the water. The red kite is more tied to forest edges and rows of trees in fields. The feet of both species show them to be pronounced small animal predators. Their food consists of fish, amphibia, reptiles, mice, rats, and hamsters. Black kites tend to work over good feeding areas in a group, and they tend to breed colonially. The red kite, on the other hand, maintains quite a large territory of 2000 to 3000 hectares in area. Both settle by preference in the vicinity of goshawks, peregrines, ospreys, and sea eagles in order to beg for some of the weighty prey of these strong predators. Hence, if one finds remains

Fig. 15-24. Snail kite pulling a snail out of its shell with its pointed beak.

Fig. 15-25. Flight silhouette of a goshawk *(Accipiter gentilis).*

of larger animals by kite nests, these will not be from prey killed by the kites, especially since both take carrion. Today, in particular, they find an abundance of animals killed by motor vehicles, mowing machines, or poisoned by water discharges or in other ways; therefore they do not suffer as much from lack of food as do other predators. The black kite often builds its nest within heron colonies in order to have easy access to food.

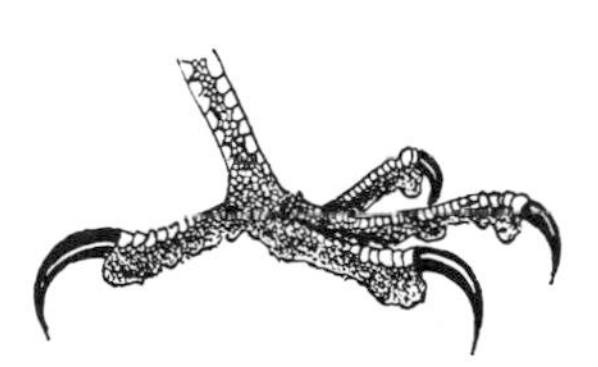

Fig. 15-26. Foot of ♀ goshawk; the spread of the foot is 9-11 cm.

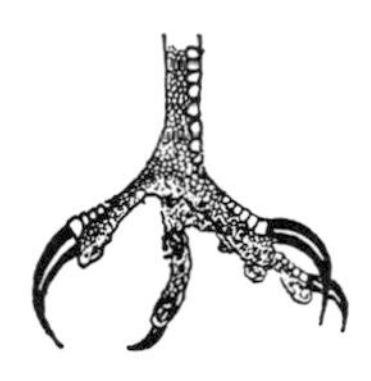

Foot of the ♂ goshawk; the spread of the foot is 7-8.5 cm.

After their return from the winter quarters, both kites often show display flights accompanied by neighing calls. Red kites circle at greater altitudes, while black kites fly in horizontal curves with sweeping wing beats. Both partners of a black kite pair may grasp one another by the feet and then tumble down together. In nest building, the kites often use old nests of raptors, crows, or gray herons. They line the nest cavity with rags and other rubbish. The female generally lays two, but if food is abundant, three or four eggs, which are probably incubated by both adults. The length of the incubation period is still unknown, although it is probably about thirty days. The young are nestlings for forty to fifty days, and are tended by the parents like in other hawk-like species (see Chapter 14). The male brings food which the female distributes to the young. After fledging, the young still continue to beg food for a while and they still feed on the nest, until finally the adults leave them the prey they have brought outside the nest. The adults moult during the breeding season in their territories. A kite may occasionally be killed by a goshawk or an eagle owl.

The ACCIPITERS (subfamily Accipitrinae) are represented in Central Europe by the goshawk and sparrow hawk. Of the six genera and fifty-two species, the following are mentioned:

Subfamily: goshawks, by H. Brüll

A. The GOSHAWKS (genus *Accipiter*), have short rounded wings and long tail feathers; there are forty-five species, among them: 1. The NORTHERN GOSHAWK *(Accipiter gentilis;* Color plate, pp. 332/333) has an L of 50-62 cm. The wingspan of the female is 110-118 cm, and that of the male is 93-101 cm; it has several subspecies, among them are: a) the EUROPEAN GOSHAWK (*Accipiter gentilis gallinarum;* Color plate p. 355); b) the SCANDINAVIAN-BALTIC GOSHAWK *(Accipiter gentilis gentilis);* c) the SIBERIAN GOSHAWK *(Accipiter gentilis buteoides),* which is lighter colored; and d) the AMERICAN GOSHAWK *(Accipiter gentilis atricapillus).* 2. The EUROPEAN SPARROW HAWK (*Accipiter nisus;* Color plates pp. 332 and 355) has an L of 25-37.5 cm. The wingspan of the male is 59-65 cm, and that of the female is 68-77 cm. It has several subspecies; among them is the AFRICAN SPARROW HAWK *(Accipiter nisus rufiventris).* 3. The AFRICAN GOSHAWK *(Accipiter tachiro),* which is the size of a sparrow hawk. 4. The AFRICAN LITTLE SPARROW HAWK (*Accipiter minullus;* Color plate p. 355), which is not quite of sparrow hawk size. 5. COOPER'S HAWK *(Accipter cooperi),* which is sparrow hawk-sized. 6. The SHARP-SHINNED HAWK *(Accipiter striatus)* which is sparrow hawk-sized. 7. The WHITE

Fig. 15-27. Flight silhouette of a Sparrow-hawk *(Accipiter nisus).*

GOSHAWK (*Accipiter novaehollandiae;* Color plate p. 355), which is goshawk-sized. 8. The BLACK SPARROW HAWK (*Accipiter melanoleucus*) which is also goshawk-sized.

B. Genus *Melierax:* 1. The PALE CHANTING GOSHAWK (*Melierax musicus;* Color plate p. 355) has an L of 42.5–45 cm. 2. The GABAR GOSHAWK (*Melierax gabar*) has an L of 27.5–35 cm.

C. Genera *Urotriorchis, Erythrotriorchis,* and *Heterospiza:* 1. The AFRICAN LONG-TAILED HAWK (*Urotriorchis macrourus;* Color plate p. 355) has an L of 55–57.5 cm. 2. The RED GOSHAWK (*Erythrotriorchis radiatus*) with an L of 60 cm, is a hunter of small birds and a nest plunderer. The nest is generally built in old crows' nests and adorned with green eucalyptus leaves. 3. The SAVANNAH HAWK (*Heterospiza meridionalis;* Color plate p. 355).

Fig. 15-28. Foot of ♀ sparrow hawk; the foot spread is 5.5–6 cm.

Foot of ♂ sparrow hawk; the foot spread is 4–5 cm.

Birds of the genus *Accipiter* and their closest relatives stalk and then pounce on their prey. They inhabit areas in which woods alternate with bush-grown areas, clearings, fields, and lakes or river shores. Their short wings and long tail make them very skillful at swerving and enable them to maintain a high speed over short distances. In areas with plenty of cover, their prey has every opportunity to disappear; hence they must take it by surprise and prevent, with the aid of their long feet, escape at the last moment. Goshawks and sparrow hawks are therefore found all over the world where they can find suitable habitats.

Fig. 15-29. 1. and 2. Goshawk (*Accipiter gentilis*); 1. American goshawk (*Accipiter gentilis atricapillus*); 2. Other subspecies; 3. African goshawk (*Accipiter tachiro*).

The northern goshawk inhabits the forest belts of the Northern Hemisphere; it is one of the most efficient hunters of its family. This has given it the reputation of being "inspired by a lust to kill and intoxicated by blood"; on the other hand, it has increased its popularity with falconers in Europe, Asia, and more recently in North America as well. The ornithologist, J. F. Naumann (1780-1857) imagined that it consumed one pheasant a day, plus some smaller birds. Actually, when a goshawk has filled its crop with a wood pigeon, it will not hunt again for at least one or two days. Brehm could not imagine how falconers could make out with such an "uncouth" bird, particularly since his own attempts to tame goshawks failed miserably. However, today, as in the Middle Ages, the goshawk is used as a very successful hunting bird (Color plate pp. 332/333) and sometimes lives twenty or more years in the falconer's care.

Fig. 15-30. 1. European sparrow hawk (*Accipiter nisus*) without the African subspecies (*Accipiter nisus rufiventris*); 2. African little sparrow hawk (*Accipiter minullus*).

As a resident of European regions, the goshawk maintains very large territories. In Central Europe, with its relatively high animal population, they may have, according to accurate determinations, areas of 3000 to 5000 hectares; however, its population density in areas with less game is considerably less. Within its territory, a resident pair contributes to the regulation of the prey populations by helping to maintain the natural balance. Such pairs have been observed more than a decade and, on the basis of moulted feathers, their territories

Fig. 15-31. 1. Cooper's hawk *(Accipiter cooperi);* 2. Bicolored sparrow hawk *(Accipiter bicolor).*

Fig. 15-32. Sharp-shinned hawk *(Accipiter striatus).*

Fig. 15-33. White goshawk *(Accipiter novaehollandiae).*

have been marked on maps. The biologically important role of these birds in their habitats was fully confirmed by these studies.

In one case three pairs of goshawks inhabited a woodland area of 400 hectares. The nest sites formed a triangle whose sides were two to three kilometers long. If one were to draw a circle of about six kilometers in diameter and divide it into three equal segments, this would result in three territories within whose borders no prey was killed (this is the nesting contrasted to the hunting territories). Within the territory of one resident pair, the remains of 3875 prey animals of 67 species were found. The prey ranged from mice to hares, from weasels to domestic cats, from red-backed shrikes to the common buzzard, from tree creepers to pheasants. Rabbits, wood and domestic pigeons, which are particularly common species, accounted for 48.52% of the total prey. The male, with a weight of 550 to 750 grams, can overcome animals up to the size of rabbits and pheasants. Only the heavier female, weighing 1100 to 1250 grams, is able to overcome large hares.

When one determines the prey taken at particular periods by resident pairs of goshawks, one obtains a good insight into their biologically regulating activity. Thus in the pair mentioned earlier, the proportion of cock and hen partridges, carrion crows, and magpies out of 1864 prey remains collected over a four year period, was assessed. It was notable that partridges were taken mainly during their courtship season by the male goshawk. There were always more cock than hen partridges, particularly in years when the former were abundant. Thus the goshawks took off the excess of cocks, several of which would gather about a single hen to win her favor. It was also noticeable that in the fall and in severe winters, the number of partridges taken declined; this coincided with the period when the partridges live in flocks and benefit from the alertness of many. The taking of crows and magpies as prey in the months of June and July aids in the increased survival of the eggs and young of many small and middle-sized birds and, beyond that, it benefits the reproduction of animals in the territory in general.

The nest of the goshawk stands in old trees, often near cuttings, ditches, or water courses which enable the hawk to fly at low level. Flying low over the ground, it approaches its prey, using all possible cover; it flies unobtrusively low above the ground to its nest. Only during courtship does the pair circle high above the breeding territory. In these flights, the male, with almost closed wings and opened bright lower tail coverts, shoots vertically upward. The nest is built in the forks of branches, or built on old crow or buzzard nests or old nests of its own species. The pair lines the depression in the nest with fresh green branches of conifers until well into the period of feeding of the young. Sometimes, particularly in the areas with an Atlantic climate, the pair uses several nests in alternation, because mites and fleas accumulate in the nests of the previous year. The clutch generally consists of three, sometimes four or five eggs, which are mainly incubated by the female. The male relieves the hen for about one to two hours a day. Incubation lasts thirty-eight to forty days. Large numbers of eggs presuppose a territory rich in prey. The fledging period of the young is also thirty-eight days, during which the female covers the young to warm them and distributes the food brought in by the male. A male feeding its young is an exception to this rule.

Fig. 15-34. 1. Black sparrow hawk *(Accipiter melanoleucus)*; 2. African long-tailed hawk *(Urotriorchis macrourus)*.

The young goshawk wears the juvenile plumage, also called the "red plumage," during its first year. Because of this, falconers call the young goshawk the "red goshawk." If a falconer releases his red goshawk in the territory of a wild pair, they will fly around it and shriek at the strange youngster. It then shows very clearly that it would like to get away as soon as possible. No doubt this also happens to wild juveniles, which, while seeking a vacant territory in the fall, enter those of strange goshawks. The resident goshawks make it clear that there is not room for the intruder and that he must move on. If the resident goshawks of a territory are trapped, it is often possible to catch others in the same area, for wandering goshawks try at once to occupy a territory which has become vacant in this way. Thus they do not congregate in a territory in large numbers to hunt pheasants, as is often believed by misinformed observers. In spring, the arrival of a goshawk, particularly of a male (who becomes sexually mature at ten months), in a territory may be fatal, as it risks being killed by the resident pair. Adult goshawks, as proven by their remains, are killed by eagle owls (ten cases), sea eagles (four cases), and members of their own species (six cases). Since its natural enemies are absent in most of Central Europe, the goshawk itself regulates its population density at about 5000 hectares per resident pair, where this is not done by man. In some parts of Europe, as in Holland, the goshawk population has declined to such an extent that the authorities pay farmers and landowners a premium for each successfully raised brood.

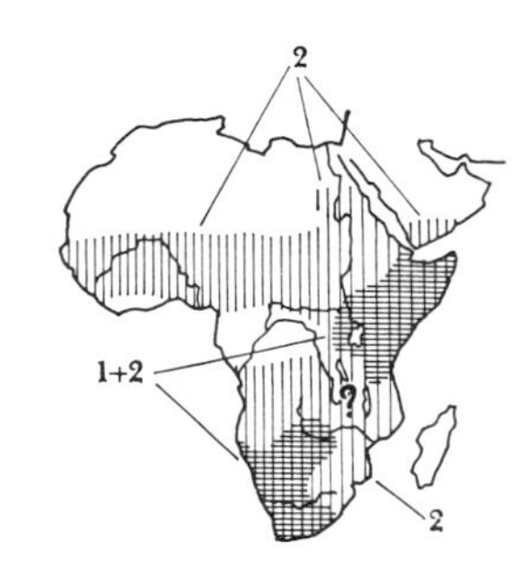

Fig. 15-35. 1. Pale chanting goshawk *(Melierax musicus)*; 2. Gabar goshawk *(Melierax gabar)*.

The sparrow hawk inhabits much the same habitats as the goshawk in the Old World. It preys mainly on small birds, and its foot has the long middle toe for grasping. The Central European sparrow hawk is a miniature of the goshawk, as far as its behavior at the nest and the effect in its territory are concerned. In both species the nest is surrounded by a so-called nest territory, in which prey is not taken. In this section only actions concerned with breeding and the rearing of the young are carried out, including the passing of the prey from the male to the female, the preparation of prey, or the search for trees on which to rest. Hence, all kinds of small birds which can breed quite undisturbed in this area, and are also safe from the enemies of their eggs or young, tend to settle in the nest territory. Meng has reported the same for the American Cooper's hawk.

Fig. 15-36. Red goshawk *(Erythrotriorchis radiatus)*.

The nest of the sparrow hawk is primarily in small conifers; it is

Fig. 15-37. Savannah hawk *(Heterospiza meridionalis).*

not, like that of the goshawk, lined with green branches. Only the female, weighing 220 to 300 grams, can incubate. The male, weighing only 110 to 130 grams, cannot do so, for it cannot cover the clutch with its small body. Thus there is a complete division of labor during the reproductive period. The female lays three to six (generally five) eggs and incubates for thirty-six days. When one considers that sparrow hawks have been found as prey of eagle owls (four times), goshawks (ninety-four times), peregrines (five times), common buzzards (four times), and tawny owls (seven times), the large clutch size and the consequent reproductive capacity is understandable. The pine martin also pursues the sparrow hawk to an appreciable extent.

Goshawks and sparrow hawks clearly show that the effect of a pair of raptors can only be judged within the context of the density of their territories. Thus, Finnish goshawks kill mainly grouse, while the Central European goshawks kill rabbits and wood pigeons. Sparrow hawks in Central Europe have territories of 700 to 1200 hectares in area. Their preferred prey depends on local conditions; in an oak-birch region it will be quite different from that in an oak-horn beam region.

Like the goshawk, the sparrow hawk takes its prey by surprise. It is a stalking and starting flight hunter. The size of the male's prey ranges from that of a wren to thrushes, while the female's ranges from thrushes to pigeons. This small raptor is a popular falconer's bird, particularly in the Near East in Persia and northern Turkey; quail in particular are hunted with it. The sparrow hawk is held in the hand and thrown like a stone. Its quick action gives it every chance of overtaking the prey and bringing it down. An old Persian proverb says, "The sparrow hawk's body is small but its heart is great."

Although the African pale chanting goshawk rarely calls outside the breeding season, its call of excitement at the nest is so remarkable that it has earned the bird both its common and scientific names. This call is composed of a modified shrill whistling, mournful notes, and an occasional loud "kek." The large flat nest of the chanting goshawk is used year after year; if a new nest is built, it will be near the old one. Feathers, clay, rags, and camel dung line the cavity. The nest is often covered with spider webs.

The sparrow hawk-sized gabar goshawk even brings in spiders with their webs, which then properly shroud the nest with webs. The prey of the chanting goshawk consists of grasshoppers and other insects, frogs, and small snakes, but mainly of lizards and only occasionally of mice and small birds. The gabar goshawk, on the other hand, is a good bird hunter and takes mainly weaver finches.

The African long-tailed hawk may be observed, although only rarely, in the forests near the equator. Members of the Bulu tribe in the Cameroon call it the "aerial leopard" because it hunts mainly on the tops of trees in pursuit of birds and the scaly-tailed squirrel. Occasionally it emerges from the forests and visits a chickenyard. There it walks about, looking for an opportunity to seize a chick. The savannah hawk in South America perches on tree trunks, low fence posts, or clumps of grass while looking for amphibia, reptiles, and small rodents; it also glides low over the ground when hunting. The rough soles on its feet facilitate its grip on slippery frogs. Its habitats are grasslands, savannahs, and fields near swamps or rivers. It also keeps its nest green with fresh foliage and incubates a clutch of one or two eggs. The nest is in the forks of branches. In their display flight, two or three of the birds rise to great heights and circle there and utter shrill calls. Then they drop down with motionless wings and extended legs.

Philippine monkey-eating eagle (*Pithecophaga jefferyi*). ▷

Subfamily: buzzards and related forms, by W. Fischer

The BUZZARDS and relatives (Buteoninae) are the most multiform subfamily of the raptors with the most species. There are great differences in their modes of life and external appearance, but ornithologists combine them on the basis of certain structural characteristics.

The L is 29-110 cm; most have long, broad wings, and a medium-long rounded tail. They have relatively short feet which are efficient for killing. There are twenty-nine genera with a total of ninety-four species.

Buzzards and eagles, large or small, are, as a rule, not such elegant hunters as are goshawks and falcons. But they are very adaptable in their methods of both obtaining prey and in their choice of prey; they can even live temporarily on ground insects, snails, or carrion. Most species, however, hunt mammals, including mice, small antelopes, young deer, or young llamas, as well as monkeys and sloths. Sea eagles kill waterfowl and fish, and many tropical species are adapted to catching the plentiful reptiles, amphibia, insects, or crabs.

The structure of their toes and claws corresponds to these needs. The toes are relatively short, but they have sharp claws. The cutting edge of the upper mandible behind the hook is curved down, and so projects like a scissor blade over the upper edge of the lower mandible.

The difference in size of the two sexes is much less pronounced in the Buteoninae than among the hawks and large falcons. In all species the female is only one-fourth to one-fifth larger than the male. The subfamily has world-wide distribution, and its members have adapted themselves to almost all areas and habitats in the most diverse forms. They inhabit forests and prairies, high plateaus and mountain forests, semideserts, deserts, steppes and jungle, sea coasts, lake shores, and swamps, as well as cultivated fields near human settlements.

Nest construction of buzzards

All species of the buzzard-eagle group build nests, but the nest sites are highly variable according to conditions in their habitats. What a contrast there is between the huge nest of a golden eagle in a rock niche on a Swiss mountain cliff and that of a steppe eagle on an eroded hillock in the Mongolian semidesert. Yet each of these sites offers the best possible chances for the preservation of the species in its particu-

1
2
8
3
9
4
7
6
5
F. Reimann

Fig. 15-38. Gray eagle buzzard *(Geranoaëtus melanoleucus).*

Fig. 15-39. 1. Common buzzard *(Buteo buteo);* 2. Himalayan buzzard *(Buteo refectus);* 3. African mountain buzzard *(Buteo oreophilus);* 4. Madagascar buzzard *(Buteo brachypterus).*

lar environment. Where buzzards and eagles, such as common buzzards, the guiana crested eagle, the slate-colored hawk, or the booted eagle, occur in forests, they all nest on trees which affords them a wider view and offers security. Some species prefer treetops; others build more or less hidden nests in forks of branches, where gaps allow free access.

The nests of the large species are of considerable size and weight as a result of being used year after year. One can clearly recognize each year's additions on these structures, which are always marked by new material. The material of an old sea eagle's nest, thrown down by a storm, filled a hay wagon and weighed over a ton; according to Stemmler, golden eagle nests of similar dimensions have also been found. There are records of territories which have been occupied for over fifty years. A sea eagle's nest on the Irtysh river was used for over eighty years, and a crowned eagle's nest in Kenya was used for seventy-five years; that of a bald eagle in America came down in a storm after sixty-three years of use.

All buzzard and eagle nests, when complete, are lined with green leaves and finally with bits of turf, grass, pieces of fur, wool, or old pellets. The "greening" is done with twigs of conifers or of deciduous trees which are either picked up or bitten off; they are pushed into the nest rim and indicate that a particular nest is being used. This is not a matter of camouflage, but an innate behavior involving the partner which is simply a part of the nest-building behavior. The species which have no access to green leaves, i.e. those which build their nests in semideserts or deserts before plant growth occurs, carry bones, bits of rags, dried fresh animal feces, and other similar materials into their nests, perhaps as a "substitute" for greenery. When the spotted eagle, normally a forest bird, nests on a slope in the steppe, it brings in pine branches from afar; the steppe eagle which nests in the same area does not.

Nest-building in the larger species takes at least two, but generally up to four, weeks. The males begin the construction, often using the nest of another species as a base; after pair formation the female stimulates the male to bring in nesting material, which she alone builds into the structure. High circling flights above the breeding area, often accompanied in many species by loud calls, denote the staking of a claim to a territory in buzzards and eagles. The population density of various species or pairs of the same species depends on the food supply. In buzzards, population density is greater in cultivated areas than in semideserts or mountain regions. When there is a real population explosion of prey animals, the population density of raptors may be considerably increased and result in larger clutches and a higher rate of reproduction.

Buzzard-like birds have existed since the Lower Tertiary, and true

◁

Crested eagles: 1. Martial eagle *(Polemaëtus bellicosus);* 2. Harpy eagle *(Harpia harpyja);* 3. Ornate hawk eagle *(Spizaëtus ornatus);* 4. Guiana crested eagle *(Morphnus guianensis);* 5. Long-crested eagle *(Lophoaëtus occipitalis);* 6. Crowned eagle *(Stephanoaëtus coronatus);* 7. New Guinea harpy eagle *(Harpyopsis novaeguineae).* Bonnelis' eagles: 8. Bonnelis' eagle *(Hieraaëtus fasciatus);* 9. Booted eagle *(Hieraaëtus pennatus).*

buzzards *(Buteo)* have occurred since the Oligocene (about thirty million years ago).

The EAGLE BUZZARD (genus *Geranoaëtus*) contains a single species, the well-known and common GRAY EAGLE BUZZARD (*Geranoaëtus melanoleucus;* Color plate p. 356). The L is 64–69 cm, while in the northern subspecies *(Geranoaëtus melanoleucus meridensis)* it is only 56 cm. The wingspan is 170–190 cm, and the weight is 1800 g for males to 2500 g for females. The nape feathers are pointed in a lancet-like manner, the wings are long, and the tail is short. Two to three eggs, which are grayish-white with brown spots, are laid from October to December.

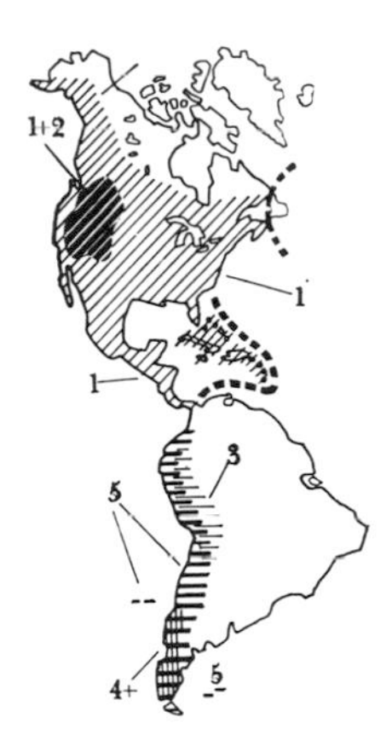

Fig. 15-40. 1. Red-tailed hawk *(Buteo jamaicensis);* 2. Ferruginous hawk *(Buteo regalis);* 3. Gurney's hawk *(Buteo poecilochrous);* 4. Red-tailed buzzard *(Buteo ventralis);* 5. Red-backed hawk *(Buteo polyosoma).*

This sizeable, and in many respects, eagle-like buzzard occupies the same ecological niches in its South American home on the coasts, in hill country, and in mountains, that in other countries are taken up by sea eagles *(Haliaeëtus)* and eagles *(Aquila);* that is, it uses the same opportunities for living. It flies singly or in pairs, generally high, carefully checking its hunting area. In swooping flight, it seizes its prey, such as a rodent or opossum, on the ground. But it can also seize pigeons or ducks, even as they take wing. Nor does it avoid carrion; sometimes it even lives on it for weeks. It is, however, a skillful raptor able to defend itself and to fight. In the large flight cage in the Berlin Zoo, a female regularly swooped down at flying Himalayan griffon vultures, held onto their backs, and let herself be carried along.

The species of the TRUE BUZZARDS (genus *Buteo*) are difficult to distinguish externally. Even within a species there are differences in the inhabitants of different areas, as well as individual differences within a locality in size, color, markings, and choice of habitat.

Fig. 15-41. 1. Long-legged buzzard *(Buteo rufinus);* 2. Upland buzzard *(Buteo hemilasius);* 3. Grasshopper buzzard *(Butastur rufipennis);* 4. Jackal or Augur buzzard *(Buteo rufofuscus).*

They are predominantly hunters of small animals; individuals tend to specialize on particular prey or to become parasitic. The head is rounder and the beak is somewhat shorter than in the closely related eagles *(Aquila).* There are twenty-seven species; among them are:

1. The COMMON BUZZARD (*Buteo buteo;* Color plates p. 332 and p. 356) has an L of 46–56 cm, a wingspan of 120–140 cm, and a weight of 600–900 g (male) and 800–1200 g (female). It lays two to four eggs in April, which are grayish-white with red or grayish-brown spots, and measure 57 x 46 mm. There are eleven subspecies; among them are the *Buteo buteo zimmermannae* in Eastern Europe and *Buteo buteo vulpinus* in western Siberia. Both are smaller with more pointed wings. They are more rusty-colored and brighter in color (falcon buzzards). 2. The RED-TAILED HAWK (*Buteo jamaicensis;* Color plate p. 356) has an L of 43-63.5 cm, a wingspan of 109–143 cm, and a weight of 880–1000 g for the male and 1250–1500 g for the female. They lay two to five eggs of the same type as the common buzzard. In the subspecies *Buteo jamaicensis borealis* the eggs measure 59.5 x 47.2 mm. The breeding season is from March to May. The smallest races are found in Alaska, while the largest are in northern Mexico. The tail feathers can also be

Fig. 15-42. 1. Rough-legged buzzard *(Buteo lagopus);* 2. Zone-tailed hawk *(Buteo albonotatus);* 3. African red-tailed buzzard *(Buteo auguralis).*

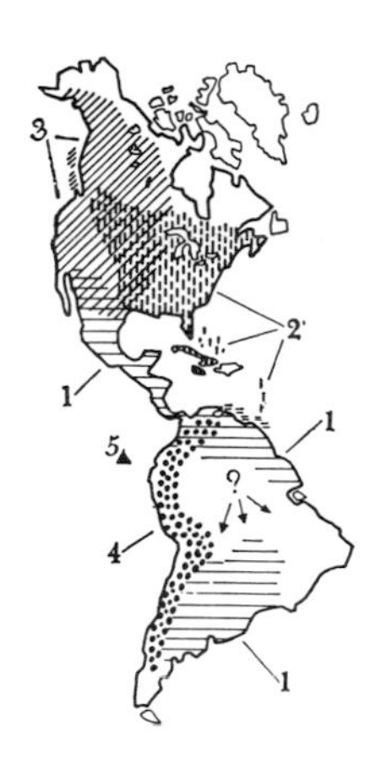

Fig. 15-43. 1. White-tailed hawk *(Buteo albicaudatus)*; 2. Broad-winged hawk *(Buteo platypterus)*; 3. Swainson's hawk *(Buteo swainsoni)*; 4. White-throated hawk *(Buteo albigula)*; 5.Galapagos hawk *(Buteo galapogoënsis)*.

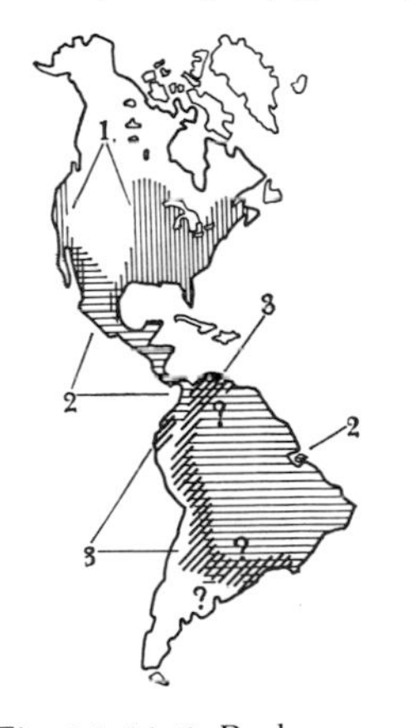

Fig. 15-44. 1. Red-shouldered hawk *(Buteo lineatus)*; 2. Gray hawk *(Buteo nitidus)*; 3. Rufous-winged hawk *(Buteo leucorrhous)*.

Fig. 15-45. Flight silhouette of the rough-legged buzzard *(Buteo lagopus)*.

gray or whitish. 3. The RED-TAILED BUZZARD *(Buteo ventralis)* is hardly distinguishable from the light phase of the red-tailed hawk, which was discovered by Charles Darwin in 1834. In winter it wanders as far north as Guiana, and meet red-tailed hawks there. 4. The LONG-LEGGED BUZZARD *(Buteo rufinus)* has tarsi with plates. 5. The UPLAND BUZZARD *(Buteo hemilasius)* has feathered tarsi. 6. The FERRUGINOUS HAWK *(Buteo regalis)* has feathered tarsi.

These three species are grouped together as EAGLE BUZZARDS, with an L of 58.5-68 cm. They are desert dwellers, and therefore they are more "pedestrians" and ground birds than other buzzards. They are long-legged. 7. The AUGUR BUZZARD *(Buteo rufofuscus)* has an L of 46-53.5 cm. It is the most common buzzard of East and South Africa. Found in foothills and steppes, it has a jackal-like call. 8. The AFRICAN RED-TAILED BUZZARD (*Buteo auguralis;* Color plate p. 356) has an L of 39-43 cm. It eats rats, mice, lizards, frogs, and insects. 9. The ROUGH-LEGGED BUZZARD (*Buteo lagopus;* Color plate p. 356) has an L of 51-61 cm, a wingspan of 130-150 cm, and a weight of 800-1100 g for males and 900-1300 g for females. The tarsi are feathered, and the flight feathers are longer and there is more gray in the plumage than in the common buzzard. The very pale subspecies *Buteo lagopus pallidus* is found in central Siberia; the darker *Buteo lagopus sanctijohannis* is found in North America. 10. The RED-SHOULDERED HAWK *(Buteo lineatus)* has an L of 47-51 cm. 11. The BROAD-WINGED HAWK *(Buteo platypterus)* has an L of 35-46 cm. It is a forest dweller; its winter quarters are mostly on the lower Amazon. As a "mass migrant," it wanders in flocks of hundreds or thousands. For other buzzard species, see distribution maps, Figures 15-39 through 15-51.

The common buzzard is the most common raptor in Europe and parts of northern Asia, as the red-tailed buzzard is in the north of the New World. Although the common buzzard breeds in woods, most of its life is spent over open fields. Here, in flapping and soaring flight, occasionally interrupted by a clumsy hovering, it searches for food. Buzzards also wait on trees, stones, or earth hillocks, and swoop onto their prey from these perches. They also take grasshoppers, beetles, snakes, worms, and snails on the ground. The research of Th. Mebs has shown that the vole *Microtus arvalis* is the principal prey of the common buzzard; where there are many voles this induces a greater population density of buzzards. Two species of voles account for 41% of the total prey; together with all voles and mice, these are 51% or just about half of all prey animals of this buzzard.

The courtship flights consist of mutual circling flights of both partners, which often utter their "heeah" call. This "weightless gliding" of the buzzard is one of the finest sights that our raptors can offer. The ornithologist Otto Kleinschmidt states: "It is quite a common raptor, and yet something magic, a poem, perhaps more beautiful than the lark's and the nightingale's song! This alone, not the mere calculation of the number of mice and young hares it catches annually, makes it worth protecting."

Fig. 15-46. Flight silhouette of the common buzzard *(Buteo buteo)*.

After thirty-two to thirty-four days of incubation, the young hatch and have an initial weight of about forty-five grams. Kleinschmidt describes a nest as follows: "A wreath of fresh oak twigs and fresh leaves surrounded it. In the middle was a little clump of charming little white buzzard children on still green twigs, while a worried white adult circled around me at close range, calling."

While adult native buzzards, as residents of Central Europe, may gather in some numbers at places where there is much food, the buzzards of Scandinavia are migrants which in part, wander through Europe, and in part, increase the Central European buzzard population in winter. In some years, the West Siberian common buzzard appears during its migration in Central Europe. Its migration route is, however, to the southeast, and in the winter this subspecies is more common in South Africa than is the native augur buzzard.

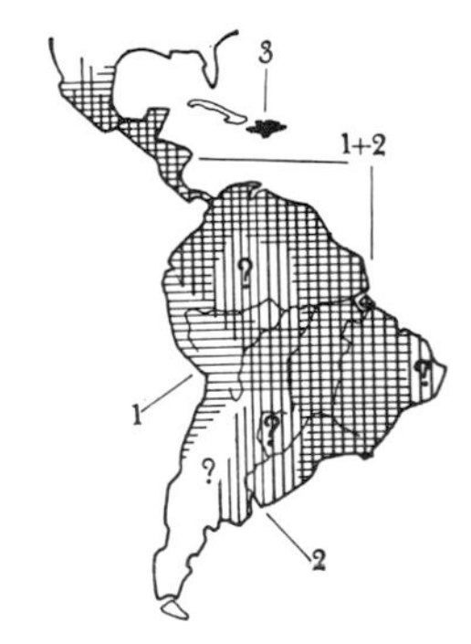

Fig. 15-47. 1. Short-tailed hawk *(Buteo brachyurus)*; 2. Roadside hawk *(Buteo magnirostris)*; 3. Ridgeway's hawk *(Buteo ridgwayi)*.

The common buzzard is not sexually mature until the third year. Human persecution is still the most common cause of death, particularly in Europe. The longest known lifespan is twenty-five years and four months.

The red-tailed hawk is one of the most common raptors of North America. It inhabits the forests of the East as well as the prairies and deserts of the West, the taiga of the North, and the tropical forests of the West Indies and Panama. This extraordinarily versatile raptor, like the falcon buzzards, combines the characteristics of buzzard and goshawk, as is evident in its more skillful and rapid flight. Its prey, corresponding to the different habitats, is varied, ranging from small mammals, snakes, lizards, frogs, and fish to insects. It kills rabbits, ground and tree squirrels, but also catches birds up to the size of prairie chickens. Its nests stand in the small trees and bushes of the taiga, on trees, in shrubs in the swamp hills of Florida, and on old cactus stems in the deserts of Mexico. It also breeds on cliffs, on the edge of deserted farms, and even on palm trees in the sugar plantations in Cuba.

The habitat of the eagle buzzard species in the dry areas of North Africa to Central Asia, and in the desert steppes of central North America is relatively uniform. In the Gobi desert there is the upland buzzard and occasionally the long-legged buzzard, which is otherwise more of a steppe bird. We found nests on rocks as well as on the ground on the edges of ravines. In the open landscape of the Gobi, upland buzzards, according to Piechocki, always raise only one youngster. The adults can only protect one against the intense sunshine; the others, which they cannot shade, perish.

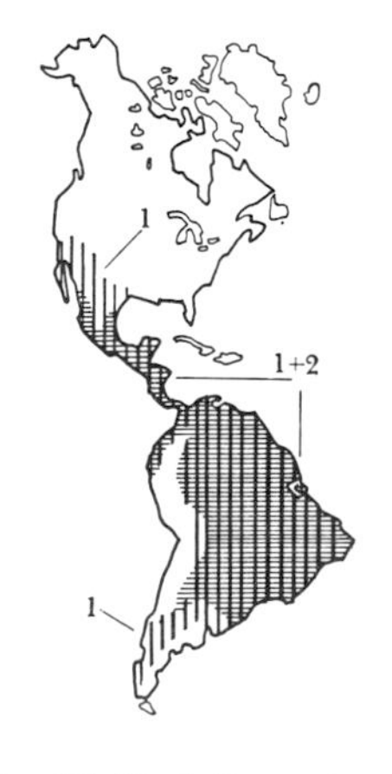

Fig. 15-48. 1. Bay-winged or Harris's hawk *(Parabuteo unicinctus)*; 2. Fishing buzzard *(Busarellus nigricollis)*.

In the northern tundra and forest tundra belt of the earth, the rough-legged buzzard is quite a common breeding bird. Its presence

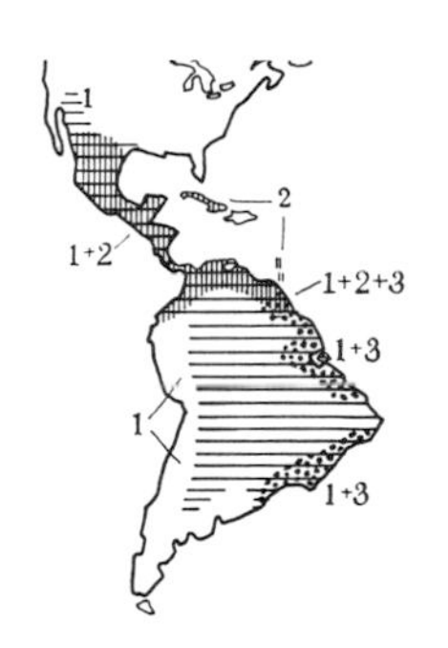

Fig. 15-49. 1. Great black hawk *(Buteogallus urubitinga)*; 2. Common black hawk *(Buteogallus anthracinus)*; 3. Rufous crab hawk *(Buteogallus aequinoctialis)*.

depends upon the abundance of its chief prey, the lemmings and other small rodents, although not to the extent of the snowy owl *(Nyctea scandiaca)*. It inhabits the entire North, up to 75° north latitude, and in years of food abundance may extend its breeding range appreciably southward, to southern Norway. The nests are on rocks, hills, and sand banks, and also in dwarf birch; because of the prevalent cold in the northern spring, these nests are very well lined with grass and wool. In years with many rodents, nests in favorable areas are only one to one and one half kilometers apart, while in less favorable areas, they are two to three or four kilometers apart. Mortality of the young is high; about 33% die as early as the first fall. Usually these are the last young to hatch, particularly when there is a sudden shortage of food. The rough-legged buzzard leaves its breeding range in September and appears from October onward in the temperate latitudes of the Northern Hemisphere, and thus also in Central Europe as a winter visitor.

The red-shouldered hawk is an inhabitant of moist and swampy areas. It also lives on farmland and catches insects very skillfully. It kills small rodents and searches for wounded birds after hunts. The blue jay *(Cyanocitta cristata)* imitates its call, "kee yoo," as readily as the common jay imitates the mewing of the common buzzard.

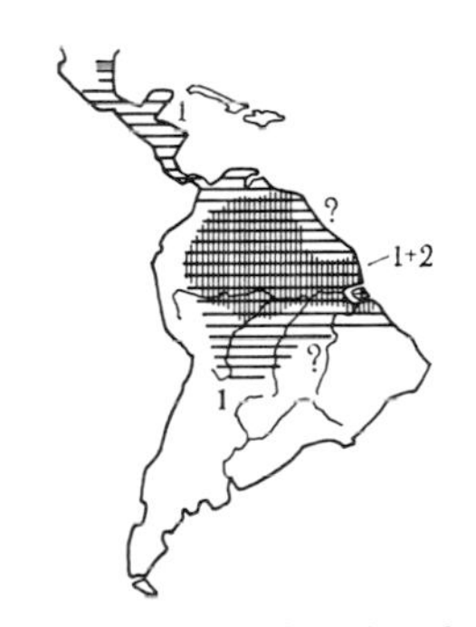

Fig. 15-50. 1. White hawk *(Leucopternis albicollis)*; 2. Slate-colored hawk *(Leucopternis schistacea)*.

DESERT BUZZARDS (genus *Parabuteo*) are very close to the true buzzards. The only species, the BAY-WINGED or HARRIS'S HAWK (*Parabuteo unicinctus*; Color plate p. 356), has long legs, and dark plumage with white and red marks. The L is 46-56 cm, and the wingspan of the male is 108 cm, and that of the female is 115 cm. Three to four eggs, measuring 54 x 41 mm, are laid. Its habitat is dry regions with scrub, but occasionally it occurs along rivers in the primeval forests of Central and South America. In Mexico it lives in the large sisal plantations, where many lizards offer it abundant food. In its style and method of its hunting flights, it is intermediate between buzzards and goshawks; its long feet enable it to seize prey, even out of the thorny scrub. It kills sizeable birds such as rails, night herons, teals, and woodpeckers.

The three species of BLACK HAWKS *(Buteogallus)* are predominantly black or dark brown. They occur in South and Central America. The wings are very broad, easily the size of the common buzzard. In flight they somewhat resemble the black vulture *(Coragyps atratus)*; they often join the latter in search of food on river banks, including frogs, reptiles, and crabs (hence they are also called crab goshawk). The nests are found on tall trees near the coast or on river shores. For the various species see distribution map Fig. 15-49.

Fig. 15-51. 1. Lizard buzzard *(Kaupifalco monogrammicus)*; 2. White-eyed buzzard *(Butastur teesa)*; 3. Gray-faced buzzard *(Butastur indicus)*; 4. Rufous-winged buzzard *(Butastur liventer)*.

The FISHING BUZZARD *(Busarellus nigricollis)*, which lives from northern Mexico to Argentina near water and on the coasts, is a peculiar bird. The L is 45-50 cm. It is a highly specialized crab and fish hunter, and is long-winged and short-tailed. The legs are long, but appear plump; the toes have strong rough pads. The plumage is red-brown with a paler head area, and black spots on the nape and throat. It lives mainly in mangrove forests. During the period of floods it moves inland and catches rodents, reptiles, snails, and small birds in flooded rice fields, often standing in the water while doing so. Its main prey, however, is fish.

In contrast to this buzzard are the hawks of the genus *Leucopternis,* which live mainly in the tropics of the New World. They are short-winged and long-tailed forest birds. The largest species is the WHITE HAWK *(Leucopternis albicollis),* with an L of 46–52 cm. The plumage, apart from the black flight feathers and tail, is white. The SLATE-COLORED HAWK (*Leucopternis schistacea;* Color plate p. 356) has an L of 41–43 cm. There are eight other species, the L being 33–51 cm.

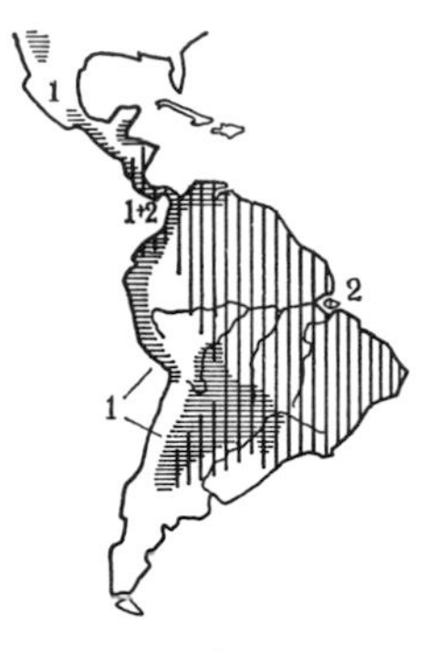

Fig. 15-52. 1. Genus *(Harpyhaliaëtus);* 2. Guiana crested eagle *(Morphnus guianensis).*

These buzzards have short feet with claws for killing. They fly from tree to tree and take their prey by surprise in swerving flights. Most species take reptiles and insects, as well as birds. Thus they appear on the displaying grounds of the cock of the rock *(Rupicola rupicola)* and seize the displaying males. The small nests of this species stand in dense forests, but are easy to find since this buzzard makes loud and shrill calls in its nesting territory. Little is known about its reproduction.

In the Old World, certain long-winged raptors live in similar habitats (see Fig. 15-51).

1. The LIZARD BUZZARD *(Kaupifalco manogrammicus),* with an L of 29–35 cm, has a gray plumage and a white throat with a black longitudinal stripe. The iris and bare areas are orange-red. It catches lizards, grasshoppers, and preying mantises, as well as mice and snakes. Its perches are in the cover of treetops; it often changes perches, gliding to another tree. Its call, a melodius sequence, is often heard. The small nest is in a fork of an isolated tree, and is often "greened" with moss and lichens.

2. The GRASSHOPPER BUZZARDS (genus *Butastur*), with an L of 30–43 cm, have four species (see Figs. 15-41 and 15-51). They use perches in open areas, often sitting on telephone poles.

The solitary, black, and crested eagles are much larger. The claws are long and strong (they show the greatest development in all raptors). The wings are rounded and the tail is long. The approach flight to prey is as swift as in goshawks. This group has horn plates on the tarsus, which is unfeathered and very strong. There are four species in Central and South America, and two others in Southeast Asia and New Guinea.

The SOLITARY EAGLES (genus *Harpyhaliaëtus)* live along river shores, including those in mountains. Apart from fish their prey consists also of mammals, such as young deer and skunks. There are two species: 1. The BLACK SOLITARY EAGLE *(Harpyhaliaëtus solitarius),* with an L of 61–66 cm; and 2. The CROWNED SOLITARY EAGLE (*Harpyhaliaëtus coronatus*), with an L of 63.5–69 cm.

One of the best known species of this group is the GUIANA CRESTED EAGLE (*Morphnus guianensis*; Color plate, p. 336) with an L of 68–80 cm, a wingspan of 150–154 cm, and a weight in the ♂ of about 2500 g. They lay one to two eggs. The guiana crested eagle is relatively common, inhabiting rain forests and open park lands. Its flights in search of prey begin from the trees on the edges of clearings. It takes its prey after a short approach flight. When hunting monkeys in the forest, it makes searching flights over and between the treetops; otherwise it usually takes iguanas, forest birds like hokkos and guans, opossums, and young llamas.

In its breeding behavior it does not differ from the HARPY EAGLE (*Harpia harpyja*; Color plate, p. 366), which is probably one of the world's strongest birds of prey. The L is 81–100 cm, and the weight is 3000 to 4400 g. They lay one to two yellowish-white eggs during the breeding season, which in Guiana, is from December to January. The legs are strong, the toes are coarse, and the claws are very large. A wide erectile crest is located on the back of the head. The tail is long, and the wings are short and wide, and do not reach the middle of the tail when the bird is resting. (Distribution, see Fig. 15-53).

There have been few birds about which there has been as much discussion since their discovery as about this mighty raptor of the South American jungles. It was described factually by d'Orbigny and Tschudi only relatively recently. The harpy eagle does not live in the interior of great forests but on the edges, on river shores, and in forested mountain valleys. Its swerving flight carries it with short wingbeats from one perch to another in the treetop region. There it seizes monkeys, particularly capuchins, squirrel monkeys, woolly monkeys, as well as coatis, tree porcupines, and sloths, which it tears off the branches with its strong feet. It appears in clearings, seizing agutis and ground birds; near human habitation it also kills dogs, baby pigs, and domestic fowl. The hunting flight of the harpy eagle is stimulated by sounds, such as the contact calls of a flock of capuchin monkeys or the calls of ara parrots clamoring in a thicket. Outside the breeding season, the partners hunt in separate territories and only meet again in the nest territory for courtship, as in the goshawks.

Fowler and Cope observed two harpy eagle families in Guiana from December 1959 to May 1960. On the basis of their report, a picture of the breeding biology of this species has been made possible. The nest is always in the top of a tall tree, often on the edge of a clearing. Here the birds are imprinted to certain landmarks in the nesting territory, and so they always fly to the nest site by the same route. The nest is richly "greened" and its depression is lined with leaves and moss. Remains of prey, such as aguti bones, are built into it as well. The nest territory extends ninety meters from the nest in all directions. Even the male, after it has relinquished its prey, is immediately chased out of this area by the female. The young have a long "branching period," which is the time they spend out of the nest on the nest tree.

Indians often take harpy eagles from the nest and keep them captive in order to easily obtain the valued feathers. According to the information of Alcide d'Orbigny (1802–1857), the owner of a living harpy eagle is highly regarded among the Indians. The feathers are among the most important objects of exchange among the natives, and successful harpy hunters thereby obtain everything they need for living.

Fig. 15-53. Harpy eagle (*Harpia harpyja*).

The "Old World harpy" is the MONKEY-EATING EAGLE (*Pithecophaga jefferyi;* Color plates, p. 327 and p. 365) which was only discovered on the Philippines in 1894. The L is 85–95 cm, and the weight is 3600–4200 g. The beak is short and high in the ♂, measuring 51 mm, but it is at most only 21 mm wide. When this "giant goshawk" erects the lancet feathers on the back of its head, which are up to 9 cm long, they form a semicircle behind the dark face with the mighty ax-shaped beak; thus the bird acquires a particularly savage aspect. In the forests of its home, where prey animals are scarce, it takes mainly macaques and often manages to catch a feeding hornbill male in front of its nest cavity. Near settlements and farms, it has concentrated on domestic animals as prey and kills smaller dogs and young pigs. Its territory has an area of thirty square kilometers. Unfortunately this impressive raptor is becoming ever rarer. It is not certain whether this is due to the natives, which persecute it as "harmful," or is due to the excessive taking of young from the nest for use in the animal trade. Presently there are hardly more than fifty breeding pairs on Luzon and Mindanao, the main Philippine islands; however, this bird was probably never common. In order to prevent its extermination, the international union of zoo directors has resolved not to buy any more monkey-eating eagles. This abolishes the profit from robbing nests and it is hoped that, without this incentive, the practice will cease. This eagle probably breeds only every two years, but it can bring two young to fledging in a single brood.

Fig. 15-54. 1. Philippine monkey-eating eagle (*Pithecophaga jefferyi*); 2. New Guinea harpy eagle (*Harpyopsis novaeguineae*).

The NEW GUINEA HARPY EAGLE (*Harpyopsis novaeguineae;* Color plate, p. 366) lives in New Guinea. The L is 79 cm. It has a head crest, long legs, and a very long tail. According to Thomas Gilliard, it is found in the forests of the central plateau between elevations of 2500 and 3100 meters; in this area it is not rare. It probably seizes mainly flying prey, such as displaying birds of paradise. It takes tree mammals such as kuskus, marsupial rats, and young tree kangaroos. It also ranges into open farming country where it takes young pigs and chickens.

In contrast to these broad-crested species, the woodland eagle species with pointed crests have feathered tarsi. They are distributed

▷

Eagles: 1. Golden eagle (*Aquila chrysaëtos*); 2. Imperial eagle (*Aquila heliaca*); 3. Steppe eagle (*Aquila nipalensis*); 4. Lesser spotted eagle (*Aquila pomarina*); 5. Greater spotted eagle (*Aquila clanga*); 6. Verreaux's eagle (*Aquila verreauxi*); 7. Indian black eagle (*Ictinaëtus malayensis*).

1
2
3
4
5
6
7

1
2
5
3
4
6
7
F. Reimann

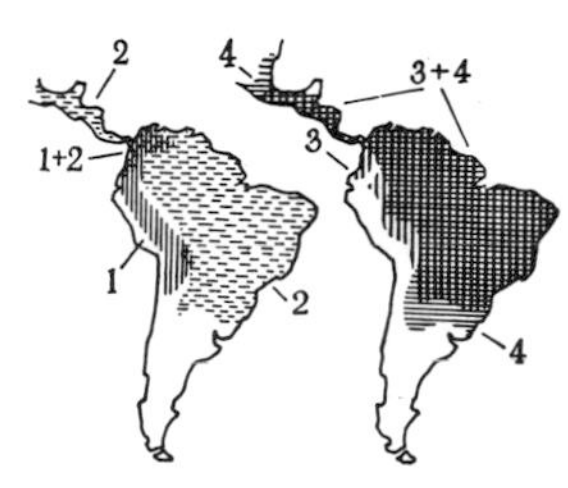

Fig. 15-55. 1. Isidor's eagle *(Oroaëtus isidori)*; 2. Black and white hawk eagle *(Spizaëtus ornatus); leucus)*; 3. Ornate hawk eagle *(Spizaëtus ornatus)*; 4. Black hawk eagle *(Spizaëtus tyrannus)*.

Fig. 15-56. Genus *Spizaëtus* in Asia: 1. Crested hawk eagle *(Spizaëtus cirrhatus)*; 2. Nepal hawk eagle *(Spizaëtus nipalensis)*.

◁
Sea eagles: 1. White tailed sea eagle *(Haliaeëtus albicilla*; immature birds are dark brown with dark head); 2. Pallas' sea eagle *(Haliaeëtus leucoryphus)*; 3. African fish eagle *(Haliaeëtus vocifer)*; 4. Bald eagle *(Haliaeëtus leucocephalus)*; 5. Steller's sea eagle *(Haliaeëtus pelagicus)*; 6. White-bellied sea eagle *(Haliaeëtus leucogaster)*. Fishing eagles: 7. Gray-headed fishing eagle *(Icthyophaga ichthyaëtus)*.

in seven species in the New World tropics, as well as in Africa, southern Asia, and partly as far as the south of the Old World northern region. The ISIDOR'S EAGLE *(Oroaëtus isidori)*, the New World mountain form of this group, is shiny black with reddish-brown underparts. It is relatively long-winged. The L is 63.5-73.5 cm. It lays one egg, which is white with brown spots. It begins to breed in April. It lives in the Andes between 600 and 3000 meters elevation and breeds in an area about 2000 meters high. There it replaces the Guiana crested eagle and the harpy eagle and, like them, feeds on monkeys, young sloths, racoons, squirrels, tree porcupines, and birds. F. C. Lehmann found in Columbia that a youngster hatched in May stayed in or near the nest for six months. The BLACK AND WHITE HAWK EAGLE *(Spizastur melano leucus)* resembles the booted eagle; it is a mobile small eagle with an L of 48-51 cm. It hunts small mammals, birds, and reptiles in the lowlands of central South America.

In the genus *Spizaëtus*, the forms with quite long crests, short wings, and long tails are combined. The habitats of two species overlap in South America; they are the ORNATE HAWK EAGLE *(Spizaëtus ornatus*; Color plates, p. 366; the L is 60-75 cm) and the dark BLACK HAWK EAGLE *(Spizaëtus tyrannus*; the L is 68-80 cm). All of the New and Old World species are paler in juvenile plumage; the lower parts are unpatterned. Originally only two Old World species were described, but the newer systematics distinguishes seven species, which occur in pale and dark color phases as well, in the range which extends from Japan to Indonesia and India. The largest range is that of the MOUNTAIN HAWK EAGLE *(Spizaëtus nipalensis)*, which is strong and densely banded below. The L is 62-79 cm. There are three subspecies. In the lower levels of southern Asia, it shares its range with the CRESTED HAWK EAGLE *(Spizaëtus cirrhatus)*, whose L is 51-61 cm, and weight is 1145 to 1800 g. There are four subspecies, and the plumage color is extraordinarily variable. The systematics of other crested eagles of Celebes, the Philippines, and Indonesia is still partially obscure. In the Berlin Zoo a dwarf crested eagle of this species is only 44 cm long and weighs 1200 g. In excitement it waves its tail to and fro in the manner of shrikes. The crested eagles lay one to three eggs, although in the South American species only one egg is incubated.

The South American crested eagle species have adjacent breeding areas. White ornate hawk eagle nests are found below the canopy of the jungle trees, while the black hawk eagle is a bird of open park landscapes. The mountain hawk eagle breeds in mountains at a height of 1500 meters, while the only slightly smaller crested hawk eagle builds its nests in the foothills, as well as the settled country near villages where it often captures chickens. The mountain hawk eagle always has two nests in its territory, using them in alternation. Its prey includes smaller mammals (rats, tree rats, squirrels, rabbits, hares, and field mice) as well as gallinaceous birds from the peacock to the quail, herons, parrots, and large reptiles.

The LONG-CRESTED EAGLE (*Lophoaëtus occipitalis;* Color plate, p. 366) also lives on smaller animals such as rodents, snakes, lizards, frogs, and insects. It is a small, long-crested eagle with the longest crest feathers of all the species related to it. The white areas of the central wing feathers and the yellow iris contrast with its brownish-black plumage. The L is 47–56 cm, and the weight is 850–1100 g. One to two eggs are laid, which are white with red spots, measuring 59.5 x 47.6 mm. The breeding season lasts from March to October.

Fig. 15-57. 1. Long-crested eagle *(Lophoaëtus occipitalis);* . Crowned eagle *(Stephanoaëtus coronatus).*

Its habitat includes forests along rivers and lakes, as well as steppe forests in areas which are not too dry. Its hunting flights are short and, in the manner of goshawks, may be carried out in dense shrubberies. It also sits and waits on trees at woodland edges, searching the ground for young birds, insects, and frogs in the manner of buzzards; it has even been observed eating wild figs. Its nest is quite small and is found in the tops or forks of trees at the forest's edge; the trees are often inhabited by stinging insects (bees and wasps). The young fledge in 65–70 days. They have pale feather edges, and their crests are not as long as those of the adults.

Africa has two large forest eagles which replace one another in the jungle and steppe. They occupy habitats similar to those of the harpy eagle and Guiana crested eagle in South America. The CROWNED EAGLE (*Stephanoaëtus coronatus;* Color plate, p. 366) often called the "African Monkey Eagle," is found in forested areas and isolated woods from Ghana to the Cape. It is long-tailed and short-winged, with a dense barring of the underparts, and yellow iris (the goshawk type). It shows a considerable size difference between the sexes. The L is 69–87.5 cm, and the weight goes from 2820 g (♂) to 4650 g (♀). Two eggs are laid during the breeding season from October to July. It hunts monkeys, particularly guenons, in the rain forests, but it also catches duikers and the calves of larger antelopes such as the reedbuck, as well as hornbills. Its powerful talons can finish off a strong male guenon when the killing grasp is properly placed on its chest and head. The pairs call a great deal; when the male utters its "kee kee week week," the female replies with "koi koi." The incubation period is about fifty days; male young fly after ninety-five days, but females only after 110 days.

Fig. 15-58. 1. Cassin's hawk eagle *(Cassinaëtus africanus);* 2. Martial eagle *(Polemaëtus bellicosus).*

The MARTIAL EAGLE (*Polemaëtus bellicosus;* Color plate, p. 366) is a steppe dweller with longer and stronger claws than the crowned eagle. The L is 72–90 cm, and the weight goes from 3550 g in the ♂ to 4740 g in the ♀. It is short-tailed, and because of its very powerful feet, the martial eagle is the strongest African eagle. Its hunting flight consists of soaring, occasionally interrupted by hovering. It flies high over the tree steppe, but when hunting monitor lizards, gallinaceous birds, hyraxes, baboons, jackals, duikers, and young impalas, it may use a

Fig. 15-59. Bonnelis' eagle *(Hieraaëtus fasciatus).*

Fig. 15-60. 1. Booted eagle *(Hieraaëtus pennatus)*; 2. *Hieraaëtus kienerii*; 3. Little eagle *(Hieraaëtus morphnoides)*; 4. Ayre's hawk eagle *(Hieraaëtus ayresii)*.

Fig. 15-61. Golden eagle *(Aquila chrysaëtos)*.

Fig. 15-62. 1. Imperial eagle *(Aquila heliaca)*; 2. Wahlberg's eagle *(Aquila wahlbergi)*.

Fig. 15-63. 1. Steppe eagle *(Aquila nipalensis)*, subspecies: a) *Aquila nipalensis nipalensis*, b) *Aquila nipalensis rapax*; 2. Gurney's eagle *(Aquila gurneyi)*; 3. Wedge-tailed eagle *(Aquila audax)*.

low-level searching flight. It takes young goats and chickens near settlements. However, the damage it and the crowned eagle do in settled areas, for which both are destroyed, is not great, since it can rarely carry off prey weighing more than three and one half kilograms.

It builds its nest high in a fork in flat crowned trees. It always lays only one egg of 185 grams, while the crowned eagle of similar size lays two eggs weighing 112 grams each. The female incubates, probably unaided, for about forty-five days; the fledging period is about one hundred days.

CASSIN'S HAWK EAGLE (*Cassinaëtus africanus;* the L is 60 cm) is in competition with the crowned eagle and long-tailed goshawk. It has no crest and, in its appearance and behavior, forms a link to Bonneli's eagle and the booted eagle.

BONNELI'S EAGLES (genus *Hieraaëtus)* exceed all other eagles in boldness and speed. In its daring and maneuverability, Bonneli's eagle (*Hieraaëtus fasciatus;* Color plate, p. 366) shows a similarity to the goshawk, but in its swooping flight, with its folded wings, it resembles the large falcons. The L is 66–74 cm, the wingspan is 145–160 cm (♂) to 165–180 cm (♀). The weight ranges from 1600 g (♂) to about 2000 g (♀). It is long-legged and short-winged. The tail projects 5–7 cm beyond the wing tips. It has relatively long claws. The hind toe measures 5–5.4 cm, and the inner toe measures 4.6–5 cm. It breeds in February in southern Europe and lays two eggs. It lives in dry areas of southern Europe and central Asia, while the darker subspecies *(Hieraaëtus fasciatus spilogaster)* is found at the edge of rain forests in Africa.

In hunting, the male and female often cooperate. Flying birds, such as jackdaws and pigeons, are seized on the wing, while medium-sized mammals, such as partridges and red-legged partridges, are seized by a rapid swoop to the ground. In display flight, this eagle plunges down steeply and then rises once more in curved figures. The nest is generally in clefts in rocks or on projections of steep cliffs. In the tropics it is also found on jungle trees. Two nests on the rock of Gibraltar were used in alternation for many years. The incubation period lasts for about forty days. Generally only one young is reared; its fledging period is up to nine weeks.

The BOOTED EAGLE (*Hieraaëtus pennatus;* Color plate, p. 366) occurs in a pale and a darker color phase. The white shoulder patch is characteristic. The L is 46–53 cm, the wingspan ranges from 110–120 cm (♂) to 122–130 cm (♀), and the weight is about 700 g (♂) to 900 g (♀). They lay two whitish, lightly reddish-brown spotted eggs, which measure 56 x 45 mm. For further species of the genus, see Fig. 15-60.

The smaller, buzzard-sized booted eagle is even more mobile than Bonneli's eagle. In display the male swoops down with folded wings, calling "yueg yueg yueg". It flies at great speed through gaps in trees, then rises again and performs loops before the circling female. All of its genus are forest steppe dwellers. Their nesting territories are on wooded mountain slopes and in wooded ravines.

The TRUE EAGLES (genus *Aquila*) live in all parts of the world except South America. The beak is strong and the tarsi are feathered. They have very strong claws and a lancet-shaped head and neck feathers. They prey mainly on small and medium-sized mammals; most species also take carrion. They soar with wing tip feathers spread widely apart, which may be bent upward a great deal by the updraft. As far as is known, up to four species of eagles may occur in the same area. The golden, imperial, and steppe eagles probably originated separately, later acquiring overlapping ranges. There are nine species:

1. The GOLDEN EAGLE (*Aquila chrysaëtos;* Color plates, pp. 327, 328 and 375). The L is 79–95 cm and the wingspan is 195–210 cm in the ♂ and 210–230 cm in the ♀. The weight is 3–4.5 kg in the ♂ and in the ♀ it is 4–6 kg. Two to three eggs of buzzard type are laid during the breeding season, beginning in March and April. There are five subspecies.

2. The IMPERIAL EAGLE (*Aquila heliaca;* Color plate, p. 375). The L is 79–84 cm and the wingspan is 185–200 cm in the ♂ and about 200–215 cm in the ♀. The weight is about 3 kg in the ♂ and about 3.5 kg in the ♀. The plumage of the juveniles is yellow to pale brown. There are two to three chalk-white eggs. Breeding begins in April and May.

3. The STEPPE EAGLE (*Aquila nipalensis;* Color plate, p. 375). The L is 66–79 cm and the wingspan is 165–180 cm in the ♂ and 180–200 cm in the ♀. The weight is 2.5 kg in the ♂ and 3 kg in the ♀. They inhabit steppes and semidesert. They remain on the ground most of the time. The tarsi are short and thick. Juveniles are paler than the illustration, with reddish-yellow feather edges on the wing coverts and secondaries. Two white faintly spotted eggs are laid. Breeding begins in April and May. The southwestern subspecies, the TAWNY EAGLE (*Aquila nipalensis rapax*) is often considered a separate species.

4. The GREATER SPOTTED EAGLE (*Aquila clanga;* Color plate, p. 375) has an L of 66–74 cm. The wingspan is 160–175 cm in the ♂ and 170–185 cm in the ♀. The weight is 1.6–2 kg in the ♂ and 2.1–2.3 kg in the ♀. The nostrils are round. They are very closely related to the next species.

5. The LESSER SPOTTED EAGLE (*Aquila pomarina;* Color plate, p. 375) has an L of 61–66 cm. The wingspan is 145–160 cm in the ♂ and 160–170 cm in the ♀. The weight is 1.5 kg in the ♂ and 1.8 kg in the ♀. The nostrils are slit-shaped. The Indian subspecies *Aquila pomarina hastata* leads over to Wahlberg's eagle.

6. WAHLBERG'S EAGLE (*Aquila wahlbergi*) has a small erectile crest on the back of the head. It is dark brown, but has a light phase with white underparts.

7. VERREAUX'S EAGLE (*Aquila verreauxi;* Color plate, p. 375) has an L of 90 cm. The weight is 3.6–4.2 kg. One to two eggs are laid; the breed-

Fig. 15-64. 1. Greater spotted eagle (*Aquila clanga*); 2. Lesser spotted eagle (*Aquila pomarina*); 3. Verreaux's eagle (*Aquila verreauxi*).

In an open savannah of 360 square km with thornbush and rocky hills (Eagle Hills in Kenya), L. Brown found the nesting territories of many raptors of five species within 10.75 square km. Each species flew over its own hunting grounds or at its own elevation; thus there was no interspecific competition:

Species	Prey
Martial eagle	wild birds, hares, small ungulates
Bonnelis' eagle	small wild birds, rodents
Wahlberg's eagle	rats, snakes, lizards
Long-crested eagle	rats, snakes, lizards
Bateleur eagle	reptiles, carrion

Three species were living and breeding in Kenya in a forest region with barren, rocky areas:

Verreaux's eagle	hyraxes, baboons birds
Crowned eagle	duikers, monkeys, forest birds
Ayre's hawk eagle	small birds, reptiles

Fig. 15-65. Flight silhouette of a golden eagle (*Aquila chrysaëtos*) and the foot of the golden eagle.

ing season lasts all year round. It is a mountain dweller, nesting on cliffs. It hunts hyraxes, baboons, hares, klipspringers, gallinaceous birds, as well as lambs and chickens, and even dares to take remains of leopard's prey. It also eats large beetles.

8. The WEDGE-TAILED EAGLE (*Aquila audax;* separate subgenus *Uraëtus;* Illus. Vol. X) has an L of 89–106 cm. The wingspan is 190–240 cm and the weight is 3.5 kg in the ♂ to 4.95 kg in the ♀. It is brownish-black. The juveniles are a dark red-brown. Two eggs are laid. Breeding begins in August and September.

9. GURNEY'S EAGLE (*Aquila gurneyi*) has an L of 76 cm. It is brownish-black, inhabiting jungles in New Guinea and the Moluccas.

The golden eagle is indisputably one of the most magnificent raptors, even one of the most magnificent birds. It has always been a symbol of nobility and fighting capacity—it is the heraldic eagle. Although slightly exceeded in size by the sea eagle, it is still the strongest raptor of the northern Old and New Worlds because of its mighty talons. Northern subspecies are lighter and have more red-brown tones. They exemplify the golden type. Southern golden eagles have gray to rusty-yellow nape feathers and represent another type of golden eagle. American golden eagles and East Siberian subspecies are darker than West Siberian and Northern European ones. Immatures have a blackish-brown plumage and a white tail with a wide black terminal band. They acquire the adult plumage at six years of age, and are then sexually mature.

Hunting Area
tree steppe
bare hills, 4.5 km away open dry areas open moist areas
wide hunting area (Specialist)
open rocky areas
forests and clearings
treetops in light forests

In many parts of its European and North American range, the golden eagle has been exterminated. In the British Isles it now breeds only in the Scottish highlands, although it has done so again in Northern Ireland since 1953. In Germany there are only about seven pairs in the Bavarian Alps; until around 1890 it still bred in northern Germany on forest trees. In the Alps, as a result of stronger application of nature conservation laws, its population is again increasing somewhat today; in all alpine areas there are now probably about 150 pairs.

In wooded areas it lives mainly in large river valleys and on the edges of forests. It avoids the dense taiga. In mountains it prefers the solitary fissured parts of steep cliffs for nesting, generally below the tree line. It also hunts at higher elevations so that it can often fly downward with heavy prey, facilitating its transport. In northern Africa, Central Asia, California, and Mexico, it also inhabits open stony deserts and ranges of medium elevation which are deficient in rainfall.

When hunting, it has to surprise its prey, which is often very strong, in order to get a suitable grip on it. For hours it searches a certain sector with the greatest exactitude, flying low over the ground, disappearing quickly in depressions, flying out of hollows, and combing through woodlots at treetop height. Prey which is surprised and scared (into motion) is pursued with a sudden swoop and killed in the air or on the ground. It can throw itself on its back, seizing the prey from below. Two eagles often hunt in cooperation, one hunting the prey and chasing it to exhaustion, the other flying along behind, then swooping down and killing the defenseless prey. When golden eagles have struck their talons into large prey, they often let themselves be dragged along for a hundred meters or more. The golden eagle hunts mainly marmots, as well as kids of the chamois and roe deer, also snow hares, grouse, foxes, and martens, particularly when rearing young. Stig Wesslen, who spent many months in northern Lapland, observed in the "Valley of the Royal Eagle," how eagles took ptarmigan out of snares in winter. They picked out the still warm, just choked birds, taking the frozen ones much less often.

Fig. 15-66. The "bold" look of eagles is created by a peculiarity of bony structure. An additional bone from the upper bony margin of the orbit extends backward and outward. It forms a flat ledge projecting as an eyebrow edge above the eye. By muscle contraction, man can similarly project the eyebrows and does so as an expression of bold decisiveness. Our inborn understanding of this expression causes us to misunderstand the unalterable "eagle look" of the bird as a sign of boldness.

A pair of golden eagles remains in its range most of the time and the partners stay together for a lifetime. The nest is placed in niches or ledges of cliff walls, often below a protecting overhang. In mountain areas, however, such as the taiga, there are tree nests. Every pair has several nests, which it uses in alternation. In courtship the golden eagle utters a fluting "giooeh;" otherwise its call is harsh "eak eak." The female does most of the incubating—incubation lasts forty-three to forty-five days. Only about 30% of the time do both newly hatched young reach maturity. They can fly at seventy-five to eighty days of age and remain over the winter with their parents, often longer when the parents do not breed the following year.

The golden eagle can kill adult roe deer or chamois only in winter. Since eagles remove mangy chamois, they contribute to the preservation of healthy members of this species. In Sweden, golden eagles only survive in areas where reindeer are reared, and then they take only the sick or dead animals. In 1941 there were still eighty-four pairs in the European north. Among the Khirgiz and Kasaks, the Berkut, as they call this eagle, is a much appreciated hunting bird. It kills foxes, roes, antelopes, and even wolves. A trained golden eagle is worth as much as two horses or camels in their areas.

The imperial eagle has moved into the park-like wooded landscapes from the Mediterranean countries to western Mongolia. Its talons are weaker than those of the golden eagle and it is smaller, plumper, and clumsier.

It is one of the chief predators of ground squirrels. It also catches rats, hamsters, and rabbits. Only rarely does it attack birds, but it regularly takes carrion. Its large nest, visible from afar, stands in the isolated wooded areas of the steppes and foothills, in the woods near rivers, or in isolated trees. In Bulgaria, it nested formerly even along country roads and in villages. The young fly nine to ten weeks after birth and often migrate in family units to Central Africa, Southwest Asia, and India. As a vagrant from Hungary, it has often reached Austria and Germany.

The steppe and tawny eagles replace the golden eagle in their ranges. In the highlands of Central Asia, they live on alpine meadows, deserts, and semideserts. North of the Caspian, the steppe eagle feeds mainly on a small ground squirrel *(Citellus pygmaeus)*. The eagles arrive in their breeding range almost at the same time as the ground squirrels come out of hibernation, at the end of March. Their often large nests are on tamarisk shrubs, sand and rubble slopes, and even on the Kasak burial mounds. In years of good food supply, they rear two young. Steppe eagles, which winter in the breeding area, feed almost entirely on carrion during this season. As late as April and May we saw dozens of steppe eagles on the corpses of dead horses. The steppe eagle is "useful" since it kills ground squirrels, which are known as carriers of certain diseases. It also pursues whistling hares and kills masses of grasshoppers during their flying season. The tawny eagle breeds in Africa on acacias and monkey-bread trees *(Baobab)*. It often scavenges from the martial eagle and, together with vultures, gathers about animal corpses.

The habitat of the greater and lesser spotted eagles consists of moist woods in swampy areas and areas rich in water, as well as lower mountain areas. An ochre-yellow color phase of the greater spotted eagle has often been found in Central Asia. The greater spotted eagle kills ducks and coots, and also catches fish and young waterfowl, and scavenges food from the sea eagles. In the breeding season, it lives mainly on frogs, reptiles, and small mammals, like the smaller species. The lesser spotted eagle hunts these mostly on foot, and so is not easily observed in the tall grass. V. Wendland found that the first hatched youngster in this species regularly lies on the smaller one, which cannot be fed and so perishes. However, in the nest of the greater spotted eagle, the two young are generally reared. In begging flight they utter their loud, long drawn calls, which at that stage show no similarity to the sonorous calls of the adults. Both species fly over the Bosphorus in flocks on migration. Their winter quarters are in the tropics of Africa and South Asia.

The wedge-tailed eagle fulfills the functions of the vulture, which does not occur in Australia. Since the settlement of whites in the country, its numbers have increased, because the large herds of cattle have provided additional food for it. As a result, the wedge-tailed eagle has largely developed into a carrion eater. John Gould (1804-1881) once saw forty of these raptors sitting about the corpse of a large ox. It also kills small kangaroos, bustards, and occasionally sheep. Because of this, it is unfortunately still vigorously pursued in the sheep grazing area of western Australia.

In the years 1928 to 1948, bounties of two, and later five shillings, were paid on the heads of 94,090 wedge-tailed eagles, mainly from the Kimberley district. About 5000 eagles were killed there annually, in

F. Reimann

Migrating birds on Lake Müritz (Germany)

Grebes: 1. Little grebe *(Podiceps ruficollis)*. □ Herons: 2. Bittern *(Botaurus stellaris)*. □ Ducks and Geese: 3. Whooper swan *(Cygnus cygnus cygnus)*. 4. Bewick's swan *(Cygnus columbianus bewickii)*. 5. White-fronted goose *(Anser albifrons)*. 6. Bean goose *(Anser fabalis fabalis)*. 7. Barnacle goose *(Branta leucopsis)*. 8. Wigeon *(Anas penelope)*. 9. Garganey *(Anas querquedula)*. 10. Golden-eye *(Bucephala clangula)*. 11. Common pochard *(Aythya ferina)*. □ Raptors: 12. White-tailed sea eagle *(Haliaeëtus albicilla)*. 13. Osprey *(Pandion haliaëtus)*. 14. Black kite *(Milvus migrans)*. □ Cranes: 15. Common crane *(Grus grus*, Vol. VIII) □ Charadriiformes: 16. Lapwing *(Vanellus vanellus)*. 17. Black-tailed godwit *(Limosa limosa)*. 18. Common snipe *(Gallinago gallinago)*. 19. Oyster catcher *(Haematopus ostralegus)*. 20. Common sandpiper *(Actitis hypoleucos)*. 21. Little ringed plover *(Charadrius dubius)*. 22. Common tern *(Sterna hirundo)*.

spite of the fact that they live mainly on refuse and rabbits. It often manages to get the prey of foxes, dingos, and other raptors, too. The mass destruction encouraged by bounties still continues. If an end is not put to it soon, the extinction of this eagle is only a matter of time.

Fig. 15-67. Indian black eagle *(Ictinaëtus malayensis).*

The lightly built INDIAN BLACK EAGLE (*Ictinaëtus malayensis;* Color plate p. 375, Fig. 15-67), with an L of 58.5–75 cm, is related to the larger eagles. It lays one to two eggs, and the breeding season is between November and March in southern India.

It uses its long rear toe as we use an ice pick, penetrating egg shells with it and then opening them with its beak. This very long-winged raptor flies in elegant, harrier-like search flight among the treetops. To see it from above in its flight through mountain ravines is a wonderful experience. In the tree tops it catches frogs, reptiles, and large insects. In Java, it catches flying foxes as well. When plundering bird nests, it does not spare the young. It also feeds on wounded pheasants and other wild gallinaceous birds.

Another group of buzzard-like birds which may, however, be more closely related to the kites, is the SEA EAGLES (genus *Haliaeëtus*). They are mighty birds with very strong beaks which are slightly curved upwards on and in front of the cere. The feet are strong, and are feathered down to one third of the tarsus. The toes have strong balls which facilitate the grasping of smooth prey, fish, and birds. They have very wide wings which give the effect of straight boards in flight. They have a pale, generally white tail, and are often white on the head, neck, breast, and bend of the wing. They call readily. All the species have high-pitched call sequences as well as shrill and vibrating gull-like calls. When calling, they stretch the head forward and, in the course of the sequence of calls, bring it back so that it almost touches the back. They are characteristic birds of the seashore and large inland lakes. There are eight species in suitable habitats in all continents except South America.

1. The WHITE-TAILED SEA EAGLE (*Haliaeëtus albicilla;* Color plate p. 376 and Color plate pp. 385/386), with an L of 77-95 cm, has a wingspan of 210–230 cm in the ♂ and 225–255 cm in the ♀. The weight is 3–4.6 kg in the ♂ and 4.3–6.7 kg in the ♀. The tail is wedge-shaped. Two to three white eggs, measuring 73 x 58 mm, are laid. Breeding begins in February or March.

2. The BALD EAGLE (*Haliaeëtus leucocephalus;* Color plate p. 376) has an L of 68–76 cm. The wingspan is 188–197 cm in the ♂ and 211 cm in the ♀. The weight is 4.1 kg in the ♂ and 5.84 kg in the ♀. They are slimmer and more mobile than the white-tailed sea eagle, and the tail is rounded. Two to three white eggs are laid, and the breeding season lasts from December to April.

3. STELLER'S SEA EAGLE (*Haliaeëtus pelagicus;* Color plate p. 376) has an L of 110 cm. The tail reaches 40 cm, and the wingspan may be up

to 280 cm. The weight of the ♂ is 5.25-6.84 kg, and may reach 8.97 kg in the ♀. The bill is massive, and often may be even higher than in the monkey-eating eagle. There are fourteen retrices (other sea eagles have twelve). The toes have longer claws but are not any stronger than in the other species. One to three white eggs, measuring 80 x 60 mm, are laid. This bird begins to breed in April or May. The North Korean subspecies *Haliaeëtus pelagicus niger,* which is black except for its white tail, is probably extinct.

Fig. 15-68. 1. White-tailed sea eagle *(Haliaeëtus albicilla)*; 2. Pallas' sea eagle *(Haliaeëtus leucoryphus)*; 3. African fish eagle *(Haliaeëtus vocifer)*; 4. Madagascar fish eagle *(Haliaeëtus vociferoides)*.

4. PALLAS' SEA EAGLE (*Haliaeëtus leucoryphus;* Color plate p. 376) has an L of only 68.5-78 cm.

5. The WHITE-BELLIED SEA EAGLE (*Haliaeëtus leucogaster;* Color plate p. 376) has an L of 63.5-69.5 cm.

6. SANFORD'S SEA EAGLE *(Haliaeëtus sanfordi)* has a dark plumage and so may be assumed at first to be the young of the above. Both species are skillful catchers of fish, and also hunt sea snakes, flying foxes while they are sleeping, and phalangers.

7. The AFRICAN FISH EAGLE (*Haliaeëtus vocifer;* Color plate p. 376) has an L of 61-72 cm and a wingspan of 168 cm in the ♂. The weight reaches 2.78 kg in the ♂. Two to three white eggs are laid; according to latitude, it breeds all year round.

Fig. 15-69. Bald eagle *(Haliaeëtus leucocephalus)*.

8. The MADAGASCAR FISH EAGLE *(Haliaeëtus vociferoides)* is darker with a white tail. It lives on the sea coast, but otherwise its habits are like the above.

The white-tailed sea eagle can hardly be confused with any other raptor because of its striking size, its mighty yellow beak, and the short, wedge-shaped tail, which is white in maturity. Because of its unusually large range, in the northern parts of the Old World, it has adapted to all possible habitats. Many pairs still inhabit the large forests of the plain and coastal areas of the Baltic in Germany. In Western Europe, however, it has been exterminated. Though mainly a coastal bird, it breeds on the tundra and forest tundra, in the taiga, in the steppes, and in low mountain ranges. It is confined everywhere to large waters, rivers, natural lakes, and dams, those being the localities where it finds sufficient food. Its nests, which are often huge, stand in trees, on coastal cliffs, in reed beds, in bushes, or on the ground. On water, sea eagles hunt waterfowl and above all fish. The adult pairs' long hunts for surface-feeding and diving ducks, mergansers, and loons have often been described. With slow wingbeats one of the birds follows the waterfowl, which has dived, while the other flies in a semicircle, keeping its eyes on the water surface and swooping in at the moment when the pursued bird reemerges.

Fig. 15-70. White-bellied sea eagle *(Haliaeëtus leucogaster)*; 2. Steller's sea eagle *(Haliaeëtus pelagicus)*; 3. Sanford's sea eagle *(Haliaeëtus sanfordi)*.

The white-tailed sea eagle's chief prey is the coot, but it also appears over reedy bays looking for young geese or moulting ducks to seize by surprise. It hovers and then swoops down upon fish. At the time of the pike's spawning in spring, it gathers in troops over the spawning

Fig. 15-71. Flight silhouette of a white-tailed sea eagle *(Haliaeëtus albicilla)*.

places. On long search flights along the seashore, it looks for stranded dead fish, dead birds, mussels, and crabs. The summer and winter diet may be different. Thus according to Scheel, Greenlandic sea eagles mostly feed on salmon in summer, while in the winter they feed on sea birds, particularly eiders. In winter, carrion and all sorts of refuse may temporarily satisfy sea eagles. It, as well as the steppe eagle, is regularly found where such things accumulate along the Caspian Sea.

The incubation period varies between thirty-five and forty days. Often one youngster dies and may be eaten if there is a food shortage, so that only fifty percent of the time are two or, even more rarely, three young raised. In western Siberia, this eagle is highly respected by the natives. The Kasym Ostyaks even regard it as holy. The Soviet fishermen on the Ob also protect it, since it removes the dead fish from their storage tanks so that they may keep the other fish alive until fall. In many other areas, however, it has been greatly persecuted, and was brought to the verge of extinction in Sweden. It is to the merit of Bengt Berg that, in his rousing book, *The Last Eagles*, he awakened the conscience of European conservationists. Since then, special protective laws have been passed in Bengt Berg's homeland and in the Central European states, while those protective laws already existing have been applied with more emphasis. Since then the sea eagle population of our latitudes can probably be considered saved.

The bald eagle is the heraldic bird of the United States. Until the last few decades it was to be found regularly throughout its whole North American range. Unfortunately, it has more recently acquired a sad fame. It was the first bird in which traces of poisonous pesticides were found in the brain and in the eggs. Probably as a result of pesticides, whole local populations of the bald eagle, such as in Florida, have broken down within a few years.

This agile sea eagle is able to catch flying birds and even flying fish. It is also a food parasite which forces turkey vultures to disgorge swallowed lumps of meat and even kills them if they do not do so quickly enough. Young bald eagles wander long distances. Many of those hatched in Florida spend the succeeding summer in southern Canada. Although its populations are declining, bounties are still paid on bald eagles in Alaska, since they are reputed to damage the salmon harvest. Between 1922 and 1940, 103,454 bald eagles were shot there, and in 1949/50, another 7455 were shot. In view of this mortality rate, the species cannot survive for long even in those remote areas, particularly if the persecution continues.

Fig. 15-72. Foot of the white-tailed sea eagle with unfeathered tarsus, (cf. foot of the golden eagle with feathered tarsus, see fig. 15-65.)

A bald eagle called "Old Abe" became famous as the mascot of the Eighth Wisconsin Regiment in the Civil War of 1861 to 1865. It carried fesses and was highly respected; it was even known to the Confederates as the "Yankee Buzzard."

As an inhabitant of barren coasts, Steller's sea eagle spends much time on the ground and on rocks. Apart from very large salmon, it takes hares, foxes, and, as has been established, small dolphins and young seals. It also eats much animal matter cast up on the seashores, and temporarily may even be restricted to such refuse. Sea eagles which have clawed themselves into large pike or carp have, according to the testimony of reliable witnesses, often been dragged down into the depths by their prey. The display flights of this eagle often take it far out to sea. In Kamchatka, nests are usually on trees. Further north they are on cliffs of the coast, while on Sakkhalin they may be in strong bushes or in low trees.

Fig. 15-73. 1. Gray-headed fishing eagle *(Icthyophaga ichthyaëtus)*; 2. Lesser fishing eagle *(Icthyophaga nana)*.

Pallas' sea eagle represents the smaller black-beaked eagle species in the high steppes and deserts of Central Asia. In part of its range it meets the white-tailed sea eagle, but generally their habitats are separate. Pallas' sea eagle also replaces the steppe eagle in areas made swampy by bends in the rivers. More recently it has moved into the steppes beyond the Volga, into northern India and Burma, and has occurred several times as a vagrant in Central Europe (Silesia, Hungary).

In the summer of 1962 on the Mongolian desert lakes Orog-Nur and Buncagan-Nur, we found Pallas' eagles of all ages. They often sat for half a day on peninsulas and on the tops of dunes, especially mornings and late afternoons when they flew along the shore in search of food. They were skillful at catching fish, which they seized often with only one foot in an elegant, swooping flight. They also settled beside dead fish on the shore, and probably often took ducks, which were abundant on the lakes, as well. Upon the appearance of a circling eagle, all the waterfowl took wing. The eagle nests, built of branches and reeds, stand on the ground, on dunes, on rocks, or in low trees and shrubs.

The tropical forms of the sea eagle group, like the white-bellied sea eagle and above all the magnificent African fish eagle, are temperamental, often vociferous, and very agile birds. Their five-syllabled, high-pitched calls can be heard over great distances. The female replies in a lower octave. These eagles call not only in flight but also when they are perched on branches or coastal coral reefs. Above all, they eat fish, and so the English consider them to be real "fish eagles." Their attacks on prey are flat and quick; they look very elegant. In the Sudan this eagle specializes in catching sheatfish, *(Synodontis schal)*, which is the bulk of its prey.

It tends to chase other fishing birds, such as pelicans and cormorants to make them drop their prey. The eagle, in its "restless, fighting spirit," attacks the large goliath heron *(Ardea goliath)*, often gripping it around the neck until the heron disgorges its food. The fish obtained in this manner are, strangely enough, often dropped again, very little having been eaten. Natives then pick up such fish and so become beneficiaries of the eagles' hunting.

Of the two species of fishing eagles, (genus *Icthyophaga*), the GRAY-HEADED FISHING EAGLE (*Ichthyophaga ichthyaëtus*; Color plate p. 376) is the larger. The L is 66–75 cm. The tarsi are long and bare. Horny tubes are found beneath the soles of the feet, similar to those of the OSPREY (*Pandion*), which facilitate the gripping of smooth fish. Two or three eggs are laid. The breeding season in Ceylon lasts from December to March.

On the lakes, pools, and rivers of Indonesia, this species replaces the white-bellied sea eagle, which only lives in coastal areas. It catches not only fish, but also reptiles, small aquatic mammals, and birds. Its nests are found in the woods of nearly every area which has pools. Where it is not pursued, it may live close to man, like the osprey in America. Near Calcutta there is a nest on a mango tree in the midst of a settlement that is used annually. A second smaller species (*Ichthyophaga nana*; the L is 54–61 cm) lives in the same area but not as far as the Sunda Islands.

Subfamily: Old World vultures by W. Fischer

Old World vultures (subfamily Aegypiinae) are not related to the New World vultures, but are placed in the family Accipitridae near the buzzards, kites, eagles, and harriers by the new systematics. Man has been familiar with the European and North American Old World vultures since the Miocene (about twenty million years ago). They disappeared from America, where eight species had been found, and the smaller black, vulture-like ancestors of the modern New World vultures took their place. The extinct genus *Neogyps*, apparently one of the original forms of the Old World vultures, resembled the present day wattled vultures to a great extent. Man has been familiar with the genera *Torgos, Gyps, Gypaëtus* since the Pleistocene (one and one half million to 20,000 years ago). The mighty malta vulture (*Gyps melitensis*) was much larger than the present day European black vulture. In the late Ice Age deposits near Salzgitter-Lebenstedt in Germany, a fossil EARED VULTURE (*Torgos tacheliotus todei*) was found which was described by Adolf Kleinschmidt. Today this species is confined to Africa. There are at present ten genera of Old World vultures with sixteen species.

Like those vultures of the New World, they have become adapted to eating the corpses of larger animals. This requires a long, hooked beak to tear open skins and abdominal walls. Since some species feed particularly on the viscera of dead animals, which soils their heads and necks, these parts are only scantily covered with feathers and down. In the griffon vultures (genus *Gyps*), the neck has seventeen vertebrae and may be forty centimeters long when extended. A neck ruff, which is sometimes collar-like, prevents soiling of the plumage. The toes of vultures are hardly suitable for seizing prey, but make most of them good pedestrians.

All vultures, in spite of their apparently "unappetizing" manner of feeding, are quite clean birds which bathe a great deal. Their voice is not varied; one generally only hears a low-pitched cackling and hissing from them. Griffon vultures in courtship utter a vibrating grunting and moaning. At the feeding place they emit a peculiar "keh keh" call. The Egyptian vulture utters a lively call resembling the trilling of kites, and it grunts at the nest. The bearded vulture has a penetrating whistling tone, which is sounded with particular strength in the mating season.

Since vultures mainly feed on the corpses of large mammals, they do not live in places like Madagascar, Ceylon, the Malayan Islands or Australia where there are no herd-forming mammals of the open country. Their habitat is open country which they fly over in high searching flight, often with other members of the same and related species, circling slowly. They observe the ground with their sharp eyes shadowing herds of game and cattle, hunting safaris, war parties, and also keeping watch on running hyenas, jackals, and flying ravens, which often lead to a source of food. When a vulture sees a corpse, it comes down in a characteristic manner. All the other vultures circling within view notice this and come down also, so that in no time there are fifty vultures or more collected at a place where previously there were none (see Color plate p. 333). Vultures probably never attack dying animals. Instead they perch nearby until all signs of life are gone.

The distensible crop and gizzard of vultures can hold large amounts of meat. Since opportunities do not rise daily, a meal has to last for several days. Vultures can voluntarily fast for a week. They can eat decaying meat which contains poisonous substances fatal to other animals. The gland system of their proventriculus secretes abundant digestive juices which make poisonous products harmless through decomposition. Only the PALM-NUT VULTURE *(Gypohierax)* and the BEARDED VULTURE *(Gypaëtus)* have a different manner of feeding.

In their first few days the young are fed with predigested "rearing broth" derived from carrion. Later the parents bring food in the crop and regurgitate it at the nest. Young vultures, particularly griffons, do not need to be fed daily. Since vultures have no flying enemies, preservation of the species is assured by the laying of one or two eggs, although usually only one youngster hatches and grows up.

Vultures have been revered in ancient cultures as embodiments of immortality and of the transmigration of souls. On the temples of ancient Egypt, representations of the long-necked griffon vulture, as well as of the Egyptian vulture, which symbolized parental love, as Pharaoh's hen, can be seen. Tibetan lamas used to dismember corpses and offer the pieces to vultures. The bones were picked up by the bearded vulture, which, as the Gourral, is holy to Tibetan Buddhists. According to myth it keeps up the eternal light during the night on a steep cliff and so guides pilgrims. On the holy Ganges and on the "towers of silence" in Bombay, the Hindus and Parsees make sure that vultures will consume the corruptible matter of man and so bring his spirit to a "new life."

Pondicherry vulture

A marked formation of cheek wattles distinguishes the PONDICHERRY VULTURE (*Sarcogyps calvus;* Color plate p. 397). The L is 68–85 cm, the weight is 4.7 kg in the ♂ and 5.4 kg in the ♀. The head has long pendant wattles. The juvenile plumage is brown, and the head and neck are covered with dark down. The ♂ has a light iris, and the ♀ a dark one. One white egg is laid between December and April.

This bird is a definite inhabitant of the tropics. It lives in the woods of high plateaus, and up to an elevation of 1700 meters. However, it does not avoid the vicinity of settlements. At feeding places it appears only in twos or threes, chasing away other vultures. Therefore, it was called the king vulture by the British in India. In contrast to this hostility is the submission gesture described by Otto Antonius: "When one of these vultures hangs its head down, with its neck slack, and sees no other way out, it lets anything happen. In this posture it can easily be caught."

Black and lappet-faced vulture

The strongest Old World vultures are the EUROPEAN BLACK or CINEREOUS VULTURE *(Aegypius monachus)* and the LAPPET-FACED VULTURE *(Torgos tracheliotus).* The downy plumage of both is prominent; on the underside of the lappet-faced vulture it is supported only by isolated, lancet-shaped feathers. The L of the European black vulture reaches 103 cm; the wingspan is 265–287 cm, and the weight is about 8–14 kg. One white, sometimes reddish spotted egg is laid. The breeding season lasts from February to April. It has a strong beak, a wide head, and a neck in the shape of a "monk's collar," which protects the front and the rear of the neck. The underlying plumage is downy. It is a cold-resistant bird. The color is dark brown to black. White areas are often found on the wing coverts. The cere is bluish. In juvenile plumage there is a black face mask of down. The L of the lappet-faced vulture is about 100 cm, the wingspan is 270–280 cm, and the weight is about 8 kg. A single cream-colored, brown spotted egg is laid. The breeding season lasts from September to April. The beak is strong. There are naked skin folds on the head and neck. Only the South African birds have a collar-like fold on the side of the head. The underparts have a grayish-white down with isolated lancet feathers (a tropical bird). The bare parts are pale pink to reddish-blue.

Fig. 15-74. 1. European black vulture *(Aegypius monachus);* 2. Pondicherry vulture *(Sarcogyps calvus);* 3. Lappet-faced vulture *(Torgos tracheliotus,* only rarely in the Sahara).

The European black or cinereous vulture inhabits the Mediterranean countries, as well as West and Central Asia. It is a characteristic bird of the Mongolian and Tibetian upland steppes and semideserts. As the forests in Morocco and the rest of North Africa disappear, it is replaced by the lappet-faced vulture. This largest vulture of African semideserts and steppes is absent only in East Africa. In the south it lives between the Zambesi and Orange rivers. Though both species have penetrated semideserts, they are "conservative" in their nesting habits, for they breed in trees wherever possible. Only in the Gobi desert did we find a vulture nest on a rock; Suschkin has seen the same in northwestern Mongolia. In Andalusia (southern Spain) the huge nests of the cinereous vulture are visible from afar in the tops of evergreen oaks. Both species occur in high mountains; the lappet-faced vulture at a height of 4000 meters in Ethiopia, and the European black vulture at 4500 meters in the Tibetan Wild Yak steppe. This is the real home of the European black vulture, which is not common anywhere else and is only seen singly or in small troops of three to five birds at feeding places. However, in Tibet, as Ernst Schäfer reports, it chases off Himalayan griffon and bearded vultures and even golden eagles with wingbeats and blows with the beak. Several of these vultures can even keep wolves at bay. They are among the most unsociable, cantankerous, and shy vultures. When disturbed from an animal corpse, they do not return at once, although Himalayan griffons may have begun to eat there again in the meantime.

Fig. 15-75. White-headed vulture *(Trigonoceps occipitalis).*

It prefers the meat of dead animals, but also takes many small bones and particularly bits of skin. Its strong beak can readily gnaw flesh off the large bones, even when the corpse has dried out. This way of feeding causes it to make pellets of hair or feathers and to disgorge them while the bones consumed are digested completely. It kills living animals regularly in the summer months when there are fewer carrion. Marmots, hares, lambs, tortoises, and lizards are included among its prey. The lappet-faced vulture is also agile enough to catch live animals. Lutz Heck saw one "cooperate" with two fan-tailed ravens in killing a hare in the highlands of Ethiopia. The lappet-faced vultures take precedence over all the other species at the carrion site. They open the abdonimal cavity with vigorous blows of their beak and devour the intestines as well as the muscle meat. Only the marabous are competitors, when they appear at the feeding place with their huge beaks, which command respect.

In the Halle Zoo, a male griffon vulture mated with a female cinereous vulture in 1928. The hybrid had the stronger body and the dark color of its mother. Reminiscent of its father was the pale head and neck, an incomplete ruff, and the pale color of the thighs. As a chick, it weighed 152 grams.

The white-headed vulture

The WHITE-HEADED VULTURE *(Trigonoceps occipitalis)* is distinguished by its white, down-covered, woolly head cap. The L is 68–76 cm, the wingspan reaches 210 cm, and the weight is 4.8 kg in the ♂ to 5.8 kg in the ♀. One egg, white with brown spots, is laid. The breeding season lasts from October to August. The general color is a dark blackish-brown; the secondaries, belly, and thighs are white. The cere is blue, and the beak and legs are red. The juveniles are a uniform black-brown.

Outside the breeding season, the white-headed vulture lives in open savannahs from the Sudan to Zululand; otherwise it lives in pairs or family units. It also feeds on carrion, but quickly learns to utilize other food sources as well. Thus these vultures watch hunters and follow wounded small game even into dense bush, killing it there. From an eagle-like hunting flight, it swoops at duikers, young ostriches, and half grown bustards, killing them with blows of the beak. Their nests are predominantly in the tops of acacias, but also may be found on cliffs or river valleys.

The griffon vultures

The GRIFFONS (genus *Gyps*) are distinguished by a slender, relatively weak beak and a long goose-like neck of even thickness. The tail is rounded, in contrast to most other genera. There are fourteen rectrices. The ruff is found only on the back of the neck and consists of dense quill and down feathers. The juvenile plumage consists of lancet-shaped feathers, also.

Fig. 15-76. 1. Griffon vulture *(Gyps fulvus)*; 2. Rüppell's griffon vulture *(Gyps rueppelli)*; 3. Cape vulture *(Gyps coprotheres)*.

There are six species: 1. The GRIFFON VULTURE (*Gyps fulvus;* Color plate p. 397) has an L of 100 cm. The sexes are almost identical. The wingspan reaches 240 cm, and the weight is 6.8-8.2 kg. A single white egg is laid. The breeding season lasts from January to March. A brown to cinnamon-red subspecies *(Gyps fulvus fulvescens)* is found from Afganistan to northern India. 2. The CAPE VULTURE *(Gyps coprotheres)* weighs 7.5-8.4 kg. Its breeding season lasts from May to July. 3. The HIMALAYAN GRIFFON VULTURE (*Gyps himalayensis;* Color plate p. 397) is larger and paler. 4. RÜPPELL'S GRIFFON VULTURE *(Gyps rueppelli)* weighs about 8 kg. Its back has pale feather edges which give the effect of barring. 5. The LONG-BILLED VULTURE *(Gyps indicus)* has a dark, almost bare head. 6. The SLENDER-BILLED VULTURE *(Gyps tenuirostris)* is found in India and Malaya. The long-billed vulture and the slender-billed vulture are often considered to be one species.

The habitats of the griffon are barren areas with few trees, preferably mountain steppes and high plateaus. It usually breeds high in the mountains where there are cliffs and deep ravines. Its main food consists of large, dead mammals of which it takes only the flesh and intestines. It leaves the bones and viscera to the Egyptian and bearded vultures. This vulture cannot survive in areas where such animal corpses are removed by man, as in Central Europe. This is also true of the cape vulture in South Africa, which is, moreover, greatly persecuted.

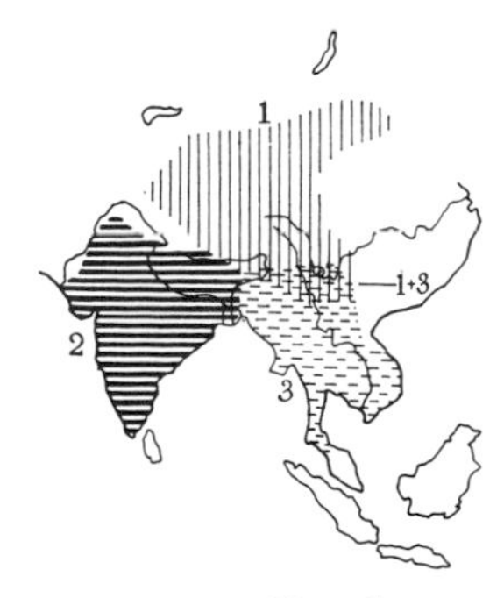

Fig. 15-77. 1. Himalayan griffon vulture *(Gyps himalayensis)*; 2. Indian griffon vulture *(Gyps indicus)*; 3. Slender-billed griffon vulture *(Gyps tenuirostris)*.

In earlier periods, however, man made it possible for griffons to migrate seasonally towards the north. Ernst Schüz has shown on the basis of finds of bones from Swabia, the griffon vulture must have been a breeding bird there up until the eighteenth century. After the Thirty-Year War, there was considerable sheep rearing in this area. This coincided with an improvement of the climate, and the herds were kept in the open except during periods of deep snow. Hence the vultures could live by eating dead sheep from early spring onward; this enabled them to breed on the Danube rocks. The Dominican naturalist, Albertus Magnus (1193-1280), reports that even in the thirteenth century, numerous griffons inhabited cavities in rockfalls in the Donnersberg, Hochwald and Hunsrück mountain ranges between Worms and Trier in Germany. Even today griffons appear annually in the spring in the High Tatra Mountains in Czechoslovakia, induced by the presence of sheep. However, they cannot breed there, for at the beginning of their breeding season in February, the sheep have not yet been driven to the high meadows.

A flock of the larger, paler Himalayan griffon vultures can consume a dead yak down to the skeleton in two hours. Ernst Schäfer found completely consumed yak corpses with skin still intact except for the entry hole in the belly. In the steppes inhabited by nomads, these vultures eat the placentas of yaks and sheep. The flight feathers of this vulture produce a humming sound during quick downbeats of the wing, which can be heard up to one hundred meters away. In their flights for food, both species of griffons often move sixty to one hundred kilometers in a straight line from their breeding places. They seek food only in the open; on search flights they avoid wooded areas.

The nests of these species are always placed on cliffs, even when trees are available. Nests are built of twigs and grass, and are found in many different situations, from well-protected rock caves to exposed ledges. In the Pyrenees, there was a breeding colony on a cliff with a southwest face, according to Terrasse and Boudoint. One group had three nests which were fifty meters apart from one another. About one hundred meters higher up, six nests stood fairly close together. Francois Levaillant (1753-1824) described "huge colonies" of the cape vulture in Cape Province, around 1820, which nested close together like rooks or gulls, often two to three pairs in one rock cave.

After a good snowfall in Central Europe, the griffons and Himalayan griffons, which breed in the flight cages of zoos in February and March, may get completely snow covered. Director Karl Max Schneider of the Leipzig Zoo reports that the breeding cleft and the nest looked like a snow heap, from which only the black wing tips of the incubating bird projected. Partners relieve one another only at two-day intervals, because the partner who is off the eggs may often have difficulties in finding food and so needs some time to do so. In the Münster Zoo, Ludwig Zukowsky found an extraordinarily long incubation period of sixty-five days in a griffon vulture. It is usually fifty-four days. The Münster chick fledged after two and one half months and was self-supporting after about one hundred days. From then on it would take meat from the ground together with other raptors.

The East African Rüppell's vulture forms a link between the griffon and cape vulture. Its nests are on trees and rocks. At Mount Marsabit in Kenya, they were at their breeding places from May to September.

▷ Old World vultures: 1. Himalayan griffon vulture *(Gyps himalayensis)*; 2. Griffon vulture *(Gyps fulvus)*; 3. Hooded vulture *(Necrosyrtes monachus)*; 4. Indian white-backed vulture *(Pseudogyps bengalensis)*; 5. Indian black vulture *(Sarcogyps calvus)*; 6. Palm nut vulture *(Gypohierax angolensis)*; 7. Bearded vulture *(Gypaëtus barbatus)*.

1
2
3
4
5
6
7
F. Reimann

1♂
3♀
2
1♀
6
4♀
♂
9
5
10
7
8

Fig. 15-78. 1. African white-backed vulture *(Pseudogyps africanus)*; 2. Indian white-backed vulture *(Pseudogyps bengalensis)*.

The slender-billed vulture ranges furthest to the southeast of all the species. In Arakan it is a common breeding bird and often nests in loose colonies of four to six pairs. Such colonies are found in trees. In the mountains, however, the southeast Asiatic species, particularly the long-billed vulture, are exclusively rock breeders.

The genus *Pseudogyps* includes the bulk of the tropical Old World carrion birds. They are about three-fifths the size of griffons. Normally they have twelve tail feathers and strongly outbulging cutting edges on the upper mandibles. There are two species: 1. The INDIAN WHITE-BACKED VULTURE (*Pseudogyps bengalensis;* Color plate p. 397). The L is 90 cm, the wingspan is 210 cm, and the weight is 4.5-5.2 kg. A single brown spotted egg is laid during the breeding season, which lasts from October to March. 2. The AFRICAN WHITE-BACKED VULTURE (*Pseudogyps africanus;* Color plates, pp. 329 and 333). The L is 95 cm, the wingspan reaches 215 cm, and the weight is 4.7-6.75 kg. They are pale and have more tawny white marks. One reddish-brown to lilac spotted egg is laid. It breeds from August to April.

These "sanitation" birds are found in the wild, in settlements, and in large cities as carrion removers. In Bombay, they sit by the hundreds on the "tower of silence" and devour the corpses of Parsees, whose religion forbids burying, cremation, or putting corpses in the water. In Africa, the white-backed vulture is characteristic of the steppes, which are teeming with game. Many of these small vultures can satiate themselves on the corpse of one large animal. At such meals they often disappear almost completely into the body cavity of large corpses. They often fight tenaciously with their competitors, hyenas and jackals. Schillings reports of an elephant corpse on which hundreds of vultures were feeding. In spite of this, two spotted hyenas emerged in the daytime from the body cavity where they had been feeding. Granvick reports on a case of nest defense by an African white-back: a male tried with outspread wings and beak open, ready to deliver blows, to defend a well-incubated egg against a man collecting eggs.

◁
Harriers: 1. Hen harrier *(Circus cyaneus)*; 2. Marsh harrier *(Circus aeruginosus)*; 3. Pallid harrier *(Circus macrourus)*; 4. Montagu's harrier *(Circus pygargus)*; 5. Pied harrier *(Circus melanoleucus)*; 6. Crane hawk *(Geranospiza caerulescens)*; 7. Madagascar harrier hawk *(Polyboroides radiatus)*. Serpent eagles: 8. Bateleur *(Terathopius ecaudatus)*; 9. Short-toed eagle *(Circaëtus gallicus)*. Ospreys: 10. Osprey *(Pandion haliaëtus)*.

The small Egyptian vultures and hooded vultures (genera *Neophron* and *Necrosyrtes*), with their weak beaks, are more than other species, refuse utilizers. However, they are also skillful hunters of small animals and insects, and good nest plunderers. There are two species: 1. The EGYPTIAN VULTURE *(Neophron percnopterus)*. The L is 66-73 cm, and the wingspan is 155-164 cm. The sexes are of identical size. The weight is 1.97-2.2 kg. The yellow face is bare. The beak is slim and the tail is wedge-shaped with fourteen rectrices. The wings are pointed. The adult plumage is "stork colored"; in juveniles it is a blackish-brown. At the back of the head and nape is a ruff of pointed, prolonged feathers, which also surround the neck. The smaller Indian subspecies *(Neophron pernopterus ginginianus)* has an entirely yellow beak. Two yellowish-white eggs, marbled with reddish-brown spots are laid during the breeding season from February to March. The eggs of the Egyptian vulture are probably the most beautiful of all raptor eggs. They inhabit deserts and desert steppes. 2. The HOODED VULTURE (*Necrosyrtes monachus*; Color plate, p. 397) has an L of 58–63 cm, and a wingspan of 157–169 cm. The ♀ is larger than the ♂; the weight is about 2 kg. The back of the head and nape is covered with down in a cap-like manner. The face and the front of the neck are bare, although the neck is bluish to reddish when the bird is excited. The wings are broad, and the tail is rounded. It lays one egg with reddish spots on a whitish base, and breeds all year round. It replaces the Egyptian vulture in the steppes and on coasts of Africa, south of the Sahara.

Fig. 15-79. 1. Egyptian vulture *(Neophron percnopterus)*, 1a) Yellow-beaked Egyptian vulture *(Neophron percnopterus ginginianus)*.

"The hooded vulture is beginning to become more common on the White Nile in the country of the Shilluks," writes A. Koenig. "Further north, where it occurs, it is found only singly." In tropical Africa, it is the only vulture which inhabits coastal areas. During the day it appears in towns, be it in Massawa or in Conakry. Like the crow, it looks for food in gutters and refuse dumps. Away from settlements, it is greatly outnumbered by the white-backed vulture. On the seashores, the hooded vulture feeds at low tide on spiny lobsters, mussels, mollusks, dead fish, and other stranded edibles. Often, however, it feeds for weeks, just like the Egyptian vulture, on human feces, and at times, feeds these to its young.

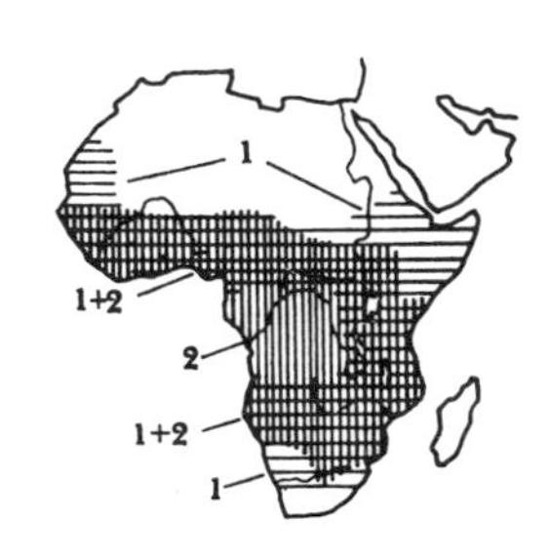

Fig. 15-80. 1. Hooded vulture *(Necrosyrtes monachus)*; 2. Palm nut vulture *(Gypohierax angolensis)*.

The Egyptian vulture has penetrated northward into Europe, but the more northern outposts, such as that on Mount St. Saleve near Geneva, have been given up again. Only in southern France, Bulgaria, and Rumania are there still outposts of this species. Often they fly, as single companions of ravens, over a large area every day and so find their food in alpine meadows, on country roads, below bearded vulture nests, or in villages. When other vulture species feed on animal corpses, they chase the smaller Egyptian vultures away. In such cases, the latter even eat blood-saturated earth. They also eat dates, palm nuts, and other plant food, and yet they are not such pronounced vegetarians as are the palm-nut vultures.

British investigators Jane van Lawick-Goodall and Hugo van Lawick observed that this agile vulture has a tendency to use tools. "When a grass fire in the East African steppe had driven ostriches from their nests, their eggs remained behind undamaged. White-headed and lappet-faced vultures tried in vain to open these eggs with blows of their beaks. Then two Egyptian vultures arrived. At first they too tried to open the eggs with their beaks. When this was in vain, they searched out stones in the vicinity weighing 100 to 300 grams, took them in their beaks, raised themselves up before the egg, and flung the stones at the egg. After four to twelve blows the egg cracked and the meal could start. At the end, a third vulture which was lowest in the peck order tossed stones at the remaining egg shells." (Color plates, pp. 330/331).

The adults often bring small animals such as ground squirrels, moles, frogs, lizards, and snakes to their young. Sometimes the young have to starve for several days when the adults cannot find food, such as during a long rain. Since the very adaptable Egyptian vulture, on the one hand, utilizes refuse and, on the other hand, can catch live animals, it is perhaps the one species of vulture which still has a chance of surviving the coming decades in Western and Southern Europe.

The palm-nut vulture

The PALM-NUT VULTURE of Africa (*Gypohierax angolensis;* Color plate p. 397) has often had its systematic position changed. The German ornithologist, Erwin Stresemann, places it not with the sea eagles, as was formerly done, but with the vultures. It seems to be the most primitive of all living Old World vultures. Its tendency to a plant diet is probably a recent adaptation to its environment. The L is 51–60 cm, the wingspan reaches 160 cm, and the weight is 1.52 kg in the ♂ and 1.83 kg in the ♀. The head is feathered except between the beak and around the eye. The beak is strong and long. The juveniles are dark brown. In the ♂, the lower parts are often pink tinted. One egg, white with brown or purple spots, is laid during the breeding season, which lasts from January to August. It lives in the mangrove belt and rain forest region. It leaves the coast at the start of the breeding season and builds its nest on or near palm trees.

Although the palm-nut vulture lives on fish, mollusks and crabs, from rivers and the sea, the fruit of the oil palm *(Elaeis guinensis)* and of the raphia palm *(Raphia ruffia)* are its main food. It is common where these palms grow in numbers. According to Braun, it is found along waters in Angola only because that is where the oil palms are. This vulture holds nuts of both palm species, which are two to three centimeters long in its foot, and pulls off the husk in such a way that only the kernel remains. It was formerly believed that this special taste was due to a high vitamin A content of the fruit. But raphia palm nuts, which this vulture likes just as much, are not rich in this vitamin. There must, therefore, be other reasons for this diet. Incidentally, it also takes insects found in the fruit of the raphia palm as well.

In the courtship period, which is during March and April, a pair of these birds in the Berlin animal park became very hostile. With a loud bleating "boooh," both threatened their attendant, and made flying attacks at the head and upper body of those people who entered the flight cage, or bit their legs. The young are nestlings for about ninety days. In Zululand, the parents fed the young in part with fruit of the very raphia palm tree in which the nest was built.

The bearded vulture

One of the most common raptors is the BEARDED VULTURE (*Gypaëtus barbatus;* Color plate p. 397), which is also a peculiar type—there are many fables and inventions about the "lamb vulture." The L reaches 115 cm, the wingspan is 280 cm, and the ♀ is slightly larger than the ♂. The weight is 4.5–7.1 kg. Two pale, reddish-brown tinted and spotted eggs are laid during the breeding season, which lasts from December to February or March, and in Africa from October to May. The tail is long and wedge-shaped. The wings are very long and pointed. The head is relatively large and the beak is long and strong. The black bristle feathers on the chin form a "beard" projecting at a forty-five degree angle. The tarsus is short and feathered in front. The thigh plumage is strongly developed. Juveniles are blackish-brown, and white below. The adult plumage is reached, after several intermediate stages, in six years. They inhabit high mountain ranges. There are four subspecies. The largest is the CENTRAL ASIATIC BEARDED VULTURE *(Gypaëtus barbatus hemalachanus)*, and the smallest is the BARE-FOOTED BEARDED VULTURE (⊕ *Gypaëtus barbatus meridionalis)* of East and South Africa.

It is significant that new feathers of this vulture are pale, but old ones are rusty colored. We know now that this change in the plumage is due to iron oxide that has traces of quartz. At the great altitudes where bearded vultures breed, there are, as a result of erosion, surface iron oxide deposits, which in moist weather discolor particularly the underside of the vulture. It is thus understandable that bearded vultures do not regain this rusty color of the feathers after the moult in zoos. P. Berthold spread iron oxide-containing sand in a flight cage in the animal park in Tripsdrill (Württemberg), Germany. The bearded vulture in this cage soon acquired a rusty yellow underside, for it would sleep stretched out on the floor which held deposits of the pigment. This resting posture on the ground of breeding cages is also an adaptation of bearded vultures in the wild.

Fig. 15-81. Bearded vulture *(Gypaëtus barbatus)*; "a" indicates that it occurs only occasionally, and "b" indicates occasional northward migrations.

In the mountains of Central Europe, bearded vultures have been exterminated. They used to nest there only at high altitudes, but came down to lower levels in winter. In the words of the Swiss naturalist, Johann Jakob von Tschudi, "In heavy snow storms their wings would almost brush the roofs of mountain villages in the Grisons." The last "German" bearded vulture was shot in 1855 near Berchtesgaden. In 1887, in the Valais, (Switzerland), an adult male which had been poisoned was found. In the Carpathians, poisoned wolf corpses killed the last bearded vultures about 1935. Only in the Pyrenees and other Spanish mountain ranges, in the southern Balkans and in the Caucasus, is this royal bird still flourishing. Terrasse and Boudoint write, "In the Pyrenees, where this bird was also on the verge of extinction, the population recovered. At any rate, one can, without exaggerated optimism, assume that the species is no longer in danger of extinction; one need only consider the many observations of bearded vultures in immature plumage." Since the 1920's, bearded vultures reappear annually in the Alps. They probably arrive with griffon vultures. Most have been seen in the Salzburg Alps in Austria.

I shall never forget my first view of a bearded vulture. It was in the Church-ul mountains in the Mongolian Gobi Desert. Late in the after-

▷

1. Caracaras: 1. Crested caracara *(Polyborus plancus)*; 2. Red-throated caracara *(Daptrius americanus)*; 3. Forster's caracara *(Phalcoboenus autralis)*; 4. Yellow-headed caracara *(Milvago chimachima)*. Laughing falcons and forest falcons: 5. Collared forest falcon *(Micrastur semitorquatus)*; 6. Laughing falcon *(Herpetotheres cachinnans)*. Pigmy falcons: 7. Red-legged falconet *(Microhierax caerulescens)*; 8. African pigmy falcon *(Polihierax semitorquatus)*; 9. Spot-winged falconet *(Spiziapteryx circumcinctus)*.

1
5
3
2
4
7
♂
♀
8
9
6
F. Reimann

a
1
b
2
6
3
4
7
5

noon an immature came sailing up over the dry stream bed. A single, strong downbeat of the wings, in which the wing tips almost met, swerved it up again and it disappeared from view behind the nearest peak. Among all large vultures, this species is the most active in its search for dead or sick animals, and it covers the greatest distances on its searching flights. Ernst Schäfer describes it in Tibet, "In the Hsifan mountain country it is rare and shy in the south, but in Tibet it is becoming quite common. It flies at all elevations from deep ravines up to the highest alpine zones. It likes to be in the vicinity of settlements, where it has become half tame and concerns itself more with consuming refuse than with waiting for food from the slaughter of domestic animals. It lives largely on bones. As an enthusiastic bone feeder, it even swallows large, whole yak bones."

Its Spanish name, Quebrantaheusos (bone breaker) indicates that it knows how to smash larger bones by letting them fall onto rocks. Then it can eat the fragments more easily. With its peculiarly built tongue, it probably extracts the marrow out of bones and brains out of skulls. R. E. Moreau found a bearded vulture "smithy" and writes, "Over an area of forty meters, white bone splinters covered the bare rocks. In depressions they lay in masses. I could have collected a dozen buckets full." In the Mediterranean area, Greek and Mauretarian tortoises *(Testudo hermanni* and *Testudo graeca)* are the bearded vulture's chief food. It shatters them on rocks, like large bones. In Bosnia and Spain it takes the placentas and dead lambs from sheep and goat herds. dogs, and chamois, with wingbeats, toppling these prey into ravines, Since the bearded vultures in the Alps often attacked goats, sheep, it was known as the Lämmergeier (lamb vulture). It was falsely accused of stealing babies. This widespread myth was the cause of the last campaign of extermination against it in the Alpine areas.

A bearded vulture nest in the lower Pyrenees is described by Terrasse and Boudoint. "The nest lies at the foot of a steep slope between a large yellow bush and a shrub. It consists of a tremendous heap of branches covered with great masses of sheep's wool. It is hidden below an overhanging rock of about five meters in the upper third of the cliff, and has a view of the whole valley, with its 500 meter high granite walls and forests."

Incubation lasts fifty-five to fifty-eight days. The incubating bird is well protected from the cold and winter storms by the thick bed of sheep wool into which it burrows. A youngster in the Sofia Zoo in Bulgaria was not fed from the parents' crop as was first assumed. The parents first cut up the meat into very small portions with the sharp edges of their beaks before they passed it to the chick. Every breeding pair has a choice of several nest niches; the nest sites which are not used are taken over by peregrines, kestrels, or ravens.

Almost all species of vultures in Europe and the Orient are endan-

◁ Gryfalcons: 1. Gyrfalcon *(Falco rusticolus),* a) dark, b) pale phase. In northern areas white gyrfalcons predominate, while in southern areas the darker birds do. 2. Prairie falcon *(Falco mexicanus);* 3. Laggar falcon *(Falco jugger),* light form; 4. Saker falcon *(Falco cherrug);* 5. Feldegg's falcon *(Falco biarmicus feldeggii).* Goshawk falcons: 6. Brown hawk *(Ieracidea berigora);* 7. New Zealand falcon *(Nesierax novaeseelandiae).*

gered at present. Poisoned bait, which is put out for wolves, is also taken by vultures. This reduces the northern species in cattle breeding areas. Modern hygiene also has, in its own way, condemned vultures, for carrion feeders are not needed when man buries dead wild and domestic animals. Only in wild, uninhabited mountain areas will our descendants probably be able to see vultures and large eagles, provided that these areas are left to maintain this biological balance, thus assuring a food supply for the large raptors.

Subfamily: harriers, by H. Brüll

The HARRIERS (subfamily Circinae) are distinguished by long legs and a striking arrangement of the head plumage. Feathers are arranged about the eye into a "mask." These feathers can be moved and so effect changes of facial expression. The goshawk also has such a "facial disc," although it projects very little. This disc may assist the harriers in finding prey by sound. For all the inhabitants of areas rich in cover, sound perception is important. Thus the goshawk pays attention to the feeping call of the female roe deer and the calls of hares and rabbits. The marsh harrier can find unhatched coot clutches by the peeping of the chicks within the eggs.

Fig. 15-82. Marsh harrier's foot *(Circus aeruginosus)*; foot spread of the male is 6.5 cm, that of the female is 8 cm.

Of the nine species of true harriers (genus *Circus*), we will mention five: 1. The HEN HARRIER (MARSH HAWK) (*Circus cyaneus;* Color plate p. 398) has an L of 45–55 cm. 2. MONTAGU'S HARRIER (*Circus pygargus;* Color plate p. 398) has an L of 40–45 cm. 3. The PALLID HARRIER (*Circus macrourus;* Color plate p. 398) has an L of 42.5–47.5 cm. The ♀♀ of these three species have white rump patches on their otherwise brown plumage. The young ♂♂ are colored like the ♀♀. Adult plumage forms only after the second moult, but they can reproduce while in the intermediate plumage. 4. The PIED HARRIER (*Circus melanoleucus;* Color plate p. 398) has an L of 37.5–45 cm. 5. The MARSH HARRIER (*Circus aeruginosus;* Color plate p. 398) has an L of 45–50 cm.

The hen harrier, one of the species which live in Germany, is distributed all over the Northern Hemisphere wherever there are moors, meadows, and large fields. Montagu's harrier needs much the same habitat. The main concentration of both these species in Central Europe was in northwestern Germany. However, their populations have been much reduced by the destruction of upland moors, the conversion of low moors to grassland and of heaths to forest. As a result, these harriers are only rarely seen today. The pallid harrier occasionally travels far to the west on its wanderings. It nested in Germany and Sweden in 1952.

Fig. 15-83. Flight silhouette and foot of a hen harrier *(Circus cyaneus)*; it glides and swoops down on its prey, grasping with its feet. The spread of a foot is 5.2 cm.

The colorful plumage of adult males is especially displayed during courtship flights, accompanied by calls of excitement. Then it can easily be seen that male hen and pallid harriers do not have a black bar along the wing, like the Montagu's harrier. In their display flights, harriers at great heights glide with their wings raised at an angle. Then they seem to tumble, rise up and down, and occasionally loop the loop.

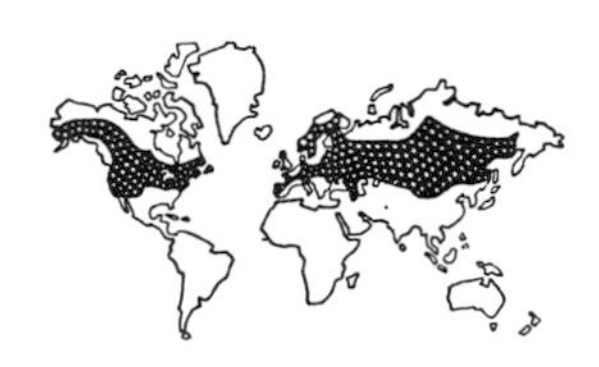

Fig. 15-84. Hen harrier *(Circus cyaneus).*

Fig. 15-85. Marsh harrier *(Circus aeruginosus).*

Fig. 15-86. 1. Montagu's harrier *(Circus pygargus);* 2. Pallid harrier *(Circus macrourus);* 3. Pied harrier *(Circus melanoleucus);* 4. Madagascar harrier hawk *(Polyboroides radiatus).*

Fig. 15-87. Crane hawk *(Geranospiza caerulescens).*

While all this is going on the male may also pass food to the female. Such passing of food on the wing is typical of harriers.

Two females mated to one male Montagus' harrier were observed on one occasion. The male was living in simultaneous bigamy.

The nest is on the ground in the low vegetation of moors, heaths, meadows, and fields. The male offers the female several nests to choose from. She lays four to five white eggs in the chosen one, and takes the main role in incubation, which lasts from twenty-eight to thirty days after the first egg is laid. The male, as with so many raptors, is the main supplier of food. Hen harriers defend their nests against all enemies, including the goshawk and sparrow hawk. This is particularly impressive when it happens about the time the young hatch.

As small animal catchers, the harriers capture insects, reptiles, mice, ground-brooding birds and their eggs with a "swooping attack" by reaching their long talons into the small coverings over low vegetation. The territories of the hen harrier and the Montagu's harrier have an area of 500 to 700 hectares. Occasionally single hen harriers spend the winter in northern Germany. In three cases one could prove that hen harriers became the prey of the goshawks.

The marsh harrier is at home in the reed beds of lakes and old river courses, and in swamps with large reed beds from Spain to the Yenissei River to western Mongolia. It has territories of about 800 to 2000 hectares. With its talons, it can overcome larger prey than its smaller relatives. It patrols reed beds and sedge meadows, as well as adjacent fields, tirelessly from late March to early October. It seeks young coot eggs as well as adult coots. It also hunts great crested grebes, ducks, and cygnets, which have become lost in the reeds and are calling for their mother. It catches reed buntings, starlings roosting in the reeds, cuckoos about to lay in reed warbler nests, as well as voles, water shrews, moles, field mice, young of hares, rabbits and pheasants, partridge, skylarks, and whatever else it can find of suitable size.

In this species too, the more colorful male shows off in display flights before the dark brown female. As a rule, the nest is located in places of limited access among the reeds. It is made of reeds, grass, water plants, and dry twigs. The female lays three to six white eggs, generally four to five, and she alone incubates them for thirty-three to thirty-six days. Incubation begins as soon as the first egg has been laid. The last-hatched youngster is inferior to those hatched earlier. It is stunted and is generally fed to the others, since the parents do not recognize such a weakling as their own. Nestlings remain in the nest for thirty-four to thirty-eight days. Even when not yet fledged, they climb about in the reeds. When fifty-five to sixty days old, they seek food from the adults in begging flight. Those from northern areas migrate to the Mediterranean countries, and also as far as tropical Africa. Those in southern Europe are resident.

A few harrier species nest in trees, including the MADAGASCAR HARRIER HAWK (*Polyboroides radiatus*; Color plate p. 398 and the CRANE HAWK (*Geranospiza caerulescens*; Color plate p. 398). Both hunt in low gliding flight over the ground, hop about in the foliage of trees, or hang on the bark or even the interior of rotted tree trunks, finding insects, lizards, bats, eggs, and young of birds. The tarsal joint can be moved backward to some degree, an arrangement which evidently favors the searching out of holes and cavities. The first species is often seen hanging on projecting house roofs. On the ground they look for ants, crabs, termites, frogs, snakes, and small rodents. In many areas of Africa they prefer a vegetarian diet in the form of palm nuts, while in other areas they feed largely on weaver finches. The crane hawk also puts its head and super mobile foot into holes to get bee larvae, ants, eggs, and young birds. It also takes lizards and small snakes, even the highly poisonous coral snakes. Its northern populations live in forests in the spring and summer, but in winter Sutton observed them over rocky areas where they hunted like Montagus' harriers. The crane hawk is one of the raptors which follows grass fires to find corpses. Evidently these two species do not always swallow their prey whole; that is, they dissect lizards neatly and remove the intestines. Both of these harriers prefer low-lying woods near rivers and swamps, although the crane hawk is also found in dry areas.

Subfamily: bateleur, harrier eagles and serpent eagles, by H. Brüll

The SERPENT EAGLES (subfamily Circaëtinae), as the name indicates, live to a great extent on snakes. There are fourteen species, five of which are in Africa. One of the African serpent eagles, because its display actions resemble circus performances, has been called the BATELEUR (*Terathopius ecaudatus*; Color plate p. 398). Its display flights are marked by sideways rolls, looping the loop, and other "artistic performances." It emphasizes these by sounds produced in flight; when the wings are beaten together, a loud clapping sound is produced which is audible for some distance. The tail feathers of the bateleur are very short.

It avoids large forests; its habitat is open land, the steppe with acacias, and other broad crowned trees. For nesting it prefers trees near native paths or elephant trails. The nest is built of coarse sticks and branches, and is "greened" with foliage. The zoologist Hoesch reports that when flying to their nest they come down to just above the ground many hundreds of meters before the nest tree; this explains their preference for trees along paths. The female lays only one rough-shelled egg, which is generally all white but occasionally has red or purple marks; it is incubated for forty-three days. The youngster hatches in ninety-three to one hundred days. Bateleurs do not have territories from which they expel members of their species; rather, they tolerate wandering adults and immatures in their vicinity.

Circling high up gives the bateleur a wide view; it searches for

Fig. 15-88. 1. Bateleur *(Terathopius ecaudatus)*; 2. Philippine serpent eagle *(Spilornis holospilus)*; 3. Crested serpent eagle *(Spilornis cheela)*.

snakes, lizards, and other small reptiles, but also hunts rats, mice, shrews, and larger mammals up to small antelopes. Tortoises, bird eggs, grasshoppers, and carrion are also on its menu. It attacks tawny eagles and vultures to induce them to regurgitate their prey. Leslie Brown observed how a bateleur repeatedly swooped at a snake which again and again tried to bite it, until the snake was exhausted and became easy prey. It is said to overcome even puffadders in this way.

To the serpent eagles in the narrower sense (genus *Circaëtus*) belongs, in addition to the BROWN HARRIER EAGLE (*Circaëtus cinereus*), a European species, the SHORT-TOED EAGLE of the Mediterranean countries (*Circaëtus gallicus*; Color plate p. 398). It is becoming even rarer, for the extensive alteration of the land by man has reduced its main prey, snakes. In the nineteenth century it was an occasional breeder in Germany, but by the middle of this century only isolated pairs were seen in Lower Silesia and East Prussia. Hibernation of snakes in northern areas forces it to migrate.

Fig. 15-89. 1. Brown harrier eagle *(Circaëtus cinereus)*; 2. Short-toed eagle *(Circaëtus gallicus)*.

In Europe it inhabits old forests near heaths and swamplands, dry steppes, and rocky areas with low scrub; in Asia it inhabits dry areas and scrubland, and in tropical Africa, savannahs and bush steppes; these are the habitats of various kinds of snakes. Here it may perch on a branch or fly high over the ground, occasionally hovering like a kestrel, while looking for prey. G. L. van Someren describes how a harrier eagle danced about a snake with its wings raised to distract it. Thus it was able to seize it at the back of the neck, aided by the rough soles of its feet; it then cut off the head and swallowed the snake's body intact. It cuts larger snakes, poisonous and non-poisonous, into pieces. These birds, like so many other snake hunting animals, are not immune to snake poisons. During reproduction the male supplies its female and young primarily with snakes; when it carries them in they often hang from its beak, or it regurgitates them from the crop. Frogs, crabs, small mammals, and ground dwelling birds which are easily accessible to it also are its prey. The small flat nests of these two species stand in the crowns of high forest trees, euphorbias and acacias. The female lays one or two rough-shelled eggs and incubates them with much perserverance, for in Africa they are much endangered by kites and one of the buzzards.

The SERPENT EAGLE (genus *Spilornis*) inhabits Southeast Asia to the Philippines. Of the six species, we mention the CRESTED SERPENT EAGLE *(Spilornis cheela)* and the PHILIPPINE SERPENT EAGLE *(Spilornis holospilus)*. Their habitats are wet jungles, forest clearings, and cultivated areas, provided there are enough snakes. The rest of their menu is also like that of the other serpent eagles. However, they do follow cultivation. Nests are in tree forks on forest edges of rivers or near swampy fields. The clutch consists of one to three eggs, which are white with brown caps at the poles and brown to red-brown spots.

Subfamily ospreys, by K. H. Moll

The OSPREYS (subfamily Pandioninae) are often placed in a separate family because of several special characters which deviate from those of other Accipitrids. There are no secondary shafts on feathers of the lower parts, the lacrimals of the skull are poorly developed, and the outer toe can be turned to the rear. There is hardly any pellet formation. There is a very large preen gland. The body shape is somewhat gull-like. The wings are narrow. There is only one species, the OSPREY (*Pandion haliaëtus;* Color plates pp. 385/386 and 398), which is distributed in six subspecies over almost the whole world. The northern subspecies *(Pandion haliaëtus haliaëtus)* occurs in Central Europe. The L is 60-66 cm, and the wing length is 48-54 cm. Other subspecies are a little smaller. The northern forms of the Old and New World, (*Pandion haliaëtus haliaëtus* and *Pandion haliaëtus carolinensis*), are migrants.

Fig. 15-90. Osprey *(Pandion haliaëtus)*, subspecies: 1. *Pandion haliaëtus haliaëtus;* 2. *Pandion haliaëtus carolinensis;* 3. *Pandion haliaëtus cristatus;* 4. *Pandion haliaëtus ridgwayi;* 5. *Pandion haliaëtus mutuus;* 6. *Pandion haliaëtus microhaliaëtus.*

Nests are, as a rule, in the crowns of free standing trees, on power line masts, on cliffs, or even on the ground. In areas with good food supply there may be a considerable accumulation of breeding pairs in a relatively small area. Thus in 1932 there were over 300 nests on Gardiner island in North America, with an area of 1200 hectares. So many ospreys in one area have not occurred in Europe for a long time. The greatest concentration of pairs in Germany was observed in 1924-1925 on the Darss (Pomerania, Germany); there were four pairs in an area of 2.4 hectares. At present the greatest concentration is on the east shore of Lake Müritz in Mecklenburg, Germany, where there were ten nests in 1967, which on the average were one and one half kilometers from the lake shore.

The body structure of the osprey is fully adapted to its method of hunting. With its short thigh plumage, particularly long and very curved claws, spiny under surface of toes, and an outer toe which can be rotated to the rear, this raptor is fully equipped for plunge-diving for fish. When hunting it cruises up to about fifty meters above a lake. When it spots prey, it hovers briefly, plunges down with folded wings and extended feet, dives, seizes the fish, and comes back to the surface by the impetus of the descent, whereby it uses the tail as a rudder to effect an upward movement, and then rises into the air again. The angle of descent varies from 45° to 90°. Although it only weighs 1.7 kg, it can bring up fish weighing 2 kg and carry them off through the air.

Fig. 15-91. Osprey flying to the nest with a large branch.

The daily food requirement is about 400 grams of fish; the average weight of the fish taken is 100 to 200 grams. Though it lives almost exclusively on fish, it keeps to the smaller species and those of least economic significance. The view that it is harmful to fishery interest has, therefore, now probably been overcome. The clutch consists of three, rarely four, eggs. The incubation period is thirty-eight days, and the fledging period is sixty days. Unfortunately this beautiful raptor has become rare as a result of strong persecution so that it is urgently necessary to strictly protect it everywhere.

Fig. 15-92. Flight silhouette of an osprey *(Pandion haliaëtus)*.

16 Falcons and Their Relatives

Family: falcons, by Th. Mebs

The FALCONS (family Falconidae) are particularly notable raptors. They are distinguished from the hawk-like raptors by certain skeletal characteristics and other features of the internal structure. Their nostrils are round or oval, and there is a longitudinal ridge on the palatal surface of the upper mandible. Eggshells show not a greenish, but a yellow or reddish tinge when light shines through them. Moult of the primaries begins not with the first (innermost) feather but with the fourth. There are four subfamilies: 1. The LAUGHING FALCONS (Herpetotherinae); 2. The CARACARAS (Polyborinae); 3. The PIGMY FALCONS (Polihieracinae); and 4. The TRUE FALCONS (Falconinae). There are thirteen genera with around sixty species.

Subfamily: laughing falcons and forest falcons

In the subfamily Herpetotherinae, the genera *Herpetotheres* and *Micrastur* are combined; both are native to tropical Central and South America and are mainly or exclusively forest dwellers. In adaptation to this habitat, they have short, round wings and a long tail which gives them great ability to swerve in flight. In their external appearance and behavior they are more like goshawks than falcons. On the basis of structural characteristics, they do, however, unmistakably belong to the falcons.

A striking feature of the LAUGHING FALCON (*Herpetotheres cachinnans;* Color plate p. 403) is the black face mask. The L is 46 cm and the weight is 570 g. They inhabit the tropical rain forests from southern Mexico to southern Brazil.

It is often seen on the outer branches of high jungle trees, only coming into the open to hunt. It preys mainly on lizards and snakes which it swoops down onto from its perch; it skillfully kills poisonous snakes by biting off their heads before carrying them to the nest. A flying laughing falcon offers a unique view when it holds a long snake dragging behind in its talons; this may have led to the scientific name of the genus, which means "snake carrier." It breeds in cavities in high trees or in rock caves; when these are lacking it uses the old nests of other raptors. Only a single dark brown egg is laid. The mother guards the youngster while the male hunts for the family. At sunrise and in the evening twilight the pair, in a duet, utters loud calls, "vah-co, vah-co, vah-co." The name, laughing falcon, refers to a far-reaching, shrill "ha ha ha" which is repeated up to fourteen times.

The COLLARED FOREST FALCON (*Micrastur semitorquatus;* Color plate p. 403; the L is about 50 cm) lives from Mexico to northern Argentina and Paraguay. The color of its underparts varies; white, light brown, and dark brown color phases are distinguished. About one-third smaller, the BARRED FOREST FALCON (*Micrastur ruficollis*) has almost the same distribution; it is banded below and has a reddish-brown throat. Two further species of laughing falcons in tropical South America, the SLATY-BACKED FOREST FALCON (*Micrastur mirandollei*) and SCLATER'S FOREST FALCON (*Micrastur plumbeus*) are rather rare.

All live in densely wooded areas, partially in almost impenetrable jungle, and show marked adaptations to this environment in shape and structure, including the short and round wings, a very long graduated tail, and long tarsi and toes. Being very fast and able to swerve like a goshawk or sparrow hawk, they fly through the branches and undergrowth, but they can also walk rapidly on the ground and jump in trees from branch to branch without opening their wings. They capture various forest birds, small mammals up to the size of rabbits, and reptiles, including snakes. They hunt both while flying or running, depending on the vegetation. While running they hold the body horizontal; their steps are long and accelerated. They have a circle of stiff feathers at the side of the head in the ear region which is similar to the facial discs of harriers or owls, and large ear openings. Evidently they hear particularly well because of this, and probably can hunt by sound in the forest twilight. The call of the collared forest falcon is a repeated "kow," which sounds peculiarly like the groaning of a man. About the reproduction of forest falcons virtually nothing is known, although they are said to nest on high trees.

Subfamily: caracaras

The CARACARAS (subfamily Polyborinae) live in the New World, mainly in South America. There are four genera with nine species; they are long-winged, rather long-legged inhabitants of steppes, high mountains, or coasts, and they like to seek food on foot, and hence do not avoid carrion (only the genus *Daptrius* is relatively short-legged and its members are jungle dwellers). They show a certain similarity to vultures in their food and way of life. The name caracara is of Indian origin and refers above all to the CARANCHO OF CRESTED CARACARA (*Polyborus plancus*), because its hard vibrant call sounds as if two pieces of wood were being vigorously rubbed together.

The two jungle species, the fifty centimeter long RED-THROATED CARACARA (*Daptrius americanus;* Color plate p. 403) of the tropical forests from Guatemala to southern Brazil, and the black, forty centimeter long YELLOW-THROATED CARACARA (*Daptrius ater*) of the Amazon area, are pronounced tree birds. They have long wings, very long tails, and relatively short legs. It was formerly believed that they preyed on small birds and snakes, but it has recently been revealed that they live mainly on insects and fruit. According to the observations of Alexander Skutch, the red-throated caracara tears open the nests of large black wasps (genus *Synoeca*) and eats the larvae and pupae from the combs. Hanging the head down, it supports itself with its claws in the lateral holes of the nest wall and puts its head far into the nest interior in order to devour the content of the combs in a single meal. The wasps swarm around it closely, but don't do anything to it. The yellow-throated caracara feeds largely on the fruit of mistletoes. These caracaras build their nests high in the gigantic trees of the tropical forest; the two or three eggs are reddish-yellow with brown spots. Their loud calls make them conspicuous, particularly in the breeding season. The red-throated species calls "kakao-ka-ka-ka-ka-kakao," and then goes on to a loud, shrill laughter which sounds like "ha ha ha."

Chimango caracaras

The CHIMANGO CARACARAS (genus *Milvago*) are among the most common raptors of the pampas and grazing lands of South America. There are two species: 1. The YELLOW-HEADED CARACARA (*Milvago chimachima*; Color plate p. 403), with an L of about 40 cm, lives from Panama to Uruguay; and 2. The CHIMANGO CARACARA (*Milvago chimango*) is somewhat smaller; it is brown above and has cinnamon-brown underparts with light bands and spots. The legs and feet of the male are blue-gray, but are yellow in the female. They inhabit southern Brazil to northern Chile and the Falkland Islands. Both species live in open country; they seek food mostly on foot and eat insects, nestlings, and other small animals, as well as carrion, eggs, and even plant material. They perch on the backs of cattle to eat ticks. They also follow the plough and play as useful a role there as do crows in Central Europe. The large nest is in trees, bushes, or on the ground, and holds two to four, usually three, reddish-brown eggs. In the open pampas they like to breed in swamps; often there are several pairs in a close neighborhood. They are in general very sociable, often being seen in flights of thirty to forty birds. Outside the breeding season they roost socially on high trees. At feeding places, however, they are cantankerous, like vultures.

The mountain caracaras

In the Andes and on the Falkland Islands there are four species of MOUNTAIN CARACARAS (genus *Phalcoboenus*), which replace one another geographically. The juvenile plumages of all four species, in contrast to the adult plumages, are very similar, which is an indication of their close relationship. The slim, long-legged MOUNTAIN CARACARA (*Phalcoboenus megalopterus*), which is about fifty centimeters long, inhabits the high elevation Puna zone from northern Peru to northern Chile and northwestern Argentina. A very similar species, the CARUNCULATED CARACARA (*Phalcoboenus carunculatus*), which has white stripes on its underside is at home in the Paramo zone of southwestern Colombia and Ecuador. The WHITE-THROATED CARACARA *(Phalcoboenus albogularis)* of Patagonia has entirely white lower parts. The largest (the L is about 56 cm) is the predominantly brownish-black FORSTER'S CARACARA (*Phalcoboenus australis*; Color plate p. 403), which has a yellowish-white striped neck and breast, of the Falkland Islands and the small islands south of Tierra del Fuego. In regard to their food, all play the role which crows play in the Northern Hemisphere. They live mainly on worms, larvae, carrion, and refuse. In the high Andes they are found on cadavers together with condors. They follow trucks in the expectation of refuse, like gulls follow ships. On rocky coasts and islands they feed almost entirely on stranded sea organisms, chase gulls to induce them to drop their prey, or eat the refuse dropped near houses. Only rarely do they attack small mammals. They build their nests on rocks by carrying in a few twigs, dry grass, or sheep's wool, often laying the eggs in a natural depression. Clutch size varies from two to four densely blackish-brown spotted eggs.

Common caracaras, and Guadalupe caracara

The CRESTED CARACARA (CARANCHO) (*Polyborus plancus*; Color plate p. 403) is marked by a small erectile crest at the back of the head. The L is about 53 cm, the tarsus is particularly long, but the toes are short. Rather than a round nostril, theirs is very small and slit-like, which is probably an adaptation to the mode of feeding. The beak has a naked red root. It is mainly a steppe dweller of the southwestern United States through Central and South America as far south as Tierra del Fuego and the Falkland Islands. The closely related GUADALUPE CARACARA *(Polyborus lutosus)* was exterminated in a few years at the end of the last century, through planned persecution by goat herds on the Island of Guadalupe off California.

Usually one sees two, three, or four of these birds soaring together or perched in the tops of isolated tall trees, which offer a wide view and which serve as communal roosts outside the breeding season. In low level flight and also while walking, they search the ground for live and dead animals of the most diverse kinds. They visit freshly plowed fields, taking road kills from country roads and floating carrion on rivers. When they find corpses of large animals, they feed more on the insect larvae living in them than on the decaying meat. Since the carancho removes dead animals it is generally tolerated rather than pursued.

In South America the crested caracara reputedly breeds twice a year, but in Florida it usually only breeds once a year. Nests are built of dry twigs in isolated trees or in swamps; it builds a large nest with a depression which is lined with dry grass, pellets, sheep's wool, and similar materials. The two to three eggs, densely dark spotted on a light brown background, are incubated by both parents. The young are fledged in about eight weeks. Their posture when calling is peculiar; the head is laid well back, almost on the back, and the bird calls "traaa," then raises the head and calls "rooo."

Subfamily: pigmy falcons

The subfamily PIGMY FALCONS (Polihieracinae) contains particularly small raptors, some of which are the smallest birds of the whole order. In structure, mode of life, and behavior, they are similar to a large degree to the true falcons. There are four genera, one in South America and three in the Old World tropics.

The barely kestrel-sized SPOT-WINGED FALCONET (*Spiziapteryx circumcinctus*; Color plate p. 403) of northern Argentina, however, has only a hint of the falcon "tooth" in the upper mandible, but it does have a round nostril with a small cone in it, like the true falcons. It is a short-winged and long-tailed forest dweller. On cool days it sits on the outer branches of densely leaved trees; in hot weather, however, it confines itself more to shade. In straight-line hunting flights with its characteristically vigorous wing beats, it captures mainly birds, some of which are as large as it is. Native Argentineans admire its courage and strength and therefore call this falconet "Rey de los pájaros," which means "king of the birds."

The AFRICAN PIGMY FALCON (*Polihierax semitorquatus*, Color plate p. 403) is appreciably smaller; the L is about 18 cm. They have the falcon's tooth on the upper mandible. Two to three dull white eggs are normally laid. It lives in East and South Africa.

It likes dry valleys fringed with acacias and thorn scrub. In mode of life and behavior it resembles the shrikes (see Vol. X). It leaves its perch with a few quick wingbeats, and then glides up to another perch. Often a pair sits very conspicuously on the branches of dead trees or thorn bushes, and from there swoops down steeply at the ground as soon as prey has been spotted. They mainly take beetles and other insects, but they also take small birds, mice, and lizards. They breed mostly in the old nests of one of the weaver birds *(Philetuirus)*, often in the middle of a colony and do not trouble their neighbors; they may also occupy deserted nests of the glossy starlings. The call is an ascending, shrill "tioo tioo tioo tioo."

The FALCONETS (genus *Microhierax*) are found in South and East Asia. The L is 14 to, at most, 18 cm. The strong beak has a sharp "tooth" and an indentation on both sides of it (often falsely designated as "two teeth"). The tarsi are short, but the toes are relatively large and strong. The nostril has no cone. 1. The RED-LEGGED FALCONET (*Microhierax caerulescens;* Color plate p. 403) inhabits the lower slopes of the Himalayas. 2. The BLACK-LEGGED FALCONET *(Microhierax fringillarius),* which is equally colorful, has black thighs. It inhabits Assam, Malacca, Sumatra, Java, and Southwest Borneo. 3. The BORNEAN FALCONET *(Microhierax latifrons)* is similar in color, but is much paler below, and has a wide white forehead band. 4. The PHILIPPINE FALCONET *(Microhierax erythrogonys)* is white only on the throat, breast and belly; otherwise it is a steely green-black.

All four species prefer open lowlands and hill country. Although they feed mainly on insects, they chase birds in a fast swoop and fearlessly seize even those that are as large, or even larger, than themselves. In courage and eagerness to attack, they are in no way inferior to the large trained falcons, in spite of their smallness. In India, the first-named falconet is therefore used by falconers to take larks and, it is said, even quail. All falconets are very sociable. They like to sit in pairs on the outer branches of dry trees to spot insects and birds and to hunt them. They also seize swarming insects in a circling soaring flight and eat them on the wing. They breed in tree cavities, mostly in old barbet or woodpecker holes, and lay three to four yellowish-white eggs. The red-legged falconet leaves its Himalayan breeding area in winter, while the others are resident.

▷ Peregrines: 1. White-cheeked peregrine *(Falco peregrinius calidus)*; 2. Central European peregrine falcon *(Falco peregrinus germanicus)* classified by some with the subspecies *Falco peregrinus peregrinus*, a) young bird, b) adult; 3. Orange-breasted falcon *(Falcon deiroleucus)*; 4. Desert falcon *(Falco pelegrinoides)*; 5. Eleonora's falcon *(Falco eleonorae)*, a) dark, and b) pale phase.

The red-legged falconet has been kept successfully in the Frankfurt Zoo since 1965. Since they are so sociable, several are always kept in one cage. In the spring of 1968, a female laid four eggs in a tree hole and incubated persistently, but without success. This is the first step in their breeding in a zoo, as far as we know.

A long graduated tail and rather rounded wings distinguish the BURMESE PIGMY FALCON *(Neohierax insignis)* from the others. It is larger; the L is about 23 cm, and the plumage is black above and white below. The head and neck of the male are pale gray with black shaft marks, while in the female they are reddish. It inhabits Burma, Siam, and Laos.

When hunting, it sits in wait for grasshoppers and other insects, and then seizes them by plunging down in forest clearings. Its flight from one tree to another is undulating, like that of a magpie. It incubates its grayish-white eggs in cavities or the old nests of other birds.

Subfamily: true falcons

The subfamily of the TRUE FALCONS (Falconinae) includes a whole series of species, some of worldwide distribution, which in structure and plumage are adapted to the pursuit of living prey in the air and in open landscapes, many of them to a great degree. In speed of flight and, above all, in the extraordinary speed of their swoops which are intensified by strong wingbeats, the so-called trained falcons (gyr, peregrine, hobby, Eleonora's falcon, and merlins) exceed all other animals. The hovering falcons, on the other hand, including the kestrel group with many species, the red-footed falcon, and gray kestrels, chiefly hunt prey on the ground by swooping down after hovering or sitting on their perch. There are great size and weight differences between the smallest and the largest members of the genus *Falco*; the small AMERICAN SPARROW HAWK *(Falco spareverius)* weighs only about 100 grams, while the female of the large GYRFALCON *(Falco rusticolus)* weighs up to 2000 grams. Yet all falcons show the following characteristics.

Fig. 16-1. Trained falcons pursue birds on the wing. Here an Eleonora's falcon chases a wagtail.

1
2a
2b
3
4
5
a
b
F. Reimann

1
2
♀
♂
3
4
5
6
F. Reimann

◁
Merlins: 1. Red-headed falcon *(Falco chiquera)*; 2. Merlin *(Falco columbarius)*; 3. Sooty falcon *(Falco concolor)*. Hobbies: 4. Hobby *(Falco subbuteo)*; 5. African hobby *(Falco cuvierii)*. Gray Kestrels: 6. Dickinson's kestrel *(Falco dickinsoni)*.

The head is relatively small; the eyes are large and dark and surrounded by an unfeathered edge of skin. There is a dark cheek mark. The beak is strong, and strongly curved, and there is a sharp projection on each side on the edge of the upper mandible, shortly behind the tip (the very characteristic "falcon's tooth"), which fits into a corresponding indentation on the edge of the lower mandible. They have a streamlined body with a long, pointed, generally short tail. They have a nostril generally with a projecting little cone in the middle. The feathers are hard and rigid, the feet are strong, and the toes, especially the middle one, are generally very long. In contrast to hawks which kill with their dagger-like claws, falcons kill by biting through the nape of their prey with the strong beak (particularly with the falcon's tooth), while the feet serve only to catch and hold the prey.

None build their own nests; all lay their eggs and incubate either in old nests of other birds, or in tree cavities, on rocks or buildings, or in a flat depression which they scratch out on rock ledges or on the ground. The eggs always show more or less densely distributed brown spots on a yellow background. The female is considerably larger and heavier than the male, particularly in the large species. Falconers therefore call the male "terzel" (German) from the Latin *tertium* meaning one-third, since it is about one-third smaller than the female. This size difference seems to play a considerable role in the division of tasks during breeding and the rearing of young. The male can provide food for the entire family due to his maneuverability, while the larger and stronger female takes over the incubation and caring for the young. Only when the young no longer need protection, but yet still need more food, does the female hunt again; she is able to master appreciably larger animals than the male. Incubation in almost all falcons takes thirty days; the fledging period in small species is barely four weeks, but in the large species it takes up to seven weeks.

Brown hawk and New Zealand falcon

Two rather primitive falcons of Australia and New Zealand take up a somewhat isolated position. Several of their characteristics suggest an early deviation from the true falcons, so they were placed in separate genera by zoologists:

1. The BROWN HAWK (*Ieracidea berigora*; Color plate p. 404) has an L of about 43 cm. It has long tarsi, but the toes are relatively short. The tail is long and rounded; their shape is somewhat like the goshawk. There are both pale and dark color phases. They inhabit Australia, Tasmania, and Southeast New Guinea. 2. The NEW ZEALAND FALCON (*Nesierax novaeseelandiae*; Color plate p. 404) is smaller; the wings are relatively short, but the tail is long.

The brown hawk is not a particularly good flier, and so it likes to hunt waiting on a perch for insects and small birds. Though it does not migrate, it arrives by the hundreds in winter in areas where there are caterpillar pests. The New Zealand falcon also generally hunts by waiting on its perch.

The falcons in a narrower sense

The remaining thirty-five falcon species all belong to the genus *Falco*. According to their different modes of life they may be divided into nine groups: A. GRYFALCONS; B. The BLACK FALCON; C. The PEREGRINES; D. The HOBBY; E. The ELEONORA'S and SOOTY FALCON; F. The MERLINS; G. The GRAY KESTRELS; H. The RED-FOOTED FALCONS; and I. The OLD WORLD KESTRELS.

A. The GYRFALCON group (subgenus *Hierofalco*) includes the largest and strongest of falcons. 1. The LANNER FALCON *(Falco biamircus)* has a reddish or yellow crown and nape and a narrow dark cheek stripe. There are four species; among them are: FELDEGG'S FALCON (*Falco biamircus feldeggi;* Color plate, p. 404) of Southeast Europe and Asia Minor, with an L of 43–49 cm, and a weight of 500–600 g in the male and 700–900 g in the female; and the NORTH AFRICAN LANNER *(Falco biamircus erlangeri).* 2. The SAKER FALCON (*Falco cherrug;* Color plate p. 404) has an L of 46–54 cm, a weight of 600–800 g in the male, and 1000–1300 g in the female; it is distinguished from the peregrine falcon by a slimmer body and a longer tail. The ALTAI FALCON *(Falco altaicus)* is, according to recent investigations by G. P. Dementiew and A. Schagdarsuren, a dark color phase of the Central Asiatic sakers. 3. The LAGGAR FALCON (*Falco jugger;* Color plate, p. 404) is found in India and Afghanistan. 4. The PRAIRIE FALCON (*Falco mexicanus;* Color plate, p. 404). 5. The GYRFALCON (*Falco rusticolus;* Color plates, p. 404 and pp. 61/62) has an L of 52–63 cm, and a weight of 900–1500 g in the male, and 1400–2100 g in the female. The cheek stripe is very weak or absent; there is a dark and a pale color phase.

Fig. 16-2. Lanner falcon *(Falco biamircus).*

The falcons inhabit bare mountain plains, deserts, steppes, or tundra, and are very fast and agile hunters. Compared to the peregrine, they have somewhat wider but less pointed wings and a decidedly longer tail; hence they can seize prey in the air as well as on the ground.

Fig. 16-3. Saker falcon *(Falco cherrug).*

In contrast to the saker, which is mainly a steppe dweller, the lanner falcon lives more in mountains and likes steep, rocky mountain masses above wide open, often semidesert plains. The pair partners keep together to a marked degree and hunt the area below from some dominating height. Often they perform an effective surprise hunt on rock doves and jackdaws, flying close to the cliffs, working together. The female, flying ahead, swerves into rock caves, flushing out the birds sitting there, while the male, flying somewhat higher, at once makes the first flying attack. The falcons, swooping in alternation, bring the hunted bird to a confused state, so that only very skillful avoidance and swift escape can save it. Even lesser kestrels and hawks are taken by lanner falcon pairs in this way, as well as other birds, from sparrows to partridge. Occasionally the lanner falcon kills animals on the ground, such as small mammals up to the size of rabbits, lizards, and beetles. It always chooses a cavity or niche on a steep cliff to breed, while the North African lanner nests in part on trees in old crows' nests.

Fig. 16-4. Gyrfalcon *(Falco rusticolus).*

The very swift and agile saker is much appreciated among Asiatic falconers, for it can kill even fairly large prey such as wild geese and hares. In the summer, the main prey of the steppe-dwelling saker are ground squirrels; at other times birds predominate. Most of the western breeding sites are in Czechoslovakia and lower Austria; in Germany it appears only rarely. Sakers usually breed on trees in lowland steppes, always in the old nests of other species, particularly in common heron or rook colonies. A pair of sakers may even take over a sea eagle's or imperial eagle's nest as a result of vigorous swooping attacks. More rarely they breed on cliffs or steep slopes, although this is usually the case in the Central Asiatic highlands. The hunting area may be very large.

The prairie falcon of the prairies and high plains of the Rocky Mountains in the western United States hunts mammals, usually rodents like prairie dogs and ground squirrels, but also larger mammals such as jack rabbits and cottontails, which weight more than twice as much as the falcon. It kills birds as well, both on the ground and in the air. It almost always nests on cliffs, often using old raven nests.

Praire falcon and gyrfalcon

The powerful gyrfalcon has been rightfully regarded, since ancient times, as the noblest and most valuable hunting hawk. In medieval diplomacy it played a great role as a valuable present. Great enthusiasm was caused not only by its outstanding hunting performance but also by the snow-white plumage (of the pale phase) of many far northern gyrs.

Keeping low over the ground in racing flight, gyrfalcons prey largely on ptarmigan, but also on all other possible species of birds, especially shore and seabirds on the coasts. They also kill many lemmings and occasionally even Arctic hares. Thanks to the versatility of their hunting methods and the wide range of prey, they can live in inhospitable habitats. Pairs use the same locality over and over, and are said to stay together as long as they live. Using old nests of ravens and rough-legged buzzards, they always breed on rocks in niches below projections, on coasts and in river valleys. In the forest tundra, however, they nest in trees, also in the old nests of other birds. Their reproduction is much influenced by recurrent fluctuations in populations of their chief prey, ptarmigan and lemmings; in years of low food supply they do not breed at all. In the fall, when there is a lack of prey, the immature gyrfalcons will travel far to the south.

B. The BLACK FALCON *(Falco subniger)* has an L of 46 cm. It is almost entirely brownish-black, with a white forehead, cheeks and throat. It is somewhat intermediate between the goshawks and the gyrfalcons, inhabiting open lowlands in Central Australia.

The peregrine and its relatives

C. 1. The PEREGRINE FALCON *(Falco peregrinus)* has subspecies varied in size and color; among them are the CENTRAL EUROPEAN PEREGRINE (*Falco peregrinus germanicus;* Color plate, p. 417) with an L of 40–48 cm, a weight of about 600 g in the male and about 900 g in the female, the NORTHERN PEREGRINE *(Falco peregrinus peregrinus)* of Northern Europe, and the WHITE-CHEEKED PEREGRINE (*Falco peregrinus calidus;* Color plate, p. 417), which is very large and pale. It breeds in the tundra, often in the immediate vicinity of wild geese, which it does not harm when they are near its nest. Southern forms are smaller and darker, like the SOUTH EUROPEAN PEREGRINE *(Falco peregrinus brookei).* 2. The DESERT FALCON (*Falco pelegrinoides;* Color plate p. 417) inhabits the desert belt from North Africa to Central Asia; it is very similar to the peregrine. The swiftest of all falcons, it has particularly large size and weight differences in the two sexes; the males average 360 g, the females 690 g. The two subspecies are: the BARBARY FALCON *(Falco pelegrinoides pelegrinoides),* which inhabits North Africa and Arabia; and the *Falco pelegrinoides babylonicus* found in Afghanistan and Turkestan; 3. The ORANGE-BREASTED FALCON (*Falco deiroleucus;* Color plate, p. 417) is fairly small, weighing from 340 g in the male to 610 g in the female. It inhabits tropical Central and South America. 4. The TAITA FALCON (◊*Falco fasciinucha)* is small, colorful, and very rare, inhabiting East Africa. The last two species hunt only birds up to pigeon size. 5. The APLOMADO FALCON *(Falco fuscocaerulescens)* inhabits southwestern United States, south to Tierra del Fuego, and leads us over toward the hobbies. Like them, it preys on small birds, insects, and reptiles.

▷ Hovering falcons: 1. Lesser kestrel *(Falco naumanni);* 2. Common kestrel *(Falco tinnunculus);* 3. Red-footed falcon *(Falco vespertinus);* 4. American sparrow hawk *(Falco sparverius);* 5. Fox kestrel *(Falco alopex).*

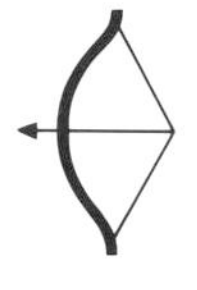

The peregrine is most completely adapted to long distance hunting on the wing. In its horizontal and its swoop dives, it reaches extraordinary speeds. The swooping speed of one which wintered on the tower of Cologne Cathedral and hunted pigeons from there was measured accurately at 70–90 meters per second, or 270–320 kilometers per hour. The name "peregrine," meaning wanderer, applies only to the northern subspecies. They leave their breeding areas every winter and migrate through the temperate zone of the Old and New World into their tropical winter quarters, sometimes even to the tropics of the Southern Hemisphere, often traveling more than 10,000 kilometers. Thus Siberian peregrines migrate as far as New Guinea and Ceylon, or even to South Africa, while those from northern Canada reach Brazil and Argentina. Those native to temperate and subtropical latitudes are, however, more or less resident; only the young, once they are independent, move away in the fall.

Fig. 16-5. Peregrine falcon *(Falco peregrinus).*

In the last twenty years both the Central European and the North American peregrine *(Falco peregrinus anatum)* have shown an alarming decline, and both subspecies are now endangered. The main cause of this striking decline is an evident decrease of fertility and breeding success. It is suspected that this is due to agricultural pesticides which are used on a large scale. These pesticides build up in the tissues of falcons through their prey, which feed on insects.

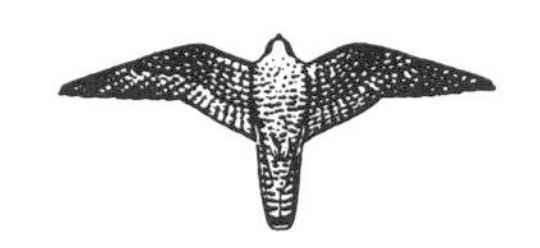

Fig. 16-6. Peregrine falcon *(Falco peregrinus)* in flight from below.

1
2
♀
♂
3
♀
♂
♂
4
♀
5
F. Reimann

1
2
3

◁
Mound birds: 1. Mariana jungle fowl *(Megapodius laperouse)*, mound building; 2. Mallee fowl *(Leipoa ocellata)*; 3. Maleo *(Macrocephalon maleo)*.

All peregrines are specifically bird hunters which like open country and, as a rule, seize only flying birds. They like to hunt along rocky coasts and over rivers, where they seize birds flying over the water. Their flat or swoop attack is often such that, because of the different speeds, they cannot grasp the flying bird and pull it along; they merely injure it by gliding the claw of their hind toe over it and knocking it down. The falcon then rises steeply, slows down, and swoops after the falling prey. In Central Europe, pigeons, starlings, thrushes, lapwings, and larks make up three-quarters of the prey. Often it only eats the breast muscles of its prey so that the wing skeleton connected through the pectoral girdle is left as typical remains of its meals.

Usually peregrines nest on cliffs, on steep banks, or in the tundra on the ground. Only in the forests of the wide plain extending from northern Germany to Eastern Europe do they also breed on trees, using the nests of other birds, generally those of common herons, sea eagles, and ospreys. In exceptional cases they breed on high buildings, such as skyscrapers, in the midst of large cities. At the nest they are conspicuous because of their loud calls, a sharp "kozick" and a long-lasting "ghe ghe ghe." It is very interesting to see the young perform their begging flight in the period when they are being "trained" by their parents in hunting. While one adult flies over the youngster with prey and drops it, the other adult flies lower and catches the prey if the youngster's swoop at it failed. This adult then climbs up to repeat the process. Finally the adults also bring living birds which the young catch and kill.

The hobbies

D. Hobbies are particularly long-winged; their flight silhouette shows narrow, pointed wings curved in a sickle shape and a short tail, and is reminiscent of a swift (see Vol. VIII). 1. The BAT FALCON *(Falco rufigularis)* occurs in the same area of tropical America as the orange-breasted falcon, but is decidedly smaller; in behavior, hunting method, and voice, it is a typical hobby. 2. The EUROPEAN HOBBY *(Falco subbuteo*; Color plate p. 418) has an L of 30–36 cm, and a weight of about 200 g (male) to about 300 g (female). 3. The AFRICAN HOBBY *(Falco cuvierii*; Color plate, p. 418) inhabits savannahs in Central and South Africa. 4. The ORIENTAL HOBBY *(Falco severus)* inhabits the forests of Southeast Asia. 5. The AUSTRALIAN HOBBY *(Falco longipennis)* has particularly long wings. It inhabits forests of Australia, Tasmania, and the lesser Sunda Islands.

Fig. 16-7. Hobby *(Falco subbuteo)*.

Fig. 16-8. Hobby *(Falco subbuteo)* in flight from below.

The habitat of the European hobby is a varied countryside where woods border on open spaces, preferably with ponds or lakes. Here it can find both food and nesting sites. In very swift flight, it catches flying prey, small birds and especially insects, exclusively. Among birds, the barn swallow, skylark, and house martin constitute two-thirds of its bird prey. Even common swifts are occasionally caught, which is thus the best proof of the hobby's speed. The major insects caught are dragonflies and beetles; it bites off and drops the wings or wing cases of these while in flight, and eats the rest. Over pools surrounded by woods, one may see several hobbies together toward a midsummer evening hunting dragonflies, flashing low over the surface of the water.

The hobby is a definite migrant, returning to its home in April or May. Therefore it starts breeding rather late, about early July. It always uses old nests of other birds, mostly those of crows. The time of rearing the young coincides with the period of greatest abundance of insects. The fledged young are trained to catch prey much as are the peregrines. As soon as the young are independent, the species migrates again to its winter quarters in Southeast Africa and South Asia.

Eleonora's and sooty falcon

E. 1. ELEONORA'S FALCON (*Falco eleonorae;* Color plate p. 417) has an L of 35-38 cm, and a weight of about 400 g (male) to about 480 g (female). In flight it resembles the hobby, but it is larger. 2. The SOOTY FALCON (*Falco concolor;* Color plate p. 418) has an L of about 33 cm. It inhabits the area from the Lybian Desert over the Somali peninsula to Madagascar. 3. The GRAY FALCON (*Falco hypoleucos*) is a little larger; the L is about 38 cm. It inhabits the semidesert of the Australian interior.

Eleonora's falcon was named after one of the few famous women of the fourteenth century, the Princess Eleonor d'Arborea, who distinguished herself as a war leader, regent, and judge of a large part of the Island of Sardinia. In 1392, she issued the codex of laws in the Sardinian language known as the "Carta de Logu." According to the desire of Gené, who first described this falcon, the name is to commemorate this woman who had issued a law for protection of goshawks and falcons and who, "in a way remarkable for the century of barbarism in which she ruled, also protected the honor, life, and property of the people."

Fig. 16-9. Eleonora's falcon *(Falco eleonorae).*

Eleonora's falcon is remarkable because of its very restricted residual distribution and its unusual breeding period. It is found only on Mediterranean Islands, notably those of the Aegean, on the coasts of Crete, Cyprus, Sardinia and the Balearic Islands, and also on the Mogador Islands off Morocco, as well as on the Canaries. On the coasts of these islands, which generally consist of steep cliffs falling down to the sea, it breeds in colonies of a few up to fifty pairs, but its population has declined, for fishermen routinely take the young from the nests to eat them. The strong tie of this falcon to the sea and the fact that it only begins to breed in August are evidently related to its specialized way of feeding when rearing the young. Migrants arriving from the north in the fall are more readily caught over the sea and so form a rich source of food for rearing the young.

The members of a breeding colony form a sort of barrier, which extends out to sea from their breeding island, by flying, facing the wind, and keeping station, awaiting the migrants coming down wind. This "barrier" which extends from the small island of Paximada con-

Fig. 16-10. Merlin *(Falco columbarius)* in flight from below.

Fig. 16-11. Merlin *(Falco columbarius).*

sists of 150 male falcons and forms a "wall" two kilometers long and 1000 meters high. They stand especially close at 800 to 1000 meters. Hunting is carried out here only in the early morning hours, since the passage of migrants slackens off later in the day. But on Mogador, where the migration continues all day, the falcons hunt for longer daily periods, catch more birds, and rear more young. Hartmut Walter, who lived among these falcons on both islands, collected thousands of remains of their kills and found sixty-two species of birds represented among them; the most common were red-backed shrikes, woodchat shrikes, willow warblers, and whitethroats, but also many other Sylvidae, nightingales, redstarts and short-toed larks. He estimates that the 2500 pairs which form this falcon's total population in the Mediterranean annually kill up to one and one quarter million migrants, almost exclusively ones which would have continued their migration over the Sahara.

Besides the kinds of bird species, which may reach the size of the rock partridge, that occur in its breeding areas or pass through as migrants, insects such as beetles, grasshoppers, and cicadas play a great role in the falcon's diet outside the breeding season. It is very sociable, for it not only breeds in colonies, but many times it hunts in company with several others until far into the evening twilight, when it often catches bats. Soon after the young are independent the species migrates to the East African coast, where it winters on Madagascar and the adjacent islands.

The merlins

F. 1. The MERLIN (*Falco columbarius;* Color plate p. 418, and Fig. 16-11) has an L of 23-29 cm, and a weight of 170 g (male) to 200 g (female). There is no cone in the round nostril. 2. The RED-HEADED FALCON (*Falco chiquera;* Color plate, p. 418) has an L of 28-33 cm. The plumage differs according to area. It inhabits India, Central, East, and South Africa, catching mainly pigmy swifts as well as weaver finches and other finches. It lays eggs in the wide leaf sheath of deleb or dum palm trees without any base.

As the hobbies appear to be miniatures of peregrines in looks, structure, and manner of hunting, so merlins are miniatures of the gyrfalcons. Like them, they catch prey in flight as well as on the ground.

The merlin inhabits open country and is found breeding on moors and dwarf shrubs of the tundra in particular. In the fall and spring it is a fairly frequent migrant in Germany, where it may also winter, preferring open country with few woodlands. In swift flight low over the ground, it is often overlooked because it is so small. But when one knows its appearance in flight, one is sure to see this elegant hunter now and then. It rarely perches on trees (this does not apply to the North American subspecies), but generally on a slight rise of the ground from which it can look around.

Its chief prey are small birds up to the size of thrushes. The most common prey species are pipits, linnets, buntings, larks, thrushes, and wheatears. It kills either in a swoop from above, or by an attack from behind and below, following an approach made low over the ground. Occasionally it seems to imitate the manner of flight of its prey in order to approach unrecognized, and only attacks when quite close. Occasionally it takes birds larger than itself such as pigeons; it also takes insects and, to a slight extent, small mammals up to the size of lemmings.

In Scotland it breeds on the ground. In Norway, according to Y. Hagen, it almost always lives in fieldfare colonies where it uses old thrush nests or those of hooded crows. In Norway, hooded crows regularly nest in fieldfare colonies if the merlins are absent. The fieldfares are protected by the merlins while breeding, for they do not take prey in the immediate areas of their nests and, moreover, very vigorously keep away crows and magpies which try to get eggs or young. The usual clutch is four eggs, but in years of lemming abundance it can rise to seven.

Gray kestrels

G. The African group of GRAY KESTRELS include three species which resemble the next two groups in structure and mode of life. They are predominantly gray, and about thirty centimeters long. 1. The GRAY KESTREL *(Falco ardosiaceus)* lives from Senegal and Eritrea to Angola and Tanzania and nests in old tree nests of other birds. 2. DICKINSON'S KESTREL (*Falco dickinsoni;* Color plate, p. 418) lives from Tanzania and Angola to the Transvaal, and prefers crowns of palms or tips of tall tree trunks as nest sites. 3. The MADAGASCAR BANDED KESTREL *(Falco zoniventris)* has a banded underside. All three species are not very agile; they hunt by sitting in wait for small animals on the ground.

The red-footed falcon

H. The RED-FOOTED FALCON (*Falco vespertinus;* Color plate, p. 423 and Fig. 16-12) stands somewhat isolated among the hovering falcons; the L is 27–30 cm, and the weight is about 200 g.

The eastern subspecies, the AMUR RED-LEGGED FALCON *(Falco vespertinus amurensis)*, lives from Lake Baikal to the Amur area, and in Manchuria and north China; it is regarded by some as a separate species. The male shows white instead of black lower wing coverts in flight, while the female has black spotted underparts.

In the west, its normal breeding range extends to Hungary and southern Poland. But occasionally it appears further west as a breeding bird, and has recently nested repeatedly in southern Germany. Many that pass regularly in spring migration in May, stay and breed if conditions are favorable.

This falcon likes open country with old trees, and is very sociable. It likes to fly and hunt with several others, including hobbies or lesser kestrels. Its food is chiefly insects, mainly grasshoppers, crickets, beetles, and dragonflies, which it seizes on the wing, or it spots them while hovering and eats them on the ground. It often hunts late in the

Fig. 16-12. Red-footed falcon *(Falco vespertinus).*

twilight to pursue flying beetles; hence, one of its vernacular German names is the evening falcon, which is also the meaning of its scientific name. Apart from insects, it takes voles and newly hatched young birds, and occasionally frogs, toads, and lizards. In September, 1925, there was an "invasion" of tens of thousands of these falcons in Hungary, coinciding with a mouse plague.

In the breeding season, which does not begin until June, pairs take nesting territories in small woodland patches on the steppe where there are rookeries, since this falcon as a rule uses old rooks' nests. By this time, the young rooks have left the nests. In Hungary over 500 pairs of these falcons have been counted in one spot; generally, however, only up to twenty pairs nest in the same stand of trees. In Asia it is said to nest in trees and ground cavities also. The incubation period of this species is strikingly short: twenty-two to twenty-three days.

Red-footed falcons begin to leave in late summer. They winter in the savannahs and steppes of East and South Africa, where, together with lesser kestrels, they follow grasshopper swarms. The eastern subspecies, *Falco vespertinus amurensis,* also winters principally in South Africa and may occur there in thousands in connection with grasshopper invasions.

The Old World kestrels

I. The last group, one with many species distributed over almost the entire world, are the kestrels and its relatives. 1. The OLD WORLD KESTREL (*Falco tinnunculus;* Color plate, p. 423) has an L of 32-34 cm, and a weight of 190-220 g. (♂) to 210-290 g (♀). 2. The LESSER KESTREL (*Falco naumanni;* Color plate, p. 423) has an L of 26-30 cm, and a weight of about 200 g. 3. The AMERICAN KESTREL (also called the sparrow hawk in America) (*Falco sparverius;* Color plate, p. 423) is the smallest of the true falcons; the L is 20-28 cm, and the weight is 100-200 g. It is found in about twenty subspecies from Alaska and Newfoundland to Tierra del Fuego. 4. The MOLUCCAN KESTREL *(Falco moluccensis),* with an L of 28-30 cm, is found in Java, Celebes, and Moluccas. 5. The AUSTRALIAN (NAUKEEN) KESTREL *(Falco cenchroides)* has an L of 30-33 cm. 6. The AFRICAN KESTREL *(Falco rupicoloides)* has an L of 28-33 cm, and inhabits upland steppes and savannahs of East and South Africa. 7. The MADAGASCAR KESTREL *(Falco newtoni)* with an L of about 27 cm, has a pale and a dark phase; it inhabits Madagascar, Aldabra, and the Komora Islands. 8. The SEYCHELLES KESTREL (⟠ *Falco araea*) has an L of 22 cm. 9. The MAURITIUS KESTREL (⟠ *Falco punctatus*) is a little larger, and is almost extinct. 10. The FOX KESTREL (*Falco alopex;* Color plate, p. 423) has particularly long wings and tail; the L is 34 to 38 cm. It is found from Ghana to Ethiopia, and likes to live on cliffs.

The kestrel is by far the most common of all raptors, at least in Europe. This is probably due to the fact that it is not choosy in its demands on its habitat. It can often be observed in the midst of cities, where it nests on church towers and other large buildings. It also breeds on the towers of castles and ruins. In mountains or other areas where there are rocks or quarries, it breeds in cliffs or cavities. But in Central Europe it breeds most often at the edges of woods or in isolated trees, where it can use old nests of crows or other birds. Occasionally it breeds in holes in trees or nest-boxes; on the North Sea coast it even breeds on the ground.

Fig. 16-13. Kestrel *(Falco tinnunculus).*

It needs country as open as possible for hunting. It can be recognized in flight from afar by its long pointed wings, long tail, and especially its characteristic hovering flight, which it uses when searching for food. When hovering, it slants its tail downward and spreads it more or less widely, depending on the strength of the wind. Thus it can observe the ground below and wait for prey to appear. Generally it hunts field mice. As soon as it spots one while hovering, it drops a little, and then at the favorable moment it swoops down steeply onto the prey. Often it rises up again with empty talons, the mouse having escaped just in time. Tirelessly, the kestrel goes on hovering elsewhere until its attack succeeds; then it flies to an elevated point to eat its prey.

Fig. 16-14. Lesser kestrel *(Falco naumanni).*

In general it swoops almost exclusively onto prey which is sitting or running on the ground; it rarely hunts flying insects or small birds. Its food, therefore, consists of 85% mice, mostly field mice. This is due to the fact that hunting by hovering is best out in the open, over fields and meadows where field mice often live in greatest numbers. It also takes moles, shrews, lizards, beetles, and grasshoppers. In Central Europe it is in part a resident, in part a migrant, depending on the food supply and weather. Those present in this area in winter are largely immigrants from further north or east.

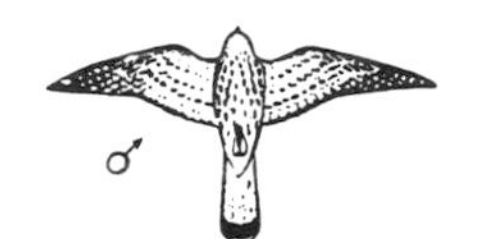

Fig. 16-15. Kestrel from below in flight *(Falco tinnunculus).*

The lesser kestrel is a "twin" species of the common kestrel, resembling it so closely in shape, color and behavior that the two may be mistaken for one another. Both species occur together from the Mediterranean to Central Asia, and often breed in close vicinity without interbreeding. But they are easily recognized by their different calls; the lesser kestrel's call is a hoarse "kechel" and is often heard near breeding colonies. As shown by the northern limits of its range, the lesser kestrel needs a warmer climate than the common kestrel. It likes the sunny slabs of rocky hills and dry meadows where it finds abundant prey, such as grasshoppers, crickets, and beetles, besides reptiles and small rodents which it seizes by swooping after hovering or by walking.

Because it is a social bird, it always breeds in colonies. Pairs use holes in cliffs, in walls, or in roofs, which are often so close together that the half-fledged young may crawl from one nest site into another and be fed there too. Occasionally it breeds in holes in trees or in nest-boxes put up especially for it, as they are in Carinthia and Styria (Austria), which are the most northern breeding areas in Central Europe. In the Balkans and Southwest Asia, it is so tied to buildings in many places that it is one of the most common and tame birds of the towns. A definite migrant, the lesser kestrel leaves its breeding area in early fall and winters mainly in the steppes and savannahs of East and South Africa, where it is just as sociable as at its breeding places. In the southern summer, flocks of thousands have been observed feeding mainly on grasshoppers.

The American kestrel also lives partly on grasshoppers and has also been called the grasshopper hawk there. In the last few years this small falcon has reached Europe a few times. In Europe, Mrs. A. Koehler first bred it in captivity. Birds raised by her came to the Frankfurt Zoo, where they have been kept since 1961 and have bred annually since 1965. A female raised in 1967 in the zoo was bred when only one year old, according to R. Faust. The four broods observed at Frankfurt were always of two or three eggs, but no more than two youngsters ever were successfully raised.

Only a few species of the large raptors have been observed closely enough to give us adequate knowledge of their biology. Man has condemned them too readily in the past, and has neither recognized nor appreciated properly the significance of individual species in their habitats. Only exact knowledge will enable us to prevent further extermination of these species which are so important for the balance of life in various habitats.

17 Gallinaceous Birds: Mound Birds, Curassows, Guans, and Chachalacas

Order: Galliformes

GALLINACEOUS BIRDS (order Galliformes) are an old group of birds whose fossils have been found as early as the Eocene (about fifty to sixty million years ago). They are mostly medium to large in size, and only a few species are small. The L is 12–235 cm, and the weight is 45–11,000 g; in domesticated forms the weight reaches 22,500 g. There are ten primaries—the outer secondaries are generally very short. The feathers often have a well developed aftershaft. Generally downy feathers are found only on the pterylae. There are no powder downs, but the preen gland is present. The males of many species are often very colorful, with widespread iridescent colors. The females generally have a protective coloration. They have very strong breast muscles which enable them to fly up quickly (except for the hoatzins). The tarsus is located anteriorly, and has horny plates (with the exception of the hoatzins) arranged in two longitudinal rows. Many species have spurs, particularly the males. They are predominantly ground birds with strong feet, often exposing their vegetable food by scratching. They have a strong beak, and almost always, a roomy, distensible crop which acts as a food reservoir. There is a very strong gizzard between whose grinding surfaces and with the help of small stones swallowed for this purpose, grains and green food are ground up. They generally have a long caecum for cellulose digestion. There is a gall bladder in all species.

There are two suborders: Galli, including the families MOUND BUILDERS, CURASSOWS, PHEASANTS and pheasant-like birds; and Opisthocomi, with the crested fowl, the HOATZIN (Vol. VIII) as the sole species. There are ninety-four genera and 263 species in total. They are distributed over most of the world, in semideserts, steppes, savannahs, forests, and cultivated country, and mountains up to far above the tree line (6000 meters). All gallinaceous birds like to bathe in dust or sand, but not in water.

The Galli are of importance to man, for they include four widely distributed domestic birds, including the domestic chicken. Fowl-like birds have always occupied a preferred position on our menus. For the modern hunter, fowl are particularly important and are therefore carefully tended in many cultivated lands. They are also very popular as ornamental fowl. No other group of birds has been transplanted and newly introduced by men to such an extent. As strict residents, the fowl-like birds are very suitable for this, for only four species are truly migratory. Man has distributed the pheasant *(Phasianus colchicus)* over most of the world. In the United States alone, sixteen to eighteen million pheasants are shot each year. The European partridge *(Perdix perdix)* and the Eurasiatic rock partridge *(Alectoris graeca)* have been introduced in North America. In Europe, many attempts at introduction and transplantations have been made often at great expense with less than twenty-two species. Most of them, however, have been without success; although, in New Zealand alone, nine gallinaceous species from America, Europe, Asia, Africa and Australia have been successfully introduced.

Evidently the great majority of all Galli can reproduce when one year old. Most species lay many eggs. In the European partridge, up to twenty-six eggs have been found in one clutch. Incubation is performed almost without exception by the hen alone. Mound-building birds do not incubate at all. Newly hatched chicks have a dense, protectively-colored down plumage and are soon able to feed themselves. They can fly in the first few weeks, sometimes even in the first few days. The wings of young of the true Galli are, however, still incomplete, having only seven short primaries. They lack secondaries. This "first wing" is much smaller than that of adults but suffices for the chicks' flight. With the increase of the bird's weight, the primaries and secondaries which were lacking grow. The inner primaries, which are too short, are replaced by longer ones. The replacements fit in with the outer primaries of later growth, which from the start are about the final length, and so are not necessarily replaced.

All true Galli not only have a "first wing" of short duration, but they also have a smaller, still incomplete "first tail" in many species. Its surface area is in accordance with the needs of the chick during the first weeks. As adults, many species moult the tail from the inside towards the outside (centrifugally). Others moult from the outside towards the inside (centripetally). Still others begin the moult in each half of the tail with a feather which lies between the central one and the outermost one.

Family: Megapodiidae (Great-footed birds), by H. J. Frith

The MOUND BIRDS (family Megapodiidae) are dark-colored Galli which differ from all other birds in their particular method of incubating their eggs. They are between the size of domestic fowl and turkey. There are seven genera with twelve species in the southwest of the Old World, with New Guinea as the focus. They are divisible into two tribes according to size:

A. 1. The mound birds proper (Megapodiini) are small dark birds with short tails, often insular. They include the SCRUB FOWL (genus *Megapodius*) with the species MICRONESIAN SCRUB FOWL (*Megapodius laperouse*; Color plate, p. 424), found on the Marianas, the NIUAFOO SCRUB FOWL (♦ *Megapodius pritchardii*) found on Niuafoo (central Polynesia), and the AUSTRALIAN SCRUB FOWL (*Megapodius freycinet*) in many subspecies from the Nicobar islands to northern Australia and Polynesia. 2. MALEO (♦ *Macrocephalon maleo*; Color plate, p. 424) is found on Celebes. 3. WALLACE'S EULIPOA (*Eulipoa wallacei*) is found on the Moluccas.

Mound builders: A pair of brush turkeys (*Alectura lathami*), with their mound.

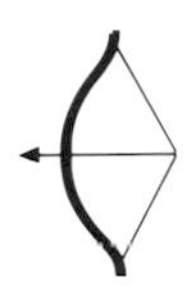

B. The large mound builders (Alecturini) are larger and are mostly tied to definite habitats. Their distribution is limited. 1. LATHAM'S BRUSH TURKEY (*Alectura lathami*) is found in eastern Australia. 2. TALEGALLAS (genus *Talegalla*), with three species on New Guinea, include the WESTERN TALEGALLA (*Talegalla cuvieri*). 3. The COMBED TALEGALLAS (genus *Aepypodius*), with two species on New Guinea, include the WATTLED BRUSH TURKEY (*Aepypodius arfakianus*). 4. The MALLEE FOWL (*Leipoa ocellata*; Color plate, p. 424) inhabits dry areas in central, southern Australia.

How the breeding mounds were discovered, by B. Grzimek

The ancient Egyptians first built ovens to incubate fowl eggs artificially and we do this today in electrically heated incubators. However, the mound birds, also called incubator birds, "discovered" this method much earlier. Many species lay their eggs near hot volcanic springs or even hotter lava. Others use the heat generated by rotting leaves and vegetation. Still others go to the seashore and lay their eggs in the sand where they can be warmed by the sun. This may almost sound like a fairy tale and it was, in fact, not believed for a long time. Some time after Magellan's unfortunate circumnavigation of the world (1519–1522), the Dominican monk Navarette brought back from Australia the story of fowl which laid eggs larger than themselves in heaps of leaves. The eggs, then, hatched without any further care. The matter of the size of the eggs was not quite correct; in regard to the second point, people believed more in mermaids and giant sea serpents than in such skills in fowl. When, centuries later, the first settlers arrived, they looked on the large mounds as toy castles made by native mothers for their children. In northern Australia they were believed to be grave mounds.

Fig. 17-1. Jungle fowl (genus *Megapodius*): N—Nicobar Islands, M—Marianas, H—New Hebrides, Nf—Niuafoo.

Only in 1840 did the naturalist John Gilbert conceive the obvious idea of digging up a mound. There were, as the natives had always claimed, eggs in it. The eggs are fairly large, weighing 185 grams. Since a mound bird is about the size of a domestic fowl which lays eggs of fifty to sixty grams, mound bird eggs represent twelve percent of the bird's body weight, while the domestic fowl's egg is only four percent. In parts of Polynesia, particular breeding mounds often have human "owners" who regularly take the megapode eggs away.

The various incubation procedures of mound birds, by H. J. Frith

Mound builders use various means to incubate their eggs. The maleo (*Macrocephalon maleo*) and Wallace's eulipoa (*Eulipoa wallacei*) use only the sun's heat. They emerge from the dark forest and dig their

1
2
3
♂
♀
4
5

◁ Cracidae: 1. White-headed piping guan *(Pipile cumanensis)*; 2. Helmeted curassow *(Pauxi pauxi)*; 3. Black curassow *(Crax alector)*; 4. Horned guan *(Oreophasis derbianus)*; 5. Razor-billed curassow *(Mitu mitu)*.

eggs into areas along the shore which are in sunshine. According to some reports, the maleo also breeds in the interior of the country and is said to bury its eggs in places warmed by volcanic heat. In the jungle fowl *(Megapodius)* the method of egg disposal varies greatly. The Australian scrub fowl *(Megapodius freycinet)* in particular, often selects places heated by volcanic action, such as New Britain and the Solomons. There it digs burrows, which may be up to a meter deep, into warm soil. In other places these birds build large brood heaps, which have a diameter of twelve meters and a height of five, of sandy soil and leaves. These are the largest structures built by birds. The incubation heat is provided partially by the sun and partially by fermentation of the leaves. In these places directly exposed to the sun, the heaps consist almost entirely of soil. In dense jungle they consist, however, almost entirely of leaves. Each egg is laid in the passage which has been dug out at a place where there is a suitable incubation temperature. After egg-laying, the birds take no further steps to control the temperature.

Talegallas and comb talegallas

The breeding methods of the genera *Alectura, Talegalla,* and *Aepypodius* are quite similar. Their brood mounds are always in dense forest and therefore are completely shaded. Thus birds depend entirely on the heat of fermentation for egg incubation. The cock scratches together leaves and other plant fragments with its large feet, making a heap three meters in diameter and one and one half meters high. Before egg-laying, the cock visits the mound daily, turns over the leaf mould, and makes sure that it gets permeated by rain and is simultaneously ventilated. As a result, fermentation is rapid and an initial "fermentation burst," with very high temperatures, occurs.

Fig. 17-2. 1. Maleo *(Macrocephalon maleo)*; 2. Wallace's eulipoa *(Eulipoa wallacei)*; 3. Talegallas (genus *Talegalla*); 4. Combed talegallas (genus *Aegypodius*); 5. Mallee fowl *(Leipoa ocellata)*; 6. Brush turkey *(Alectura lathami)*.

When this initial "burst" is over, the cock lets the hens come to the mound to lay their eggs into holes dug into it. First the hens test the decaying leaves with open beaks to make sure that the temperature is about right. Each hen lays ten to thirteen or more eggs. In Latham's brush turkey (*Alectura lathami;* Color plate, p. 435), these need seven to twelve weeks of incubation. During egg-laying and incubation, the cock continues to bring in new plant material and to dig it into the mound. This insures a constant production of heat. The cock must work all the time in order to keep the material around the eggs at the right temperature.

How does he do this without a thermometer? Now and then he scratches holes into his incubator from above and sticks his head inside. I have also observed, in another species, that the cock takes sand out of the depth of the mound with his beak. Mound birds probably have a temperature sense in the beak, perhaps on the tongue or on the palate.

The mallee fowl

The MALLEE FOWL (*Leipoa ocellata;* Color plate, p. 424) must overcome much more difficult problems than its relatives. In contrast to the other mound builders, it inhabits the dry Mallee Brush areas of Central Australia, where there are large temperature fluctuations during the day as well as during the year (on many days temperatures vary by more than 40°C). The cock also finds little organic material on the ground in these areas and when he heaps it up, it does not ferment but dries out and blows away, or is eaten by termites. Therefore, the male mallee fowl builds his mound in quite a special way. It rather resembles the wood pile of the brush turkey, but is removed from the drying, dessicating action of the dry air by being buried deep in the ground. The whole business takes a long time. For ten months in a year, the birds are busy almost from morning to evening in maintaining the right temperature inside their incubation mounds.

During the winter, the cock digs out a pit which is two meters wide and one meter deep. It gathers together all the dry leaves and twigs it can find within a circle of fifty meters, and brings them into the pit, where the light winter rains then provide moisture. When this has happened, the cock covers the moistened plant material with sand. His activity finally produces a heap above the pit five meters in diameter and one and one half meters in height. The leaves and twigs that are buried, decompose and produce the necessary heat. Four months pass until the cock is ready for eggs to be laid. Then the hen lays at intervals of five to seventeen days, placing the eggs in a hole inside the structure.

However, the cock's work is not yet over. He attends the mound during the spring, summer, and fall, for he must insure that its temperature never deviates much from 33.5°C. Since the development of the embryos in the mound takes seven or more months, the external temperature naturally varies a great deal with the season.

In the spring the mound must be kept warmer than the average air and soil temperatures. In summer it must be cooler than its environment, and in the fall, once more it must be warmer. The fermentation of the buried material is extremely rapid in spring, while in the height of summer only little heat is produced by fermentation. Thus in the spring, the cock digs holes daily to let off heat, in order to prevent overheating of the eggs. In midsummer, however, he brings more solid material to the top of the mound, for the sun is very hot (the external temperature reaches 46°C) and the eggs could get too hot. In addition, during hot weeks, the cock opens up the pit at dawn every few days and fills it up again, mixing in soil which has become cooled by the morning air. Thus he reduces heat conduction from the surface to the interior. In the fall the temperature of the egg chamber could readily fall below 33.5°C, so the cock makes the mound flatter so that only a few centimeters of soil cover the eggs. At the warmest time of each day, he comes along and scratches soil away from the mound into a shallow layer which gets warmed by the sun. Later he covers the egg chamber once with layers of the sand which has been warmed. When the shorter and cooler days of fall make this activity impossible and the temperature of the mound decreases, the incubation period ends.

The mallee fowl and the brush turkey are strictly limited to territories. The territory is up to fifty hectares in area. It is selected during the winter and contains a spot suitable for a mound. The cock does most of the work of mound-building and defends his structure against other cocks. He also prevents the hen from approaching the mound before a suitable temperature has been reached in the spring by attacking her and chasing her off. The other mound birds are less territorial. In those species which lay eggs into sun-warmed or volcanically heated sand, several birds may even place their clutches in closed colonies. The cock and hen of the ordinary mound builders build the mound together and form a closely knit pair. It has often been asserted that occasionally more than one hen of this and other species may deposit eggs in one mound. But there is no proof of this so far.

Egg-laying

Behavior before and during egg-laying differs in the various species. The cock mallee fowl opens the mound and the egg chamber before the laying of each egg by scratching away the sand above it. To do this, he must move one to two cubic meters of sand, which takes several hours. Only then is the hen allowed to approach. She checks the exposed pit with her beak and chooses a spot in the egg chamber where she digs a small niche. After deposition of the egg, the hen usually withdraws and the cock rebuilds the heap to its original height. In the brush turkey, the hen digs out a niche in the surface of the mound, lays the egg in it, and covers it herself. Since different parts of the mound are at different temperatures depending on composition and the progress of fermentation, the hen sometimes digs several pits and checks their temperature before selecting one. In the ordinary mound bird and the maleo, cock and hen work together at digging and at the subsequent filling in of the egg pit.

Evidently the beginning and the duration of incubation depends largely on the supply of necessary heat to the eggs. The brush turkey and the mallee fowl lay their first eggs in September, which is the Australian spring, because at that time the temperature in the mound reaches the level required for development of the embryo. If it is such an unusually dry spring that the materials in the mound will not ferment, the birds postpone egg-laying or give it up altogether. In hot, dry summers they stop egg-laying earlier than normal. In mild summers the period of egg-laying is more extended. Some species of jungle fowl, however, live in areas where the temperature remains high and fairly equable all the year round and the precipitation also is evenly distributed. Hence these birds lay eggs throughout the year. In a mild breeding season with a favorable laying period, a female mallee fowl may lay up to thirty-five eggs.

The chicks hatch deep underground

Mound bird chicks emerge from the eggs as deep as one meter

below the surface. It may take them several hours to work their way up and out. In some species the young have a minute egg tooth before hatching, but it is not used. They destroy the egg shell completely by the pressure of their legs and necks. After hatching, they rise to the surface of the mound. This can be dangerous for mallee fowl chicks. In very hot weather, some chicks suffocate before reaching the top. Once they arrive at the top, having dug through about a meter of loose sand, they are so exhausted that they can usually only just manage to drag themselves to the nearest cover. There they rest for an hour or more before wandering further into the brush.

The chicks of brush turkeys have a rather simpler task. They only have to burrow out of loose leaves into the open. Therefore they are not so out of breath when they arrive on the surface, and they can quickly run away. I have watched them hatch through a glass screen with which I had divided an artificial mound. When such a little bird shows its head on the surface, it looks into a rather hostile world. Its father and mother pay no attention to it; it avoids them just as it does all other living things it meets. Fortunately it can already flutter and within at most twenty-four hours it flies to a low branch. The chicks are still downy, yet their wings already have closed feather vanes. Every flight feather is enclosed in a slimy gelatin-like sheath at first, which gets stripped off during the climb up to the surface of the mound. Thus, when the little brush turkey first reaches daylight, its flight feathers are already fully formed, dry, and suitable for flight.

Breeding of mound builders in zoos, by B. Grzimek

Many questions on the mound birds could only be properly clarified when the birds bred in captivity. At first it was not expected that young brush turkeys *(Alectura lathami)* would actually hatch in rainy weather at the Frankfurt Zoo. But as early as 1872 one was reared in the Berlin Zoo, and another in 1932. Success was not frequent. But one morning one of our keepers found a little gray creature sitting under some wooden steps fairly far from the brush turkey enclosure. At first sight he took it for a rat. It had wandered a long way through wire fences. Altogether we have reared thirty brush turkeys in the succeeding six years. These mound birds are already fully grown within a year after hatching and so set about building mounds themselves. This is not so simple in a zoo, because they need huge masses of leaves for this purpose. If one drops a wagon load of leaves for the cock through the cage door in the early morning, he has spread it all a few meters around the whole enclosure and built it into a brood mound by the late morning.

S. Baltin and Dr. Faust and his wife tried to hatch brush turkey eggs in an incubator in Frankfurt. At first they failed. Not a single egg showed even the beginnings of embryonic development. Evidently the temperature corresponding to that used for domestic hens' eggs was too high. Next year the researchers set it at 33.6 to 34.4°C and placed the eggs in a glass aquarium in leaf mould, which was moistened regularly. The air was 78% saturated. One egg which lay in a natural mound for thirty-three days hatched after fifteen more days in the incubator. A second newly-laid egg placed in the incubator immediately developed for a while, but the embryo died at twenty-one days. The first chick hatched in the Frankfurt incubator came out of the third egg. Its development took forty-seven days.

The Frankfurt incubator-hatched chicks were, at first, unable to stand upright, and could only walk properly after about twenty-four hours. Only then did they look like newly emerged chicks hatched naturally in the mound.

When one first hears of the "invention" of the mound birds, one wonders why all birds do not use such "incubators." It was formerly believed that mound birds were a primitive group of birds, for their method of incubation seemed similar to that of reptiles. However, now that we are more familiar with the mound birds and have studied them, we know that there are actually many differences between their breeding process and that of reptiles. Their peculiar method of incubation is surely not a primitive characteristic, but was developed by birds belonging to the gallinaceous group. When one watches these "hard workers" scratching leaves and soil back and forth for months, from early morning to late evening; digging holes and fiercely chasing away any creature that looks even approximately fowl-like, it becomes clear that the whole business was not really a matter of "progress." The other gallinaceous birds, as it were, "surmised" that the good old way was more comfortable: one might as well settle on the eggs and sit there quietly for a few weeks.

Family curassows, guans and chachalacas (Cracidae), by A. Skutch

The second family of the gallinaceous birds, the Cracidae, are confined to subtropical and tropical America. They range in size from scarcely as large as a black grouse *(Lyrurus tetrix)* to almost turkey-sized tree fowl. The L is 42–100 cm. They are slim, long-legged, and have short, rounded wings and a fairly long tail which may be a little shorter or a little longer than the wings. The beak is strong but fairly short. It is lightly curved, often with a conspicuous cere at the root. It also often has a knob of variable shape. The feet are like those of the mound birds and, in contrast to the Phasianidae, they have long, well-developed hind toes on the tarsus at the same level as the other toes ("pigeon-footed"). It has a long caecum; in many species the trachea is prolonged. The plumage is mainly a shiny black or olive-brown to reddish-brown, often with white marks, which sometimes form a helmet-like crest on the head of many species. The feathers lack an aftershaft.

They are mainly plant eaters, but eat insects and other small animals to a lesser extent. The nests are generally found on trees. They are small, and built with twigs and leaves. At present they are only distributed in Central and South America, from southern Texas and Mexico to northern Argentina. Fossils also have been found in North America. There are eleven (7–11) genera with forty-six (37–47) species.

Fig. 17-3. Range of the family Cracidae.

A. The tribe of large CURASSOWS (Cracini) has recently been classed by some investigators as a single genus *(Crax)*: the head generally has an erectile crest of stiff, forward-curled feathers. In some species the males have a fleshy knob or hump on the root of the beak. Sometimes there are bare, brightly colored areas of skin on the head, and the cere of the beak is also often colorful. 1. The CRESTED CURASSOWS *(Nothocrax)* have one species, the NOCTURNAL CURASSOW *(Nothocrax urumutum)*, which is found from southwestern Venezuela to northeastern Peru. 2. The HELMETED CURASSOWS *(Pauxi)* have a bony outgrowth on the forehead. There are two species: among them are the HELMETED CURASSOWS (*Pauxi pauxi*; Color plate, p. 436), found in northern Venezuela and northeastern Colombia. 3. The MITU CURASSOWS *(Mitu)* have a shorter head crest and a tall, narrow red comb running along the forehead and upper mandible of the beak. There are three species, including the RAZOR-BILLED CURASSOW (*Mitu mitu*; Color plate, p. 436) found from the Amazon area to Bolivia and Mato Grosso, and the CRESTLESS CURASSOW *(Mitu tomentosa)*, found in South America north of the Amazon. 4. The ORDINARY CURASSOWS (*Crax*, in a narrower sense) have eight species; among them are the GREAT CURASSOW (MEXICAN) *(Crax rubra)*, with a bright yellow knob on the forehead, found from southern Mexico to western Ecuador, and the WATTLED CURASSOW *(Crax globulosa)*, which has a reddish-yellow cere. The males are black with a white belly, while the females have reddish-brown belly feathering; they are found from northern Brazil to northeastern Peru, Bolivia and Mato Grosso; the YELLOW-KNOBBED CURASSOW *(Crax daubentoni)*, with a yellow forehead, is found from Venezuela to Colombia; and the BLACK CURASSOW (*Crax alector*; Color plate, p. 436) is found from eastern Colombia and northern Brazil to Guiana.

B. Medium-sized to small Cracidae, tribe GUANS and CHACHALACAS (Penelopini): 1. The PLAIN GUANS *(Chamaepetes)* have two species, including the BLACK-BELLIED GUAN *(Chamaepetes unicolor)* found in Costa Rica and western Panama. 2. The ABURRIS *(Aburria)* have a naked throat with a pendant skin wattle. There is one species, the WATTLED GUAN *(Aburria aburri)* found from northwestern Venezuela and Colombia to Peru. 3. The PIPING GUANS *(Pipile)* have four species, including the WHITE-HEADED PIPING GUAN (*Pipile cumanensis*; Color plate, p. 436), which is found from Colombia to Paraguay. 4. The GUANS proper *(Penelope)* have thirteen species, including the BAND-TAILED GUAN *(Penelope argyrotis)* which is found in Venezuela, Colombia, western Ecuador and northwestern Peru, and the CRESTED GUAN *(Penelope purpurascens)* found from Mexico to Venezuela, northern Colombia and southwestern Ecuador. 5. The *Penelopina* has one species, the GUATEMALAN BLACK CHACHALACA *(Penelo-*

pina nigra), which is found from southern Mexico to Nicaragua. 6. The CHACHALACAS proper *(Ortalis)* have ten species, including the CHESTNUT-WINGED CHACHALACA *(Ortalis garrula),* which is found from Honduras to northern Colombia, the PLAIN CHACHALACA *(Ortalis vetula),* found from southeastern Mexico to Nicaragua, and the RUFOUS-VESTED CHACHALACA *(Ortalis ruficauda)* which inhabits northeastern Colombia, northern Venezuela, and Tobago. 7. The HORNED GUANS *(Oreophasis)* are red and have a somewhat backward slanting spur-like bone of five centimeters in length on the crown. There is one species DERBY'S MOUNTAIN PHEASANT or HORNED GUAN (⊕ *Oreophasis derbianus;* Color plate p. 436) found in Guatemala and Chiapas (the most southern province of Mexico).

The Cracidae are very different from the mound birds in their reproductive behavior; otherwise they resemble the mound birds in some aspects of structure. They replace the pheasants of Eurasia in tropical America and are in some respects also reminiscent of the American turkeys. Spanish-speaking Latin Americans therefore call them "Pavos" or pavones (turkeys), or faisanes (pheasants).

The large members of the family are birds of dense tropical forest country. Some live in areas with a long dry season where the trees periodically shed their leaves, or in the "fringing forests" which extend along waterways in an otherwise treeless country. The smaller species, on the other hand, avoid the interior of dense forests and are at home in light secondary growth, like woods newly grown on formerly cultivated land. When the smaller species are not pursued too much, they also live on plantations and near houses as long as some trees and shrubs are there. Most species inhabit warmer lowlands. Only a few like the horned guan *(Oreophasis derbianus)* and the black-bellied guan *(Chamaepetes unicolor)* occur in cool mountain forests up to an altitude of at least 3000 meters.

All are eminently adapted to tree life. They walk about lightly and skillfully on thin branches in the tops of trees. When seen as silhouettes against the sky, their slim bodies appear attractive and dainty. When they have to cross a woodland clearing, they rise by jumping, and flying from branch to branch until they have reached the top of the highest tree at the edge of the clearing. From there they launch themselves into the air and as soon as they have gained enough speed in their fall, they spread their wings and glide downwards, often over distances far greater than one hundred meters. Only when this glide will not enable them to reach their goal do they beat their wings to gain height. However, occasionally even heavy species like the great curassow *(Crax ruba)* undertake more distant aerial journeys. In Guatemala, I once watched one beating its wings with effort as it flew over a high, bare hill to reach the forest beyond. As far as is known, all Cracidae roost on trees overnight.

They feed mainly on fruit and seeds which ripen during the course of the year in the tropical forests. They swallow berries and other small fruits singly and in one piece. However, they bite into larger fruit like mangoes and guavas *(Psidium guayava).* They also bite off soft leaves and opening buds. In Costa Rica, I saw three crested guans *(Penelope purpurascens)* spend half an hour stuffing themselves with the tender, young leaves of a climbing plant which had overgrown a dead tree at the forest's edge. One of these birds tore a piece from a leaf which the other held in its beak. The original "owner," however, did not get excited. Guans love sprouting beans and can do much damage to bean fields. The smaller species sometimes eat flowers. Most species seek food in trees and bushes where one can even see them hanging upside down while eating. Now and then they come down from the trees to eat. The large curassows *(Crax, Pauxi)* gather great amounts of fallen fruit on the forest floor. Insects, snails, and other small animals form only very small parts of the menu. When feeding on the ground, Cracidae, in contrast to many other Galliformes, do not appear to scratch the ground. Strangely enough, they sometimes fill their stomachs with soil.

A crop is present as a dilatation of the gullet in curassows and in the horned guan. Other members of the family which lack a crop have a distensible gullet, so that food can be stored in it before digestion begins. The stomach is emptied first, and then food enters it from the crop or the gullet. A few species regurgitate indigestible seeds.

The strong voice of curassows

The vocalizations of the Cracidae are loud and sometimes arousing, but rarely pleasant to the human ear. In a number of species, the vocal power of the cocks, and sometimes of the hen as well, is increased by a prolonged trachea. It runs far back between skin and breast muscles, and then turns and runs forward to the point of entry into the chest cavity. The call of the helmeted curassow *(Pauxi pauxi)* is a prolonged, low-pitched grunting or groaning which sounds like "mm-mm-mm-mm." The cock produces it by breathing out with a closed beak. The great curassow *(Crax rubra)* also utters a low-pitched call, "boo boo boo," which is audible over a distance. The yellow-knobbed curassow *(Crax daubentoni),* on the other hand, sounds a soft whistle, "yeeeeeee," which lasts four to six seconds. Crested guans are particularly noisy. When one enters their forest home, they perch high up in trees and continually protest with a very loud prolonged shrieking which sounds peculiarly high for such relatively large birds. Probably these calls warn others of stalking tree mammals, such as cats. It is, however, suicidal when uttered at the approach of a hunter, for it emphatically draws his attention to the birds. When disturbed, horned guans give off a throaty (guttural) shriek, which in its suddenness and intensity, has the effect of a loud explosion. Then they threaten the intruder from a high perch by clattering their yellow beaks like castanets.

▷

Grouse: 1. Spruce grouse *(Canachites canadensis);* 2. Ruffed grouse *(Bonasa umbellus);* 3. Blue grouse *(Dendragapus obscurus);* 4. Greater prairie chicken *(Tympanuchus cupido).*

1
2
3
4

4
6
1 ♂
3 ♂
5 b
5 a
2 ♂

The morning chorus of the plain chachalaca

The full morning chorus of the plain chachalaca *(Ortalis vetula)* is unforgettable. One of these birds, sitting in a tree above dense secondary growth, calls with a rough, unmelodic, but remarkably strong voice, "cha cha lack, cha cha lack." The neighbors take part and a real din of loud calls arises. When those nearby have become quiet, one hears other more distant voices. The chorus seems to decline until from a distance of one kilometer, one can hardly hear any more. Then the noise surges back with increasing strength and finally an ear-splitting din is produced by a group of six to eight of the birds situated vertically above the observer. The rufous-vested chachalaca *(Ortalis ruficauda)* of Venzuela gives a similar performance. Its native name, "guacharaca," is a good reproduction of its temperamental call. The bird seems to be trying to say "ch-chlaka" with a full "mouth." It calls loudly during moonlight, but its fullest choruses are to be heard at daybreak. It keeps calling on and off all morning during the rainy season of April and May. In the last two species mentioned, the voices of some "singers" are conspicuously softer and higher pitched than others. No doubt these are the hens. Strangely enough, the chestnut-winged chachalacas *(Ortulis gurrula),* which I know well in Costa Rica, have never behaved in this way.

Apart from vocal utterances, several species of this family also produce drumming or clattering sounds with the wings. Thus a crested guan may climb to the top of a high tree at the edge of a clearing and fly with slow measured beats over the open space. When it has gained enough speed, it will beat its wings much more rapidly and so produce a loud drumming noise. Then it may glide for a stretch, drum again, and continue its flight across the clearing into the trees on the opposite side. This peculiar drumming is heard only rarely, just at dawn or dusk and on moonlit nights.

◁ Forest grouse: 1. Capercaillie *(Tetrao urogallus)*; 2. Black grouse *(Lyrurus tetrix)*; 3. Rackel grouse (Hybrid between black cock and hen capercaillie); 4. The red grouse *(Lagopus lagopus scoticus)* does not assume a white winter plumage and thus has brown instead of white wings; 5. The willow ptarmigan *(Lagopus lagopus)* in summer (a) also has white wings; the winter plumage (b) is all white with a black tail; 6. Hazel grouse *(Tetrastes bonasia)*.

The black-bellied guan's *(Chamaepetes unicolor)* wing sound is quite different. When this bird, in its long glide, has reached the middle of a clearing, it beats its wings rapidly over a short stretch in such a way that the longer feathers alternately separate and beat together. Thus it produces a wooden-sounding clatter of surprising loudness, which can be imitated by drawing a thin, narrow piece of wood over an iron grating or by holding it against the spokes of a wheel. This instrumental sound of the black-bellied guan sounds "sharper" than the crested guan's drumming, because in the latter the ends of the two outermost primaries have almost no vanes, but only bare shafts.

Since the Cracidae are so shy and generally avoid man, not too much is known as yet about their social life and their breeding behavior. There are indications that at least some species live in polygamy. Thus guans are found at all seasons more or less socially in flocks, and their nests sometimes stand together in groups. Male yellow-knobbed curassows may be found in the breeding season with three, four, or occasionally more hens. Black-bellied guans, on the other hand, live singly outside the breeding season and are only found in pairs during the breeding season. With them, therefore, a temporary monogamy seems to apply. Pairs of crested guans own territories in which they may remain with their young until the next breeding season. Possibly they maintain a permanent pair bond. After the breeding season, guans sometimes join in flocks at spots particularly rich in food.

Cracidae nest in trees

The Cracidae build their nests in trees or in bushes in the forest or in thickets. The rufous-vested chachalaca and the black chachalaca occasionally nest on the ground, and the horned guan is also reported to do this. In all species the nest is a rough, disorderly structure shaped like a flat dish or a platform with a depression, which is often longer than it is wide. It is built from twigs, climbing plant stems, leaves, grass, palm frond pieces, and similar items. The larger species may use branches of two to three centimeters in diameter for the nest base. Often they pluck leaf-bearing twigs or grasses which they bring to the nest while still fresh and green.

The hens generally appear to lay only two eggs. One often sees three or four eggs in a guan nest. They may be from two hens, since guans are polygamous. A clutch of nine eggs of the plain chachalaca found by R. J. Fleetwood was evidently derived from three hens. As far as is known, all Cracidae lay white eggs. The thick shell is usually rough and grainy or has markings which look like pin pricks. In some species, such as the chestnut-winged guan, the creamy yellow surface has white spots. Often the eggs are surprisingly large. Those of the helmeted curassow may be 8½ to 9½ cm long and 6 to 6½ cm wide. During incubation, particularly in wet weather, the white eggs get stained from the leaves on which they lie.

According to the most recent observations, only hens incubate the eggs. The helmeted curassow hen leaves her eggs once a day, usually between eight and ten in the morning, and she stays off them from one to two and a half hours. If it rains the whole day, she may omit this "outing." A chestnut-winged chachalaca that I observed in Costa Rica interrupted her incubation twice a day, once in the early morning, and again late in the afternoon, staying off the eggs from one to one and one-quarter hours each time. Incubation periods vary according to species and size. The eggs of plain chachalacas, when placed under domestic hens, hatched in twenty-one days or less. In the chestnut-winged chachalaca, incubation by the mother lasted as long. But in the much larger helmeted curassow, incubation lasted thirty-four to thirty-six days.

Chicks of the Cracidae are well developed

Like the mound birds, the Cracidae chicks are well developed and precocial at hatching. At hatching, the larger wing feathers begin to widen out, so that the young are very soon able to fly or at least to flutter over short distances. They leave the nest very soon after their down is dry, sometimes even before. Schäfer saw a helmeted curassow chick leave the nest when it was scarcely two hours old. Guan chicks may stay in the nest a little longer. A chestnut-winged chachalaca that I observed left the nest twice to fetch berries from a nearby bush during the morning on which its three young hatched. It was back after four to five minutes, its red throat bulging with the berries. It squashed a berry in its beak and offered it to the young, which first pecked at and then took the food. After the last chick had become dry, it was another three hours before the mother got down to the ground from the nest, which was one and one half meters high. The young left the nest with her. Through binoculars, I tried to determine how they got out, but could ascertain no details. Possibly they clung to the mother's leg, as was variously reported for the plain chachalaca.

The hen carries the chicks out of the tree

Once I saw a chestnut-winged chachalaca leading its young on the ground, while a great many adults moved about in the trees and bushes nearby. At the approach of a person, the hen carried three chicks up into a tree one after the other, where she left them perched on a branch five meters above ground. The young of curassows and guans do not seem to need such parental care. They can move, when only a few days old, fly, hop, and walk along twigs at quite a respectable height. G. K. Cherrie has reported on crestless curassows hatched under a hen. As early as the day of hatching, the chicks climbed to a higher perch and showed no sign of wanting to crouch under the hen.

Young chachalacas may join a flock when only a few days old. But the young of curassows and guans are guarded by both parents. In the presence of danger, both parents make a great din and attract attention to themselves while the young quickly disappear. A great curassow hen, which evidently was leading chicks, ran close up to a human intruder, and grunted loudly as she limped along over the ground with outspread wings and shaking tail movements, a sort of distraction display (instinctive injury feigning often lures away the enemy). Otherwise such injury feigning is only rarely observed in birds of this family, and perhaps never in such a typical manner as is seen among many other groups of birds. Young guans grow slowly and are led by their parents for a long time.

Young Cracidae do not seem to reproduce until they are two years old. As far as is presently known, they breed only once a year, each hen rearing two or at most three young. Compared to other gallinaceous birds, the Cracidae thus have a very low rate of reproduction which is, however, quite appropriate to their condition of life. Only when man appears on the scene and pursues them with firearms does their low rate of reproduction become insufficient and consequently dangerous. Wherever these birds are not protected by sensible and strictly enforced laws, they are in danger of extermination. Only the small chachalacas, which can adapt to the plants of cultivated country

Cracidae are in danger of extinction

and are less desirable as food for people, seem to be able to flourish in more densely settled and cultivated areas.

As domestic birds, Cracidae are far less suitable than turkeys or Old World Galliformes. Though they are often kept and tamed in their native countries, they do not reproduce enough so that they can, with any utility, be made into domesticated birds. Guans and others of the group which are kept in tropical and subtropical America by amateurs are usually derived from eggs found in the wild and hatched under domestic hens. Often they become tamer and friendlier with their human friends than domestic fowl do. They are also more intelligent, more venturesome, more devoted and they move around more than domestic fowl. L. Griscom, in his work on bird distribution in Guatemala, reports on a plain chachalaca which lived in Ocos on the Pacific coast. It was allowed to move freely about the village. Its chosen task was to keep peace and order among the domestic fowl. Whenever two cockerels began to fight, the chachalaca came racing up to separate them. From "fear of punishment," the cockerels would run away as soon as the "feathered policeman" appeared on the scene. This was enough for the latter. It never chased the cockerels to tyrannize them.

18 Grouse

Family: pheasants, quails and peacocks

All other Galliformes are united in the large family Phasianidae. The size and weight are quite variable, ranging from only 45 g (Chinese painted quail) to 22.5 kg (domestic turkey). There are primitive species and highly specialized ones, as well as many intermediate ones. In species which have remained primitive, males and females both have a uniform, camouflaging plumage. In highly specialized species, males have bright plumage colors, decorative ornaments, excessively large decorative feathers, and colorful distensible structures on the head and neck. These decorative feathers are important in courtship display. Except in the peacocks and the Congo peacock *(Afropavo congensis)*, the feathers have a long aftershaft, and the preen gland (absent in the argus pheasant) is feathered. The tarsus is fairly short to long, with or without spurs. The hind toe, in contrast to the Cracidae, is inserted higher on the tarsus than the remaining toes.

Distinguishing characteristics

They are ground dwellers. Their food consists mainly of vegetation, grains, berries, roots, conifer needles, etc., but many insects and other small animals are also eaten. The union of the sexes is extraordinarily variable, including monogamy, polygamy, or virtually no bond at all. As with birds in general, the more complicated the male's decoration and courtship behavior, the less is its participation in the rearing of offspring. Their nests are built on the ground and rarely, with the exception of the tragopans, in trees. In most cases only females incubate. The young are precocial. They are distributed over most of the world, but are absent on many islands. In America they are represented by grouse, one tribe or group of Perdicinae, the toothed quails, and the turkeys.

Of the nine subfamilies (grouse, tragopans, pheasants, turkeys, argus pheasants, and peafowl) the first three are dealt with in this volume, and the rest in the following volume. Altogether there are seventy-five genera with 204 species.

Subfamily: grouse, by G. Niethammer

The GROUSE (subfamily Tetraoninae) live only in the Northern

Distinguishing characteristics

Hemisphere in areas with temperate or cold climates. They are medium to large in size, weighing 350 to 6500 g. The plumage is dense and with heavy down in the under plumage (an adaptation to the cold winters of their habitat). The nostrils are covered with feathers and the tarsi are partially or entirely covered also. Hence, the German name rough-legged fowl. In ptarmigan, even the toes are densely covered by feathers in winter (the feet therefore stand, as it were, in "fur" and snow-shoes). Instead of feathered toes, all other grouse have two combs of laterally projecting horny platelets on each toe (i.e., a row of platelets on each side of the toe), which perform the function of snowtires, and which are shed each year in the spring, growing slowly again until fall. There are three tribes with eleven genera and eighteen species, including ten in North America, six in northern Eurasia, and two around the Old and New World north pole regions.

Fig. 18-1. 1. and 2. Spruce grouse *(Canachites canadensis)*; 2. Franklin's grouse, subspecies *(Canachites canadensis franklinii)*.

A. FOREST GROUSE with the following genera: 1. CAPERCAILLIES *(Tetrao)*. 2. BLACK GROUSE *(Lyrurus)*. 3. PTARMIGAN *(Lagopus)*. 4. The NORTH AMERICAN SPRUCE GROUSE *(Canachites)* has one species, the SPRUCE GROUSE (*Canachites canadensis;* Color plate p. 445). The weight is 400-550 g. The display consists of short flights from a branch of one tree to a lower branch of a neighboring tree, or the reverse, with a moderately loud wing-whirring sound. They show no escape response towards man, and therefore are endangered or exterminated near settlements. They inhabit the coniferous forest areas of North America. 5. The SHARP-WINGED GROUSE *(Falcipennis)* has one species, the SHARP-WINGED GROUSE *(Falcipennis falcipennis)*. It is the size of the spruce grouse. The males and females are not very different in plumage, and the outermost primaries are narrow and pointed, producing a peculiar sound in flight. They inhabit northeastern Asia, and are little known. 6. The BLUE GROUSE *(Dendragapus)* has a weight of 900-1500 g. The courtship call consists of five to six low "ventriloquistic" hoots. They inhabit the forest belt in western North America. Today these grouse are located principally in large national parks. There are two species, including the BLUE GROUSE (*Dendragapus obscurus;* Color plate p. 445) (in North America only one species is recognized: translator), which are separated only by slight plumage and behavior differences. All species are inhabitants of forests, forest tundras or mountains.

Fig. 18-2. Blue grouse *(Dendragapus obscurus)*.

B. The PRAIRIE CHICKENS (genera *Centrocercus, Tympanuchus,* and *Pedioecetes*) are inhabitants of steppes and semideserts in North America.

C. The HAZEL GROUSE (genera *Tetrastes* and *Bonasa*) are inhabitants of mixed forests in Eurasia and North America.

The life of grouse

Phylogenetically, the grouse are a young crop of birds, known to man only since the Middle Tertiary (Lower Miocene, twenty to twenty-five million years ago) from ten fossil species. They probably developed in the area which once reached from Northeast Asia over the then-existing land bridge in the area of the present Bering Straits as far as Alaska. All species of grouse are highly valued as game birds by hunters both in the Old and New Worlds. Grouse are hunted by many different methods. They may be stalked, or flushed from cover into an area where concealed hunters lie in wait. They may be shot and retrieved by dogs, or they may be attacked by various modes of enticement. Since ancient times in Central Europe, the capercaillies and black grouse have been hunted during their display periods. In all of northern Eurasia grouse are looked upon as first class game and therefore are of economic significance. At the same time, these birds are an important source of food for many raptors of the tundra and taiga.

Most species are restricted to dry, brittle food in winter. For example, they eat the needles of coniferous trees in quantity. In the severe Russian winter, a single capercaillie consumes twenty-four kilograms of pine needles. Grouse have a large crop serving as a food reservoir, and most of them have a very strong muscular gizzard in which many small quartz stones (grit) aid the grinding down of the food. These small stones (grit) are taken in with the food by the birds. In the capercaillies' gizzard, one always finds fifteen to twenty grams of them. Sometimes they are colorful and often they are well polished. In some districts they are used as "pearls."

However, grouse also eat buds, leaves, insects, and other small animals. Their food is quite varied. The needles of pines and other conifers, which some species must depend upon in winter, are difficult to digest. Usually, the cellulose of the food is not digested until it reaches the caeca. In grouse these are up to fifty-two centimeters long, a length only exceeded in ostriches and rheas. The caeca are emptied every morning and the caecal feces are readily distinguished by their tough, sticky consistency and black color from the ordinary "end gut" feces which are excreted several times daily as brownish, curved little "sausages."

Generally, grouse that live in the forest are darker than those which are steppe or prairie dwellers. Only the ptarmigan of the tundra and high mountains assume a white winter camouflage plumage (which is a phenomenon restricted among birds to this genus, although it is well known in weasels and some other mammals). They are engaged in moulting more than other birds, for the cock ptarmigan moults four times a year, and the hen, three. The willow ptarmigan constantly changes its feathers from spring to fall without any real cessation of moulting; most of the time it carries feathers of several plumages and is, except in winter, almost always in a transitional plumage.

Many species have a proper moult of the beak sheath and claws. The capercaillie casts off its old beak sheath in early summer when it changes to its more tender summer food. By the time it turns once more to brittle conifer needles, its beak has again grown hard. In the same seasonal rhythm, ptarmigan cast off the long winter claws early in summer. In the fall the claws become longer again and the toes become feathered, in preparation for winter.

In some species, there is little difference in sexes. This is true in the GREATER PRAIRIE CHICKEN *(Tympanuchus cupido)*, the RUFFED GROUSE *(Bonasa umbellus)* and the SHARP-TAILED GROUSE *(Pedioecetes phasianellus)*. On the other hand, sexual differences are very prominent in the CAPERCAILLIE *(Tetrao urogallus)* and the BLACK GROUSE *(Lyrurus tetrix)*. Here the cocks not only differ completely from the hens in color, but are also much larger. Mature capercaillie cocks may weigh up to six kilograms, twice as much as the hens. The spurs which are so characteristic of male domestic fowl and of many other galliform birds are lacking in all grouse. In general, the males have red or orange patches of skin ("the combs") above the eye, which look somewhat like warts and can swell up. These combs are merely suggested in the hens. The cocks of some species inflate the air sacs of the neck tremendously in display, so that they swell out at each side of the neck feathering, almost like balloons. As a result of this swelling, bare patches of bright orange, purple or yellowish-green skin become visible. The air sacs also serve as resonating chambers and strengthen the voice. Anyone who has ever experienced the display of black cocks knows the great distance their calls can be heard. The wide variety of display postures and "songs" are further accompanied by other sounds: the cocks beat their wings strongly while standing, which produces characteristic clattering sounds.

In black grouse, prairie chickens, and some other species, there is no pair formation. Instead there is a highly organized social display. The cocks exhibit their "dances" and utter their "songs" on collective display grounds. The hens go to the grounds for mating, then leave the males and look after the remaining phases of the reproductive process alone. Other species, like the ptarmigans *(Lagopus)*, the hazel grouse *(Tetrastes bonasia)*, and the ruffed grouse *(Bonasa umbellus)*, live in monogamy. Collective display is customary in the species of open country, and takes place on the ground, while individual display is performed in trees by forest dwellers. It is notable that Old World grouse species, at least, occasionally interbreed in the wild. Thus hybrids are recognized between the black cock and hazel hen, the hazel grouse cock and hen willow ptarmigan, the male willow ptarmigan and gray hen, between the black cock and the capercaillie hen (many times), and (more rarely), between the cock capercaillie and the gray hen. German hunters call the black grouse-capercaillie hybrids "rackel" fowl.

Nests of grouse are always on the ground and contain various numbers of eggs, according to species. Generally the clutch size is from six to twelve. The incubation period is shortest in the prairie chicken and hazel grouse groups, lasting twenty-one days, and longest in the capercaillie, twenty-six to twenty-eight days. In the first few days the chicks eat predominantly animal food. Initially, they are unable to maintain their body temperature, and so at night, or in bad weather, they are absolutely dependent on being kept warm by the mother. They are very susceptible to infectious diseases. This is probably the reason why the maintenance of grouse in captivity is more difficult than that of most other gallinaceous birds.

Grouse have suffered more at the hands of man than all other gallinaceous birds, for man has constantly interfered with their habitats. Moors were drained and cultivated. Forestry has destroyed the natural woodlands and replaced them with stands of a single type of tree in many places. Forests have also been replaced by fields in which woodland grouse cannot thrive. Perhaps only a few of our children and grandchildren will ever see a living capercaillie, for the grouse population in Central Europe is steadily decreasing. In 1964, D. Popp estimated that in West Germany there were 6002 capercaillies, 14,708 black grouse, and 4120 hazel grouse. These populations are annually decreasing in many areas. At the turn of the century the yearly harvest of capercaillies in northern European Russia was 65,000, but now it is only about 1800. In eastern North America, an eastern subspecies of the prairie chicken *(Tympanuchus cupido cupido)* has been extinct for several decades.

The capercaillie, by B. Grzimek and D. Müller-Using

The CAPERCAILLIE (*Tetrao urogallus;* Color plate p. 446) is the largest of all the grouse. The L is 110 cm, and the weight is 4–5 (occasionally 6) kg. The comb above the eyes of males swells during display. Females are smaller, weighing 2.5–3 kg. The upper parts are brown with black and whitish markings. The under parts are whitish, with a brown crop band. The tail is rounded in both sexes. The capercaillie is now extinct in many areas of its originally large range from Western and Northern Europe to North Asia.

In Germany everyone knows what a displaying capercaillie looks like. Pictures of them are common, and often one may see a stuffed bird hung on the wall above the sofa in a home. However, less than 100,000 of presently living Germans have ever seen a living cock caper. This is because zoos rarely keep them. In the wild as well, they are becoming more and more scarce. Capercaillies prefer primeval mixed forests with swampy areas and moors, although in the crops of cocks shot in display one often finds only pine needles, fir and spruce, their winter food. This is probably because they did not take time to look for other food during the demanding weeks of display and mating. Generally, capercaillies require a varied diet. One soon finds this out when keeping them in captivity. They want buds, young twigs, and a variety of berries.

Chicks and the young particularly live almost entirely on beetles, caterpillars, larvae, flies, earthworms, and snails. There are not enough of these in our organized forests. An abundance of the red wood ant is particularly important for the welfare of the capercaillie. Where there are enough of these ants, the chicks are assured of the necessary insect food in their first few days. Where these ants have disappeared, capercaillie populations have also declined markedly. In many places an excessive harvest of cocks has also contributed to the disappearance of the species.

Fig. 18-3. Capercaillie (*Tetrao urogallus*).

Unfortunately nearly all attempts to introduce capercaillies artificially into their former habitats have failed. Only in Scotland was an attempt successful. In 1837 and 1838, forty-eight capercaillies were brought to Scotland from Sweden. In addition, caper eggs were placed into the nests of black grouse and hatched out there. Though there is the danger of producing black cock-hen caper hybrids with this procedure, the number of capercaillies in Scotland was estimated at over 1000 twenty years after the introduction, and they have maintained themselves well since then.

Probably most other attempted introductions were made with an insufficient number of birds. Generally only mature capercaillies were released, often only a single pair. In Poland, two to three month old youngsters were liberated. However, it is not so easy to rear young capercaillies. The Heinroths discovered this when their five chicks died of a crop disease at seven weeks of age. Dietland Müller of the Max Planck Institute at Seewiesen also lost almost all of his young to various diseases. One surviving chick attached itself very closely to the zoologist who looked after it, perching at the foot end of his bed every evening. If he worked too long, or late, the chick would lie close to him on his table or desk, and sleep. If it was left alone, it gave the lost call "dee" of its species. Another group of fifteen chicks which Müller had kept the preceeding year could be made to "weep" in this manner when he remained quiet and motionless. As soon as one of the chicks uttered the lost call, the others, which until then had been looking for food without concern, joined in. If they heard Müller's voice, the whole flock would come running and fluttering, and busily look for insects, blossoms, and stalks near him.

The display of the capercaillie

Capercaillies have become known primarily as a result of their special, conspicuous display. The displaying cock caper takes up a posture much like that of the gobbling turkey. The large bird often utters its surprisingly soft "song" before the first light of dawn. The song consists of sounds like those made by sharpening a tool, as well as snapping and smacking sounds. A capercaillie's trachea is a full one-third longer than its neck and makes a large loop in the crop region. This strengthens the sound of the song as does the inflation of air sacs under the skin of the neck. The complicated muscle apparatus found in song birds is completely lacking in the syrinx of the caper. The investigators

Wurm and von Schumacher have ascertained that the display calls are made without use of the beak or of the vocal apparatus. The cock produces them in another, quite peculiar way.

"The double beat of the snapping," says Wurm, "which can be imitated by pressing slightly moistened lips together, rapidly separating them, then pressing them together, is the result, I am convinced, of similar motions in the cock. In his case the very mobile throat and tongue, in conjunction with the deeply indented palate, play the role of human lips. Before snapping, the cock opens his beak, then draws his tongue into the triangular palatal groove and lets it drop rapidly once more. The forceful entry and exit of air with this motion produce the ringing double sound." If, as Wurm determined, one hangs a freshly-killed cock caper by its upper mandible and then taps it on the neck where the steel-green plumage begins, one clearly hears sounds which resemble the snapping, and even a weak version of the "main beat." The finger tapping strikes and raises the tracheal loop and when that drops back, the tongue, which accompanies these movements, produces the sounds. This is confirmed by von Schumacher. He was even able to produce the snapping sounds when the syrinx had first been removed.

The snapping consists of a hard, double sound repeated two to ten times with pauses of varying duration. It can be fairly well imitated by knocking two pieces of hard wood together. The cock begins his "song" with a few snaps, "ko lupp, ko lupp, ko kupp, ko lupp, ko lupp, kaluppkalupp, kalupupup." With the last few sounds, the "trill," arises as a result of a running together of the snapping sounds. This trill ends in a loud "klack," which sounds like the pulling of a cork from a bottle, and is called the main beat. Throughout this process the beak is not closed and it remains open during the final phase, the "grinding" or "whetting," which actually sounds like the stroking of a scythe with a whetstone.

During the "whetting," the cock is believed to be deaf. The stalking hunter uses this phase of the song to take a few steps toward his victim, or so it is said in a number of books. There are many hypotheses concerning the bird's deafness in these moments; the ear passage is supposed to be compressed by blood engorgement, or a particular position of the head is assumed to kink the ear passage. However, no anatomical finds support for such suggestions. The caper probably does not hear while "whetting," because he is producing loud sounds and is also very excited in this phase of the "song." Photographs show that the nictitating membrane flips over the eye at this moment, so that the cock is also temporarily blind.

The morning display usually begins at early dawn. The display song is repeated 200 to 300 times (even 600 times at the peak of the season) every morning. The whole song is repeated eight times per minute.

The "concert," however, is usually interrupted by intervals, occasionally up to an hour long, during which the cock looks about. The morning display lasts about one and one half hours. The evening display is much shorter and occurs only during the peak period. When the cocks have taken a perch on their sleeping tree in the evening, they often utter a coarse rough sound, sounding like a person vomiting. In German this is called "worgen." This sound has, certainly incorrectly, been looked upon as a sort of throat clearing.

The cocks maintain display territories and generally chase younger cocks away from them. At the peak of the display season, they descend to the ground very early in the day and accompany their "whetting" with a jump into the air, so that the rushing sound of the wingbeats almost drowns out the "whetting" sound. When no other cock is to be heard or seen, near or far, the performer presumably lacks a rival from whom he can claim his territory. Other animals or even people may then be taken as substitute rivals. This is perhaps the explanation of the "mad" behavior of some cock capers. In a forest near Gablonz in Bohemia one April, a cock capercaillie suddenly blocked the path on which a forest worker was walking. It evidently objected to human entry into its territory, but hardly bothered to notice the hen capers. When approached closer than fifty meters, it would withdraw. At the end of May, after the mating season, it was seen no more. The next spring, it was there again, recognizable by one abnormally short tail feather. This time it was bolder and would jump at a person's leg. When it was caught, moved ten kilometers by truck, and then released, it was back in its territory by the next morning.

Capercaillies also display at humans

A few years later, another displaying capercaillie in Upper Bavaria came down from the trees straight towards two people taking a walk. Though they tried to tease it with sticks and stones, it took little notice. When these people had walked on, the cock ran toward another person and came within about five meters of him. The man fended it off with a stick. He could have killed the bird but it took off without pushing the matter to such extremes.

Such "mad" capercaillies continue to occur. They jump down from the tree as soon as a person approaches and display at him. They perch on the hoods of cars, and even land on the shoulders of intruders. They always become local "sights," and usually somebody turns up sooner or later who kills them. The caper who lives with us at the Frankfurt Zoo shows no apparent concern for his brown hen, which shares the enclosure with him. She is mostly hidden among the bushes. He, however, angrily approaches all visitors. Since the flight cages of the gallinaceous birds in the Frankfurt Zoo can be entered, a low fence had to be placed in the caper's area to keep it away from people. Fortunately this was enough, for he does not try to jump at their heads. The self-confidence of this particular bird is probably due to the fact that it was reared on a farm in Sweden. Such capers readily look upon people as either companions, or as rivals for the rest of their lives.

Incubation and rearing of the young

The cock participates very little in reproduction of his species. He tries to "inspire" hens with his display, and mates with those that enter his territory, but that is all. The hen alone incubates on the ground for twenty-six to twenty-eight days, laying six to ten eggs about the size of chicken's eggs.

When leading chicks, the hen can become surprisingly bold. Two hunters in the Lichtenstein foothills sat down for a snack beside a huge fallen spruce and chatted there for half an hour. At last they noticed the tail feathers of a hen capercaillie beneath the tree trunk. Chicks, while hiding under the hen, had caused her tail feathers to move. I. Notar reports, "After five minutes of tense waiting, I became impatient and gave the hen a gentle little push with my pipe. Now she became upset. She felt discovered and rushed off between my legs, with at least ten cute, little brownish-gray "balls of wool" behind her. The chicks let themselves be seen for only seconds; like their mother, they disappeared with astonishing speed into the thicket."

Although rearing capers is difficult in captivity, it has been successfully achieved in a few game parks. In the Frankfurt Zoo our hen, acquired in 1964, laid ten eggs under a bush and incubated them carefully. To avoid all risks, we moved half of the eggs to an incubator. Four chicks hatched under the hen and four hatched in the incubator. One chick died, and since the hen unfortunately proved to be a poor mother, we took the remaining chicks from her and tried to rear them in our breeding station. After much care, we were successful, and at the end of July, 1965, six half-grown, healthy, round caper chicks could be transferred into the pheasant enclosure. However, only one hen survived. The other five died of some intestinal disease of unknown cause.

Once capers have been reared to maturity, they may become quite old in captivity. A cock lived for eighteen years with a Mr. Sterger in Krainsburg, Germany. The longevity record in the wild was of a cock near Fulda, Germany, which was known by its individual tail feather pattern and which maintained a specific display territory. It reached an age of twelve years, as F. Müller informed me. Another one, marked in Finland in 1950, was shot in the same district in 1960. It had reached a weight of four and one half kilograms.

The other forest fowl, by D. Müller-Using

The black grouse

The BLACK GROUSE (male black cock, female gray hen, *Lyrurus tetrix;* Color plates, p. 446 and p. 467) is probably the most popular of our native game birds. The weight of the males is 1200–1300 g, and the females weigh between 750–1000 g. The wing length of the males is 270 mm, and of the females, 240 mm. The combs above the eyes of the males are strongly developed. There is a white wing speculum which is noticeable at a distance in flight. Closely related is the CAUCA-

SIAN BLACK GROUSE *(Lyrurus mlokosiewiczi)* which lacks the wing band, and has much shorter under tail coverts. It is black. The tail feathers are only slightly curved, and the combs only slightly developed.

This is probably a primitive species which survived only at high altitudes in the Caucasus and northeastern Asia Minor. Display consists essentially of jumps into the air, while the cocks loudly beat their wings, and an imposing strut in an upstretched attitude. There are no vocalizations with this display.

Fig. 18-4. Black grouse *(Lyrurus tetrix).*

The black cock's display is quite different. The "Schuhplattler" (after a Bavarian peasant's dance) is the only dance derived from animal behavior which has been kept up in Germany. It represents a well-observed imitation of black grouse display. The men behave quite wildly and imitate the strong wing-beating, which goes with the flutter jumps of the cocks, by beating their knees, jumping like black cocks, and shouting a ringing "yoohooee" which in the cock is a mere hissing "tshchooee." The girls, on their own initiative, do nothing but rotate on the spot. However, they are picked up by the men, thrown up, caught again, and whirled about, in a purely human development of the courtship dance.

The elegantly curved outer three or four tail feathers of the black cock were used in the uniform cap of the well-known Tyrolean Kaiserjäger, a mountain unit of the army of the old Hapsburg empire, which had its garrison in the Tyrol. Black cock tail feathers were also used as part of the hunter's garb throughout the German and Austrian alpine area. Sometimes, as in the Chiemgau (Bavaria), they were part of the local costume. Even after the last war young peasants in that area would poach upon the black cock in order to give its tail to their fiancees as a bridal decoration. Earlier, when such a poacher was caught, he lost his weapon but was left with the bird, since custom declared that the "wedding cock" must be hunted by the bridegroom himself.

Today the populations of black grouse, like all grouse, have declined seriously. Almost everywhere there are agricultural authorities who convert moors into grazing land, or worthless and unwanted rye and potato fields. Reforestation is then carried out, but by that time the natural character of the landscape is lost, and with it, the black grouse too, has disappeared.

In the plains, display begins in March and reaches its peak in April. However, in the high mountains many times it does not begin until early May and then goes until late June. Like the caper, the black cock displays singly or in groups of five to over fifty cocks. According to the observation of Brüll, there are two to seven younger cocks to each mature one in the social display. The hens will generally only let the strongest ones mate with them. The onset of the display is recognized in the mountains by finding the imprints not only of the feet but also of the dragging wings in a snow field.

When several grouse species occur in the Alps, one hears the caper calling first, when it is still dark. Next one hears the reeling song of the cock rock ptarmigan above the tree line, and lastly at dawn, the black cocks arrive on their lek (as the collective display grounds are known in Britain). After arrival they generally look about with raised heads for quite a while. Then they utter their "crowing" without accompanying it by a flutter jump. Finally they begin their "gobbling." While giving this call, the tail is fanned out, the wings are rapidly beaten, and the head and neck are extended horizontally. All the body plumage is puffed out so that the cock looks much larger than before or after the display.

Fig. 18-5. This is how the black cocks display:

At the start of display the cock fans the tail out, and the combs swell. Then he makes jerky, fluttery jumps. This evidently stimulates other cocks to do the same.

"Gobbling" sounds like a serial repetition of "curroo curroo currooo," which may be uttered continuously for half an hour. Often, however, the cock interrupts itself by crowing, usually combined with a flutter jump, particularly when other cocks fly in and approach the owner of the display territory (a section of the lek) too closely. When greatly excited, the cock may add a harsher rattling "ca carr" to his call. The display song, particularly the "gobbling," may also be heard regularly on fine autumn days. Though it does not sound so loud when one stands close to a cock, it can be heard for about three kilometers on quiet nights with a soft favorable breeze.

Oskar Heinroth, one of the founders of comparative ethology, explains the "meaning" of the black cock's song, and those of other grouse, with the apt words, "Here there can be mating." Actually, "gobbling" is a signal to those hens ready to mate. They fly into the lek, are courted intensely by the cock in whose territory they happen to be, and finally crouch and so release copulation. Consequently the mating behavior of the male grouse is different from that of the barnyard cockerel, which "rapes" every hen which crosses its path. However, the intensity of the sex drive in domestic animals is often excessive, as a result of unnatural conditions of life and maintenance, in comparison to conditions in the wild.

Fig. 18-6. With inflated neck the cock utters his long lasting gobbling calls...

The gray hen lays about eight eggs. Young cocks begin to sprout the first blue-black feathers of the adult neck plumage in August. At the same time, the initial feathers of the juvenile wing are replaced by the second generation of primaries. As a result, the birds cannot fly well for a while. In Finland, I was able to pick up a young cock in the second half of August after a short pursuit.

Fig. 18-7. ...and walks toward a neighboring cock. At the border of the individual display territories there are often playful-looking attacks and withdrawals. Only rarely are injuries inflicted.

In October the young already have the full plumage colors. Then young cocks can be distinguished from mature ones only by a brown tint on their back feathers and their shorter, less curved tail feathers. They generally join the all-cocks flocks. Both sexes spend the winter in more open areas where they satisfy their decreased needs for food mainly with birch buds and buds of other deciduous trees and bushes. They need less food in winter than in spring and summer because they keep much more quiet. During the day they even sit still for hours.

Hybrids between capercaillie and black grouse

Strangely enough, capers and black grouse occasionally interbreed. These hybrids are called "rackel" fowl in German (Color plate, p. 446) because of the rough, vibrant display call uttered by the rackel cock. They are not always sterile, as used to be assumed. They mate with either of the parent species. Thus Bergmann was able to get young from a rackel cock and a gray hen in a flight cage.

In the display season, rackel cocks often seek leks of black cocks (probably the larger caper cocks are too strong for them as rivals). According to the observations of Jost Straubinger, rackel cocks display more fervently than black cocks,"...and their calls are feebly reminiscent of those of the capercaillie. The black cocks were dismayed by this. Whenever, within a 300 meter radius, a black cock dared to 'gobble,' the rackel cock would rush up and cause its possible father to beat a hasty retreat."

Ptarmigans, by G. Niethammer

Of all land birds, the PTARMIGANS (genus *Lagopus*) are distributed furthest to the north. There are three species: 1. The ROCK PTARMIGAN (*Lagopus mutus;* see Vol. XIII), which weighs about 450 g, has a shorter and wider beak than does the willow ptarmigan. The winter plumage is snow-white and only the tail feathers are black with white tips. Males also have a black stripe in front of and behind the eye. It inhabits stony, rocky, treeless areas on the tundra and mountains in the north of the Old and New World, and in the south, the Pyrennees, Alps, and areas in Central and East Asia. 2. The WILLOW PTARMIGAN (*Lagopus lagopus;* Color plate, p. 446) weighs about 600 g (sometimes up to 800 g). It has a longer, slimmer beak. Males in winter plumage lack the black stripe on the side of the head. Distribution ranges from Britain over the north of Europe, Asia, and North America. It inhabits lower and milder areas than the rock ptarmigan. 3. The WHITE-TAILED PTARMIGAN (*Lagopus leucurus*) has a completely white tail in the winter plumage and differs from both of the other species in this respect. It inhabits the mountains of western North America from Alaska to New Mexico.

Fig. 18-8. Rock ptarmigan (*Lagopus mutus*).

Fig. 18-9. Willow ptarmigan (*Lagopus lagopus*).

Ptarmigan are protected against the cold by their very dense plumage and against detection by enemies by their white winter plumage. They can also feed themselves through the long polar night of the far northern winter. When seeking food they burrow long passages through the snow to reach the tips of twigs and leafbuds which form their food. They can endure the winter even in northern Greenland and on Spitsbergen. Lockwood found rock ptarmigan in May, 1882, in northern Greenland at 83°24′ N latitude where temperatures regularly reach minus 40°C. Whether ptarmigan were able to survive in such Arctic areas during the last Ice Age (fifteen to fifty thousand years ago) is questionable. Johansen believes that rock ptarmigan have only penetrated so far north since the last Ice Age. At that time, they also distributed themselves widely over broad areas in Europe. When the climate improved, they withdrew to Northern Europe and into the

The rock ptarmigan

highest mountains. They still live in high mountains, widely separated from their main northern area of distribution, as typical "Ice Age relics" above the tree line, generally between 1800 and 2000 meters, sometimes even higher than 3000 meters.

Food of the ptarmigan

In the Alps, they feed on the buds and tips of almost all alpine herbs, as well as leaves, blossoms and seeds. In Scandinavia, they feed on reindeer moss. The crop of five rock ptarmigan taken from Switzerland in March contained the remains of no less than thirty-eight plant species, mostly various willows, as well as *Polygonum* (knotweed), *Cardamine* (bitter cress), and seven species of saxifrage. Later in the year, ptarmigan eat blueberries, cranberries, moorberries and bearberries and, in winter, tips of spruce needles.

Besides this plant material, ptarmigan also eat insects, which they snap up in the air or pick up from the snow and ice. They remain near their breeding grounds even in unfavorable weather. They seek out steep slopes from which the snow is cleared early and, if necessary, dig deep snow passages to reach their food, as they do in the Arctic.

Territory formation and display

In spring, cocks leave the winter flocks and seek out territories. "Here," as Bodenstein describes the display, "they stand almost motionless on a rock slab or other projecting point in their territory, looking about with a somewhat upslanted head and bright red combs. Then the cock rises with loud, repeated, woodenly-ringing "carr" calls, and with a loud whirring, flies several meters into the air at an angle of about 30°. He glides a bit, sometimes swerves aside, rises again, and follows a snake-like course over his territory, landing on another projecting perch. When landing he always calls again and the wing whirring is heard once more. In these flights, distances of at least forty meters are covered, and in longer flights, further calling spells are included in the flight. The cock is very conspicuous in flight with his contrasting pattern (white wings, reddish-brown body, black tail), but after perching he blends into the background."

Cocks that often display within sight of one another show that they claim their own territories by these display flights. Hens, in their simple breeding plumage, remain quite inconspicuous. After mating they seek out a suitable spot below stones or dwarf shrubs, scratch out a shallow depression, and often line it with rootlets, grass or stems.

Incubation and rearing of young

The six to ten, rarely up to eighteen, eggs are not laid in the Alps until June. The eggs, like those of all other grouse, have dark spots on a rusty yellow background and measure 43.5 x 31 mm. The hen incubates very steadily, alone, for twenty-one to twenty-four days. The cock generally stays on an elevated perch close by.

When the young are dry after hatching, the hen heads them off to feed. She utters a warning call in case of danger, at which the young immediately scatter to all sides, crouching so still and silently behind stones or plants that they are almost impossible to find. The cock, too, joins them, differing from most other grouse in this respect, and takes part in guiding, guarding, and keeping them warm.

The willow ptarmigan

The larger willow ptarmigan is aptly named, for it is a typical inhabitant of lightly wooded moorlands, or tundra with many dwarf shrubs. In summer it feeds on the leaves and blueberries, cranberries, swampberries, crowberries, and bearberries. In winter it eats the tips of twigs and buds of the dwarf birch. From early May on, one can hear the display call of the males, which sounds somewhat like "carr ak ak ak."

After the breeding season, several families gather into larger packs. Willow ptarmigan crouch on the ground when endangered by ground enemies like foxes or wolves. This makes them favorite game for hunting with pointer dogs. In northern Russia they used to be caught in snares in great numbers. The willow ptarmigan was once common in East Prussia, but disappeared completely within a decade around 1880 without any evident cause.

The British subspecies, the RED GROUSE (*Lagopus lagopus scoticus;* Color plate p. 446) differs from all other willow ptarmigan in its lack of a white winter plumage. This is to be attributed to the relatively mild climate of its home. It lives on moors and heathlands in Scotland, northern England, Wales, and Ireland, and on the Orkney islands and the Hebrides. It has long been one of the most favored gamebirds of the British and was, therefore, introduced in several countries in which it did not originally occur.

The introduction of this grouse in the High Venn on the moors of the German-Belgium border was very successful. There, the textile manufacturer, Scheibler, from Monschau, released over seventy pairs in the early 1890's. They multiplied so well that by 1904, three hunters could bag forty birds in a day. Their total population in the Venn was estimated at 1000 at that time. In order to safeguard the new gamebird, the Prussian House of Lords, without debate, passed a proposed closed season law for the red grouse on January 9, 1902.

Unfortunately there were large fires in the central areas of the High Venn in 1911 and, according to witnesses, grouse perished in flocks in the flames. Yet the population soon recovered and the value of hunting rights in this area, particularly in the Belgian sector, rose tremendously. However, from 1930 on, there was a decline. Red grouse first disappeared from the marginal areas of the High Venn and then became even scarcer in the Belgian central areas about Baraque Michel. Only near Botrange are there a few left. From 1960 to 1962 I could always observe them at a certain location in the central Venn, where they also bred. The draining of swampy areas and the forestation of open moorlands with pines have pushed these birds back to residual areas. Perhaps they will be able to hold out there for a long time, because in 1957 the Venn was declared a Belgian National Park.

Fig. 18-10. Cock sage grouse in display.

The prairie chickens, by D. Müller-Using

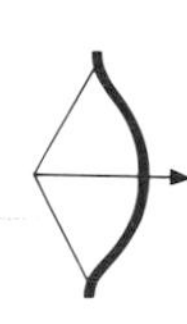

Four species of grouse of the American grass steppes are grouped together as prairie chickens: 1. The SAGE GROUSE *(Centrocercus urophasianus)* has an L of 72 cm, and a weight of 3.5 kg. The males become almost as large as capercaillies. They have yellow combs, a black belly, a black and white marked brown back, grayish-white dark spotted neck feathers, and very long pointed tail feathers. The area over the cervical air sacs is almost entirely feathered. They inhabit sage brush steppes. 2. The GREATER PRAIRIE CHICKEN (⊖ *Tympanuchus cupido*; Color plate p. 445), has an L of 47 cm, and a weight of about 800 g. They are blackish-brown, and have light tawny color above, and are white with brown barring underneath. They have a long erectile feather "ear" on each side of the neck. There are several subspecies on the prairies. 3. The LESSER PRAIRIE CHICKEN *(Tympanuchus pallidicinctus)* is a third smaller, and is also paler. It inhabits semideserts. 4. The SHARP-TAILED GROUSE *(Pediocetes phasianellus)* has an L of 44 cm, and a weight of 550-950 g. They are pale gray with darker bars and black and white spotting above, while the lower parts are white with dark bars on the lower neck and flanks. The voice is higher pitched than that of the prairie chicken. The combs of the cock are bright yellow, and the bare skin over the neck air sac is purple. This is the most common bird of this group in Canada. It inhabits grasslands and herb steppes.

Fig. 18-11. 1. Sage grouse *(Centrocercus urophasianus)*; 2. Greater prairie chicken *(Tympanuchus cupido)*. Both species have now been exterminated in large parts of the ranges, as shown here.

The display of the sage grouse is particularly interesting. The cocks have huge air sacs which are inflated during the display and then suddenly emptied with a whipping, snap-like sound which can be heard, in favorable circumstances, from 300 meters away. A rustling noise is produced when the air sacs are inflated, for the cock vibrates its stiff white neck and breast feathers. This is reminiscent of the feather vibration of the peacock when it erects and fans out its tail. Such an inflated cock looks quite grotesque. The lively color of the throat and upper breast, and particularly the bright yellow bare skin neck patches which are visible from afar but become exposed only at the peak of inflation, contribute to the effect.

As I. W. Scott observed, the cocks in the North American sagebrush lands display in societies with a peculiar rank order. Besides a head cock, there is a second cock, three to six other ranking cocks, and an even greater number of low-ranking cocks which are more or less scattered singly over the area. The hen solicits mating. According to Scott's observations, 74% of all matings were performed by the head cock, 13% by the second cock, 3% by the third to sixth ranked cocks, and only 10% by all the remaining low-ranking cocks. The latter are thus not completely excluded from reproduction.

Fig. 18-12. Sharp-tailed grouse *(Pedioecetes phasianellus)*.

The food of the sage grouse consists almost exclusively of buds, twiglets, leaves, and fruit of the North American sage *(Artemisia tridentata)*, as well as insects during the summer. The bird is quite dependent on the presence of the plant which gives it its vernacular name. Since the sage is relatively soft, the sage grouse is the only grouse to have a gizzard with a soft lining. It also does not need to take up grit. The birds are sociable and gather into larger flocks in winter. Formerly these flocks consisted of hundreds of birds, but unfortunately, they have become much scarcer nowadays. Bent writes, "Natural enemies merely kept their increase within bounds, but when man came it meant extermination."

The fate of the prairie chicken has been much the same. Once it was the most common grouse of the United States, occurring in the extensive prairies in such numbers that, as Audubon wrote, "Its meat was not valued anymore than ordinary meat and no real hunter looked on it as worthy of pursuit." Today the eastern subspecies is totally extinct. Those other subspecies which have survived were able to adapt here and there to agricultural cultivation, or have penetrated westward where they found favorable habitats beyond their original range.

Prairie chickens are definitely ground dwellers and only rarely do they fly into trees or bushes (this hardly applies to the sharp-tailed grouse). Like partridges, they roost on the ground. The cock has two bare yellow-red patches over its air sacs which are exposed when the air sacs are inflated in social display. It then looks as if it had a mandarin orange on each side of its neck (Color plate p. 445). In display the cock lowers its beak toward the ground and utters howling and drum-like sounds. At the same time, it erects the tuft of feather ears on each side of its neck and moves the tail into an upright position so that, as in the black cock, the white under tail coverts stand out in contrast to the dark tail feathers. The prairie chicken somewhat resembles the European black grouse, in its food preference as well, but in areas of grain cultivation it takes a greater extent of grain. In some areas it has practically become a follower of cultivation and approaches farm settlements without fear. It is easily kept in captivity.

▷ Gray partridge *(Perdix perdix)* at the dust bath. Gallinaceous birds use dust or sand baths but never bathe in water (upper picture). Displaying black cock *(Lyrurus tetrix)* (lower picture).

▷▷ Fully erect, a gray partridge cock *(Perdix perdix)* shows off his black breast patch.

▷▷▷ Incubating gray partridge hen (enlarged).

The hazel grouse, by D. Müller-Using

The HAZEL GROUSE (*Tetrastes bonasia;* Color plate p. 446), weighs up to 400 to 450 g, and is distributed in mixed forests with rich undergrowth in Europe and North Asia. Within the last century, it has declined markedly for reasons similar to those that have affected the capercaillie. Apparently the ever-increasing noise level is also a factor, for the hazel grouse seems to be extremely sensitive to noise. Wherever the din of motor saws, bulldozers, trucks, and motor highways penetrates into remote woodlands, the hazel grouse does not feel at home anymore.

Hazel grouse need forests with variety and much undergrowth. They do inhabit pure deciduous or pure coniferous woods, but only where many bushes, layers of dwarf shrubs, or small grassy areas are available to them. They like richly varied tracts, particularly southernly slopes with dense plant growth, and fresh springs. Thus they are highly demanding in their choice of environment. They live in pairs, possibly

Fig. 18-13. 1. Hazel grouse *(Tetrastes bonasia)*; 2. Black-breasted hazel grouse *(Tetrastes severtzowi)*.

♀
1
♂
2♀
3 a ♂
4♂
5♂
3b
♂

◁

Francolins: 1. *Francolinus hartlaubi bradfieldi*, a subspecies of Hartlaub's francolin. 2. Cameroun mountain francolin *(Francolinus camerunensis).* 3. Orange River francolin *(Francolinus levaillantoides);* Subspecies: a. *Francolinus levaillantoides pallidior,* b. *Francolinus levaillantoides stresemanni.* 4. Swainson's francolin *(Francolinus swainsoni chobiensis).* 5. Francolin *(Francolinus francolinus).*

in permanent pairs, in territories of twelve to eighteen hectares in size, and they tolerate no members of their own species within these areas. The spring display begins in April. The cock then utters a call which sounds like "tsee, tseetseree, tseetsee tswee." It displays both on the ground as well as on bushes, the lowest branches of trees, tree trunks, or on small rocks, by drooping its wings and whipping its expanded tail up and down in a lively manner.

The hen lays seven to ten eggs, rarely more. She incubates them alone and is a very "tight" sitter. It is a notable fact that all the chicks hatch at the same time. During the approximately three week incubation period, the embryos of the eggs laid first develop more slowly than those of the eggs laid last. Furthermore, hatching of the first-laid eggs lasts somewhat longer than that of the later ones. Chicks peep while still within the egg. After drying them by covering them with her plumage, the hen leads them away at once, and warms them a great deal during the first few days. She feeds them by holding out insects to them. Already on their fourth day, the chicks make little flights of about forty centimeters long at a height of twenty-five centimeters. Young reared by Krätzig could fly a distance of six meters at half a meter high, and could even swerve, when only eight days old. At two weeks of age they can fly fully. At this age they switch gradually from an insect to a plant diet by beginning to eat the tips of grasses and herbs, buds, leaves, conifer needles, and berries. From their very first days of birth the chicks pick up bits of grit to grind food in the stomach. Initially these are only little stones of pinhead size, but later they include lentil-sized stones as well.

Fig. 18-14. Ruffed grouse in normal posture...

At three months of age the young reach their final weight of about 400 grams. They are already in the adult plumage and become hostile toward one another. Consequently, the cohesion of the family, which has been joined by the cock soon after the hatching, soon ends. Young cocks seek out their own territories in the fall, and with the courtship song mentioned above, and display flights, and wing whirring they seek to attract a hen. When a pair has found one another they maintain a sort of "engagement" until the next spring, as is the case with many ducks and anseriform birds.

Fig. 18-15. ...and in display.

The closely related BLACK-BREASTED HAZEL GROUSE *(Tetrastes severtzowi)* replaces the hazel grouse in the Chinese provinces of Kansu and Szechwan, and in North America it is replaced by the RUFFED GROUSE (*Bonasa umbellus;* Color plate, p. 445). The L is 42 cm. The males have a tuft of feathers at the side of the neck which are erected to a ruff in display (see Figures 18-14 and 18-15).

Fig. 18-16. Ruffed grouse *(Bonasa umbellus).*

While displaying, the cock produces a wing whirring sound which carries quite a distance, and the so-called drumming (which sounds very much like a somewhat distant motor cycle starting up). The hazel grouse does this too, but not as loudly. Hens lay nine to twelve eggs and generally rear the young alone. Only in a few cases does the cock take a share in guiding the young. The American naturalist Ernest Thompson Seton has described a cock which, after the death of its hen, led the young alone until they became independent. Recently, American wildlife biologists have placed very small radio transmitters under the wings of ruffed grouse. This has enabled them to study the daily rhythm of activity and the size of the territories of this grouse.

19 Partridges, Quails, and Tragopans

Subfamily: Perdicinae, by S. Raethel

The Perdicinae, with 132 species, are the largest subfamily of the Phasianidae. Their shape is rather compact. The beak is short, and the tarsi are relatively short, with or without spurs. The weight is 45 g in the Chinese painted quail and reaches 3 kg in the snowcocks. The plumage is generally of a camouflage color, and is rarely colorful. The tail is generally short or fairly short. The moult of the tail begins with the central feathers (centrifugal moult). They are omnivorous, with the exception of the snow partridge and the snowcock (which are predominantly vegetarian) and the wood partridge (which mainly take animal food). There are three tribes: the PERDICINI, the QUAILS, and the AMERICAN or TOOTHED QUAILS.

Tribe: Perdicini

Among the Perdicini, some species approach the grouse in their appearance: 1. The SNOW PARTRIDGES *(Lerwa)* have tarsi which are feathered in the upper part and a small horny prolongation of the plates of the toes along the sides of the toes. The sexes are similar in color. The ♂♂ have short spur knobs. 2. The SNOWCOCKS *(Tetraogallus)* are very large, the weight reaching 3 kg. Their plumage is extremely dense (they are inhabitants of areas with severe climate). The tarsi are short and robust. The ♂♂ have blunt spurs. The sexes are similar in color. There are 20–22 tail feathers, and the crop is merely suggested. The gizzard is strongly developed and the caecum is very long. 3. The MONAL PARTRIDGES *(Tetraophasis)* greatly resemble the snowcocks, but are smaller. 4. VERREAUX'S MONAL PARTRIDGE *(Tetraophasis obscurus)*.

The snow partridge

The SNOW PARTRIDGE (*Lerwa lerwa:* Color plate, p. 480) inhabits the Himalayas east as far as Yunhan. Ernst Schäfer observed these birds in eastern Tibet. There they are the most common gallinaceous birds found in rugged high mountains above the tree line at 5000 meters. They prefer stony slopes, often with scattered snow patches. Their plumage offers such good camouflage that one only notices them upon hearing their alarm call, which is a very loud cackling. When escaping, they plunge downhill with skillful swerves, often repeating shrill whistles. They can curve about corners of rocks with the utmost exactitude. Except during the few hours of strong sun at noon, they walk to the alpine meadows to feed. Their food consists mainly of grass tips. The breeding season begins in May; then they are particularly noisy. Their nest, a hollow which is scraped out beneath an overhanging rock, is well lined with moss and leaves. It holds three to five cream-colored, light reddish spotted eggs.

The snowcocks

The SNOWCOCKS *(Tetraogallus)* are real giants among the Perdicini, for they range from black grouse to capercaillie size. There are five species in high Asiatic mountains.

1. The CAUCASIAN SNOWCOCK *(Tetraogallus caucasicus)* is the most western species, and is found in the Caucasus. 2. The CASPIAN SNOWCOCK *(Tetraogallus caspicus)* inhabits mountain chains of Taurus, Armenia, northern Iran, and Southwest Transcaspia. 3. The TIBETAN SNOWCOCK *(Tetraogallus tibetanus)* inhabits Pamir, Tibert, West Kansu, Szechwan, and Sikkim. 4. The ALTAI SNOWCOCK *(Tetraogallus altaicus)* inhabits the Altai and Sajan mountains. 5. The HIMALAYAN SNOWCOCK (*Tetraogallus himalayensis;* Color plate, p. 481) is found in the western Himalayas, Ladak, Gharval, Tianshan, Pamir, Alatau, western Kwenlun, Altyndag, Humboldt, and the South Kokonor mountains. The strictly residential habits, in association with isolated breeding areas, have led to the formation of numerous subspecies.

Niethammer says, "The snowcocks are the kings of the birds of the Asiatic high mountains, for they are as large and as powerful as the capercaillie, as swift and as enduring on foot as the rock partridge, and in their arrow-like flight they have no equal. They live high above many other mountain birds and most large mammals, near the permanent snow, and only rarely do they descend to below 4000 meters or into levels of 2000 meters or less. Only after climbing to 4700 meters did Ernst Schäfer enter the range of the Tibetan snowcock, and I first heard the melodious whistle of the Himalayan snowcock in the Pandjir Hindukush when the altimeter showed 3900 meters."

As Schäfer reports, the snowcocks are the most vocal of all high mountain birds. They cackle loudly in the morning and evening all year round, often for five minutes without interruption. These calls are taken up by one bird after another, so that soon the whole rocky slope rings with their calls. "In flight they are the fastest and most agile gallinaceous bird I know," says Schäfer of the Tibetan snowcock. "Almost like birds of prey, they launch forth, calling loudly from the talus slopes, beating their wings for only the first twenty to thirty meters, to produce a rapid whirring sound, and then gliding with their pointed wings angled a little towards the rear in an elegant flight, reminiscent of a falcon's swoop, into the depths. During these flights they develop unbelievable speeds. They can make sudden swerves around rock projections and, while doing so, slant themselves in the

Fig. 19-1. 1. Snowcocks (genus *Tetraogallus*); 2. Snow partridge *(Lerwa lerwa)*.

air so that when seen from above it often looks as if they must crash into the rock." Upon taking flight, they utter shrill whistles which begin slowly and then become increasingly faster, until the individual whistles merge into one another. After landing they utter a monotonous contact call.

The display begins generally in April, with the onset of spring in the mountains. During the morning hours, the cocks emit whistles which can be heard for kilometers, at short intervals from high rock perches, while they throw back their heads. They erect their neck feathers and tail in such a way as to display the white under tail coverts. In this posture a cock will slowly circle around the hen with its head lowered.

Snowcocks are monogamous. The hen lays her three to nine eggs in a shallow depression under an overhanging rock. The eggs are from creamy-white to clay yellow, reddish spotted and blotched. The cock stays on watch near the nest and warns the hen of danger with a loud whistle. The young are led by both parents. When endangered, the whole family walks uphill in Indian file with vigorous up and down wingbeats. When the young have become independent, neighboring families gather into larger groups. They like to associate with the ungulates of their mountain home, with the ibex in the Caucasus and the blue sheep in Tibet. Probably they search the droppings of these mammals for insects. However, they mainly eat plants.

Rock partridges

The ROCK PARTRIDGES (genus *Alectoris*) are quite similar in size and shape to ordinary partridges. They have brown and gray plumage colors without fine vermiculations of predominate striped patterns. The sides of the body are always conspicuously barred with black, brown, and white. The beak, eyecombs, and legs are bright red. Both sexes are similarly colored. The tarsi in the ♂♂ have short spur knobs. There are five species, including: 1. The ROCK PARTRIDGE *(Alectoris graeca)* with twenty-two subspecies; 2. *Alectoris melanocephala;* 3. The RED-LEGGED PARTRIDGE *(Alectoris rufa)* which is not as dependent upon mountains as the rock partridge; 4. The BARBARY PARTRIDGE (*Alectoris barbara;* Color plate p. 481).

The best known European subspecies is the ALPINE ROCK PARTRIDGE *(Alectoris graeca saxatilis)* which prefers sunny southern slopes of mountain chains with low vegetation and shrubs of the alpine rose, as well as lightly wooded mountain slopes and plant-free areas up to the snow border. Rock partridge are residents which only seek lower elevations at the onset of winter. In spring the cocks, holding themselves very upright, call "tchert tseereet tshee tshee." The female normally scrapes out two nest depressions about a hundred meters apart and lays a clutch of nine to fifteen eggs in each. The eggs are thick-shelled and brown spotted on a sand-colored background. As long ago as Aristotle's time (384 to 322 B.C.), it was known that one of the two clutches is incubated by the cock alone. The chicks hatch after twenty-four days of incubation and at four months of age have practically acquired the adult plumage.

The rock partridge reared in European zoos are mainly from the Himalayas; they are CHUKAR PARTRIDGE *(Alectoris graeca chucar).* They inhabit the Himalayan chain from East Dalak to Nepal at heights of 3000 to 4600 meters. They are distinguished from the alpine rock partridge by chestnut-brown ear coverts and pale yellow wings. They received their name from their call, "chuk-chukor." They have been introduced successfully into the mountains of the southwestern United States. In Armenia, Iran, Afghanistan, and West Pakistan, young rock partridges are often reared and kept free in houses and yards. The cocks attack visitors and dogs courageously. The combativeness of the cock rock partridges is often used by people in these countries who want to stage cock fights.

Fig. 19-2. 1. Rock partridge *(Alectoris graeca);* 2. Red-legged partridge *(Alectoris rufa);* 3. Barbary partridge *(Alectoris barbara);* 4. Arabian red-legged partridge *(Alectoris melanocephala).*

The RED-LEGGED PARTRIDGE *(Alectoris rufa)* likes shrubby barren areas, fields, and vineyards. In 1770 it was introduced into England by the Marquis of Hertford and Lord Rendlesham. Both obtained large numbers of eggs from France, had them hatched under domestic hens, and liberated the fully-grown young on their estates. This partridge has probably reached the Atlantic Islands as the result of introductions. Early attempts to introduce this attractive bird into southwest Germany were also successful. Up to the end of the sixteenth century it lived in the middle Rhine area near Bacharach.

Partridges

The TRUE PARTRIDGE (genus *Perdix*) are distinguished from the rock partridge by sixteen to eighteen tail feathers and spurless tarsi. There is a sex difference in color, and the plumage shows fine, dark vermiculation. There are three species in temperate Eurasia: 1. The COMMON PARTRIDGE *(Perdix perdix),* with eight subspecies; 2. The DAURIAN PARTRIDGE *(Perdix dauuricae),* with elongated throat feathers and a large black belly patch, has five subspecies. 3. The TIBETAN PARTRIDGE *(Perdix hodgsoniae),* with a reddish-brown crown and nape band feathers, a white throat, a black spot below the eye, a black-and-white crop and breast barring, and a large black patch on the belly. It has three subspecies and is found in the Tibetan Yak meadows and dwarf shrub areas.

The common or GRAY PARTRIDGE owes its German name, "Rebhuhn" (vinehen), not to its presence in vineyards but to its shrill alarm call, "rep rep rep." Originally it lived on heaths and moorlands. However, it has long become a follower of cultivation and has settled cultivated fields of the lowlands in numbers. In the Pyrenees it also occurs on mountain meadows. In general, partridge are residents, but in the fall, they may gather into larger flocks which wander about widely. In Eastern Europe flocks of the local, pale subspecies *(Perdix perdix lucida)* wander to the south and west when there is a lack of food, particularly

Fig. 19-3. 1. Common or gray partridge *(Perdix perdix);* 2. Daurian partridge *(Perdix dauuricae);* 3. Tibetan partridge *(Perdix hodgsoniae).*

Fig. 19-4. 1. Sand partridge *(Ammoperdix hayi);* 2. See-see partridge *(Ammoperdix griseogularis);* 3. Stone partridge *(Ptilopachus petrosus).*

when the snow depth exceeds fifty centimeters. In the winter of 1937 large flights of "eastern wandering partridge" reached the eastern shores of the Kurische Haff (East Prussia) from the northeast. In the moors of the Ems area and the Netherlands there is an interesting dark subspecies *(Perdix perdix sphagnetorum).* As a result of progressive moorland drainage, it is condemned to extinction and has largely interbred with ordinary partridges which have immigrated into its former range.

Partridge eat animal and plant food. They are particularly beneficial to man because of their consumption of injurious insects. Investigations have shown that their diet is of 63% plant origin. In spring the cock displays to the hen with muttering, an open beak, fluffed out flank feathers, and fanned out tail. Nests are found from May to June. They are depressions in the earth, scraped out by the hen below shrubs or among grass. They contain ten to twenty olive-brown, spotted eggs which hatch after twenty-three to twenty-five days of incubation. The cock stands on guard near the nest but takes no part in the incubation. He covers the newly hatched chicks beneath his plumage at once. At sixteen days the young can fly well. In the fall, neighboring families join together in flocks which separate only in the spring.

This partridge has been successfully introduced into wide areas of southern Canada and from coast to coast in the northern United States.

Other Perdicini

The SAND PARTRIDGE (genus *Ammoperdix)* is found in Northeast Africa and Southwest Asia. There are two species: the SAND PARTRIDGE *(Ammoperdix hayi)* and the SEE-SEE PARTRIDGE *(Ammoperdix griseogularis).* These partridge are sand-colored inhabitants of barren, stony, desert mountains. They run and jump among rocks with noticeable agility. Cocks make themselves known with a loud clear double whistle.

Fig. 19-5. Hill partridges (genus *Arborophila*).

The STONE PARTRIDGE *(Ptilopachus petrosus)* is a small peculiar galliform bird. Its systematic position is still unsettled. It is bantam-sized and has fourteen tail feathers which are longer than those of the partridge or francolins. The tail is carried upright as with domestic fowl. The tarsi have no spurs, and the sexes are similar in color.

They live on rocky hills and in the cliffs of steppe areas on the southern edge of the Sahara. They move by hops among the rock debris and jump with great agility. In their shape and tail carriage they remind one of bantam fowl. Outside the breeding season, they gather in small family groups of six to eight birds. In display, and while breeding, the cocks utter a flute-like call. The nest is well hidden in the grass and is merely a depression lined with leaves and dry stalks. It holds four to six ochre-yellow eggs.

Fig. 19-6. Roulroul *(Rollulus roulroul).*

The HILL PARTRIDGE (genus *Arborophila*) are almost the size of the common partridge, but they are short-tailed. The area around the eyes and over the throat is bare of feathers, and the tarsi are unspurred.

There are eleven species, one of which is the HILL PARTRIDGE *(Arborophila torqueola)*. The upper head and cheeks of the hill partridge are red-brown and the eyestripe, chin, and throat patch are black. The upper breast and nape region are black striped; otherwise they are predominantly ochre-yellow and reddish-brown. They inhabit southwestern Himalayan slopes from Kashmir to Assam. The TREE PARTRIDGES (genus *Tropicoperdix*) are closely related. They are inconspicuously colored with tufts of white down-like feathers behind the shoulders. They inhabit dense tropical forests rich in undergrowth in Burma, Malaya, Sumatra, and Borneo.

▷ Perdicini: 1. Mountain quail *(Oreortyx picta)*; 2. Spotted wood quail *(Odontophorus guttatus)*; 3. California quail *(Lophortyx californica)*; 4. Scaled quail *(Callipepla squamata)*; 5. Montezuma quail *(Cytronyx montezumae)*; 6. Bobwhite *(Colinus virginianus)*.

Hill partridge prefer mountain forests with numerous ravines and dense evergreen undergrowth at altitudes from 1200 to 3600 meters. Outside the breeding season, they are found in flocks, scratching the soil for insect food. The flock crowds tightly together in the collective sleeping places. Their call is a double whistle which sounds like "whee hoo." It can be heard a long way and is sounded mainly in the morning and evening. The nest is a depression, well lined with grass and leaves and is closely interwoven with the surrounding grass tufts to form a proper chamber. It holds three to five milk-white eggs. The chicks are reared by both parents and are independent when six weeks old.

A number of small, often brightly colored species of galliforms have the vernacular name of bush quails, but they do not belong to the true quails. Instead, they belong to the partridge-like birds.

A. Indian species: 1. The JUNGLE BUSH QUAIL *(Perdicula asiatica)* has a short, high beak, a spur knob on the tarsus, and twelve tail feathers. The upper parts are pale reddish-brown and there are closely barred black and white stripes on the head and lower body. The call of the males is a clear, high-pitched, long drawn trill. Outside the breeding season they form small flocks. They spend the night in a "hedgehog" formation on the ground. 2. The PAINTED BUSH QUAIL *(Cryptoplectron erythrorhynchum)* is unspurred. It has long, very dense feathering about the rump and upper tail coverts, and twelve tail feathers. The beak and legs are red; the crown stripe is white. The face mark and upper head are black, the back is blue-gray, and the underparts are rusty-brown.

▷▷

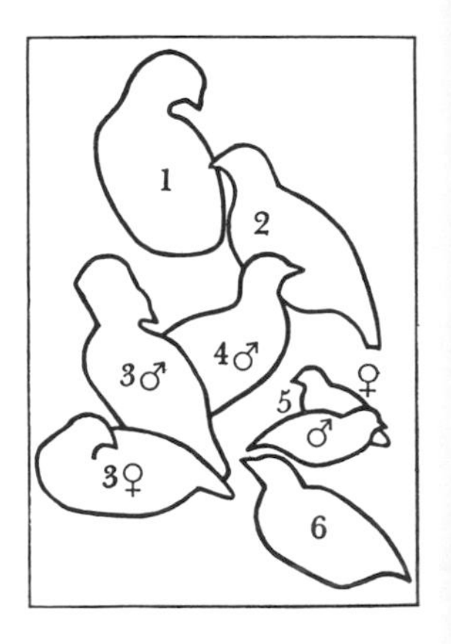

1. Blood pheasant *(Ithaginis cruentus)*; 2. Snow partridge *(Lerwa lerwa)*; 3. Roulroul *(Rollulus roulroul)*; 4. Crimson-headed partridge *(Haematortyx sanguiniceps)*; 5. Chinese painted quail *(Excalfactoria chinensis)*; 6. European quail *(Coturnix coturnix coturnix)*.

B. Malayan and Indonesian species: 1. The BLACK WOOD PARTRIDGE *(Melanoperdix nigra)* has a short thick beak. The legs are spurless, but there are short claws on its rear toes. The males are uniformly black, and the females are red-brown with a central white abdomen and banded wings. It inhabits dense forests of southern Malaya and in Sumatra up to 1200 meters in the mountains. 2. The CRIMSON-HEADED PARTRIDGE *(Haematortyx sanguiniceps*; Color plate p. 480) has a weak, pigeon-like beak. The males have sharp double spurs on the tarsus. The head and longest under tail covert are red, and in males the crop area is also red. It inhabits forests of North Borneo, especially the sand flats

1
2
3
4
5
6

2
1
5
4♂
4a
3

♂
1
♀
2 ♂
3 ♂

◁

Tragopans: 1. Satyr tragopan *(Tragopan satyra)*; 2. Western tragopan *(Tragopan melanocephalus)*; 3. Blyth's tragopan *(Tragopan blythii blythii)*.

◁◁

Perdicini: 1. Himalayan snowcock *(Tetraogallus himalayensis)*; 2. Barbary partridge *(Alectoris barbara)*; 3. Red-legged partridge *(Alectoris rufa)*; 4. Gray partridge *(Perdix perdix)*; 5. See-see partridge *(Ammoperdix griseogularis)*.

of valley floors. As protection against floods, the nest is built as a well-shaped structure in tall grass tufts. 3. The FERRUGINOUS WOOD PARTRIDGE *(Caloperdix oculea)* has a degenerate hind toe, and the males are spurred. The head, neck, breast, and lower parts are reddish-brown. The wing coverts and flanks are decorated with black drop-shaped spots. The male's call is repeated eight to nine times in increasingly rapid sounds and is answered by the female with a series of even faster whistles. The nest is said to be roofed over and to have a side entrance. It is found in Malaya, Sumatra, and Borneo. 4. The CROWNED WOOD PARTRIDGE (ROULROUL) (*Rollulus roulroul*; Color plate, p. 480) has a tarsi without spurs and the hind toe is clawless. The male has a chestnut-red hood and in front of this, on the forehead, is a tuft of black bristle feathers. The female has only the latter as a head decoration.

The roulroul is doubtlessly the most beautiful representative of the Indonesian wood partridges. These pretty birds like deforested areas with shrubs, as well as dry, open jungle and bamboo copses. In such habitats they are found in small flocks of seven to fifteen birds, all of which are looking for food. They scratch busily for insects and worms, moving the feet far forward before scratching. They are among the few gallinaceous birds which live mainly on an animal diet. Besides this, they also like to eat fruit. On Borneo, Pepper saw a feeding association between wild pigs and roulrouls; both, close together, fed on fallen *Lithocarpus* fruit.

Like many of the true quails, roulrouls also live in monogamy. The hen builds a completely roofed-over nest, of twigs and leaves, with a small entry hole in the front. After eighteen days of incubation the four yellowish-white eggs hatch. Observations of captive birds have shown that, until the chicks are twenty-five days old, the hen leads them to the nest every evening and then carefully closes the nest entrance with twigs. The cock takes no part in rearing the young. During the first few days the young only take food from the mother's beak. From the fifth day on they execute the characteristic sweeping, scratching movements of the adults. At ten days of age they can flutter, and at twenty-five days they can perch on trees. When one month old, the males show an indication of the red-brown hood.

The partridge-sized LONG-BILLED PARTRIDGE *(Rhizothera longirostris)* is regarded as a link between the south Asiatic wood partridge and the francolins. The beak is fairly long and curved, and both sexes have a sharp spur on the tarsus. The males are quail-colored above and rusty-yellow below. The crown is dark brown while the head is red-brown with a wide gray neck band. They inhabit hill country plentiful with dry bush and bamboo in south Tenasserim, southwest Thailand, Malaya, Sumatra, and Borneo. In the morning and evening they make themselves noticeable by calling shrilly.

Francolins

FRANCOLINS (genus *Francolinus*) are partridge-like fowl. The tarsi are strong and have one or two sharp spurs in the males. The tail is short and faintly rounded, with fourteen tail feathers. In many species, the sexes are similar in color, while in others they differ. There are thirty-four species in Africa and five in West and South Asia. Francolins inhabit steppes, savannahs, primeval forests, and mountains. The preferred habitats of the RED-BILLED FRANCOLIN (*Francolinus adspersus*) of southern Angola, Southwest Africa, Botswana, and Zambia are moist areas in dry bush and extensive river courses. The HARTLAUB'S FRANCOLIN (*Francolinus hartlaubi*; Color plate, p. 470) can be found among cliffs and also in roosts on the ground. Many African mountain chains have their own species of francolins. For example, the Ethiopian mountains are inhabited by the CHESTNUT-NAPED FRANCOLIN (*Francolinus castaneicollis*), while the CAMEROUN MOUNTAIN FRANCOLIN (*Francolinus camerunensis*; Color plate, p. 470) occurs only in the mountain country of the Camerouns. Some other francolin species are shown on page 470.

Fig. 19-7. Francolin (*Francolinus francolinus*).

All francolins live in monogamy and aggregate in family units outside the breeding season. During the breeding season cocks utter loud, rough calls from some elevated perch, such as a termite hill, a rock, or a branch. One of the most beautiful species, the FRANCOLIN (*Francolinus francolinus*; Color plate, p. 470), was introduced into southern Spain, Sicily, and Greece by the Moors and Saracens from West Asia, during the Middle Ages. However, it soon became extinct in Europe as a result of unrestricted killing. It prefers cultivated land with irrigation channels and much cover. In spring, the cocks call during the morning and evening hours and before rain storms, with a very audible "tshik-tshik-tshik keeraikek." The breeding season lasts from April to July. Nests are made between grass tussocks and are lined with grass stalks. The clutch consists of six to eight hard, thick-shelled eggs, which are olive to chocolate-brown in color, and covered with round, white, chalky spots. As in many Perdicinae, the cock warns the incubating hen of danger. After twenty-one to twenty-three days of incubation, the young hatch and are led by both parents.

Northern Transcaucasia forms the northern limit for the distribution of the francolins. The climate there is especially severe, and the population shrinks repeatedly as a result of the lack of food during the cold, snowy winters. But the francolins recover again after a few mild winters. The successful introduction of francolins in the Nucha area of Kachetia (Caucasus) shows how productive this bird is. In 1932, three cocks and two hens were released in Agri-tschai Valley. In 1947 there were already francolins all over this valley, and they had settled the Alasan river valley over a distance of one hundred kilometers as well.

The ANGOLA RED-NECKED FRANCOLIN (*Francolinus afer*) is a common gallinaceous bird of the bush steppes and cultivated lands of Angola, the southern Congo, and other parts of South Africa. Cocks sound their loud "koraki koraki" from the top of termite hills in the morning and evening. Nests are well-concealed with grass. There are five to nine cream-colored or light brown eggs, which have white pore spots, but are sometimes unspotted. They are incubated for eighteen to twenty days. The hen, when leading her young, is so concerned for the safety of her brood that she may even attack animals the size of dogs.

The BAMBOO PARTRIDGE (genus *Bambusicola*) inhabit southern China, Taiwan, and parts of Malaya. They are partridge-sized with a fairly long tail, which is graduated with fourteen feathers. Males have a pointed spur. Among the species in this genus is the BAMBOO PARTRIDGE *(Bambusicola thoracica)* which has a red-brown face, a stripe above the eye, and a gray throat. The lower parts are pale ochre-yellow; and there are red-brown drop-shaped spots on the flanks. It has long been successfully introduced in southern Japan.

Fig. 19-8. Bamboo partridges (genus *Bambusicola*).

Bamboo partridges are forest dwellers, but they have successfully adapted to the cultivated lands in southern China. They frequently inhabit hills with low bamboo shrubbery or field copses. The cock's call sounds very loud and consists of a series of calls extended for half a minute. These calls sound like "gigigigigigigeroigigeroi," and is heard at all times of the year. The hen lays three to seven tawny-colored eggs with fine brownish-red dots. These eggs are distinguished by their particularly hard and thick shells. Both parents rear the young together. The Chinese often keep these birds in small cages and enjoy their shrill cries, which sound anything but pleasant to European ears.

The MOUNTAIN QUAIL *(Ophrysia superciliosa)* and the three species of SPURFOWL are pheasant-like, but are much smaller. The mountain quail lacks spurs and is gray and black with a white band at the side of the head. There are three species of spurfowl (genus *Galloperdix*). Among them is the PAINTED SPURFOWL *(Galloperdix lunulata)* which is a little over partridge-size and has one or more sharp spurs on the tarsi of both the males and females. There is a bare area around the eye, and the tail is fairly long. The male has a black and white spotted crown, a black throat with a wide rusty-yellow edge, a black-spotted crop, breast, and upper belly feathers, and shining green tips on the shoulder and tail feathers. Otherwise the plumage is red-brown and is covered with white and black edged spots. The female is uniformly brown.

The mountain quail inhabits, or used to inhabit, grassy areas on the Himalayan slopes in northwestern India. However, it does not seem to have been observed or reported there since more than ninety years ago. Spurfowl prefer dense bush on a rocky base and avoid true jungles. They always escape on foot and are not at all eager to fly. In the morning and evening hours they cackle as loudly as domestic hens. The nest, hidden among grass tussocks, is a very shallowly scraped area. It holds four white eggs which are incubated for twenty-one days.

Fig. 19-9. Blood pheasant *(Ithaginis cruentus)*.

The BLOOD PHEASANTS (genus *Ithaginis*) are included with the true pheasants by many zoologists, yet they have little in common with the latter. Externally they resemble the francolins. The moult of the fourteen feathers proceeds centrifugally as in the Perdicinae. Being high mountain dwellers, they have a dense soft plumage, most of which, in the cocks, consists of long lancet-shaped feathers. The hen's plumage is an inconspicuous brown with fine blackish vermiculations and stripes.

There is only one species, the BLOOD PHEASANT *(Ithaginis cruentus)*. It has a dense feather hood in both sexes which can be erected. The beak is short, thick, and curved. The male's tarsus has one to three spurs. Fourteen subspecies inhabit the Central Asiatic mountains from Nepal over northern Burma to northwest China at elevations of 3000 to 4500 meters. The largest known subspecies is the NEPAL BLOOD PHEASANT *(Ithaginis cruentus cruentus;* Color plate p. 480) which lives in moist, subalpine mixed forests with a dense undergrowth of bamboo, rhododendrons, and elderberry.

Blood pheasants regularly feed in the alpine meadows above the tree line during the daytime. Ernst Schäfer reports on the mode of life of the southeast Tibetan subspecies: "In cultivated areas they come to the harvested fields where they feed on weed seeds, and in the winter months they often associate with flocks of eared pheasants. The latter, which hear better and are shyer, always take over the leadership of such mixed flocks. Blood pheasants are agile runners and jumpers. Only rarely do they perch in trees. They search for food almost exclusively on the ground. They are sociable and keep together in flocks of ten to forty birds until early May. In large flocks, there is a striking predominance of males in the ratio of three to one. This is also noticeable during the breeding season, when the excess cocks do not disturb the monogamous pairs but, instead, form pairs of their own and display to one another. At this time, two or three pairs of nonbreeding cocks often join together. Blood pheasants are good at escaping. When they run from people, they are easily mistaken for mammals in the dense undergrowth because of their loud piping. When pursued by dogs they generally take to trees. However, they never fly far. Instead, they sit close to the branches in the middle of dense trees so that they are difficult to recognize. As soon as the disturbance is over, they come down from the trees, raise their hoods, and, by piping, call one another to reassemble. They often walk directly behind one another in single file.

In the breeding season cocks show off with erected hood, fluffed out plumage, drooping wings, and fanned-out tail, strutting repeatedly around the hen. Both partners take part in raising the young. Blood pheasants live predominantly on plant food.

Tribe: quails

The smallest Perdicinae are found among the true quails, tribe Coturnicini. Some of them are the only migrants among the gallinaceous

Fig. 19-10. 1-4. Quail *(Corturnix coturnix)* subspecies: 1. European quail *(Coturnix coturnix coturnix)*; 2. Ussuri quail *(Coturnix coturnix ussuriensis)*; 3. Japanese quail *(Coturnix coturnix japonica)*; 4. African quail *(Coturnix coturnix africana)*; 5. Rain quail *(Coturnix coromandelica)*; 6. New Zealand quail *(Coturnix novaezelandiae)*.

Fig. 19-11. Harlequin quail *(Coturnix delegorguei)*.

Fig. 19-12. Chinese painted quail *(Excalfactoria chinensis)*.

birds. Their wings are generally long and pointed. They are good fliers. The tail, as a rule, is quite short, having ten to twelve feathers, and consequently is quite hidden by the upper tail coverts. Males and females differ in color, and have no spurs. There are eight species in five genera. Among them are: 1. The QUAILS in the narrow sense, genus *Coturnix*, with four species: The COMMON QUAIL (*Coturnix coturnix*; Color plate p. 480), the RAIN QUAIL *(Coturnix coromandelica)*, the NEW ZEALAND QUAIL *(Coturnix novaezelandiae)*, and the HARLEQUIN QUAIL *(Coturnix delegorguei)*. The harlequin quail is particularly colorful. Males have a white throat surrounded by a black, anchor-shaped pattern. The centers of the crop and breast are black. The lower parts are reddish-brown and black spotted and striped. Females are inconspicuously brown. 2. The CHINESE PAINTED QUAIL (*Excalfactoria chinensis*; Color plate p. 480) is the world's smallest gallinaceous bird. It is sparrow-sized, with an L of 12 cm and a weight of 45 g. They inhabit grassy steppes and swamp grasslands. There are ten subspecies. 3. The NEW GUINEA MOUNTAIN QUAIL *(Anurophasis)* has one species. 4. The MADAGASCAR PARTRIDGE *(Margaroperdix)* has one species, the MADAGASCAR PARTRIDGE *(Margaroperdix madagascariensis)*. It has a short, high beak and twelve tail feathers. The male has a black face stripe and throat, a white eyebrow and temple stripe, a white cheek band, a reddish-brown crown, gray sides of the head, neck, and breast, a black lower body with white pearly dots, reddish-brown flanks with a white, black-edged shaft stripe, and a reddish-brown white-striped upper part. It is found in Madagascar.

The COMMON QUAIL *(Coturnix coturnix coturnix)* inhabits a particularly wide range as a breeding bird (see Fig. 19-10). It is a late European migrant, arriving in Germany only in mid-May. It begins to breed about the end of May. Originally it was an inhabitant of grass and dwarf bush steppes, but it has long adapted to farmland and settles grain fields, clover, lucerne and pea fields, dry meadows, and herb-grown fallow land. The cocks sound their melodious "pick werwick" call especially during the night. They fight with determination over their territories, which they inhabit with several females. The cock attracts the hen with a morsel of food in his beak and then struts about her with his plumage fluffed-out.

Only hens concern themselves with incubation and the rearing the chicks. They lay eight to fourteen thick-shelled eggs which are yellowish with black-brown spots. The nest depression is quickly scratched out and incubation lasts seventeen days. The young are capable of flying in fourteen days and are fully fledged in nineteen days. The mother leads them for about seven weeks. They are sexually mature when a year old.

Quail migrate at night from late August to late October, flying low over the ground in dense flocks towards the south. The routes of migration for the populations are quite complicated and have only been partially clarified. Banding recoveries have shown that quail from eastern France, Romania, and western Russia meet in Sardinia, Sicily, Tunisia, and Algeria. Many quail which breed in North Africa migrate as early as April or June, together with their two-month-old young, north from their breeding areas (which are then in summer drought) as far as Albania. Great flocks of quail which, after an exhausting crossing of the Mediterranean, reached the Sinai Desert are reported in the Bible. Such gigantic swarms were seen until the beginning of this century. As late as 1920, Egypt exported three million of these small fowl. Since the 1930's, however, they have arrived on the southern Mediterranean shores only in small flocks. Ceaseless pursuit in the south of their range, complete weed control, and the excessive use of pesticides in Europe have caused the quail to become rare in Central Europe. The case of this small, once common fowl, shows how modern pest control agents destructively interfere with the highly sensitive balance of nature and endanger a harmless, "useful" species which people like to see.

Fig. 19-13. Range of American quail (tribe Odontophorini).

Many quail populations extend their migration as far as the steppes of the Sudan and the northeast of East Africa. There are two quite similar subspecies in Africa. Eastwards, the range of the European quail borders on that of the USSURI QUAIL *(Corturnix corturnix ussuriensis)*. The JAPANESE QUAIL *(Coturnix cotrunix japonica)* is very similar and is distinguished from the European quail mainly by its much rougher call, and also by its throat feathers which are prolonged and pointed after the fall moult.

This quail has long been domesticated in Japan. The beginning of Japanese quail breeding can be traced back to 1595. Originally the Japanese kept the quail for pleasure, mainly because of its melodious call. Today it is appreciated as a supplier of eggs and meat. One to two million quail a year are hatched in incubators in the large Japanese hatcheries. The chicks weigh seven grams. Immediately after hatching they are segregated according to sex. Most of the cocks are killed. Japanese quail develop three and one half times as fast as domestic fowl of small-laying breeds. At thirty days they are almost fully grown. They are then placed singly in cages with a floor area of 15 x 15 centimeters, which are stacked in five-layered laying batteries. At six weeks of age the young quail hens lay their first eggs. These weigh nine to eleven grams. The hens proceed, like machines, to lay an egg every sixteen to twenty-four hours for eight to twelve months. Finally they are killed and replaced by young hens.

Fig. 19-14. Cock Chinese painted quail circles his hen with wings spread outward.

In Europe, particularly in Italy, Japanese quail have been introduced as domestic breeders. Their eggs and meat are already of economic importance and are popular with gourmets. They were also in use more recently as experimental animals for medical research.

Fig. 19-15. The frontal display of the cock Chinese painted quail effectively emphasizes the black-white face pattern and the reddish-brown lower parts.

The wanderings of Indian, Australian, and African species of quail depend upon the rainy seasons. Thus the colorful HARLEQUIN QUAIL *(Corturnix delegorguei)* breed wherever grass seed and insects are plentiful enough to assure raising of the young. They appear suddenly in such areas, breed, and then disappear with the fully grown young. Meinertzhagen has reported: "Soon harlequin quail flocks ran all over the camp, walked over our feet, and inhabited our tents. The steppe grass was bubbling with quail, as it were, which uttered their mournful contact calls in chorus and quickly migrated towards the north. The passage began about six p.m. and lasted until midnight. The migrating birds were in a trance-like state and took no notice of people."

The CHINESE PAINTED QUAIL *(Excalfactoria chinensis)* lives in monogamy. The cock attracts the hen with a morsel of food in his beak and then displays his side towards her by raising the side of his body facing her and drooping his wing on the opposite side. The hen lays four to six, olive-brown, dark brown, or black spotted eggs in a ground depression which she quickly scrapes out. The cock stands guard by the nest and rushes at every enemy, even ones as large as dogs, with lowered head and trailing wings. The chicks, which are no larger than bumble bees, hatch after sixteen days of incubation. They are reared by both parents. They can fly towards the end of their second week. Young cocks are sexually mature at five months while hens mature at seven to eight months of age. Like the Japanese quail, the Chinese painted quail has also become a popular cage bird. It has lived up to ten years under human care.

Rand found the MADAGASCAR PARTRIDGE *(Margaroperdix madagascariensis)* particularly common in the heath bush areas of the central mountainous lands of Madagascar. The male's call sounds like "koo koo koo." Both parents tend the young and later form small flocks with them. These flocks remain together until the next breeding season.

Tribe: American quails

In America, where there are no true quail, partridge, or francolins, the AMERICAN OR TOOTHED QUAIL (tribe Odontophorini) occupies the habitats of these small fowl. Varying in size from that of a quail to a partridge, these birds have a compact build with spurless tarsi and have a high, short beak with cutting "teeth" behind the tip of the lower mandible which are often worn down. Crests are common. In most species, the sexes are of different colors. They are monogamous. Nests are always on the ground, and both partners take part in rearing the young. They are adapted to a variety of habitats from southern Canada to northern Argentina. There are ten genera with thirty-four species (Color plate p. 479).

Fig. 19-16. Bobwhite often spend the night secure towards all directions in a "hedgehog" formation.

The WOOD PARTRIDGE (genus *Dendrortyx*) inhabits the mountain forests of Mexico and Central America. It is almost partridge-sized and thus is the largest of the American quail. Its long crown feathers form a short hood at the back of the head. The tail is as long as that of a domestic pigeon and is therefore quite long for this group. The BUFFY-CROWNED WOOD PARTRIDGE *(Dendrortyx leucophrys)* inhabits the uplands of Guatemala, Honduras, Nicaragua, and Costa Rica. Van Rossem observed wood partridge in El Salvador. There they were common in the impenetrable secondary growth of the mountains (newly grown forest on formerly cultivated land) and lived in small flocks. They perched in high trees during the night. When they thought themselves unobserved, they walked about on their long legs. At the slightest sound, they flattened themselves and disappeared silently into the shrubbery in a manner usually only observed in rails.

North America also has a large quail, the beautiful MOUNTAIN QUAIL (*Oreortyx picta;* Color plate p. 479), which occurs in the western coastal mountains from the Columbia River in the north to lower California in the south. On its head it has two feathers which may be up to six centimeters long. It lives on canyon slopes, which are overgrown with hardwoods, and in light pine and elderberry forests on mountain ridges. It remains at high elevations only from late spring to fall. With the onset of the cold season, it wanders regularly down the valleys to warmer areas. Distances of up to sixty kilometers are covered in these journeys and they are made exclusively on foot.

The only quail of the arid desert steppes of Texas, Kansas, Colorado, and central Mexico is the SCALED QUAIL (*Callipepla squamata;* Color plate p. 479). It has a short, wide crest and a scale-like pattern of contour feathers. The rounded feathers have black edges. In its desert-like ranges, the scaled quail regularly flies to distant watering places in flocks of 6 to 200 birds. It generally breeds in June and July, the only rainy months in its habitat. If the rains fail or are unusually excessive no young are reared. However, if the weather in succeeding years is favorable, two to three broods are raised in a season, and so the population soon returns to normal.

The CRESTED QUAIL (genus *Lophortyx*) live in the dry areas of Mexico and southwestern United States. The best known species are the CALIFORNIA QUAIL (*Lophortyx californica;* Color plate p. 479), and GAMBEL'S QUAIL *(Lophortyx gambelii).* The BANDED QUAIL *(Philortyx fasciatus),* with its head plume of narrow feathers directed towards the rear, is similar to the two above-mentioned species.

The California quail is common in the chaparral (oak and hardwood forests) of the slopes and valleys in the Pacific coastal mountains of North America. In many places it has become a follower of cultivation and inhabits vineyards, gardens, and city parks. Hence, man has been able to introduce it successfully in other states and continents, such as Utah, New Mexico, British Columbia, Hawaii, New Zealand, and Chile. From late March to early April, the cock's "kaah" call is heard everywhere. The cock displays in front of the hen by bowing before her and dancing around her, uttering "giggling" and chattering sounds.

Crests in American quail:

Fig. 19-17. California quail.

Fig. 19-18. Mountain quail.

Fig. 19-19. Montezuma quail.

Fig. 19-20. Scaled quail.

The nest is a ground depression fitted with a few leaves and stalks. It holds nine to seventeen whitish, brown-spotted eggs which are incubated for twenty-two days. The cock stands guard by the nest and if the hen perishes, he takes over the breeding duties alone. There is only one brood per year. Both parents lead the young which are fledged at fifteen days of age and are independent when four weeks old. In the fall five to six families, with an average of nine young each, gather into winter flocks. Each such flock claims an area of about three kilometers in diameter and defends it against intruding members of the species. California quail are long-lived in human care. They are easily reared, and are, consequently, often kept by European bird fanciers.

The TREE QUAIL (genus *Colinus*), with their very short hoods, are small American quail. They occur in four species in eastern and central United States, throughout Mexico, Central America, Venezuela, Guiana, and northern Brazil. The name of the United States species, bobwhite, comes from its call. The BOBWHITE (*Colinus virginianus;* Color plate p. 479) lives on fallow fields, meadowlands rich in bushes, and open woodlands from the Canadian border southwards to Mexico and Cuba. While many other species have suffered greatly from the cultivation of the North American landscape, the bobwhite has been able to adapt to the altered circumstances. Deforestation, the planting of roadside shelter belts, and the vast extension of grain cultivation have even furthered its increase to a great extent. It has also wandered into city parks.

In spring, those cocks not yet paired sound their loud, tuneful "bobwhite" call from some elevated point such as a fence post. The cock courts the hen by displaying from the side and from in front. The clutch consists of up to fifteen white eggs, which the hen, and more rarely the cock as well, incubates for twenty-three days. Chicks can fly within a few days of hatching and are fully grown when eight weeks old. In the fall, neighboring pairs gather into winter flocks. The birds then spend the night together in some protected depression in the ground. They lie in a circle with their heads pointing outwards. Thus they can soon spot an enemy from whatever side it may approach and then whirr off in all directions. Even one-day-old chicks assume this star pattern.

Bobwhites are reared today on quail farms in the United States for hunting purposes. Generally they are kept in very small wire cages, each of which holds a pair. A hen may lay up to eighty eggs a year. As a result of this production, the bobwhite has almost become a domestic bird. Selective breeding has produced white, tawny, reddish-brown, and black races. A captive bobwhite once lived for nine years.

The smallest American quail is the HARLEQUIN OR MONTEZUMA QUAIL (*Cyrtonyx montezumae;* Color plate p. 479) which is found from the highlands of New Mexico south to Nicaragua. It is compact, short-tailed, and has a thick short hood lying along the back of its head. This colorful little bird lives in light oak or pine woods with a grass undergrowth at elevations of 1200 to 2700 meters. It regularly ploughs over the ground in search of its main food, the bulbs of a carex sedge. Its nest is, for a gallinaceous bird, quite skillfully built. It consists of a ground depression lined with grasses and down, which is roofed over with grass for concealment. Both partners incubate the six to sixteen white eggs for twenty-five to twenty-six days, and together they rear the young. Vigorous cattle grazing in their home range has already made this quail rather rare.

The WOOD QUAILS (genus *Odontophorus*) inhabit tropical forests from southern Mexico to northern Argentina in fifteen species. They are almost as large as a partridge and are marked by strong thick beaks, large thick feet, and short wings. The head is decorated by a wide, short nape hood. The SPOTTED WOOD QUAIL (*Odontophorus guttatus*; Color plate p. 479) lives in Central America from southern Mexico to Panama. Wood quails are characterized by their loud calls which sound like "ooroo." They build nests which are roofed over and lay four white eggs which are incubated for twenty-seven to twenty-eight days.

The only songster in the Galliformes is the SINGING QUAIL (*Dactylortyx thoracicus*) from the mountain forests of Vera Cruz to Honduras. This weak-billed, short-tailed bird sings a short song, particularly at the onset of twilight. It begins with three low-pitched whistles which are given with steadily increasing intensity. Then it utters the syllables "tshee wa leeo," which are uttered three to six times.

Subfamily: tragopans, by S. Raethel

The TRAGOPANS (subfamily Tragopaninae) are fairly large, compact, and very colorful Galliformes. The tail is short and rounded. There are eighteen tail feathers. The tarsi have a short spur in the male. In the cock, the throat and side of the face are bare. The cock also has a distensible skin flap at the throat which can be inflated in excitement along with horn-like, stalk-shaped erectile bodies on the back of its head. The tail moult is centrifugal, as in the Perdicinae. It breeds in old crow nests or in nests which it builds in trees. There is one genus (*Tragopan*) with five closely related species.

1. The WESTERN TRAGOPAN (◊ *Tragopan melanocephalus*; Color plate p. 482) is found in Himalayan mountain forests from western Kashmir to Garchal at 2000 to 4000 meters. 2. The SATYR TRAGOPAN (*Tragopan satyra*; Color plate p. 482) inhabits mountain forests of deodar cedars, oaks, and rhododendrons with rich bushy undergrowth in the Himalayas from Garvhal to northern Assam, at 1800 to 3900 meters. 3. TEMMINCK'S TRAGOPAN (*Tragopan temminckii*) has a bare face of cobalt blue. The plumage is predominantly red with white pearly spots on the back and wings. The belly is silver-gray with fire-red feather edges. It inhabits cool mountain forests with high rainfall from northern Assam to Hupeh at 900 to 2700 meters. 4. BLYTH'S TRAGOPAN (*Tragopan blythii*;

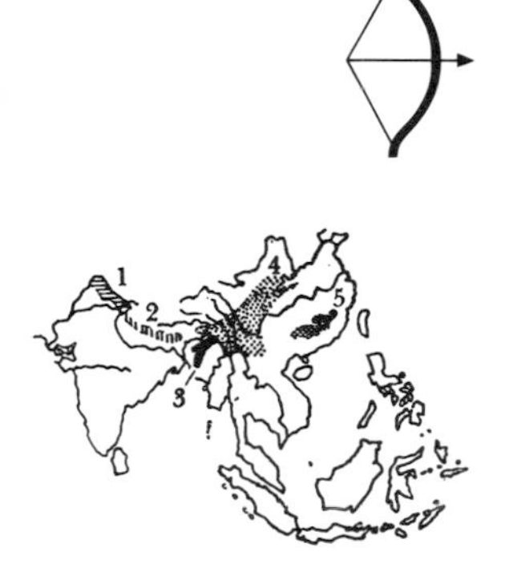

Fig. 19-21. 1. Western tragopan (*Tragopan melanocephalus*); 2. Satyr tragopan (*Tragopan satyra*); 3. Blyth's tragopan (*Tragopan blythii*); 4. Temminck's tragopan (*Tragopan temminckii*); 5. Cabot's tragopan (*Tragopan caboti*).

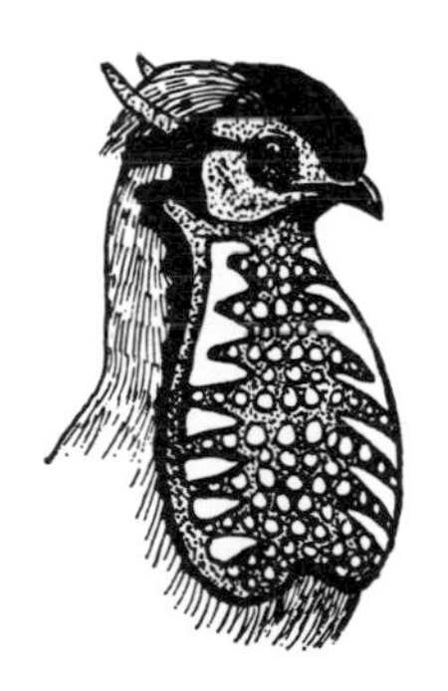

Fig. 19-22. The bare, colorful throat-skin of cock tragopans swells up in the excitement of display to a "display shield." Temminck's tragopan is above, the satyr tragopan is below.

Color plate p. 482) inhabits dense, moist mountain forests with streams in Assam and northwestern Burma at altitudes of 1800 to 2700 meters. 5. CABOT'S TRAGOPAN (⊖ *Tragopan caboti*) is the smallest species. The sides of the head and throat are bare and orange-yellow. The neck and undersides are light ochre-yellow, while the upperside is chestnut-brown covered with pale brown drop-shaped spots. It inhabits mountains in Fukien and Kwantung at 900 to 1500 meters.

Like the curassows and guans of tropical America, the tragopans spend a great deal of time on the branches in the crowns of trees where they find much of their food. They mainly eat tender leaves, buds, berries, fruit, and a lesser amount of animal food. They are unbelievably shy and have a very acute sense of hearing. At the slightest suspicious sound they slink off like cats, using every bit of cover. Only rarely do they leave the protective darkness of shady forests and, at the most, take a sunbath in lightly wooded areas at noon. Their flight is swift and accompanied by a whirring noise. But in general, tragopans are rather disinclined to fly.

Tragopans are solitary. In summer they live in pairs and join up in loose family groups only during the cold season. From April to May the cocks begin to display. Only during this season does one hear a loud series of calls from these normally very quiet birds. The initial sounds of the call are distinct, but these disappear toward the end of the phrase into a long drawn-out call, which sounds like a sheep bleating from a distance. While calling, the cock stands very erect and partially unfolds its colorful breast lappet. The hen then replies with a soft "quack quack."

The Brehm illustrator Gustav Mützel described the very impressive display of the tragopans at the Berlin Zoo as long ago as 1878: "Slowly the horns become erect. Like the jerky motions of the head, the throat-skin lowers itself in jerks, and as it becomes longer it also increases in width. The waves of excitement rise higher. The head flings wildly back and forth so that the still lax throat lappets, and the only half-erect "horns" fly about the bird's head. The wings are raised and stretched, and the tail feathers are lowered to form a wheel, the edges of which touch the ground. The ankle joint is flexed so that the cock, "raging with love," almost lies on the ground on its breast. The wings drag on the ground accompanied by hissing and puffing sounds. Suddenly all movement ceases. Crouched down low, breathing deeply, the plumage fluffed out, wings and tail pressed against the ground and eyes closed, the bird remains in full ecstasy. One can only see the beak and forehead crest of its head. The turquoise-blue horns, needle-like, stand erect, stiff, and vertical, and all parts of the throat shield are swollen and are now unfolded to the full extent. The skin is transparent sky-blue, deep cornflower-blue, and fiercest blood-red of colors; it is a peculiar, indescribably beautiful sight."

One of the few researchers who has watched a hen SATYR TRAGOPAN

(Tragopan satyra) building a nest in the wild was the American zoologist William Beebe. "The nest," he writes, "was right up against the trunk of a silver spruce and was partially roofed over by branches of this tree. The bird climbed about the needle-grown branchy wilderness from branch to branch around the tree, without a sound, its beak full of leaves. It came to the nest just as silently, climbing down the resinous "stairs." It was an old nest, probably that of a raven. It had a strong base of twigs and blanched grass which showed that it had endured a great deal. The lining of green, juicy oak leaves and herbs was still fresh, as were the torn off twigs from which an aromatic odor spread."

The hen is the sole incubator of the three to four cream-colored, brown spotted eggs. After twenty-six days the chicks hatch with tiny, well developed wings. As early as forty-eight hours of age, they can flutter from branch to branch and spend the night beneath their mother's wing in the crown of a tree.

Young tragopans grow more slowly than other gallinaceous birds. Though the sexes can be recognized as early as three months, young cocks do not acquire their full nuptial plumage until the fall of the second year and only then can they reproduce. A particularly striking feature of captive hens is that they can breed in their first year of life. A Temminck's tragopan cock lived for fourteen years with a French breeder.

At present there is only a single Blyth's tragopan in captivity, since its range has been inaccessible for political reasons for years. The cock arrived on April 14, 1963, at the Frankfurt Zoo, directly from India. Since March 22, 1967, it has been in the care of the internationally known pheasant breeder Dr. Steenbeck (Willemsstad/Rotterdam) in order that it might breed with the seven-eighths pure hens present there. This was successful within the first year.

Systematic Classification

The opinions of ornithologists diverge about the ordering of avian families into a natural system. The co-editors of the bird volumes, W. Meise, G. Niethammer and J. Steinbacher, agreed to use of the system used by Peters, Mayr and Greenway in the Check List of Birds of the World with a few changes. This system goes back to Fürbringer, Gadow, Wetmore, Mayr and Amadon. One of the co-editors, W. Meise, would have preferred the system, also going back to Fürbringer, used by Stresemann, or the system proposed by himself and used by Berndt/Meise. We give here a comparative review of these four systems. The numbers indicate the sequence of the orders of the system used here; several units which have the same figure have been joined into single orders in Grzimek's Animal Life.

Grzimek's Animal Life (1968)	**Peters (1932-1967 ff)**	**Stresemann (1959)**	**Meise (1960-1962)**
1. Tinamous (Tinamiformes)			Crypturi (1)
2. Ratites (Struthioniformes)	Ostriches (to 2)	Ostriches (to 2)	Ratites (2)
	Rheas (to 2)	Rheas (to 2)	
	Cassowaries (to 2)	Cassowaries (to 2)	
	Aepyornithiformes (to 2)	Aepyornithes (to 2)	
	Apterygiformes (to 2)	Apteryges (to 2)	
	Tinamous (1)	Crypturi (1)	
	Penguins (5)		
3. Grebes (Podicipediformes)	Colymbiformes (3)		
4. Loons (Gaviiformes)	Gaviiformes (4)		
5. Penguins (Sphenisciformes)			
6. Tubenoses (Procellariiformes)	Procellariformes (6)		
7. Pelecaniformes	Pelecaniformes (7)		
8. Ciconiiformes	Ciconiiformes (8+9)		
9. Flamingos (Phoenicopteriformes)			
10. Anseriformes	Anseriformes (10)		
11. Raptors (Falconiformes)	Falconiformes (11)		
12. Gallinaceous Birds (Galliformes)	Galliformes (12)	Galli (to 12)	Galli (12)
		Opisthocomi (to 12)	
13. Gruiformes	Gruiformes (13)	Turnices (to 13)	Grues (13)
		Columbae (to 15)	
		Pterocletes (to 15)	
		Ralli (to 13)	
		Heliornithes (to 13)	
		Mesoenades (to 13)	
		Jacanae (to 14)	
		Thinocori (to 14)	
		Rhynocheti (to 13)	
		Eyrypygae (to 13)	
		Cariamae (to 13)	
		Psophiae (to 13)	
		Grues (to 13)	
		Otides (to 13)	
14. Shorebirds and Gulls (Charadriiformes)	Charadriiformes (14)	Laro-Limicolae (to 14)	Limicolae-Lari (14)
		Alcae (to 14)	
			Anseres (10)
			Phoenicopteri (9)
			Gressores (8)
			Accipitres (11)
			Steganopodes (7)
			Tubinares (6)
			Sphenisci (5)

Grzimek's Animal Life (1968)	Peters (1932-1967 ff)	Stresemann (1959)	Meise (1960-1062)
		Gaviae (4)	Pygopodes (3+4)
		Podicipedes (3)	
		Sphenisci (5)	
		Tubinares (6)	
		Anseres (to 10)	
		Anhimae (to 10)	
		Steganopodes (7)	
		Phoenicopteri (9)	
		Gressores (8)	
		Accipitres (11)	
15. Pigeons (Columbiformes)	Columbiformes (15)		
16. Parrots (Psittaciformes)	Psittaciformes (16)		
17. Cuckoos (Cuculiformes)	Cuculiformes (17)	Musophagae (to 17)	Cuculi (17)
		Cuculi (to 17)	
			Columbae (15)
		Psittaci (16)	Psittaci (16)
18. Owls (Strigiformes)	Strigiformes (18)	Striges (18)	Striges (18)
19. Nightjars (Caprimulgiformes)	Caprimulgiformes (19)	Caprimulgi (19)	Caprimulgi (19)
20. Swifts (Apodiformes)	Apodiformes (20)		
21. Hummingbirds (Trochiliformes)			
22. Mousebirds (Coliiformes)	Coliiformes (22)		
23. Trogons (Trogoniformes)	Trogoniformes (23)		Trogones (23)
24. Rollers (Coraciiformes)	Coraciiformes (24)	Coraciae (to 24)	Coracii (24)
		Halcyones (to 24)	
		Meropes (to 24)	
		Momoti (to 24)	
		Todi (to 24)	
		Upupae (to 24)	
		Trogones (23)	
		Colii (22)	Colii (22)
		Apodes (20)	Apodes (20)
		Trochili (21)	Trochili (21)
25. Woodpeckers (Piciforms)	Piciformes (25)	Pici (25)	Pici (25)
26. Passerine Birds (Passeriformes)	Passeriformes (26)	Passeres (26)	Passeres (26)

In the review below of species and groups described in this volume, fossil forms are not listed. Page numbers refer to the main article; numbers in parentheses refer to illustrations or distribution maps of species not mentioned in the text. Species and subspecies without page numbers are not mentioned in the text, nor illustrated or shown on maps. Species and subspecies marked with) are endangered, those marked with + are extinct.

Class Birds (Aves)

Subclass Recent Birds (Neornithes)

Order Tinamous (Tinamiformes)

Family Tinamous (Tinamidae) 81
Subfamily Tinamous (Tinaminae) 82
Genus Rough Taos *(Tinamus)* 82
Great Tinamous, *T. major* (Gmelin, 1789) 82
Gray Tinamou, *T. tao* Temminck, 1815 82
Genus Scaled Taos *(Nothocercus)* 82
Bonaparte's Tinamou, *N. bonapartei* (Gray, 1867) 82
Black-capped Tinamou, *N. nigrocapillus* (Gray, 1867) 82

Genus Smooth Taos *(Crypturellus)* 82
Little Tinamou, *C. soui* (Hermann, 1783) 82
Variegated Tinamou, *C. variegatus* (Gmelin, 1789) 82
Brushland Tinamou, *C. cinnamomeus* (Lesson, 1842) 82
Slaty-breasted Tinamou, *C. boucardi* (Sclater, 1859) 82
Tataupa, *C. tataupa* (Temminck, 1815) 82

Subfamily Steppe Tinamous (Rhynchotinae) 82
Genus Red-winged tinamous *(Rhynchotus)* 80
Red-winged tinamou, *Rh. rufescens* (Temminck, 1815) 80
Genus Crested Tinamous *(Eudromia)* 82
Martineta tinamou, *Eu. elegans* Is. Geoffroy, 1832 82
Genus Three-toed Tinamous *(Tinamotis)* 82
Puna Tinamou, *T. pentlandii* Vigors, 1837 82
Genus Partridge Tinamous *(Nothoprocta)*
Ornate Tinamou, *N. ornata* (Gray, 1867) 82
Brushland Partridge Tinamou, *N. cinerascens* (Burmeister, 1860) 82
Genus Quail Tinamous *(Nothura)* 82
Spotted Nothura, *N. maculosa* (Temminck, 1815) 82
Genus Peacock Tinamous *(Taoniscus)* 82
Dwarf Tinamou, *T. nanus* (Temminck, 1815) 82

Order Ratites (Struthioniformes)

Suborder Rheas (Rheae)

Family Rheas (Rheidae) 90
Genus Rheas *(Rhea)* 90
Common Rhea, *Rh. americana* (Linné, 1758) 90
Rh. americana americana (Linné, 1758) (91)
Rh. americana intermedia Rothschild & Chubb, 1914 (91)
Rh. americana albescens Arribálzaga and Holmberg, 1878 (91)
Rh. americana araneipes Brodkorb, 1938 (91)
Rh. americana nobilis Brodkorb, 1939 –
Genus *Pterocnemia*
Darwin's Rhea, *P. pennata* (d'Orbigny, 1834) 90
P. pennata pennata (d'Orbigny, 1834) (91)
P. pennata garleppi Chubb, 1913 90

Suborder Ostriches (Struthiones)

Family Ostriches (Struthionidae) 91
Genus Ostrich *(Struthio)* 91
African Ostrich, *St. camelus* Linné, 1758 91
North African Ostrich, *St. camelus camelus* Linné, 1758 (92)
Somali Ostrich, *St. camelus molybdophanes* Reichenow, 1883 (92)
Massay Ostrich, *St. camelus massaicus* Neumann, 1898 (92)
South African Ostrich, *St. camelus australis* Gurney, 1868 (92)
Arabian Ostrich, +*St. camelus syriacus* Rothschild, 1919 92

Suborder Cassowaries (Casuarii)

Family Emus (Dromaiidae) 101
Genus *Dromaius* 101
Emu, *D. novaehollandiae* (Latham, 1790) 101
Tasmanian Emu, +*D. novaehollandiae diemenensis* Le Souëf, 1907 –
Black Emu, +*D. minor* Spencer, 1906 102

Family Cassowaries (Casuariidae) 104
Genus *Casuarius* 104
Australian Cassowary, *C. casuarius* (Linné, 1758) 105
One-wattled Cassowary, *C. unappendiculatus* Blyth, 1860 105
Bennett's Cassowary, *C. bennetti* Gould, 1857 105
Papuan Cassowary, *C. bennetti papuanus* Schlegel, 1871 105

Suborder Kiwis (Apteryges)

Family Kiwis (Apterygidae) 107
Genus *Apteryx* 107
Common Kiwi, *A. australis* Shaw, 1813 108
Southern Common Kiwi, *A. australis australis* Shaw, 1813 108
Northern Common Kiwi, *A. australis mantelli*

Order Grebes (Podicipediformes)

Order Loons (Gaviiformes)

Order Penguins (Sphenisciformes)

Order Tubenoses (Procellariiformes)

Tribe Hydrobatini 156
- Genus *Hydrobates* 156
  - British Storm Petrel, *H. pelagicus* (Linné, 1758) 156
- Genus *Oceanodroma* 156
  - Leach's Storm Petrel, *O. leucorhoa* (Vieillot, 1817) 156
  - Hornby's Storm Petrel, *O. hornbyi* (Gray, 1854) 156
  - Fork-tailed Storm Petrel, *O. furcata* (Gmelin, 1789) 156
  - Black Storm Petrel, *O. melania* (Bonaparte, 1854) 156
- Genus *Halocyptena* 156
  - Least Storm Petrel, *H. microsoma* Coues, 1864 15

Family Pelecanoididae 16
- Genus *Pelecanoides* 16
  - Peruvian Diving Petrel, *P. garnotii* (Lesson, 1828) 16
  - Magellan Diving Petrel, *P. magellani* (Mathews, 1912) 16
  - Georgian Diving Petrel, *P. georgicus* Murphy & Harper, 1916 16
  - Common Diving Petrel, *P. urinatrix* (Gmelin, 1789) 16
  - Kerguelen Diving Petrel, *P. exsul* Salvin, 1896 16

Order Pelecaniformes

Family Phaëthontidae 162
- Genus Tropic-Birds *(Phaëthon)*
  - Red-billed Tropic-Bird, *Ph. aethereus* Linné, 1758 162
  - White-tailed Tropic-Bird, *Ph. lepturus* Daudin, 1802 162
  - Red-tailed Tropic-Bird, *Ph. rubricauda* Boddaert, 1783 163

Family Pelecanidae 163
- Genus *Pelecanus* 164
  - Eastern White Pelican, *P. onocrotalus* Linné, 1758 164
  - Dalmatian Pelican, *P. crispus* Bruch, 1832 164
  - American White Pelican, *P. erythrorhynchos* Gmelin, 1789 169
  - Pink-backed Pelican, *P. rufescens* Gmelin, 1789 169
  - Gray Pelican, *P. philippensis* Gmelin, 1789 169
  - Australian Pelican, *P.. conspicillatus* Temminck, 1824 169
  - Brown Pelican, *P. occidentalis* Linné, 1766 169
  - *P. occidentalis occidentalis* Linné, 1766 169
  - Chilean Pelican, *P. occidentalis thagus* Molina, 1782 169

Family Cormorants (Phalacrocoracidae) 169
- Genus *Phalacrocorax* 169
  - Common Cormorant, *Ph. carbo* (Linné, 1758) 171
  - White-bellied African Cormorant, *Ph. lucidus* (Lichtenstein, 1823) 171
  - Japanese Cormorant, *Ph. capillatus* (Temminck & Schlegel, 1850) 171
  - Double-crested Cormorant, *Ph. auritus* (Lesson, 1831) 171
  - Bigua Cormorant, *Ph. olivaceus* (Humboldt, 1805) 171
  - Long-tailed Shag, *Ph. africanus* (Gmelin, 1798) 170
  - Pygmy Cormorant, *Ph. pygmaeus* (Pallas, 1773) 171
  - Little Pied Cormorant, *Ph. melanoleucus* (Vieillot, 1817) 17
  - Little Black Cormorant, *Ph. sulcirostris* (Brandt, 1837) 17
  - Flightless Cormorant,) *Ph. harrisi* Rothschild, 1898 17
  - Shag Cormorant, *Ph. aristotelis* (Linné, 1761) 17
  - Red-faced Cormorant, *Ph. urile* (Gmelin, 1789) 17
  - Spectacled Cormorant, + *Ph. perspicillatus* Pallas, 1811 17
  - Pelagic Cormorant, *Ph. pelagicus* Pallas, 1811 17
  - Red-legged Cormorant, *Ph. gaimardi* (Lesson, 1828) 17
  - Magellan Cormorant, *Ph. magellanicus* (Gmelin, 1789) 17
  - King Cormorant, *Ph. albiventer* (Lesson, 1831) 17
  - Blue-eyed Cormorant, *P. atriceps* King, 1828 17
  - Rough-faced Cormorant, *Ph. carunculatus* (Gmelin, 1789) 17
  -) *Ph. carunculatus carunculatus* (Gmelin, 1789) -
  - White-breasted Cormorant, *Ph. fuscescens* (Vieillot, 1817) 17
  - Guanay Cormorant, *Ph. bougainvillei* (Lesson, 1837) 17
  - Brandt's Cormorant, *Ph. penicillatus* (Brandt, 1837) 17
  - Spotted Cormorant, *Ph. punctatus* (Sparrman, 1786) 17
  - Chatham Cormorant, *Ph. punctatus featherstoni* Buller, 1873 17
  - Socotra Cormorant, *Ph. nigrogularis* Og.-Grant & Forbes, 1899 17

Family Anhingas (Anhingidae) 18
- Genus Anhingas, *(Anhinga)* 18
  - American Anhinga, *A. anhinga* (Linné, 1766) 18
  - Old World Anhinga, *A. rufa* (Daudin, 1802) 18
  - African Anhinga, *A. rufa rufa* (Daudin, 1802) -
  - Indian Anhinga, *A. rufa melanogastra*

Order Ciconiiformes

Order Flamingos (Phoenicopteriformes)

Order Anseriformes

Gray Duck, *A. superciliosa* Gmelin, 1789 —
Meller's Duck, *A. melleri* Sclater, 1864 —
Spotbill, *A. poecilorhyncha* Forster, 1781 (306)
Philippine Duck, *A. luzonica* Frazer, 1839 —
Yellow-billed Duck, *A. undulata* Du Bois, 1837 (307)
Chestnut-breasted Teal, *A. castanea* (Eyton, 1838) (307)
Flightless Teal, *A. aucklandica* (Gray, 1844) —
Auckland Island Flightless Teal, ◊ *A. aucklandica aucklandica* (Gray, 1844) —
New Zealand Brown Teal, ◊ *A. aucklandica chlorotis* Gray, 1845 —
Campbell Island Flightless Teal, ◊ *A. aucklandica nesiotis* (Fleming, 1935) —
Gray Teal, *A. gibberifrons* S. Müller, 1842 (307)
Madagascar Teal, ◊ *A. bernieri* (Hartlaub, 1860) (307)
Salvadori's Duck, *A. waigiuensis* (Rothschild & Hartert, 1894) (307)
Green-winged Teal, *A. crecca* Linné, 1758 314
Chilean Teal, *A. flavirostris* Vieillot, 1816 315
Gadwall, *A. strepera* Linné, 1758 315
Coue's Gadwall, + *A. strepera couesi* (Streets, 1876) 315
Falcated Teal, *A. falcata* Georgi, 1775 315
Baikal Teal, *A. formosa* Georgi, 1775 315
Wigeon, *A. penelope* Linné, 1758 315
American Wigeon, *A. americana* Gmelin, 1789 315
Chiloë Wigeon, *A. sibilatrix* Poeppig, 1829 315
Yellow-billed pintail, *A. georgica* Gmelin, 1789 (307)
Pintail, *A. acuta* Linné, 1758 315
Common Pintail, *A. acuta acuta* Linné, 1758 —
Kerguelen Pintail, *A. acuta eytoni* (Sharpe, 1875) —
Crozet Pintail, *A. acuta drygalskii* Reichenow, 1904 —
Bahama Pintail, *A. bahamensis* Linné, 1758 (303)
Red-billed Pintail, *A. erythrorhyncha* Gmelin, 1789 (307)
Silver Teal, *A. versicolor* Vieillot, 1816 (303)
Hottentot Teal, *A. punctata* Burchell, 1822 (307)
Cape Teal, *A. capensis* Gmelin, 1789 (307)
Marbled Teal, *A. angustirostris* Ménétriès, 1832 (304)
Garganey, *A. querquedula* Linné, 1758 316
Blue-winged Teal, *A. discors* Linné, 1766 316
Cinnamon Teal, *A. cyanoptera* Vieillot, 1816 316
Shoveller, *A. clypeata* Linné, 1758 316
Cape Shoveller, *A. smithi* Hartert, 1891 —
Red Shoveller, *A. platalea* Vieillot, 1816 (307)
Australian Shoveller, *A. rhynchotis* Latham, 1801 —

Genus *Rhodonessa* 316
Pink-headed Duck, + *R. caryophyllacea* (Latham, 1790) 316

Genus *Merganetta* 316
Torrent Duck, *M. armata* Gould, 1841 316

Genus *Malacorhynchos* 316
Blue-eared Duck, *M. membranaceus* (Latham, 1801) 316

Genus *Hymenolaimus* 316
Blue Duck, *H. malacorhynchos* (Gmelin, 1789) 316

Genus *Stictonetta* 316
Freckled Duck, ◊ *St. naevosa* (Gould, 1840) 316

Tribe Eiders (Somateriini) 321

Genus Eiders *(Somateria)* 321
Common Eider, *S. mollissima* (Linné, 1758) 321
King Eider, *S. spectabilis* (Linné, 1758) 321
Spectacled Eider, *S. fischeri* (Brandt, 1847) 322

Genus *Polysticta* 321
Steller's Eider, *P. stelleri* (Pallas, 1769) 321

Tribe Diving Ducks (Aythyini) 322

Genus *Netta* 322
South American Rosybill, *N. peposaca* (Vieillot, 1816) 322
Red-crested Pochard, *N. rufina* (Pallas, 1773) 322
South American Pochard, *N. erythrophthalma* (Wied, 1832) 322

Genus *Aythya* 323
European Pochard, *Ay. ferina* (Linné, 1758) 323
Canvasback, *Ay. valisneria* (Wilson, 1814) 323
Red-head, *Ay. americana* (Eyton, 1838) (307)
Ferruginous Duck, *Ay. nyroca* (Güldenstädt, 1769) 323
Madagascar White-eye, *Ay. innotata* (Salvadori, 1894) —
Australian White-eye, *Ay. australis* (Eyton, 1838) —
Baer's Pochard, *Ay. baeri* (Radde, 1863) (307)
Tufted Duck, *Ay. fuligula* (Linné, 1758) 323
Ring-necked Duck, *Ay. collaris* (Donovan, 1809) (307)
New Zealand Scaup, *Ay. novaeseelandiae* (Gmelin, 1789) (307)
Greater Scaup, *Ay. marila* (Linné, 1761) (323)
Lesser Scaup, *Ay. affinis* (Eyton, 1838) —

Tribe Cairinini 323

Genus *Amazonetta* 323
Brazilian Teal, *A. brasiliensis* (Gmelin, 1789) 323

Genus *Calonetta* 323
Ringed Teal, *C. leucophrys* (Vieillot, 1816) 323

Genus Pigmy Geese *(Nettapus)* 323
Indian Pigmy Goose, *N. coromandelianus* (Gmelin, 1789) 323
Green Pigmy Goose, *N. pulchellus* Gould, 1842 323
African Pigmy Goose, *N. auritus* (Boddaert, 1783) 323

Genus *Plectropterus* 323
Spur-winged Goose, *P. gambiensis* (Linné,

Order Raptors (Falconiformes)

Temminck, 1829 422
Russet-headed Desert Falcon, *F. pelegrinoides babylonicus* P. L. Sclater, 1861 422
Orange-breasted Falcon, *F. deiroleucus* Temminck, 1825 422
Taita Falcon, ⟠ *F. fasciinucha* Reichenow & Neumann, 1895 422
Aplomado Falcon, *F. fuscocaerulescens* Vieillot, 1817 422
Bat Falcon, *F. rufigularis* Daudin, 1800 425
European Hobby, *F. subbuteo* Linné, 1758 425
African Hobby, *F. cuvierii* A. Smith, 1830 425
Oriental Hobby, *F. severus* Horsfield, 1821 425
Australian Hobby, *F. longipennis* Swainson, 1837 425
Eleonora's Falcon, *F. eleonorae* Géné, 1839 426
Sooty Falcon, *F. concolor* Temminck, 1825 426
Gray Falcon, *F. hypoleucos* Gould, 1840 426
Merlin, *F. columbarius* Linné, 1758 427
Red-headed Falcon, *F. chiquera* Daudin, 1800 427
Gray Kestrel, *F. ardosiaceus* Vieillot, 1823 428
Dickinson's Kestrel, *F. dickinsoni* P. L. Sclater, 1864 428
Madagascar Banded Kestrel, *F. zoniventris* Peters, 1853 428
Red-footed Falcon, *F. vespertinus* Linné, 1766 428
Amur Red-legged Falcon, *F. vespertinus amurensis* Radde, 1863 428
Old World Kestrel, *F. tinnunculus* Linné, 1758 429
Lesser Kestrel, *F. naumanni* Fleischer, 1818 429
American Kestrel or Sparrow Hawk, *F. sparverius* Linné, 1758 429
Guadalupe Kestrel, ⟠ *F. sparverius guadalupensis* Bond, 1943 —
Moluccan Kestrel, *F. moluccensis* (Bonaparte, 1850) 429
Australian Kestrel, *F. cenchroides* Vigors & Horsfield, 1827 429
African Kestrel, *F. rupicoloides* A. Smith, 1830 429
Madagascar Kestrel, *F. newtoni* (Gurney, 1863) 429
Aldabra Kestrel, ⟠ *F. newtoni aldabranus* Grote, 1928 —
Seychelles Kestrel, ⟠ *F. araea* (Oberholser, 1917) 429
Mauritius Falcon, ⟠ *F. punctatus* Temminck, 1821 429
Fox Kestrel, *F. alopex* (Heuglin, 1861) 429

Order Gallinaceous Birds (Galliformes)

Suborder Galli

Family Mound Birds (Megapodiidae) 433
Tribe Megapodiini 433
Genus *Megapodius* 434
Micronesian Scrub Fowl, *M. laperouse* Gaimard, 1823 434
⟠ *M. laperouse laperouse* Gaimard, 1823 —
⟠ *M. laperouse senex* Hartlaub, 1867 —
Niuafoo Megapode, ⟠ *M. pritchardii* Gray, 1864 434
Australian Scrub Fowl, *M. freycinet* Gaimard, 1823 434
Genus *Macrocephalon* 434
Maleo, ⟠ *M. maleo* S. Müller, 1846 434
Genus *Eulipoa* 434
Wallace's Eulipoa, *Eu. wallacei* (Gray, 1860) 434

Tribe Alecturini 434
Genus *Alectura* 434
Latham's Brush Turkey, *A. lathami* J. E. Gray, 1831 434
Genus *Talegalla* 434
Western Tallegalla, *T. cuvieri* Lesson, 1828 434
Genus *Aepypodius* 434
Wattled Brush Turkey, *Ae. arfakianus* (Salvadori, 1877) 434
Genus *Leipoa* 434
Mallee Fowl, *L. ocellata* Gould, 1840 434

Family Curassows (Cracidae) 441
Tribe Cracini 442
Genus *Nothocrax* 442
Nocturnal Curassow, *N. urumutum* (Spix, 1825) 442
Genus *Pauxi* 442
Helmeted Curassow, *P. pauxi* (Linné, 1766) 442
Genus *Mitu* 442
Razor-billed Curassow, *M. mitu* (Linné, 1766) 442
Crestless Curassow, *M. tomentosa* (Spix, 1825) 442
Genus *Crax* 442
Great Curassow, *C. rubra* Linné, 1758 442
Cozumel Curassow, ⟠ *C. rubra griscomi* Nelson, 1926 —
Wattled Curassow, *C. globulosa* Spix, 1825 442
Yellow-knobbed Curassow, *C. daubentoni* Gray, 1867 442
Black Curassow, *C. alector* Linné, 1766 442

Tribe Penelopini 442
Genus *Chamaepetes* 442
Black-bellied Guan, *Ch. unicolor* Salvin, 1867 442
Genus *Aburria* 442
Wattled Guan, *A. aburri* (Lesson, 1828) 442
Genus *Pipile* 442
White-headed Piping Guan, *P. cumanensis* (Jacquin, 1784) 442
Trinidad Piping Guan, ⟠ *P. pipile pipile*

Other subfamilies of the Phasianidae as well as the second suborder of the Galliformes are considered in Vol. VIII.

ANIMAL DICTIONARY

I. English—German—French—Russian

For scientific names of species see the German-English-French-Russian section of this dictionary or the index.

In most cases names of subspecies are formed by putting an adjective or geographical specification before the name of species. These English names of subspecies will, as a rule, not appear in this part of the zoological dictionary.

ENGLISH NAME	GERMAN NAME	FRENCH NAME	RUSSIAN NAME
Abdim's Stork	Regenstorch	Cigogne d'Abdim	Абдимский аист
Abyssinian Blue-winged Goose	Blauflügelgans	Bernache aux ailes bleues	Голубокрылый гусь
Adelie Penguin	Adeliepinguin		Синий пингвин
African Black Duck	Schwarzente	Canard noir africain	
– Hooby	Afrikanischer Baumfalk	Hobereau africain	
– Pygmy Goose	Afrikanische Zwergglanzente	Sarcelle de Madagascar	
– Red-tailed Buzzard	Felsenbussard	Buse à queue rousse	
– Sea Eagle	Schreiseeadler	Aigle pêcheur	Крикливый орлан
– Sparrowhawk	Afrikanischer Habicht	Autour tachiro	
– Spoonbill	Schmalschnabel-Löffler	Spatule d'Afrique	Африканская белая колпица
– Swallow-tailed Kite	Schwalbengleitaar	Naucler d'Afrique	
– White-backed Vulture	Zwerggänsegeier	Griffon africain à dos blanc	Белоспинный сип
Albatrosses and Mollymawks	Albatrosse	Diomédeidés	Альбатросы
American Bittern	Nordamerikanische Rohrdommel	Butor d'Amérique	Североамериканская выпь
– Black Duck	Dunkelente	Canard noir de l'Amérique	
– Comb Duck	Südamerikanische Höckerglanzente		Бразильянский шишконосый гусь
– Darter	Amerikanischer Schlangenhalsvogel	Anhinga d'Amérique	Американская ахинга
– Flamingo	Roter Flamingo	Flamant rouge	Красный фламинго
– Goshawk	Amerikanischer Habicht		Североамериканский тетеревятник
– Rhea	Nandu	Nandou américain	Обыкновенный нанду
– Sparrow Hawk	Buntfalk	Crécerelle americaine	Американская пустельга
– White Pelican	Nashornpelikan		Пеликан-носорог
– Wigeon	Amerikanische Pfeifente	Canard siffleur	Американская свиязь
Andean Condor	Anden-Kondor	Condor des Andes	Андский кондор
– Flamingo	Andenflamingo		Андский фламинго
– Goose	Andengans	Bernache des Andes	
Angola Red-necked Partridge	Rotkehlfrankolin	Pterniste à cou nu	
Antarctic Fulmar	Antarktischer Eissturmvogel		Антарктический глупыш
– Petrel	Antarktissturmvogel		Антарктический буревестник
Arctic Fulmar	Eissturmvogel	Pétrel glacial	Атлантический глупыш
Argentine Ruddy Duck	Argentinische Schwarzkopfruderente	Erismature d'Argentine	
Ascension Frigate Bird	Adlerfregattvogel	Frégate aigle	Орлиный фрегат
Ashy-headed Goose	Graukopfgans	Bernache à tête grise	
Asian Open Bill	Indien-Klaffschnabel	Bec-Ouvert asiatique	Индийский аист-разиня
Australasian Gannet	Australischer Tölpel		Австралийская олуша
– White-eye	Australische Moorente	Milouin d'Australie	
Australian Cassowary	Helmkasuar	Casoar à casque	Племоносный казуар
– Magpie Goose	Spaltfußgans	Oie semi-palmée	Расщепнолапый гусь
– Musk Duck	Lappenente	Canard à membrane	
– Pelican	Brillenpelikan		Австралийский пеликан
– Shelduck	Australische Kasarka	Casarca d'Australie	Австралийская пеганка
Ayres' Eagle	Haubenzwergadler	Aigle Autour d'Ayres	
Baer's Pochard	Schwarzkopf-Moorente		Нырок Бэра
Bahama Pintail	Bahama-Ente	Canard de Bahama	
Baikal Teal	Gluckente	Sarcelle élégante	Чирок-клоктун
Bald Eagle	Weißkopf-Seeadler	Aigle à tête blanche	Белоголовый орлан
Barbary Partridge	Felsenhuhn	Perdrix de Barbarie	Берберская каменная куропатка
Bar-headed Goose	Streifengans	Oie à tête barrée	Индийский горный гусь
Barnacle Goose	Weißwangengans	Bernache nonnette	Белощёкая казарка
Barred Quail	Bandwachtel		
Barrow's Golden-eye	Spatelente	Garrot islandais	Исландский гоголь
Bateleur	Gaukler	Bateleur	Орел-скоморох
Bat Hawk	Fledermaus-Gleitaar	Faucon des chauves-souris	Индийский совиный сарыч
Bean Goose	Saatgans	Oie des moissons	Гуменник
Bearded Vulture	Bartgeier	Gypaëte barbu	Бородач

ENGLISH NAME	GERMAN NAME	FRENCH NAME	RUSSIAN NAME
Bennett's Cassowary	Bennettkasuar	Casoar de Bennett	Казуар Беннетта
Bewick's Swan	Zwergschwan	Cygne de Bewick	Западный тундровый лебедь
Birds	Vögel	Oiseaux	Птицы
– of Prey	Greifvögel	Falconiformes	Хищные птицы
Bishop Stork	Afrikanisch-Indischer Wollhalsstorch	Cigogne épiscopale	Африканско-индийский пуховый аист
– Storks	Wollhalsstörche	Cigognes épiscopales	Пуховые аисты
Bittern	Rohrdommel	Butor étoilé	Большая выпь
Black Cormorant	Kormoran	Grand Cormoran	Обыкновенный баклан
– Goshawk	Trauerhabicht	Épervier pie	
– Grouse	Birkhühner, Birkhuhn	Tétras lyre	Тетерева, Тетерев
– Heron	Glockenreiher	Héron ardoisé	
– Ibis	Warzenibis	Ibis noir oriental	
– Kite	Schwarzmilan	Milan noir	Черный коршун
– Stork	Schwarzstorch	Cigogne noire	Черный аист
– Swan	Trauerschwan	Cygne noir	Черный лебедь
– Vulture	Mönchsgeier, Rabengeier	Vautour moine, Vautour noir	Чёрный гриф, Американский черный гриф
Black-bellied Storm Petrel	Schwarzbauch-Sturmschwalbe	Pétrel des tropiques	Чернобрюхая качурка
Black-billed Whistling Duck	Kuba-Pfeifgans		Кубинская древесная утка
Black-breasted Buzzard Kite	Haubenmilan	Buse à poitrine noire	
Black-browed Albatross	Schwarzbrauenalbatros	Albatros à sourcil noir	Чернобровый альбатрос
Black-capped Petrel	Schwarzkappensturmtaucher	Pétrel diablotin	Черношапочный тайфунник
Black-crowned Night Heron	Nachtreiher	Héron bihoreau à couronne noire	Кваква
Black-faced Spoonbill	Kleiner Löffler	Petite Spatule	Малая колпица
Black-footed Albatross	Schwarzfußalbatros	Albatros à pieds noirs	Черноногий альбатрос
– Penguin	Brillenpinguin	Manchot du Cap	Очковый пингвин
Black-headed Duck	Kuckucksente	Canard à tête noire de l'Argentine	Кукутковая утка
– Heron	Schwarzhalsreiher	Héron à tête noire	
Black-necked Grebe	Schwarzhalstaucher	Grèbe à cou noir	Черношейная поганка
– Stork	Indien-Großstorch	Jabirou asiatique	Индийский исполинский аист
– Swan	Schwarzhalsschwan	Cygne à cou noir	Черношеий лебедь
– Scoter	Trauerente	Macreuse noire	Синьга
Black-throated Diver	Prachttaucher	Plongeon lumme	Чернозобая гагара
Black-winged Kite	Schwarzflügel-Gleitaar	Faucon blanc	Чернокрылый коршун
Blood Pheasant	Blutfasan	Ithagine sanguine	
Blue Petrel	Blausturmvogel		Голубой тайфунник
Blue-billed Pintail	Spießente	Canard pilet	Шилохвость
Blue-winged Teal	Blauflügelente	Sarcelle soucrourou	Синекрылый чирок
Blyth's Tragopan	Blyth-Satyrhuhn	Tragopan de Blyth	
Boat-billed Heron	Kahnschnabel	Savacou huppé	Цапля-челноклюв
Bonelli's Eagle	Habichtsadler	Aigle de Bonelli	Ястребиный орел
Boobies	Tropische Tölpel	Fous	
Booted Eagle	Zwergadler	Aigle botté	Орел-карлик
Brahming Kite	Brahminenweih		Браминский коршун
Brandt's Cormorant	Pinselkormoran		Кисточковый баклан
Brazilian Teal	Amazonas-Ente	Sarcelle du Brésil	
Brent Goose	Ringelgans	Bernache cravant	Черная казарка
Brents	Meergänse	Bernaches	Казарки
Broad-winged Hawk	Breitschwingenbussard	Petite Buse	Ширококрылый канюк
Brown Bittern	Australische Rohrdommel		Австралийская выпь
– Booby	Brauntölpel	Fou à ventre blanc	
– Harrier Eagle	Afrikanischer Schlangenadler	Circaète brun	
– Pelican	Brauner Pelikan		Бурый пеликан
Brush Turkey	Buschhuhn		Кустовая курица
Bufflehead	Büffelkopfente	Garrot albéole	Гоголь-головастик
Buller's Albatross	Bullers Albatros		Альбатрос Буллера
Bulwer's Petrel	Weichnasen-Sturmvogel	Pétrel de Bulwer	Тайфунник Бульвера
Burmese Pigmy Falcon	Langschwanz-Zwergfalk	Fauconnet à pattes jaunes	
Buzzards	Bussarde	Buses	Сарычи
Cabot's Tragopan	Cabot-Satyrhuhn	Tragopan de Cabot	
California Condor	Kalifornischer Kondor	Condor de Californie	Калифорнийский кондор
Cameroun Mountain Francolin	Kamerunbergwald-Frankolin	Francolin du Mont Cameroun	
Canada Goose	Kanadagans	Bernache du Canada	Канадская казарка
Canadian White-tailed Ptarmigan	Amerikanisches Alpenschneehuhn	Lagopède à queue blanche	Американская белая куропатка
Canvas-back	Riesentafelente	Milouin aux yeux rouges	
Cape Barren Goose	Hühnergans	Céréopse de Nouvelle-Hollande	Куриный гусь
– Gannet	Kaptölpel		Капская олуша

ENGLISH NAME	GERMAN NAME	FRENCH NAME	RUSSIAN NAME
Capercaillie	Auerhuhn	Grand Tétras	Глухарь
Cape Shoveler	Kap-Löffelente	Souchet du Cap	Капская широконоска
– Teal	Kapente	Sarcelle du Cap	
Carolina Wood Duck	Brautente	Canard Carolin	Американская брачная утка
Caspian Snowcock	Kaspisches Königshuhn	Perdrix des neiges caspienne	Каспийский улар
Cassin's Hawk Eagle	Schwarzachseladler	Aigle Autour de Cassin	
Cassowaries	Kasuare	Casuariidés, Casoars	Казуары
Cathartines	Neuweltgeier	Vulturidés	Американские грифы
Cattle Egret	Kuhreiher	Héron garde-boeufs	Египетская цапля
Caucasian Black Grouse	Kaukasisches Birkhuhn	Tétras lyre de Géorgie	Кавказский тетерев
Chestnut Teal	Kastanienente	Sarcelle d'Australie	
Chestnut-winged Chachalaca	Rotflügelguan	Ortalide babillarde	Краснокрылый шуан
Chilean Flamingo	Chilenischer Flamingo		Чилийский фламинго
Chiloë Wigeon	Chile-Pfeifente	Siffleur du Brésil	Чилийская свиязь
Chimango Caracara	Chimango		Химанго
Chinese Egret	China-Seidenreiher		Желтоклювая чепура-нужда
– Little Bittern	Chinesendommel	Blongios chinois	Китайский волгок
– Merganser	Schuppensäger	Harle chinois	Чемуйчатый крохаль
– Painted Quail	Zwergwachtel	Caille peinte de Chine	
– Pond Heron	Bacchusreiher	Crabier chinois	Белокрылая цапля
Christmas Frigate Bird	Weißbauch-Fregattvogel		Белобрюхий фрегат
Cinnamon Bittern	Zimtdommel	Blongios cannelle	
– Teal	Zimtente	Sarcelle cannelle	
Comb Duck	Höckerglanzente	Sarcidiorne à crète	Шишконосый гусь
Common Buzzard	Mäusebussard	Buse variable	Обыкновенный сарыч
– Diving Petrel	Pinguin-Sturmtaucher		Новозеландский нырцовый буревестник
– Eider	Eiderente	Eider à duvet	Обыкновенная гага
– Golden-eye	Schellente	Canard garrot	Обыкновенный гоголь
– Kiwi	Streifenkiwi	Kiwi austral	Обыкновенный киви
– Quail	Wachtel	Caille des blés	Обыкновенный перепел
– Shelduck	Brandgans	Tadorne de Belon	Пеганка
Cooper's Hawk	Rundschwanzsperber	Épervier de Cooper	Ястреб Купера
Cormorants	Kormorane	Phalacrocoracidés, Cormorans	Бакланы
Coscoroba Swan	Koskorobaschwan	Cygne coscoroba	Гигаитская утка коскороба
Cotton Pygmy Goose	Indische Zwergglanzente	Sarcelle de Coromandel	Индийская карликовая утка
Crested Caracara	Carancho		Каранхо
– Curassow	Glattschnabelhokko	Hocco de la Guiane	Гладкоклювый гокко
– Honey Buzzard	Malayen-Wespenbussard		Хохлатый осоед
– Screamer	Weißwangenwehrvogel	Chauna chavaria	Белощекая паламедея
– Serpent Eagle	Indischer Schlangenhabicht		Орел-хеела
– Tinamou	Perl-Steißhuhn	Tinamou huppé	
Crowned Eagle	Kronenadler	Blanchard	Венценосный орел
– Wood Partridge	Straußwachtel	Roulroul	Страусовый перепел
Curassows, Guans and Chachalacas	Hokkos	Cracidés	Гокко
Dalmatian Pelican	Krauskopfpelikan	Pélican frisé	Кудрявый пеликан
Darter	Altwelt-Schlangenhalsvogel	Oiseau-Serpent	Ахинга старого света
Darters	Schlangenhalsvögel	Anhingidés, Anhingas	Эмеешейки, Анхинги
Darwin's Rhea	Darwin-Nandu	Nandou de Darwin	Нанду Дарвина
Derby's Mountain Pheasant	Bergguan		Горный шуан
Diurnal Birds of Prey	Greifvögel	Falconiformes	Хищные птицы
Divers	Seetaucher	Plongeons	Гагары
Diving Petrels	Tauchsturmvögel	Pélécanoïdidés, Pétrels plongeurs	Нырцовые буревестники
Domestic Duck	Hausente	Canard doméstique	Домашняя утка
– Goose	Hausgans	Oie doméstique	Домашний гусь
– Swan Goose	Haus-Höckergans	– cygnoïde doméstique	Домашний китайский гусь
Ducks and Geese	Entenvögel	Anatidés	Утиные
Dusky Grey Heron	Sumatrareiher	Héron de Sumatra	Тифонова цапля
– Grouse	Felsengebirgshuhn	Tétras sombre	Дымчатый тетерев
Eastern Bob White	Virginiawachtel	Colin de Virginie	Виргинская куропатка
– Reef Heron	Riffreiher	Aigrette des récifs	
– Ruffed Grouse	Kragenhuhn	Gelinotte huppée	Воротничковый рябчик
– White Pelican	Rosapelikan	Pélican blanc	Розовый пеликан
Egyptian Goose	Nilgans	Oie d'Égypte	Нильский гусь
– Vulture	Schmutzgeier	Percnoptère d'Égypte	Стервятник
Eleonora's Falcon	Eleonorenfalk	Faucon d'Eléonore	Чеглок Элеоноры
Emperor Goose	Kaisergans	Oie empereur	Белошей
– Penguin	Kaiserpinguin	Manchot impérial	Императорский пингвин
Emu	Emu	Emeu d'Australie	Эму

ENGLISH NAME	GERMAN NAME	FRENCH NAME	RUSSIAN NAME
Emus	Emus	Dromicéiidés	Эму
Erect-crested Penguin	Gelbschopfpinguin		Желточубый пингвин
Eurasian Short-toed Eagle	Schlangenadler	Circaète Jean-le-Blanc	Змееяд
– Sparrowhawk	Sperber	Épervier d'Europe	Ястреб-перепелятник
European Flamingo	Rosaflamingo	Flamant rose	Розово-красный фламинго
– Pochard	Tafelente	Canard milouin	Красноголовый нырок
– Wigeon	Pfeifente	– siffleur	Свиязь
Fairy Prion	Feenwalvogel		Буревестник-горлица
Falcated Teal	Sichelente	Sarcelle à faucilles	Касатка
Falcons	Falken, Falken i. e. S.	Falconidés, Faucons	
Falkland Flightless Steamer Duck	Falkland-Dampfschiffente	Canard à ailes courtes	
Feldegg's Falcon	Feldeggsfalk	Faucon de Feldegg	Средиземноморский рыжеголовый балобан
Ferruginous White-eye	Moorente	Canard nyroca	Белоглазый нырок
– Rough-legged Hawk	Königsbussard	Buse rouilleuse	
Fiordland Penguin	Dickschnabelpinguin		Толстоклювый пингвин
Flamingoes	Flamingos	Flamants, Phoenicoptéridés	Фламинго
Fork-tailed Storm Petrel	Gabelschwanz-Wellenläufer		Серая качурка
Fowllike Birds	Hühnervögel	Galliformes	Куриные
Fox Kestrel	Fuchsfalk	Faucon-Renard	
Francolin	Halsbandfrankolin	Francolin d'Europe	Турач
Francolins	Frankoline	Francolins	
Frigate Birds	Fregattvögel	Frégatidés, Frégates	Фрегаты
– Petrel	Fregattensturmschwalbe	Pétrel frégate	Морская качурка
Fulmars	Möwensturmvögel		Глупыши
Fulvous Whistling Duck	Fahlpfeifgans		Пегая древесная утка
Gabar Goshawk	Gabar-Habicht	Autour gabar	
Gadfly Petrels	Hakensturmtaucher		Тайфунники
Gadwall	Schnatterente	Canard chipeau	Серая утка
Galapagos Penguin	Galapagospinguin		Галапагосский пингвин
Gambel's Quail	Helmwachtel	Colin de Gambel	Калифорнийский шлемоносный перепел
Gannets	Tölpel i. e. S.	Fous	
Garganey	Knäkente	Sarcelle d'été	Чирок-трескунок
Gentoo Penguin	Eselspinguin	Manchot Gentoo	Ослиный пигнвин
Giant Ibis	Riesenibis	Ibis géant	
Glossy Ibis	Brauner Sichler	– luisant	Каравайка
Golden Eagle	Steinadler	Aigle doré	Беркут
Goliath Heron	Goliathreiher	Héron goliath	
Goosander	Gänsesäger	Harle bièvre	Большой крохаль
Goshawk	Habicht	Autour des palombes	Ястреб-тетеревятник
Goshawks	Habichte i. e. S.	Éperviers	Ястреба
Grasshopper Buzzard	Heuschreckenbussard	Busard des sauterelles	
Great Blue Heron	Amerikanischer Graureiher	Grand Héron	
– Crested Grebe	Haubentaucher	Grèbe huppé	Большая поганка
– Frigate Bird	Bindenfregattvogel		Большой фрегат
– Northern Diver	Eistaucher	Plongeon imbrin	Полярная гагара
– Shearwater	Großer Sturmtaucher	Puffin majeur	Большой пестробрюхий буревестник
– Tinamou	Großtao	Grand Tinamou	Большой тинаму
– White Egret	Silberreiher	Grande Aigrette	Большая белая цапля
Greater Eider Ducks	Eiderenten	Eiders	Гаги
– Flamingo	Flamingo	Flamant rose	Обыкновенный фламинго
– Marabou	Argala-Marabu	Grand Marabout	Индийский марабу-аргал
– Scaup	Bergente	Canard milouinan	Морская чернеть
– Spotted Eagle	Schelladler	Aigle criard	Большой подорлик
Grebes	Lappentaucher, Taucher	Podicipédiformes, Podicipitidés, Grèbes	Поганковые, Поганки
Green Cormorant	Krähenscharbe	Cormoran huppé	Длинноносый баклан
– Heron	Grünreiher	Héron vert	
– Pygmy Goose	Grüne Zwergglanzente	Sarcelle verte d'Australie	
Green-backed Heron	Mangrovereiher	Héron à dos vert	Зеленая кваква
Grey Duck	Augenbrauenente	Canard à sourcil blanc	
– Heron	Graureiher	Héron cendré	Серая цапля
– Kestrel	Graufalk	Faucon ardoisé	
– Partridge	Rebhuhn	Perdrix grise	Серая куропатка
– Pelican	Graupelikan		Серый пеликан
– Teal	Weißkehlente	Sarcelle grise des Indes	
Grey-faced Buzzard	Graugesichtbussard		Ястребиный сарыч
Grey-headed Albatross	Graukopfalbatros	Albatros à tête grise	Сероголовый альбатрос
Greylag Goose	Graugans	Oie cendrée	Серый гусь
Grey-winged Petrel	Langflügelsturmtaucher		Длиннокрылый тайфунник

ENGLISH NAME	GERMAN NAME	FRENCH NAME	RUSSIAN NAME
Griffon Vulture	Gänsegeier	Vautour fauve	Белоголовый сип
Grouse	Rauhfußhühner	Tétraoninés	
Guadalupe Caracara	Guadalupe-Karakara		Гваделупский каракара
Guatemalan Black Chachalaca	Mohrenguan		Черный шуан
Gyrfalcon	Gerfalk	Faucon gerfaut	Кречет
Hadada	Hagedasch	Ibis hagedash	
Hamerkop	Hammerkopf	Ombrette	Молотоглав
Hamerkops	Hammerköpfe	Scopidés, Ombrettes	Теневые птицы, Молотоголовые цапли
Harlequin Duck	Kragenente	Garrot harlequin	Каменутка
– Quail	Harlekinwachtel	Caille harlequine	
Harpy Eagle	Harpyie	Harpye	
Harriers and Crane Hawks	Weihen	Circinés	Луни
Hasting's Tragopan	West-Satyrhuhn	Tragopan de Hasting	
Hawaiian Goose	Hawaiigans	Bernache des îles Sandwich	Гавайская казарка
– Petrel	Hawaiisturmvogel		Гавайский тайфунник
Hawks, Old World Vultures and Harriers	Habichtartige	Accipitridés	
Hazel Hen	Haselhuhn	Gelinotte des bois	Рябчик
Heath Hen	Präriehuhn	Poule des prairies	Большой луговой тетерев
Helmeted Curassow	Helmhokko	Pauxi pierre	Шлемоносный гокко
Hen Harrier	Kornweihe	Busard Saint-Martin	Полевой лунь
Herons	Reiher	Ardéidés	Цапли
Hermit Ibis	Waldrapp	Ibis chauve	Горный ибис
Himalayan Griffon	Schneegeier		Снежный гималайский сип
Honey Buzzard	Wespenbussard	Bondrée apivore	Осоед
Hooby	Baumfalk	Faucon hobereau	Чеглок
Hooded Merganser	Kappensäger	Harle couronné	Американский крохаль
– Vulture	Kappengeier	Charognard	
Horned Screamer	Hornwehrvogel	Kamichi cornu	Рогатая анхима
Hottentot Teal	Hottentotten-Ente	Sarcelle Hottentote	
Hudsonian Spruce Grouse	Tannen-Waldhuhn	Tétras des savanes	Канадский тетерев
Ibises	Sichler	Threskiornithinés, Ibis	Ибисы
– and Spoonbills	Ibisvögel	Threskiornithidés	Ибисы
Imperial Eagle	Kaiseradler	Aigle impérial	Орел-могильник
Indian Whistling Duck	Indien-Pfeifgans	Dendrocygne siffleur	Яванская древесная утка
– White-backed Vulture	Bengalgeier	Griffon indien à dos blanc	Индийский гриф
Jabiru	Jabiru	Jabiru	
James' Flamingo	James-Flamingo		Короткоклювый фламинго
Jankowski's Swan	Jankowski-Schwan		Восточный тундровый лебедь
Japanese Cormorant	Japanischer Kormoran		Японский баклан
– Crested Ibis	– Ibis		Китайский красноногий ибис
– Quail	Japanische Wachtel	Caille du Japon	Немой перепел
Jardine's Chachalaca	Rotschwanzguan		Краснохвостый шуан
Javanese Pond Heron	Prachtschopfreiher	Crabier malais	
Kelp Goose	Tanggans	Bernache antarctique	
King Eider	Prachteiderente	Eider à tête grise	Гага-гребенутка
– Penguin	Königspinguin	Manchot royal	Королевский пингвин
– Vulture	Königsgeier	Vautour royal	Королевский гриф
Kites	Gleitaare	Élaninés	
Kiwis	Kiwivögel	Kiwis, Aptérygidés	Бескрылые, Киви
Korean Crested Shelduck	Schopfkasarka		Хохлатая пеганка
Laggar Falcon	Laggarfalk		Индийский балобан-лаггар
Lanner Falcon	Lannerfalk	Faucon lanier	Рыжеголовый балобан
Lappet-faced Vulture	Ohrengeier	Vautour oricou	Ушастый гриф
Leach's Petrel	Wellenläufer	Pétrel cul-blanc	Северная качурка
Lesser Bittern	Indianerdommel	Petit Butor	
– Flamingo	Zwergflamingo	– Flamant	Африканский карликовый фламинго
– Flamingoes	Zwergflamingos	Petits Flamants	
– Frigate Bird	Kleiner Fregattvogel		Белобокий фрегат
– Grey-headed Chachalaca	Braunflügelguan	Ortalide du Mexique	Бурокрылый шуан
– Kestrel	Rötelfalk	Faucon crécerellette	Степая пустельга
– Marabou	Sunda-Marabu	Petit Marabout	Яванский марабу
– Prairie Hen	Kleines Präriehuhn		Малый луговой тетерев
– Razor-billed Curassow	Samthokko	– Hocco à bec de rasoir	
– Spotted Eagle	Schreiadler	Aigle pomarin	Малый подорлик
– White-fronted Goose	Zwerggans	Oie naine	Пискулька
Letter-winged Kite	Schwarzachsel-Gleitaar	Elanion écrit	
Light-mantled Sooty Albatross	Südlicher Rußalbatros	Albatros fuligineux	Аантарктический дымчатый альбатрос
Little Bittern	Zwergdommel	Blongios nain	Малая выпь
– Blue Heron	Blaureiher	Petit Héron bleu	
– Egret	Seidenreiher	Aigrette garzette	Малая белая цапля

ENGLISH NAME	GERMAN NAME	FRENCH NAME	RUSSIAN NAME
– Grebe	Zwergtaucher	Grèbe castagneux	Малая поганка
– Penguin	Zwergpinguin		Карликовый пингвин
– Pied Cormorant	Australische Zwergscharbe		Австралийский малый баклан
– Tinamou	Brauntao	Soui	
Lizard Buzzard	Kehlstreifbussard	Buse unibande	
Long-billed Vulture	Indischer Geier	Vautour à bec long	Индийский белоголовый сип
Long-crested Eagle	Schopfadler	Aigle huppé	Гребневый орел
Long-legged Hawk	Adlerbussard	Buse féroce	Канюк-курганник
Long-tailed Cormorant	Gelbschnabel-Zwergscharbe	Cormoran africain	
– Duck	Eisente	Canard de Miquelon	Морянка
– Hawk	Langschwanzhabicht	Autour à longue queue	Длиннохвостый ястреб
Loons	Seetaucher	Gaviiformes, Gaviidés	Гагаровые, Гагары
Louisiana Heron	Dreifarbenreiher	Héron à ventre blanc	
Macaroni Penguin	Goldschopfpinguin		Золоточубый пингвин
Madagascan White-eye	Madagaskar-Moorente	Milouin de Madagascar	
Magellan Goose	Magellangans	Bernache de Magellan	
Magellanic Penguin	Magellanpinguin	Manchot de Magelhaen	
Magnificent Man-o'-War Bird	Prachtfregattvogel	Frégate superbe	Малый фрегат
Magpie Geese	Spaltfußgänse	Anseranatinés, Oies semi-palmées	
Malayan Wood Ibis	Malaien-Nimmersatt	Tantale blanc	Малайский тантал
Mallee Fowl	Thermometerhuhn	Leipoa ocellé	Лейпоа
Mandarin Duck	Mandarinente	Canard mandarin	Мандаринка
Maned Wood Duck	Mähnengans	Bernache à crinière	
Manx Shearwater	Schwarzschnabel-Sturmtaucher	Puffin des Anglais	Обыкновенный буревестник
Marabou	Afrika-Marabu	Marabout	Африканский марабу
Marabous	Marabus	Marabouts	Марабу
Marbled Teal	Marmelente	Sarcelle marbrée	Мраморный чирок
Marsh Harrier	Rohrweihe	Busard harpaye	Камышевый лунь
Martial Eagle	Kampfadler	Aigle Martial	Орел-боец
Masked Booby	Maskentölpel	Fou masqué	Голуболицая олуша
Mediterranean Shearwater	Gelbschnabel-Sturmtaucher	Puffin cendré	Большой белобрюхий буревестник
Meller's Duck	Madagaskar-Ente	Canard de Meller	Мадагаскарская утка
Mergansers	Säger	Harles	Крохали
Merlin	Merlin	Faucon émerillon	Дербник
Mexican Curassow	Tuberkelhokko	Grand Hocco	
– Scaled Quail	Schuppenwachtel	Quaglia azzurra	
Moas	Moas	Dinornithidés	
Monkey-eating Eagle	Affenadler		Обезьяноед
Montagu's Harrier	Wiesenweihe	Busard de Montagu	Луговой лунь
Montezuma's Quail	Montezumawachtel	Colin de Masséna	
Moscovy Duck	Moschusente	Canard musqué	Мускусная утка
Moundfowl	Großfußhühner	Mégapodiidés	Сорные куры
Mountain Caracara	Anden-Karakara		Андский каракара
– Hawk Eagle	Nepal-Haubenadler		Хохлатый орел
Mute Swan	Höckerschwan	Cygne muet	Лепедь-шипун
Naukeen Kestrel	Australischer Turmfalk		Австралийская пустельга
Night Herons	Nachtreiher	Hérons bihoreaux	
Nocturnal Curassow	Schopfhokko		Хохлатый гокко
North American Wood Ibis	Amerika-Nimmersatt	Cigogne américaine	Американский ярибу
Northern Gannet	Baßtölpel	Fou de Bassan	Атлантическая олуша
– Giant Fulmar	Nördlicher Riesensturmvogel		Северный гигантский буревестник
– Mallard	Stockente	Canard colvert	Кряква
– Rough-legged Buzzard	Rauhfußbussard	Buse pattue	Мохноногий канюк
– Sharp-tailed Grouse	Schweif-Waldhuhn	Gelinotte à queue fine	Хвостатый тетерев
– Shoveler	Löffelente	Canard souchet	Широконоска
Old World Vultures	Altweltgeier	Aegypiinés	Настоящие грифы
– – Goshawk	Europäischer Habicht		Среднеевропейский тереревятник
– – Kestrel	Turmfalk	Faucon crécerelle	Обыкновенная пустельга
– – Quails	Feldhühner	Perdricinés	
Olive Ibis	Olivgrüner Ibis	Ibis olivâtre	
One-wattled Cassowary	Goldhalskasuar	Casoar unicaronculé	Лоскутный казуар
Open Bill	Afrika-Klaffschnabel	Bec-Ouvert	Африканский аист-разиня
– Bills	Klaffschnäbel	Becs-Ouverts	Аисты-разини
Oriental Hooby	Indischer Baumfalk	Hobereau à poitrine rousse	
– Ibis	Schwarzkopfibis	Ibis à tête noire	
– White Stork	Schwarzschnabelstorch		Черноклювый аист
Orinoco Goose	Orinokogans	Oie de l'Orinoque	
Osprey	Fischadler	Balbuzard fluvatile	Скопа
Ospreys	Fischadler		Скопы

ENGLISH NAME	GERMAN NAME	FRENCH NAME	RUSSIAN NAME
Ostrich	Strauß	Autruche	Африканский страус
Ostriches	Strauße	Autruches, Struthionidés	Африканские страусы, Страусы
Owen's Kiwi	Fleckenkiwi	Kiwi d'Owen	Киви Оуэна
Painted Stork	Indien-Nimmersatt	Tantale indien	Индийский тантал
Pale Chanting Goshawk	Singhabicht		Певчий ястреб
Pale-footed Shearwater	Blaßfußsturmtaucher		Бледноногий буревестник
Pallas Cormorant	Brillenkormoran		Очковый баклан
Pallas' Sea Eagle	Bandseeadler	Pygargue à queue blanche	Орлан-долгохвост
Pallid Harrier	Steppenweihe	Busard pâle	Степной лунь
Palm-nut Vulture	Palmgeier	Vautour palmiste	Грифовый орлан
Paradise Shelduck	Paradieskasarka	Casarca de Paradis	Райская пеганка
Pearl Kite	Perlenweih		Карликовый лунь
Pelagic Cormorant	Nordpazifischer Kormoran		Берингов баклан
Pelicans	Pelikane	Pélécanidés, Pélicans	Пеликаны
Penguins	Pinguine	Sphénisciformes, Sphéniscidés	
Peregrine Falcon	Wanderfalk	Faucon pèlerin	Сокол-сапсан
Peruvian Penguin	Humboldtpinguin		Пингвин Гумбольдта
Pheasants, Quails and Peacocks	Fasanenartige	Phasianidés	
Philippine Duck	Philippinen-Ente	Canard des Philippines	Филиппинская утка
– Serpent Eagle	Philippinen-Schlangenhabicht		Филиппинский змеиный орел
Pied Harrier	Schwarzweißweihe	Busard pie	Чернопегий лунь
Pigmy Falcons	Zwergfalken		Карликовые соколы
Pink-backed Pelican	Rötelpelikan	Pélican à dos rosé	Красноспинный пеликан
Pink-footed Goose	Kurzschnabelgans	Oie à bec court	Короткоклювый гуменник
Pintado Petrel	Kapsturmvogel	Pétrel damier	Капский буревестник
Plumed Quail	Berghaubenwachtel	Colin des montagnes	
Pochards	Tauchenten		
Pondicherry Vulture	Lappengeier	Vautour de Pondichéry	
Prairie Falcon	Präriefalk	Faucon des prairies	Мексиканский сокол
– Hens	Eigentliche Präriehühner	Cupidons	
Procellariids	Sturmvögel	Procellariidés	Буревестники
Ptarmigan	Alpenschneehuhn	Lagopède muet	Тундряная куропатка
Purple Heron	Purpurreiher	Héron pourpré	Рыжая цапля
Pygmy Cormorant	Zwergscharbe	Cormoran pygmée	Малый баклан
– Geese	Zwergglanzenten		Карликовые утки
Pygopodes	Steißfüße	Pygopodes	Поганки и гагары
Quails	Wachteln i. e. S.	Cailles	
Radjah Shelduck	Radjahgans		Пеганка-раджа
Rain Quail	Regenwachtel	Caille du Coromandel	Коромандельская перепелка
Ratites	Laufvögel	Ratites	Бегающие
Razor-billed Curassow	Mitu	Grand Hocco à bec de rasoir	
Red Goshawk	Australischer Habicht	Autour à ventre rouge	Австралийский ястреб
– Kite	Rotmilan	Milan royal	Красный коршун
– Shoveler	Südamerikanische Löffelente	Souchet roux	Южноамериканская широконоска
Red-billed Teal	Rotschnabelente	Canard à bec rouge	
– Tropic Bird	Rotschnabel-Tropikvogel	Paille-en-queue à bec rouge	Красноклювый фаэтон
Red-breasted Goose	Rothalsgans	Bernache à cou roux	Краснозобая казарка
– Merganser	Mittelsäger	Harle huppé	Средний крохаль
Red-crested Duck	Kolbenente	Brante roussâtre	Красноносый нырок
Red-faced Cormorant	Aleuten-Kormoran		Краснолицый баклан
Red-footed Booby	Rotfußtölpel	Fou aux pieds rouges	Красноногая олуша
– Falcon	Rotfußfalk	Faucon kobez	Кобчик
Red-head	Rotkopfente	Milouin à tête rousse	Краснолобый нырок
Red-headed Falcon	Rotkopfmerlin	Faucon shikra	
Red-legged Partridge	Rothuhn	Perdrix rouge	Красная каменная куропатка
Red-necked Grebe	Rothalstaucher	Grèbe jougris	Серощёкая поганка
Red-shouldered Hawk	Rotschulterbussard	Buse à épaulettes rousses	
Red-tailed Hawk	Rotschwanzbussard	Buse à queue rousse	Краснохвостый канюк
– Tropic Bird	Rotschwanz-Tropikvogel		Краснохвостый фаэтон
Red-throated Caracara	Rotkehl-Karakara		Красношеий каракара
– Diver	Sterntaucher	Plongeon catmarin	Краснозобая гагара
Reef Heron	Küstenreiher	Dimorphe	
Rheas	Nandus	Nandous, Rhéidés	Американские страусы, Нанду
Ring-necked Duck	Halsringente	Morillon à collier	
Rockhopper Penguin	Felsenpinguin	Gorfou sauteur	
Rock Partridge	Steinhuhn	Perdrix bartavelle	Кеклик
Ross' Goose	Zwergschneegans	Oie de Ross	Карликовый белый гусь
Rosy-billed Pochard	Peposakaente	Canard à bec rosé	
Royal Albatross	Königsalbatros	Albatros royal	Королевский альбатрос
Ruddy Duck	Schwarzkopfruderente	Erismature à tête noire	

ENGLISH NAME	GERMAN NAME	FRENCH NAME	RUSSIAN NAME
– Shelduck	Rostgans		Огарь
Ruddy-headed Goose	Rotkopfgans	Bernache à tête rousse	
Rufous Tinamou	Pampashuhn	Tinamou roussâtre	
Rufous-thighed Falconet	Indischer Zwergfalk	Fauconnet à collier	Красноногий карликовый сокол
Rüppell's Griffon	Sperbergeier	Vautour de Rüppell	
Sacred Ibis	Heiliger Ibis	Ibis sacré	Священный ибис
Sage Grouse	Beifußhuhn	Gelinotte des armoises	Полынный тетерев
Saker Falcon	Würgfalk	Faucon sacre	Обыкновенный балобан
Sand Partridge	Arabisches Sandhuhn	Perdrix de Hay	Аравийская пустынная курочка
Satyr Tragopan	Rot-Satyrhuhn	Tragopan satyre	
Scarlet Ibis	Roter Sichler	Ibis rouge	Красный ибис
Schrenck's Little Bittern	Mandschurendommel	Blongios de Schrenck	Амурский волгок
Scoters	Trauerenten	Macreuses	Турпаны
Screamers	Wehrvögel	Anhimidés	Паламедеи
Scrub Fowl	Freycinet-Großfußhuhn		Большеног Фрейсинэ
Secretary Bird	Sekretär	Secrétaire	Секретарь
– Birds	Sekretäre	Sagittaridés	Африканские секретари
Seddle Bill	Afrika-Sattelstorch	Jabirou du Sénégal	Африканский ярибу
Shark-shinned Hawk	Eckschwanzsperber	Épervier brun	
Shearwaters	Sturmtaucher		
Sheldgeese and Shelducks	Entenverwandte	Anatinés	
Shelducks	Halbgänse	Tadornes et Casarcas	Пеганки
Shoebill	Schuhschnabel	Bec-en-Sabot	Китоглав
Shoebills	Schuhschnäbel	Balaenicipitidés, Baleniceps	Китоглавы
Short-tailed Shearwater	Millionensturmtaucher		Тонкоклювый буревестник
Silver Teal	Kappenente	Sarcelle versicolore	
Slavonian Grebe	Ohrentaucher	Grèbe esclavon	Рогатая поганка
Smew	Zwergsäger	Harle piette	Луток
Snow Partridge	Haldenhuhn	Perdrix lerwa	Гималайская куропатка
– Petrel	Schneesturmvogel		Снежный буревестник
– Goose	Schneegans	Oie des neiges	Белый гусь
Snowy Egret	Schmuckreiher	Aigrette neigeuse	
Sooty Albatross	Nördlicher Rußalbatros		Северный дымчатый альбатрос
– Falcon	Schieferfalk	Faucon concolore	
– Shearwater	Rußsturmtaucher	Puffin fuligineux	Серый буревестник
South African Shelduck	Graukopfkasarka	Casarca du Cap	Сероголовая пеганка
– American Green-winged Teal	Chile-Krickente	Sarcelle du Chili	Чилийский чирок
– – Sheldgeese	Spiegelgänse	Bernaches sud-américaines	
Southern Giant Fulmar	Südlicher Riesensturmvogel		Южный гигантский буревестник
Spectacled Eider	Plüschkopfente	Eider de Fischer	Очковая гага
Spoonbills	Löffler	Plataléinés, Spatules	Колпицы
Spot-billed Duck	Fleckschnabelente	Canard à bec tacheté	Черная желтоносая кряква
Spotted Cormorant	Tüpfelkormoran		Пятнистый баклан
– Partridge	Tropfenzahnhuhn		
– Whistling Duck	Tüpfelpfeifgans		Крапчатая древесная утка
Spur-wing	Sporengans	Oie de Gambie	Обыкновенный шпорцевый гусь
Spur-winged Geese	Sporengänse	Oies armées	Шпорцевые гуси
Squacco Heron	Rallenreiher	Héron crabier	Желтая цапля
Square-tailed Kite	Schopfmilan		Хохлатый коршун
Steamer Ducks	Dampfschiffenten	Canards-vapeurs	
Steller's Albatross	Kurzschwanzalbatros	Albatros de Steller	Белоспинный альбатрос
– Eider	Scheckente	Eider de Steller	Сибирская гага
– Sea Eagle	Riesenseeadler		Белоплечий орлан
Steppe Eagle	Steppenadler		Восточный степной орел
Stiff-tailed Ducks	Ruderenten	Erismatures	
Stiff-tails	– i. e. S.	Erismatures	
Storks	Störche, Eigentliche Störche	Ciconiidés, Cigognes	Аисты
Storm Petrel	Sturmschwalbe	Pétrel tempête	Малая прямохвостая качурка
– Petrels	Sturmschwalben	Hydrobatidés	Качурки
Striated Caracara	Südlicher Karakara		Южный каракара
Surf Scoter	Brillenente	Macreuse à lunettes	Пестроносый турпан
Swainson's Hawk	Präriebussard	Buse de Swainson	
Swallow-tailed Kite	Schwalbenweih	Milan à queue fourchue	Вилохвостый лунь
Swan Goose	Schwanengans	Oie cygnoïde	Сухонос
Swans	Schwäne	Cygnes	Лебеди
Tataupa Tinamou	Tataupa	Tinamou tataupa	
Tawny Eagle	Raubadler	Aigle impérial	Африканский степной орел
Teal	Krickente	Sarcelle d'hiver	Чирок-свистунок

ENGLISH NAME	GERMAN NAME	FRENCH NAME	RUSSIAN NAME
Temminck's Tragopan	Temminck-Satyrhuhn	Tragopan de Temminck	
Tiger Bittern	Schwarzschopfreiher	Butor malais	
Tinamous	Steißhühner, Wald-Steißhühner	Tinamous	Скрытохвосты, Скрытохвостые куры, Лесные скрытохвостые куры
Totipalmate Swimmers	Ruderfüßer	Pélécaniformes	Веслоногие
Tropic Birds	Tropikvögel	Phaëthontidés, Phaëthons	Фаэтоны
Trumpeter Swan	Trompeterschwan	Cygne trompette	Лебедь-трубач
Tube-Nosed Swimmers	Röhrennasen	Procellariiformes	Трубконосые
Tufted Duck	Reiherente	Canard morillon	Хохлатая чернеть
Turkey Vulture	Truthahngeier	Vautour à tête rouge	Индюшачий гриф
Typical Birds	Neuvögel		Веерохвостые
Upland Buzzard	Hochlandbussard		Мохноногий курганник
Valley Quail	Kalifornische Schopfwachtel	Colin de Californie	Калифорнийский хохлатый перепел
Variable Goshawk	Weißbrauenhabicht		Белый ястреб
Variegated Tinamou	Rotbrusttao	Tinamou varié	
Velvet Scoter	Samtente	Macreuse brune	Черный турпан
Verreaux's Eagle	Kaffernadler	Aigle de Verreaux	Капский орел
Wahlberg's Eagle	Silberadler	Aigle de Wahlberg	
Wandering Albatross	Wanderalbatros	Albatros hurleur	Странствующий альбатрос
Waterfowl and Screamers	Gänsevögel	Ansériformes	Пластинчатоклювые
Wedge-tailed Eagle	Keilschwanzadler		Клинохвостый орел
Whistling Ducks	Pfeifgänse	Canards siffleurs	Древесные утки
Whistling Swan	Pfeifschwan	Cygne américain	
White Ibis	Weißer Sichler	Ibis blanc	
– Spoonbill	Löffler	Spatule blanche	Колпица
– Stork	Weißstorch	Cigogne blanche	Белый аист
White backed Duck	Weißrückenente	Canard à dos blanc	
White-billed Diver	Gelbschnabel-Eistaucher	Plongeon à bec blanc	Белоносая гагара
White-crested Bittern	Weißnackenreiher	Butor à crête blanche	
White-faced Ibis	Brillensichler	Ibis à face blanche	
– Whistling Duck	Witwenpfeifgans	Dendrocygne veuf	Монашенка
White-fronted Goose	Bleßgans	Oie rieuse	Белолобый гусь
White-headed Duck	Weißkopfruderente	Canard à tête blanche	Савка
– Piping Guan	Schakutinga	Pénélope siffleuse	
– Vulture	Wollkopfgeier	Vautour huppé	
White-tailed Sea Eagle	Seeadler	Pygargue à queue blanche	Орлан-белохвост
White-throated Caracara	Weißkehl-Karakara		Белогорлый каракара
– Petrel	Brustbandsturmtaucher	Pétrel à ailes blanches	Белолобый тайфунник
Whooper		Cygne sauvage	Лебедь-кликун
– Swan	Singschwan	– sauvage	Лебедь-кликун
Willow Grouse	Moorschneehuhn	Lagopède des Saules	Белая куропатка
Wilson's Petrel	Buntfüßige Sturmschwalbe	Pétrel océanite	Качурка Вильсона
Wood Grouse	Blauhühner		
– Ibis	Afrika-Nimmersatt	Tantale ibis	Африканский тантал
– Ibises	Nimmersatte	Tantales	Танталы
Yellow-billed Duck	Gelbschnabelente	Canard à bec jaune	
– Egret	Edelreiher	Aigrette à bec jaune	Средная белая цапля
– Pintail	Spitzschwanzente	Pilet du Chili	
– Tropic Bird	Weißschwanz-Tropikvogel	Paille-en-queue à bec jaune	Белохвостый фаэтон
Yellow-headed Caracara	Gelbkopf-Chimachima		Химахима
Yellow-nosed Albatross	Gelbnasenalbatros	Albatros à bec jaune	Желтоклювый альбатрос
Yellow-throated Caracara	Gelbkinn-Karakara		Желтозобый каракара

II. German—English—French—Russian

Unterartnamen werden meist aus den Artnamen durch Voranstellen von Eigenschaftswörtern oder geographischen Bezeichnungen gebildet. In diesem Teil des Tierwörterbuchs sind so gebildete deutsche Unterartnamen sowie die wissenschaftlichen Unterartnamen in der Regel nicht aufgeführt.

GERMAN NAME	ENGLISH NAME	FRENCH NAME	RUSSIAN NAME
Abdimstorch	Abdim's Stork	Cigogne d'Abdim	Абдимский аист
Accipiter	Goshawks	Éperviers	Ястреба
– *cooperi*	Cooper's Hawk	Épervier de Cooper	Ястреб Купера
– *gentilis*	Goshawk	Autour des palombes	Ястреб-тетеревятник
– *melanoleucus*	Black Goshawk	Épervier pie	
– *nisus*	Eurasian Sparrowhawk	– d'Europe	Ястреб-перепелятник
– *novaehollandiae*	Variable Goshawk		Белый ястреб

GERMAN NAME	ENGLISH NAME	FRENCH NAME	RUSSIAN NAME
– striatus	Shark-shinned Hawk	– brun	
– tachiro	African Sparrowhawk	Autour tachiro	
Accipitridae	Hawks, Old World Vultures and Harriers	Accipitridés	
Adeliepinguin	Adelie Penguin		Синий пингвин
Adlerbussard	Long-legged Hawk	Buse féroce	Канюк-курганник
Adlerfregattvogel	Ascension Frigate Bird	Frégate aigle	Орлиный фрегат
Aegypiinae	Old World Vultures	Aegypiinés	Настоящие грифы
Aegypius monachus	Black Vulture	Vautour moine	Чёрный гриф
Affenadler	Monkey-eating Eagle		Обезьяноед
Afrika-Klaffschnabel	Open Bill	Bec-Ouvert	Африканский аист-разиня
Afrika-Marabu	Marabou	Marabout	Африканский марабу
Afrika-Nimmersatt	Wood Ibis	Tantale ibis	Африканский тантал
Afrikanischer Baumfalk	African Hooby	Hobereau africain	
– Habicht	– Sparrowhawk	Autour tachiro	
– Kormoran	White-breasted Cormorant		
– Schlangenadler	Brown Harrier Eagle	Circaète brun	
Afrikanische Zwergglanzente	African Pygmy Goose	Sarcelle de Madagascar	
Afrikanisch-Indischer Wollhals-storch	Bishop Stork	Cigogne épiscopale	Африканско-индийский пуховый аист
Afrika-Sattelstorch	Seddle Bill	Jabirou du Sénégal	Африканский ярибу
Aix galericulata	Mandarin Duck	Canard mandarin	Мандаринка
– sponsa	Carolina Wood Duck	– Carolin	Американская брачная утка
Albatrosse	Albatrosses and Mollymawks	Diomédeidés	Альбатросы
Alectoris barbara	Barbary Partridge	Perdrix de Barbarie	Берберская каменная куропатка
– graeca	Rock Partridge	– bartavelle	Кеклик
– rufa	Red-legged Partridge	– rouge	Красная каменная куропатка
Alectura lathami	Brush Turkey		Кустовая курица
Aleuten-Kormoran	Red-faced Cormorant		Краснолицый баклан
Alopochen aegyptiacus	Egyptian Goose	Oie d'Égypte	Нильский гусь
Alpenschneehuhn	Ptarmigan	Lagopède muet	Тундряная куропатка
Altweltgeier .	Old World Vultures	Aegypiinés	Настоящие грифы
Altwelt-Schlangenhalsvogel	Darter	Oiseau-Serpent	Ахинга старого света
Amazonas-Ente	Brazilian Teal	Sarcelle du Brésil	
Amazonetta brasiliensis	Brazilian Teal	Sarcelle du Brésil	
Amerika-Nimmersatt	North American Wood Ibis	Cigogne américaine	Американский ярибу
Amerikanische Pfeifente	American Wigeon	Canard siffleur	Американская свиязь
Amerikanischer Graureiher	Great Blue Heron	Grand Héron	
– Schlangenhalsvogel	American Darter	Anhinga d'Amerique	Американская ахинга
Amerikanisches Alpenschneehuhn	Canadian White-tailed Ptarmigan	Lagopède à queue blanche	Американская белая куропатка
Ammoperdix hayi	Sand Partridge	Perdrix de Hay	Аравийская пустынная курочка
Anas acuta	Blue-billed Pintail	Canard pilet	Шилохвость
– americana	American Wigeon	– siffleur	Американская свиязь
– angustirostris	Marbled Teal	Sarcelle marbrée	Мраморный чирок
– aucklandica	Brown Teal		
– bahamensis	Bahama Pintail	Canard de Bahama	
– capensis	Cape Teal	Sarcelle du Cap	
– castanea	Chestnut Teal	– d'Australie	
– clypeata	Northern Shoveler	Canard souchet	Широконоска
– crecca	Teal	Sarcelle d'hiver	Чирок-свистунок
– cyanoptera	Cinnamon Teal	– cannelle	
– discors	Blue-winged Teal	– soucrourou	Синекрылый чирок
– erythrorhyncha	Red-billed Teal	Canard à bec rouge	
Anas falcata	Falcated Teal	Sarcelle à faucilles	Касатка
– flavirostris	South American Green-winged Teal	– du Chili	Чилийский чирок
– formosa	Baikal Teal	– élégante	Чирок-клоктун
– georgica	Yellow-billed Pintail	Pilet du Chili	
– gibberifrons	Grey Teal	Sarcelle grise des Indes	
– luzonica	Philippine Duck	Canard des Philippines	Филиппинская утка
– melleri	Meller's Duck	– de Meller	Мадагаскарская утка
– penelope	European Wigeon	– siffleur	Свиязь
– platalea	Red Shoveler	Souchet roux	Южноамериканская широконоска
– platyrhynchos	Northern Mallard	Canard colvert	Кряква
– poecilorhyncha	Spot-billed Duck	– à bec tacheté	Черная желтоносая кряква
– punctata	Hottentot Teal	Sarcelle Hottentote	
– querquedula	Garganey	– d'été	Чирок-трескунок
– rubripes	American Black Duck	Canard noir de l'Amérique	
– sibilatrix	Chiloë Wigeon	Siffleur du Brésil	Чилийская свиязь

GERMAN NAME	ENGLISH NAME	FRENCH NAME	RUSSIAN NAME
– *smithi*	Cape Shoveler	Souchet du Cap	Капская широконоска
– *sparsa*	African Black Duck	Canard noir africain	
– *strepera*	Gadwall	– chipeau	Серая утка
– *superciliosa*	Grey Duck	– à sourcil blanc	
– *undulata*	Yellow-billed Duck	– à bec jaune	
– *versicolor*	Silver Teal	Sarcelle versicolore	
Anastomus	Open Bills	Becs-Ouverts	Аисты-разини
– *lamelligerus*	– Bill	Bec-Ouvert	Африканский аист-разиня
– *oscitans*	Asian Open Bill	– asiatique	Индийский аист-разиня
Anatidae	Ducks and Geese	Anatidés	Утиные
Anatinae	Sheldgeese and Shelducks	Anatinés	
Andenflamingo	Andean Flamingo		Андский фламинго
Andengans	– Goose	Bernache des Andes	
Anden-Karakara	Mountain Caracara		Андский каракара
Anden-Kondor	Andean Condor	Condor des Andes	Андский кондор
Anhima cornuta	Horned Screamer	Kamichi cornu	Рогатая анхима
Anhimidae	Screamers	Anhimidés	Паламедеи
Anhinga	Darters	Anhingas	Анхинги
– *anhinga*	American Darter	Anhinga d'Amerique	Американская ахинга
– *rufa*	Darter	Oiseau-Serpent	Ахинга старого света
Anhingidae	Darters	Anhingidés	Змеешейки
Anser albifrons	White-fronted Goose	Oie rieuse	Белолобый гусь
– *anser*	Greylag Goose	– cendrée	Серый гусь
– *caerulescens*	Snow Goose	– des neiges	Белый гусь
– *canagicus*	Emperor Goose	– empereur	Белошей
– *cygnoides*	Swan Goose	– cygnoïde	Сухонос
– *erythropus*	Lesser White-fronted Goose	– naine	Пискулька
– *fabalis*	Bean Goose	– des moissons	Гуменник
– *indicus*	Bar-headed Goose	Oie à tête barrée	Индийский горный гусь
– *rossii*	Ross' Goose	Oie de Ross	Карлковый белый гусь
Anseranas	Magpie Geese	Oies semi-palmées	
– *semipalmata*	Australian Magpie Goose	Oie semi-palmée	Расщепнолапый гусь
Anseranatinae	Magpie Geese	Anseranatinés	
Anseriformes	Waterfowl and Screamers	Ansériformes	Пластинчатоклювые
Anserinae	Whistling Ducks, Swans and Geese	Ansérinés	
Antarktischer Eissturmvogel	Antarctic Fulmar		Антарктический глупыш
Antarktissturmvogel	– Petrel		Антарктический буревестник
Aptenodytes forsteri	Emperor Penguin	Manchot impérial	Императорский пингвин
– *patagonica*	King Penguin	– royal	Королевский пингвин
Apteryges	Kiwis	Kiwis	Бескрылые
Apterygidae	Kiwis	Aptérygidés	Киви
Apteryx	Kiwis	Kiwis	Киви
– *australis*	Common Kiwi	Kiwi austral	Обыкновенный киви
– *owenii*	Owen's Kiwi	– d'Owen	Киви Оуэна
Aquila audax	Wedge-tailed Eagle		Клинохвостый орел
– *chrysaëtos*	Golden Eagle	Aigle doré	Беркут
– *clanga*	Greater Spotted Eagle	– criard	Большой подорлик
– *gurneyi*	Gurney's Eagle		
– *heliaca*	Imperial Eagle	– impérial	Орел-могильник
– *nipalensis*	Steppe Eagle		Восточный степной орел
– *pomarina*	Lesser Spotted Eagle	– pomarin	Малый подорлик
– *verreauxi*	Verreaux's Eagle	– de Verreaux	Капский орел
– *wahlbergi*	Wahlberg's Eagle	– de Wahlberg	
Arabisches Sandhuhn	Sand Partridge	Perdrix de Hay	Аравийская пустынная курочка
Arborophila torqueola		– percheuse à collier	Древесная куропатка
Ardea cinerea	Grey Heron	Héron cendré	Серая цапля
– *goliath*	Goliath Heron	– goliath	
– *herodias*	Great Blue Heron	Grand Héron	
Ardea melanocephala	Black-headed Heron	Héron à tête noire	
– *purpurea*	Purple Heron	– pourpré	Рыжая цапля
– *sumatrana*	Dusky Grey Heron	– de Sumatra	Тифонова цапля
Ardeidae	Herons	Ardéidés	Цапли
Ardeola ralloides	Squacco Heron	Héron crabier	Желтая цапля
– *bacchus*	Chinese Pond Heron	Crabier chinois	Белокрылая цапля
– *ibis*	Cattle Egret	Héron garde-boeufs	Египетская цапля
– *speciosa*	Javanese Pond Heron	Crabier malais	
Argala-Marabu	Greater Marabou	Grand Marabout	Индийский марабу-аргал
Argentinische Schwarzkopf-ruderente	Argentine Ruddy Duck	Erismature d'Argentine	
Aucklandente	Brown Teal		
Auerhuhn	Capercaillie	Grand Tétras	Глухарь
Augenbrauenente	Grey Duck	Canard à sourcil blanc	
Australische Kasarka	Australian Shelduck	Casarca d'Australie	Австралийская пеганка

GERMAN NAME	ENGLISH NAME	FRENCH NAME	RUSSIAN NAME
– Moorente	Australasien White-eye	Milouin d'Astralie	
– Rohrdommel	Brown Bittern		Австралийская выпь
– Zwergscharbe	Little Pied Cormorant		Австралийский малый баклан
Australischer Habicht	Red Goshawk	Autour à ventre rouge	Австралийский ястреб
– Tölpel	Australasian Gannet		Австралийская олуша
– Turmfalk	Naukeen Kestrel		Австралийская пустельга
Aves	Birds	Oiseaux	Птицы
Aythya americana	Red-head	Milouin à tête rousse	Краснолобый нырок
– *australis*	Australasian White-eye	-- d'Australie	
– *baeri*	Baer's Pochard		Нырок Бэра
– *collaris*	Ring-necked Duck	Morillon à collier	
– *ferina*	European Pochard	Canard milouin	Красноголовый нырок
– *fuligula*	Tufted Duck	– morillon	Хохлатая чернеть
– *innotata*	Madagascan White-eye	Milouin de Madagascar	
– *marila*	Greater Scaup	Canard milouinan	Морская чернеть
-- *nyroca*	Ferruginous White-eye	– nyroca	Белоглазый нырок
– *valisneria*	Canvas-back	Milouin aux yeux rouges	
Bacchusreiher	Chinese Pond Heron	Crabier chinois	Белокрылая цапля
Bahama-Ente	Bahama Pintail	Canard de Bahama	
Balaeniceps	Shoebills	Baleniceps	Китоглавы
– *rex*	Shoebill	Bec-en-Sabot	Китоглав
Balaenicipitidae	Shoebills, Whale-headed Storks	Balaenicipitidés	Китоглавы
Bandseeadler	Pallas's Sea Eagle	Pygargue à queue blanche	Орлан-долгохвост
Bartgeier	Bearded Vulture	Gypaëte barbu	Бородач
Baßtölpel	Northern Gannet	Fou de Bassan	Атлантическая олуша
Baumfalk	Hooby	Faucon hobereau	Чеглок
Beifußhuhn	Sage Grouse	Gelinotte des armoises	Полынный тетерев
Bengalgeier	Indian White-backed Vulture	Griffon indien à dos blanc	Индийский гриф
Bennettkasuar	Bennett's Cassowary	Casoar de Bennett	Казуар Беннетта
Bergente	Greater Scaup	Canard milouinan	Морская чернеть
Bergguan	Derby's Montain Pheasant, Horned Guan		Горный шуан
Berghaubenwachtel	Plumed Quail	Colin des montagnes	
Bindenfregattvogel	Great Frigate Bird		Большой фрегат
Birkhuhn	Black Grouse	Tétras lyre	Тетерев
Birkhühner	Black Grouse		Тетерева
Biziura lobata	Australian Musk Duck	Canard à membrane	
Blaßfußsturmtaucher	Pale-footed Shearwater		Бледноногий буревестник
Blauflügelente	Blue-winged Teal	Sarcelle soucrourou	Синекрылый чирок
Blauflügelgans	Abyssinian Blue-winged Goose	Bernache aux ailes bleues	Голубокрылый гусь
Blaureiher	Little Blue Heron	Petit Héron bleu	
Blausturmvogel	Blue Petrel		Голубой тайфунник
Bleßgans	White-fronted Goose	Oie rieuse	Белолобый гусь
Blutfasan	Blood Pheasant	Ithagine sanguine	
Blyth-Satyrhuhn	Blyth's Tragopan	Tragopan de Blyth	
Bonasa umbellus	Eastern Ruffed Grouse	Gelinotte huppée	Воротничковый рябчик
Bonin-Albatros	Steller's Albatross	Albatros de Steller	Белоспинный альбатрос
Botaurus lentiginosus	American Bittern	Butor d'amérique	Североамериканская выпь
– ***poiciloptilus***	Brown Bittern		Австралийская выпь
– *stellaris*	Bittern	– étoilé	Большая выпь
Brahminenweih	Brahming Kite		Браминский коршун
Brandgans	Common Shelduck	Tadorne de Belon	Пеганка
Branta	Brents	Bernaches	Казарки
– *bernicla*	Brent Goose	Bernache cravant	Черная казарка
– *canadensis*	Canada Goose	– du Canada	Канадская казарка
– *leucopsis*	Barnacle Goose	– nonnette	Белощёкая казарка
– *ruficollis*	Red-breasted Goose	– à cou roux	Краснозобая казарка
-- *sandvicensis*	Hawaiian Goose	– des îles Sandwich	Гавайская каварка
Brauner Pelikan	Brown Pelican		Бурый пеликан
– Sichler	Glossy Ibis	Ibis luisant	Каравайка
Braunflügelguan	Lesser Grey-headed Chachalaca	Ortalide du Mexique	Бурокрылый шуан
Brauntao	Little Tinamou	Soui	
Brauntölpel	Brown Booby	Fou à ventre blanc	
Brautente	Carolina Wood Duck	Canard Carolin	Американская брачная утка
Breitschwingenbussard	Broad-winged Hawk	Petite Buse	Ширококрылый канюк
Brillenente	Surf Scoter	Macreuse à lunettes	Пестроносый турпан
Brillenkormoran	Pallas Cormorant		Очковый баклан
Brillenpelikan	Australian Pelican		Австралийский пеликан
Brillenpinguin	Black-footed Penguin	Manchot du Cap	Очковый пингвин
Brillensichler	White-faced Ibis	Ibis à face blanche	
Brustbandsturmtaucher	White-throated Petrel	Pétrel à ailes blanches	Белолобый тайфунник
Bucephala	Golden-eyes		Гоголи
– *albeola*	Bufflehead	Garrot albéole	Гоголь-головастик

GERMAN NAME	ENGLISH NAME	FRENCH NAME	RUSSIAN NAME
– *clangula*	Common Golden-eye	Canard garrot	Обыкновенный гоголь
– *islandica*	Barrow's Golden-eye	Garrot islandais	Исландский гоголь
Büffelkopfente	Bufflehead	– albéole	Гоголь-головастик
Bullers Albatros	Buller's Albatross		Альбатрос Буллера
Bulweria bulwerii	Bulwer's Petrel	Pétrel de Bulwer	Тайфунник Бульвера
Buntfalk	American Sparrow Hawk	Crécerelle americaine	Американская пустельга
Buntfüßige Sturmschwalbe	Wilson's Petrel	Pétrel océanite	Качурка Вильсона
Buschhuhn	Brush Turkey		Кустовая курица
Bussarde	Buzzards	Buses	Сарычи
Butastur indicus	Grey-faced Buzzard		Ястребиный сарыч
– *rufipennis*	Grasshopper Buzzard	Busard des sauterelles	
Buteo	Buzzards	Buses	Сарычи
– *auguralis*	African Red-tailed Buzzard	Buse à queue rousse	
– *buteo*	Common Buzzard	– variable	Обыкновенный сарыч
– *hemilasius*	Upland Buzzard		Мохноногий курганник
– *jamaicensis*	Red-tailed Hawk	– à queue rousse	Краснохвостый канюк
– *lagopus*	Northern Rough-legged Buzzard	– pattue	Мохноногий канюк
– *lineatus*	Red-shouldered Hawk	– à épaulettes rousses	
– *platypterus*	Broad-winged Hawk	Petite Buse	Ширококрылый канюк
– *regalis*	Ferruginous Rough-legged Hawk	Buse rouilleuse	
– *rufinus*	Long-legged Hawk	– féroce	Канюк-курганник
– *swainsoni*	Swainson's Hawk	– de Swainson	
Butorides striatus	Green-backed Heron	Héron à dos vert	Зеленая кваква
– *virescens*	Green Heron	– vert	
Cabot-Satyrhuhn	Cabot's Tragopan	Tragopan de Cabot	
Cairina moschata	Moscovy Duck	Canard musqué	Мускусная утка
Callipepla squamata	Mexican Scaled Quail	Quaglia azzurra	
Canachites canadensis	Hudsonian Spruce Grouse	Tétras des savanes	Канадский тетерев
Carancho	Crested Caracara		Каранхо
Casmerodius albus	Great White Egret	Grande Aigrette	Большая белая цапля
Cassinaëtus africanus	Cassin's Hawk Eagle	Aigle Autour de Cassin	
Casuarii	Cassowaries and Emus	Casoars et Emeus	Австралийские страусы
Casuariidae	Cassowaries	Casuariidés	Казуары
Casuarius	Cassowaries	Casoars	Казуары
– *bennetti*	Bennett's Cassowary	Casoar de Bennett	Казуар Беннетта
– *casuarius*	Australian Cassowary	– à casque	Шлемоносный казуар
– *unappendiculatus*	One-wattled Cassowary	– unicaronculé	Лоскутный казуар
Cathartes aura	Turkey Vulture	Vautour à tête rouge	Индюшачий гриф
Cathartidae	Cathartines	Vulturidés	Американские грифы
Centrocercus urophasianus	Sage Grouse	Gelinotte des armoises	Полынный тетерев
Cereopsis novaehollandiae	Cape Barren Goose	Céréopse de Nouvelle-Hollande	Куриный гусь
Chauna chavaria	Crested Screamer	Chauna chavaria	Белощекая паламедея
Chelictinia riocourii	African Swallow-tailed Kite	Naucler d'Afrique	
Chenonetta jubata	Maned Wood Duck	Bernache à crinière	
Chile-Krickente	South American Green-winged Teal	Sarcelle du Chili	Чилийский чирок
Chilenischer Flamingo	Chilean Flamingo		Чилийский фламинго
Chile-Pfeifente	Chiloë Wigeon	Siffleur du Brésil	Чилийская свиязь
Chimango	Chimango Caracara		Химанго
China-Seidenreiher	Chinese Egret		Желтоклювая чепура-нужда
Chinesendommel	Chinese Little Bittern	Blongios chinois	Китайский волгок
Chloephaga	South American Sheldgeese	Bernaches sud-americaines	
– *hybrida*	Kelp Goose	Bernache antarctique	
– *melanoptera*	Andean Goose	– des Andes	
– *picta*	Magellan Goose	– de Magellan	
– *poliocephala*	Ashy-headed Goose	– à tête grise	
– *rubidiceps*	Ruddy-headed Goose	– à tête rousse	
Ciconia	Storks	Cigognes	
– *abdimii*	Abdim's Stork	Cigogne d'Abdim	Абдимский аист
– *boyciana*	Oriental White Stork		Черноклювый аист
– *ciconia*	White Stork	– blanche	Белый аист
– *nigra*	Black Stork	– noire	Черный аист
Ciconiidae	Storks	Ciconiidés	Аисты
Circaëtinae	Bateleur, Harrier Eagles and Serpent Eagles	Circaetinés	
Circaëtus cinereus	Brown Harrier Eagle	Circaète brun	
Circaetus gallicus	Eurasian Short-toed Eagle	Circaète Jean-le-Blanc	Змееяд
Circinae	Harriers and Crane Hawks	Circinés	Луни
Circus aeruginosus	Marsch Harrier	Busard harpaye	Камышевый лунь
– *cyaneus*	Hen Harrier	– Saint-Martin	Полевой лунь
– *macrourus*	Pallid Harrier	– pâle	Степной лунь
– *melanoleucus*	Pied Harrier	– pie	Чернопегий лунь
– *pygargus*	Montagu's Harrier	– de Montagu	Луговой лунь
Clangula hyemalis	Long-tailed Duck	Canard de Miquelon	Морянка
Cochlearius cochlearius	Boat-billed Heron	Savacou huppé	Цапля-челноклюв

GERMAN NAME	ENGLISH NAME	FRENCH NAME	RUSSIAN NAME
Colinus virginianus	Eastern Bob White	Colin de Virginie	Виргинская куропатка
Coragyps atratus	Black Vulture	Vautour noir	Американский черный гриф
Coscoroba coscoroba	Coscoroba Swan	Cygne coscoroba	Гигантская утка коскороба
Coturnix	Quails	Cailles	
– coromandelica	Rain Quail	Caille du Coromandel	Коромандельская перепелка
– coturnix	Common Quail	– des blés	Обыкновенный перепел
– delegorguei	Harlequin Quail	– harlequine	
Cracidae	Curassows, Guans and Chachalacas	Cracidés	Гокко
Crax alector	Crested Curassow	Hocco de la Guiane	Гладкоклювый гокко
– rubra	Mexican Curassow	Grand Hocco	
Crypturellus soui	Little Tinamou	Soui	
– tataupa	Tataupa Tinamou	Tinamou tataupa	
– variegatus	Variegated Tinamou	– varié	
Cyanochen cyanopterus	Abyssinian Blue-winged Goose	Bernache aux ailes bleues	Голубокрылый гусь
Cygnus	Swans	Cygnes	Лебеди
– atratus	Black Swan	Cygne noir	Черный лебедь
– columbianus bewickii	Bewick's Swan	– de Bewick	Западный тундровый лебедь
– – columbianus	Whistling Swan	– américain	
– – jankowskii	Jankowski's Swan		Восточный тундровый лебедь
– cygnus	Whooper	– sauvage	Лебедь-кликун
– – buccinator	Trumpeter Swan	– trompette	Лебедь-трубач
– – cygnus	Whooper Swan	– sauvage	Лебедь-кликун
– melanocoryphus	Black-necked Swan	– à cou noir	Черношейный лебедь
– olor	Mute Swan	– muet	Лебедь-шипун
Cyrtonyx montezumae	Montezuma's Quail	Colin de Masséna	
Dampfschiffenten	Steamer Ducks	Canards-vapeurs	
Daption capensis	Pintado Petrel	Pétrel damier	Капский буревестник
Daptrius americanus	Red-throated Caracara		Красношейный каракара
– ater	Yellow-throated Caracara		Желтозобый каракара
Darwin-Nandu	Darwin's Rhea	Nandou de Darwin	Нанду Дарвина
Dendragapus obscurus	Dusky Grouse	Tétras sombre	Дымчатый тетерев
Dendrocygna	Whistling Ducks	Canards siffleurs	Древесные утки
– arborea	Black-billed Whistling Duck		Кубинская древесная утка
– bicolor	Fulvous Whistling Duck		Пегая древесная утка
– guttata	Spotted Whistling Duck		Крапчатая древесная утка
– javanica	Indian Whistling Duck	Dendrocygne siffleur	Яванская древесная утка
– viduata	White-faced Whistling Duck	– veuf	Монашенка
Dendrocygnini	Whistling Ducks	Canards siffleurs	
Dickschnabelpinguin	Fiordland Penguin		Толстоклювый пингвин
Dinornithidae	Moas	Dinornithidés	
Diomedea albatrus	Steller's Albatross	Albatros de Steller	Белоспинный альбатрос
– bulleri	Buller's Albatross		Альбатрос Буллера
– chlororhynchos	Yellow-nosed Albatross	– à bec jaune	Желтоклювый альбатрос
– chrysostoma	Grey-headed Albatross	– à tête grise	Сероголовый альбатрос
– epomophora	Royal Albatross	– royal	Королевский альбатрос
– exulans	Wandering Albatross	– hurleur	Странствующий альбатрос
– melanophris	Black-browed Albatross	– à sourcil noir	Чернобровый альбатрос
– nigripes	Black-footed Albatross	– à pieds noirs	Черноногий альбатрос
Diomedeidae	Albatrosses and Mollymawks	Diomédéidés	Альбатросы
Dissoura	Bishop Storks	Cigognes épiscopales	Пуховые аисты
– episcopus	– Stork	Cigogne épiscopale	Африканско-индийский пуховый аист
Dreifarbenreiher	Louisiana Heron	Héron à ventre blanc	
Dromaiidae	Emus	Dromicéiidés	Эму
Dromaius novaehollandiae	Emu	Emeu d'Australie	Эму
Dunkelente	American Black Duck	Canard noir de l'Amérique	
Eckschwanzsperber	Shark-shinned Hawk	Épervier brun	
Edelreiher	Yellow-billed Egret	Aigrette à bec jaune	Средняя белая цапля
Egretta eulophotes	Chinese Egret		Желтоклювая чепура-нужда
– garzetta	Little Egret	– garzette	Малая белая цапля
– gularis	Reef Heron	Dimorphe	
– sacra	Eastern Reef Heron	Aigrette des récifs	
– thula	Snowy Egret	– neigeuse	
Eiderente	Common Eider	Eider à duvet	Обыкновенная гага
Eiderenten	Greater Eider Ducks	Eiders	Гаги
Eigentliche Falken	Gyrfalcons, Falcons and Kestrels	Falconinés	Соколы
– Präriehühner	Prairie Hens	Cupidons	
– Störche	Storks	Cigognes	
Eisente	Long-tailed Duck	Canard de Miquelon	Морянка
Eissturmvogel	Arctic Fulmar	Pétrel glacial	Атлантический глупыш

GERMAN NAME	ENGLISH NAME	FRENCH NAME	RUSSIAN NAME
Eistaucher	Great Northern Diver	Plongeon imbrin	Полярная гагара
Elaninae	Kites	Élaninés	
Elanoides forficatus	Swallow-tailed Kite	Milan à queue fourchue	Вилохвостый лунь
Elanus caeruleus	Black-winged Kite	Faucon blanc	Чернокрылый коршун
– scriptus	Letter-winged Kite	Élanion écrit	
Eleonorenfalk	Eleonora's Falcon	Faucon d'Eléonore	Чеглок Элеоноры
Emu	Emu	Emeu d'Australie	Эму
Emus	Emus	Dromicéiidés	Эму
Entenverwandte	Sheldgeese and Shelducks	Anatinés	
Entenvögel	Ducks and Geese	Anatidés	Утиные
Ephippiorhynchus senegalensis	Seddle Bill	Jabirou du Sénégal	Африканский ярибу
Erythrotriorchis radiatus	Red Goshawk	Autour à ventre rouge	Австралийский ястреб
Eselspinguin	Gentoo Penguin	Manchot Gentoo	Ослиный пингвин
Eudocimus albus	White Ibis	Ibis blanc	
– ruber	Scarlet Ibis	– rouge	Красный ибис
Eudromia elegans	Crested Tinamou	Tinamou huppé	
Eudyptes atratus	Erect-crested Penguin		Желточубый пингвин
– chrysolophus	Macaroni Penguin		Золоточубый пингвин
– crestatus	Rockhopper Penguin	Gorfou sauteur	
– pachyrhynchus	Fiordland Penguin		Толстоклювый пингвин
Eudyptula minor	Little Penguin		Карликовый пингвин
Excalfactoria chinensis	Chinese Painted Quail	Caille peinte de Chine	
Fahlpfeifgans	Fulvous Whistling Duck		Пегая древесная утка
Falco	Falcons	Faucons	
– alopex	Fox Kestrel	Faucon-Renard	
– ardosiaceus	Grey Kestrel	– ardoisé	
– biarmicus	Lanner Falcon	– lanier	Рыжеголовый балобан
– – feldeggii	Feldegg's Falcon	– de Feldegg	Средиземноморский рыжеголовый балобан
– cenchroides	Naukeen Kestrel		Австралийская пустельга
– cherrug	Saker Falcon	– sacre	Обыкновенный балобан
– chiquera	Red-headed Falcon	– shikra	
– columbarius	Merlin	– émerillon	Дербник
– concolor	Sooty Falcon	– concolore	
– cuvierii	African Hooby	Hobereau africain	
– eleonorae	Eleonora's Falcon	Faucon d'Eléonore	Чеглок Элеоноры
– jugger	Laggar Falcon		Индийский балобан-лаггар
– mexicanus	Prairie Falcon	– des prairies	Мексиканский сокол
– naumanni	Lesser Kestrel	– crécerellette	Степная пустельга
– peregrinus	Peregrine Falcon	– pèlerin	Сокол-сапсан
– rusticolus	Gyrfalcon	– gerfaut	Кречет
– severus	Oriental Hooby	Hobereau à poitrine rousse	
– sparverius	American Sparrow Hawk	Crécerelle americaine	Американская пустельга
– subbuteo	Hooby	Faucon hobereau	Чеглок
– tinnunculus	Old World Kestrel	– crécerelle	Обыкновенная пустельга
– vespertinus	Red-footed Falcon	– kobez	Кобчик
Falconidae	Falcons	Falconidés	
Falconiformes	Diurnal Birds of Prey	Falconiformes	Хищные птицы
Falconinae	Gyrfalcons, Falcons and Kestrels	Falconinés	Соколы
Falken	Falcons	Falconidés	
– i. e. S.	Falcons	Faucons	
Falkland-Dampfschiffente	Falkland Flightless Steamer Duck	Canard à ailes courtes	
Falkland-Kormoran	King Cormorant		
Fasanenartige	Pheasants, Quails and Peacocks	Phasianidés	
Feenwalvogel	Fairy Prion		Буревестник-горлица
Feldeggsfalk	Feldegg's Falcon	Faucon de Feldegg	Средиземноморский рыжеголовый балобан
Feldhühner	Old World Quails	Perdricinés	
Felsenbussard	African Red-tailed Buzzard	Buse à queue rousse	
Felsengebirgshuhn	Dusky Grouse	Tétras sombre	Дымчатый тетерев
Felsenhuhn	Barbary Partridge	Perdrix de Barbarie	Берберская каменная куропатка
Felsenpinguin	Rockhopper Penguin	Gorfou sauteur	
Fischadler	Ospreys, Osprey	Balbuzard Fluviatile	Скопы, скопа
Flamingo	Greater Flamingo	Flamant rose	Обыкновенный фламинго
Flamingos	Flamingoes	Phoenicoptéridés, Flamants	Фламинго
Fleckenkiwi	Owen's Kiwi	Kiwi d'Owen	Киви Оуэна
Fleckschnabelente	Spot-billed Duck	Canard à bec tacheté	Черная желтоносая кряква
Fledermaus-Gleitaar	Bat Hawk	Faucon des chauves-souris	Индийский совиный сарыч
Florida caerulea	Little Blue Heron	Petit Héron bleu	
Francolinus	Francolins	Francolins	
Francolinus afer	Angola Red-necked Partridge	Pterniste à cou nu	
– camerunensis	Cameroun Montain Francolin	Francolin du Mont Cameroun	
– francolinus	Francolin	– d'Europe	Турач

GERMAN NAME	ENGLISH NAME	FRENCH NAME	RUSSIAN NAME
Frankoline	Francolins	Francolins	
Fregata	Frigate Birds	Frégates	Фрегаты
– andrewsi	Christmas Frigate Bird		Белобрюхий фрегат
– aquila	Ascension Frigate Bird	Frégate aigle	Орлиный фрегат
– ariel	Lesser Frigate Bird		Белобокий фрегат
– magnificens	Magnificent Man-o'-War Bird	– superbe	Малый фрегат
– minor	Great Frigate Bird		Большой фрегат
Fregatidae	Frigate Birds	Frégatidés	Фрегаты
Fregattensturmschwalbe	– Petrel	Pétrel frégate	Морская качурка
Fregattvögel	– Birds	Frégatidés, Frégates	Фрегаты
Fregetta tropica	Black-bellied Storm Petrel	Pétrel des tropiques	Чернобрюхая качурка
Freycinet-Großfußhuhn	Scrub Fowl		Большеног Фрейсинэ
Fuchsfalk	Fox Kestrel	Faucon-Renard	
Fulmarinae	Fulmars		Глупыши
Fulmarus glacialis	Arctic Fulmar	Pétrel glacial	Атлантический глупыш
– glacialoides	Antarctic Fulmar		Антарктический глупыш
Gabar-Habicht	Gabar Goshawk	Autour gabar	
Gabelschwanz-Wellenläufer	Fork-tailed Storm Petrel		Серая качурка
Galapagospinguin	Galapagos Penguin		Галапагосский пингвин
Galliformes	Fowllike Birds	Galliformes	Куриные
Gampsonyx swainsonii	Pearl Kite		Карликовый лунь
Gänsegeier	Griffon Vulture	Vautour fauve	Белоголовый сип
Gänsesäger	Goosander	Harle bièvre	Большой крохаль
Gänseverwandte	Whistling Ducks, Swans and Geese	Ansérinés	
Gänsevögel	Waterfowl and Screamers	Ansériformes	Пластинчатоклювые
Gaukler	Bateleur	Bateleur	Орел-скоморох
Gavia	Divers	Plongeons	Гагары
– adamsii	White-billed Diver	Plongeon à bec blanc	Белоносая гагара
– arctica	Black-throated Diver	– lumme	Чернозобаян гагара
– immer	Great Northern Diver	– imbrin	Полярная гагара
– stellata	Red-throated Diver	– catmarin	Краснозобая гагара
Gaviidae	Loons	Gaviidés	Гагары
Gaviiformes	Loons	Gaviiformes	Гагаровые
Geierseeadler	Palm-nut Vulture	Vautour palmiste	Грифовый орлан
Gelbkinn-Karakara	Yellow-throated Caracara		Желтозобый каракара
Gelbkopf-Chimachima	Yellow-headed Caracara		Химахима
Gelbnasenalbatros	Yellow-nosed Albatross	Albatros à bec jaune	Желтоклювый альбатрос
Gelbschnabel-Eistaucher	White-billed Diver	Plongeon à bec blanc	Белоносая гагара
Gelbschnabelente	Yellow-billed Duck	Canard à bec jaune	
Gelbschnabel-Sturmtaucher	Mediterranean Shearwater	Puffin cendré	Большой белобрюхий буревестник
Gelbschnabel-Zwergscharbe	Long-tailed Cormorant	Cormoran africain	
Gelbschopfpinguin	Erect-crested Penguin		Желтчубый пингвин
Gerfalk	Gyrfalcon	Faucon gerfaut	Кречет
Geronticus eremita	Hermit Ibis	Ibis chauve	Горный ибис
Glattschnabelhokko	Crested Curassow	Hocco de la Guiane	Гладкоклювый гокко
Gleitaare	Kites	Élaninés	
Glockenreiher	Black Heron	Héron ardoisé	
Gluckente	Baikal Teal	Sarcelle élégante	Чирок-клоктун
Goldhalskasuar	One-wattled Cassowary	Casoar unicaronculé	Лоскутный казуар
Goldschopfpinguin	Macaroni Penguin		Золоточубый пингвин
Goliathreiher	Goliath Heron	Héron goliath	
Gorsachius melanolophus	Tiger Bittern	Butor malais	
Graufalk	Grey Kestrel	Faucon ardoisé	
Graugans	Greylag Goose	Oie cendrée	Серый гусь
Graugesichtbussard	Grey-faced Buzzard		Ястребиный сарыч
Graukopfalbatros	Grey-headed Albatross	Albatros à tête grise	Сероголовый альбатрос
Graukopfgans	Ashy-headed Goose	Bernache à tête grise	
Graukopfkasarka	South African Shelduck	Casarca du Cap	Сероголовая пеганка
Graupelikan	Grey Pelican		Серый пеликан
Graureiher	– Heron	Héron cendré	Серая цапля
Greifvögel	Diurnal Birds of Prey	Falconiformes	Хищные птицы
Großer Sturmtaucher	Great Shearwater	Puffin majeur	Большой пестробрюхий буревестник
Großfußhühner	Moundfowl	Mégapodiidés	Сорные куры
Großtao	Great Tinamou	Grand Tinamou	Большой тинаму
Grüne Zwergglanzente	Green Pygmy Goose	Sarcelle verte d'Australie	
Grünreiher	– Heron	Héron vert	
Guadalupe-Karakara	Guadalupe Caracara		Гваделупский каракара
Gymnogyps californianus	California Condor	Condor de Californie	Калифорнийский кондор
Gypaëtus barbatus	Bearded Vulture	Gypaète barbu	Бородач
Gypohierax angolensis	Palm-nut Vulture	Vautour palmiste	Грифовый орлан
Gyps fulvus	Griffon Vulture	– fauve	Белоголовый сип
– himalayensis	Himalayan Griffon		Снежный гималайский сип

GERMAN NAME	ENGLISH NAME	FRENCH NAME	RUSSIAN NAME
Gyps indicus	Long-billed Vulture	– à bec long	Индийский белоголовый сип
– rueppelli	Rüppell's Griffon	– de Rüppell	
Habicht	Goshawk	Autour des palombes	Ястреб-тетеревятник
Habichtartige	Hawks, Old World Vultures and Harriers	Accipitridés	
Habichte i. e. S.	Goshawks	Eperviers	Ястреба
Habichtsadler	Bonelli's Eagle	Aigle de Bonelli	Ястребиный орел
Hagedasch	Hadada	Ibis hagedash	
Hagedashia hagedash	Hadada	Ibis hagedash	
Hakensturmtaucher	Gadfly Petrels		Тайфунники
Halbgänse	Shelducks	Tadornes et Casarcas	Пеганки
Haldenhuhn	Snow Partridge	Perdrix lerwa	Гималайская куропатка
Haliaeëtus albicilla	White-tailed Sea Eagle	Pygargue à queue blanche	Орлан-белохвост
– leucocephalus	Bald Eagle	Aigle à tête blanche	Белоголовый орлан
– leucoryphus	Pallas' Sea Eagle		Орлан-долгохвост
– pelagicus	Steller's Sea Eagle		Белоплечий орлан
– vocifer	African Sea Eagle	Aigle pêcheur	Крикливый орлан
Haliastur indus	Brahming Kite		Браминский коршун
Halobaena caerulea	Blue Petrel		Голубой тайфунник
Halsbandfrankolin	Francolin	Francolin d'Europe	Турач
Halsringente	Ring-necked Duck	Morillon à collier	
Hamirostra melanosterna	Black-breasted Buzzard Kite	Buse à poitrine noire	
Hammerkopf	Hamerkop	Ombrette	Молотоглав
Hammerköpfe	Hamerkops	Scopidés, Ombrettes	Теневые птицы, Молотоголовые цапли
Harlekinwachtel	Harlequin Quail	Caille harlequine	
Harpia harpyja	Harpy Eagle	Harpye	
Harpyie	Harpy Eagle	Harpye	
Haselhuhn	Hazel Hen	Gelinotte des bois	Рябчик
Hauben-Karakara	Crested Caracara		Каранехо
Haubenmilan	Black-breasted Buzzard Kite	Buse à poitrine noire	
Haubentaucher	Great Crested Grebe	Grèbe huppé	Большая поганка
Haubenzwergadler	Ayres' Eagle	Aigle-Autour d'Ayres	
Hausente	Domestic Duck	Canard doméstique	Домашняя утка
Hausgans	– Goose	Oie doméstique	Домашний гусь
Haus-Höckergans	– Swan Goose	– cygnoïde doméstique	Домашний китайский гусь
Hawaiigans	Hawaiian Goose	Bernache des îles Sandwich	Гавайская казарка
Hawaiisturmvogel	– Petrel		Гавайский тайфунник
Heiliger Ibis	Sacred Ibis	Ibis sacré	Священный ибис
Helmhokko	Helmeted Curassow	Pauxi pierre	Шлемоносный гокко
Helmkasuar	Australian Cassowary	Casoar à casque	Шлемоносный казуар
Helmwachtel	Gambel's Quail	Colin de Gambel	Калифорнийский шлемоносный перепел
Heteronetta atricapilla	Black-headed Duck	Canard â tête noire de l'Argentine	Кукутковая утка
Heuschreckenbussard	Grasshopper Buzzard	Busard des sauterelles	
Hieraaëtus ayresii	Ayres' Eagle	Aigle Autour d'Ayres	
– fasciatus	Bonelli's Eagle	Aigle de Bonelli	Ястребиный орел
– pennatus	Booted Eagle	– botte	Орел-карлик
Histrionicus histrionicus	Harlequin Duck	Garrot harlequin	Каменутка
Hochlandbussard	Upland Buzzard		Мохноногий курганник
Höckerglanzente	Comb Duck	Sarcidiorne à crète	Шишконосый гусь
Höckerschwan	Mute Swan	Cygne muet	Лебедь-шипун
Hokkos	Curassows, Guans and Chachalacas	Cracidés	Гокко
Hornwehrvogel	Horned Screamer	Kamichi cornu	Рогатая анхима
Hottentotten-Ente	Hottentot Teal	Sarcelle Hottentote	
Hügelhuhn		Perdrix percheuse à collier	Древесная куропатка
Hühnergans	Cape Barren Goose	Céréopse de Nouvelle-Hollande	Куриный гусь
Hühnervögel	Fowllike Birds	Galliformes	Куриные
Humboldtpinguin	Peruvian Penguin		Пингвин Гумбольдта
Hydranassa tricolor	Louisiana Heron	Héron à ventre blanc	
Hydrobates pelagicus	Storm Petrel	Pétrel tempête	Малая прямохвостая качурка
Hydrobatidae	– Petrels	Hydrobatidés	Качурки
Ibis	Wood Ibises	Tantales	Танталы
– cinereus	Malayan Wood Ibis	Tantale blanc	Малайский тантал
– ibis	Wood Ibis	– ibis	Африканский тантал
– leucocephalus	Painted Stork	– indien	Индийский тантал
Ibisse	Ibises	Threskiornithinés, Ibis	Ибисы
Ibisvögel	Ibises and Spoonbills	Threskiornithidés	Ибисы
Indianerdommel	Lesser Bittern	Petit Butor	
Indien-Großstorch	Black-necked Stork	Jabirou asiatique	Индийский исполинский аист
Indien-Klaffschnabel	Asian Open Bill	Bec-Ouvert asiatique	Индийский аист-разиня

GERMAN NAME	ENGLISH NAME	FRENCH NAME	RUSSIAN NAME
Indien-Nimmersatt	Painted Stork	Tantale indien	Индийский тантал
Indien-Pfeifgans	Indian Whistling Duck	Dendrocygne siffleur	Яванская древесная утка
Indischer Baumfalk	Oriental Hooby	Hobereau à poitrine rousse	
Indischer Geier	Long-billed Vulture	Vautour à bec long	Индийский белоголовый сип
– Schlangenhabicht	Crested Serpent Eagle		Орел-хеела
– Zwergfalk	Rufous-thighed Falconet	Fauconnet à collier	Красноногий карликовый сокол
Indische Zwergglanzente	Cotton Pygmy Goose	Sarcelle de Coromandel	Индийская карликовая утка
Ithaginis cruentus	Blood Pheasant	Ithagine sanguine	
Ixobrychus cinnamoneus	Cinnamon Bittern	Blongios cannelle	
– eurhythmus	Schrenck's Little Bittern	– de Schrenck	Амурский волгок
– exilis	Lesser Bittern	Petit Butor	
– minutus	Little Bittern	Blongios nain	Малая выпь
– sinensis	Chinese Little Bittern	– chinois	Китайский волгок
Jabiru	Jabiru	Jabiru	
Jabiru mycteria	Jabiru	Jabiru	
James-Flamingo	James' Flamingo		Короткоклювый фламинго
Jankowski-Schwan	Jankowski's Swan		Восточный тундровый лебедь
Japanischer Ibis	Japanese Crested Ibis		Китайский красноногий ибис
Japanischer Kormoran	– Cormorant		Японский баклан
Kaffernadler	Verreaux's Eagle	Aigle de Verreaux	Капский орел
Kahnschnabel	Boat-billed Heron	Savacou huppé	Цапля-челноклюв
Kaiseradler	Imperial Eagle	Aigle impérial	Орел-могильник
Kaisergans	Emperor Goose	Oie empereur	Белошей
Kaiserpinguin	– Penguin	Manchot impérial	Императорский пингвин
Kalifornischer Kondor	California Condor	Condor de Californie	Калифорнийский кондор
Kalifornische Schopfwachtel	Valley Quail	Colin de Californie	Калифорнийский хохлатый перепел
Kamerunbergwald-Frankolin	Cameroun Mountain Francolin	Francolin du Mont Cameroun	
Kampfadler	Martial Eagle	Aigle Martial	Орел-боец
Kanadagans	Canada Goose	Bernache du Canada	Канадская казарка
Kapente	Cape Teal	Sarcelle du Cap	
Kap-Löffelente	– Shoveler	Souchet du Cap	Капская широконоска
Kappenente	Silver Teal	Sarcelle versicolore	
Kappengeier	Hooded Vulture	Charognard	
Kappensäger	– Merganser	Harle couronné	Американский крохаль
Kapsturmvogel	Pintado Petrel	Pétrel damier	Капский буревестник
Kaptölpel	Cape Gannet		Капская олуша
Kaspisches Königshuhn	Caspian Snowcock	Perdrix des neiges caspienne	Каспийский улар
Kastanienente	Chestnut Teal	Sarcelle d'Australie	
Kasuare	Cassowaries	Casuariidés, Casoars	Казуары
Kasuarvögel	– and Emus	Casoars et Emeus	Австралийские страусы
Kaukasisches Birkhuhn	Caucasian Black Grouse	Tétras lyre de Géorgie	Кавказский тетерев
Kaupifalco monogrammicus	Lizard Buzzard	Buse unibande	
Kehlstreifbussard	Lizard Buzzard	Buse unibande	
Keilschwanzadler	Wedge-tailed Eagle		Клинохвостый орел
Kiwis	Kiwis	Aptérygidés, Kiwis	Киви
Kiwivögel	Kiwis	Kiwis	Бескрылые
Klaffschnäbel	Open Bills	Becs-Ouverts	Аисты-разини
Kleiner Adjutant	Lesser Marabou	Petit Marabout	Яванский марабу
– Fregattvogel	– Frigate Bird		Белобокий фрегат
– Löffler	Black-faced Spoonbill	Petite Spatule	Малая колпица
Kleines Präriehuhn	Lesser Prairie Hen		Малый луговой тетерев
Klippenhuhn	Barbary Partridge	Perdrix de Barbarie	Берберская каменная куропатка
Knäkente	Garganey	Sarcelle d'été	Чирок-трескунок
Kolbenente	Red-crested Duck	Brante roussâtre	Красноносый нырок
Königsalbatros	Royal Albatross	Albatros royal	Королевский альбатрос
Königsbussard	Ferruginous Rough-legged Hawk	Buse rouilleuse	
Königsgeier	King Vulture	Vautour royal	Королевский гриф
Königspinguin	– Penguin	Manchot royal	Королевский пингвин
Kormoran	Black Cormorant	Grand Cormoran	Обыкновенный баклан
Kormorane	Cormorants	Phalacrocoracidés, Cormorans	Бакланы
Kornweihe	Hen Harrier	Busard Saint-Martin	Полевой лунь
Koskorobaschwan	Coscoroba Swan	Cygne coscoroba	Гигантская утка коскороба
Kragenente	Harlequin Duck	Garrot harlequin	Каменутка
Kragenhuhn	Eastern Ruffed Grouse	Gelinotte huppée	Воротничковый рябчик
Krähenscharbe	Green Cormorant	Cormoran huppé	Длинноносый баклан
Krauskopfpelikan	Dalmatian Pelican	Pélican frisé	Кудрявый пеликан
Krickente	Teal	Sarcelle d'hiver	Чирок-свистунок
Kronenadler	Crowned Eagle	Blanchard	Венценосный орел

GERMAN NAME	ENGLISH NAME	FRENCH NAME	RUSSIAN NAME
Kuba-Pfeifgans	**Black-billed Whistling Duck**		**Кубинская древесная утка**
Kuckucksente	**Black-headed Duck**	**Canard à tête noire de l'Argentine**	**Кукутковая утка**
Kuhreiher	**Cattle Egret**	**Héron garde-boeufs**	**Египетская цапля**
Kurzschnabelflamingo	**James' Flamingo**		**Короткоклювый фламинго**
Kurzschwanzalbatros	**Steller's Albatross**	**Albatros de Steller**	**Белоспинный альбатрос**
Küstenreiher	Reef Heron	Dimorphe	
Laggarfalk	Laggar Falcon		Индийский балобан-лаггар
Lagopus lagopus	Willow Grouse	Lagopède des Saules	Белая куропатка
– leucurus	Canadian White-tailed Ptarmigan	– à queue blanche	Американская белая куропатка
– mutus	Ptarmigan	– muet	Тундряная куропатка
Lampribis olivacea	Olive Ibis	Ibis olivâtre	
Langflügelsturmtaucher	Grey-winged Petrel		Длиннокрылый тайфунник
Langschwanzhabicht	Long-tailed Hawk	Autour à longue queue	Длиннохвостый ястреб
Langschwanz-Zwergfalk	Burmese Pigmy Falcon	Fauconnet à pattes jaunes	
Lannerfalk	Lanner Falcon	Faucon lanier	Рыжеголовый балобан
Lappenente	Australian Musk Duck	Canard à membrane	
Lappengeier	Pondicherry Vulture	Vautour de Pondichéry	
Lappentaucher	Grebes	Podicipédiformes, Podicipitedés	Поганковые, поганки
Laufvögel	Ratites	Ratites	Бегающие
Leipoa ocellata	Mallee Fowl	Leipoa ocellé	Лейпоа
Leptoptilos	Marabous	Marabouts	Марабу
– crumeniferus	Marabou	Marabout	Африканский марабу
– dubius	Greater Marabou	Grand Marabout	Индийский марабу-аргал
– javanicus	Lesser Marabou	Petit Marabout	Яванский марабу
Lerwa lerwa	Snow Partridge	Perdrix lerwa	Гималайская куропатка
Löffelente	Northern Shoveler	Canard souchet	Широконоска
Löffler	Spoonbills, White Spoonbill	Plataleinés, Spatules, Spatule blanche	Колпицы, колпица
Lophoaëtus occipitalis	Long-crested Eagle	Aigle huppé	Гребневый орел
Lophoictinia isura	Square-tailed Kite		Хохлатый коршун
Lophortyx californica	Valley Quail	Colin de Californie	Калифорнийский хохлатый перепел
– gambelii	Gambel's Quail	– de Gambel	Калифорнийский шлемоносный перепел
Lyrurus	Black Grouse		Тетерева
– mlokosiewiczi	Caucasian Black Grouse	Tétras lyre de Géorgie	Кавказский тетерев
– tetrix	Black Grouse	Tetras lyre	Тетерев
Machaerhamphus alcinus	Bat Hawk	Faucon des chauves-souris	Индийский совиный сарыч
Macronectes giganteus	Southern Giant Fulmar		Южный гигантский буревестник
– halli	Northern Giant Fulmar		Северный гигантский буревестник
Madagaskar-Ente	Meller's Duck	Canard de Meller	Мадагаскарская утка
Madagaskar-Moorente	Madagascan White-eye	Milouin de Madagascar	
Magellangans	Magellan Goose	Bernache de Magellan	
Magellanpinguin	Magellanic Penguin	Manchot de Magelhaen	
Mähnengans	Maned Wood Duck	Bernache à crinière	
Malaien-Nimmersatt	Malayan Wood Ibis	Tantale blanc	Малайский тантал
Malayen-Wespenbussard	Crested Honey Buzzard		Хохлатый осоед
Mandarinente	Mandarin Duck	Canard mandarin	Мандаринка
Mandschurendommel	Schrenck's Little Bittern	Blongios de Schrenck	Амурский волгок
Mangrovereiher	Green-backed Heron	Héron à dos vert	Зеленая кваква
Marabus	Marabous	Marabouts	Марабу
Marmelente	Marbled Teal	Sarcelle marbrée	Мраморный чирок
Maskentölpel	Masked Booby	Fou masqué	Голуболицая олуша
Massenawachtel	Montezuma's Quail	Colin de Masséna	
Mäusebussard	Common Buzzard	Buse variable	Обыкновенный сарыч
Meergänse	Brents	Bernaches	Казарки
Megapodiidae	Moundfowl	Mégapodiides	Сорные куры
Megapodius freycinet	Scrub Fowl		Большеног Фрейсинэ
Melanitta	Scoters	Macreuses	Турпаны
– fusca	Velvet Scoter	Macreuse brune	Черный турпан
– nigra	Black Scoter	– noire	Синьга
– perspicillata	Surf Scoter	– à lunettes	Пестроносый турпан
Melanophoyx ardesiaca	Black Heron	Héron ardoisé	
Melierax gabar	Gabar Goshawk	Autour gabar	
– musicus	Pale Chanting Goshawk		Певчий ястреб
Mergus	Mergansers	Harles	Крохали
– albellus	Smew	Harle piette	Луток
– cucullatus	Hooded Merganser	– couronné	Американский крохаль
– merganser	Goosander	– bievre	Большой крохаль
– serrator	Red-breasted Merganser	– huppé	Средний крохаль
– squamatus	Chinese Merganser	– chinois	Чемуйчатый крохаль

GERMAN NAME	ENGLISH NAME	FRENCH NAME	RUSSIAN NAME
Merlin	Merlin	Faucon émerillon	Дербник
Mesophoyx intermedia	Yellow-billed Egret	Aigrette à bec jaune	Средняя белая цапля
Mexiko-Stockente	Mexican Duck		Мексиканская кряква
Microhierax caerulescens	Rufous-thighed Falconet	Fauconnet à collier	Красноногий карликовый сокол
Milane i. e. S.		Milvinés	Коршуны
Millionensturmtaucher	Short-tailed Shearwater		Тонкоклювый буревестник
Milvago chimachima	Yellow-headed Caracara		Химахима
– chimango	Chimango Caracara		Химанго
Milvus		Milvinés	Ко̂ршуны
– migrans	Black Kite	Milan noir	Черный коршун
– milvus	Red Kite	– royal	Красный коршун
Mittelsäger	Red-breasted Merganser	Harle huppé	Средний крохаль
Mitu	Razor-billed Curassow	Grand Hocco à bec de rasoir	
Mitu mitu	Razor-billed Curassow	Grand Hocco à bec de rasoir	
– tomentosa	Lesser Razor-billed Curassow	Petit Hocco à bec de rasoir	
Moas	Moas	Dinornithidés	
Mohrenguan	Guatemalan Black Chachalaca		Черный шуан
Mönchsgeier	Black Vulture	Vautour moine	Чёрный гриф
Montezumawachtel	Montezuma's Quail	Colin de Masséna	
Moorente	Ferruginous White-eye	Canard nyroca	Белоглазый нырок
Moorschneehuhn	Willow Grouse	Lagopède des Saules	Белая куропатка
Morus	Gannets	Fous	
– bassanus	Northern Gannet	Fou de Bassan	Атлантическая олуша
– capensis	Cape Gannet		Капская олуша
– serrator	Australasian Gannet		Австралийская олуша
Moschusente	Moscovy Duck	Canard musqué	Мускусная утка
Möwensturmvögel	Fulmars		Глупыши
Mycteria americana	North American Wood Ibis	Cigogne américaine	Американский ярибу
Nachtreiher	Night Herons, Black-crowned Night Heron	Hérons bihoreaux, Héron bihoreau à couronne noire	Кваква
Nandu	American Rhea	Nandou américain	Обыкновенный нанду
Nandus	Rheas	Nandous, Rhéidés	Американские страусы, нанду
Nashornpelikan	American White Pelican		Пеликан-носорог
Necrosyrtes monachus	Hooded Vulture	Charognard	
Neochen jubatus	Orinoco Goose	Oie de l'Orinoque	
Neohierax insignis	Burmese Pigmy Falcon	Fauconnet à pattes jaunes	
Neophron percnopterus	Egyptian Vulture	Percnoptère d'Égypte	Стервятник
Neornithes	Typical birds		Веерохвостые
Nepal-Haubenadler	Mountain Hawk Eagle		Хохлатый орел
Netta peposaca	Rosy-billed Pochard	Canard à bec rosé	
Nettapus	Pygmy Geese		Карликовые утки
– auritus	African Pygmy Goose	Sarcelle de Madagascar	
– coromandelianus	Cotton Pygmy Goose	– de Coromandel	Индийская карликовая утка
– pulchellus	Green Pygmy Goose	– verte d'Australie	
Netta rufina	Red-crested Duck	Brante roussâtre	Красноносый нырок
Neuvögel	Typical Birds		Веерохвостые
Neuweltgeier	Cathartines	Vulturidés	Американские грифы
Nilgans	Egyptian Goose	Oie d'Égypte	Нильский гусь
Nimmersatte	Wood Ibises	Tantales	Танталы
Nipponia nippon	Japanese Crested Ibis		Китайский красноногий ибис
Nordamerikanische Rohrdommel	American Bittern	Butor d'Amérique	Североамериканская выпь
Nördlicher Riesensturmvogel	Northern Giant Fulmar		Северный гигантский буревестник
– Rußalbatros	Sooty Albatross		Северный дымчатый альбатрос
Nordpazifischer Kormoran	Pelagic Cormorant		Берингов баклан
Nothocrax urumutum	Nocturnal Curassow		Хохлатый гокко
Nycticorax	Night Herons	Hérons bihoreaux	
– nycticorax	Black-crowned Night Heron	Héron bihoreau à couronne noire	Кваква
Oceanites oceanicus	Wilson's Petrel	Pétrel océanite	Качурка Вильсона
Oceanodroma furcata	Fork-tailed Storm Petrel		Серая качурка
– leucorhoa	Leach's Petrel	– cul-blanc	Северная качурка
Ohrengeier	Lappet-faced Vulture	Vautour oricou	Ушастый гриф
Ohrentaucher	Slavonian Grebe	Grèbe esclavon	Рогатая поганка
Olivgrüner Ibis	Olive Ibis	Ibis olivâtre	
Oreophasis derbianus	Derby's Montain Pheasant		Горный шуан
Oreortyx picta	Plumed Quail	Colin des montagnes	
Orinokogans	Orinoco Goose	Oie de l'Orinoque	
Ortalis garrula	Chestnut-winged Chachalaca	Ortalide babillarde	Краснокрылый шуан
– ruficauda	Jardine's Chachalaca		Краснохвостый шуан
– vetula	Lesser Grey-headed Chachalaca	– du Mexique	Бурокрылый шуан
Oxyura	Stiff-tails	Erismatures	
– jamaicensis	Ruddy Duck	Erismature à tête noire	

GERMAN NAME	ENGLISH NAME	FRENCH NAME	RUSSIAN NAME
– leucocephala	White-headed Duck	Canard à tête blanche	Савка
– vittata	Argentine Ruddy Duck	Erismature d'Argentine	
Oxyurini	Stiff-tailed Ducks	Erismatures	
Pachyptila turtur	Fairy Prion		Буревестник-горлица
Pagodroma nivea	Snow Petrel		Снежный буревестник
Palmgeier	Palm-nut Vulture	Vautour palmiste	Грифовый орлан
Pampashuhn	Rufous Tinamou	Tinamou roussâtre	Бразильянский степной скрытохвост
Pampasstrauße	Rheas	Nandous, Rhéidés	Американские страусы, нанду
Pandion haliaëtus	Osprey	Balbuzard fluvatile	Скопа
Pandioninae	Ospreys		Скопы
Paradieskasarka	Paradise Shelduck	Casarca de Paradis	Райская пеганка
Pauxi pauxi	Helmeted Curassow	Pauxi pierre	Шлемоносный гокко
Pedioecetes phasianellus	Northern Sharp-tailed Grouse	Gelinotte à queue fine	Хвостатый тетерев
Pelagodroma marina	Frigate Petrel	Pétrel frégate	Морская качурка
Pelecanidae	Pelicans	Pélécanidés	Пеликаны
Pelecaniformes	Totipalmate Swimmers	Pélécaniformes	Веслоногие
Pelecanoides	Diving Petrels	Pétrels plongeurs	
Pelecanoididae	Diving Petrels	Pélécanoïdidés	Нырцовые буревестники
Pelecanus	Pelicans	Pélicans	Пеликаны
– conspicillatus	Australian Pelican		Австралийский пеликан
– crispus	Dalmatian Pelican	Pélican frisé	Кудрявый пеликан
– erythrorhynchos	American White Pelican		Пеликан-носорог
– occidentalis	Brown Pelican		Бурый пеликан
– onocrotalus	Eastern White Pelican	– blanc	Розовый пеликан
– philippensis	Grey Pelican		Серый пеликан
– rufescens	Pink-backed Pelican	– à dos rosé	Красноспинный пеликан
Pelecanoides urinatrix	Common Diving Petrel		Новозеландский нырцовый буревестник
Pelikane	Pelicans	Pélécanidés, Pélicans	Пеликаны
Penelopina nigra	Guatemalan Black Chachalaca		Черный шуан
Peposakaente	Rosy-billed Pochard	Canard à bec rosé	
Perdicinae	Old World Quails	Perdricinés	
Perdix perdix	Grey Partridge	Perdrix grise	Серая куропатка
Perlenweih	Pearl Kite		Карликовый лунь
Perl-Steißhuhn	Crested Tinamou	Tinamou huppé	
Pernis apivorus	Honey Buzzard	Bondrée apivore	Осоед
– ptilorhynchus	Crested Honey Buzzard		Хохлатый осоед
Pfeifente	European Wigeon	Canard siffleur	Свиязь
Pfeifgänse	Whistling Ducks	Canards siffleurs	Древесные утки
Pfeifschwan	– Swan	Cygne américain	
Phaëthon	Tropic Birds	Phaëthons	Фаэтоны
– aethereus	Red-billed Tropic Bird	Paille-en-queue à bec rouge	Краснохвостый фаэтон
– lepturus	Yellow-billed Tropic Bird	– à bec jaune	Белохвостый фаэтон
– rubricauda	Red-tailed Tropic Bird		Красноклювый фаэтон
Phaëthontidae	Tropic Birds	Phaëthontidés	Фаэтоны
Phalacrocoracidae	Cormorants	Phalacrocoracidés	Бакланы
Phalacrocorax	Cormorants	Phalacrocoracinés	Бакланы
– africanus	Long-tailed Cormorant	Cormoran africain	
– aristotelis	Green Cormorant	– huppé	Длинноносый баклан
– capillatus	Japanese Cormorant		Японский баклан
– carbo	Black Cormorant	Grand Cormoran	Обыкновенный баклан
– melanoleucus	Little Pied Cormorant		Австралийский малый баклан
– pelagicus	Pelagic Cormorant		Берингов баклан
– penicillatus	Brandt's Cormorant		Кисточковый баклан
– perspicillatus	Pallas Cormorant		Очковый баклан
– punctatus	Spotted Cormorant		Пятнистый баклан
– pygmaeus	Pygmy Cormorant	Cormoran pygmée	Малый баклан
– urile	Red-faced Cormorant		Краснолицый баклан
Phalcoboenus albogularis	White-throated Caracara		Белогорлый каракара
– australis	Striated Caracara		Южный каракара
– megalopterus	Mountain Caracara		Андский каракара
Phasianidae	Pheasants, Quails and Peacocks	Phasianidés	
Philippinen-Ente	Philippine Duck	Canard des Philippines	Филиппинская утка
Philippinen-Schlangenhabicht	– Serpent Eagle		Филиппинский змеиный орел
Phoebetria fusca	Sooty Albatross		Северный дымчатый альбатрос
– palpebrata	Light-mantled Sooty Albatross	Albatros fuligineux	Антарктический дымчатый альбатрос
Phoeniconaias	Lesser Flamingoes	Petits Flamants	
– minor	– Flamingo	Petit Flamant	Африканский карликовый фламинго
Phoenicoparrus andinus	Andean Flamingo		Андский фламинго
– jamesi	James' Flamingo		Короткоклювый фламинго

GERMAN NAME	ENGLISH NAME	FRENCH NAME	RUSSIAN NAME
Phoenicopteri	Flamingoes	Flamants	Фламинго
Phoenicopteridae	Flamingoes	Phoenicoptéridés	Фламинго
Phoenicopterus	Flamingoes	Flamants	Фламинго
– *chilensis*	Chilean Flamingo		Чилийский фламинго
– *ruber*	Greater Flamingo	Flamant rose	Обыкновенный фламинго
– – *roseus*	European Flamingo	Flamant rose	Розово-красный фламинго
– – *ruber*	American Flamingo	– rouge	Красный фламинго
Pinguine	Penguins	Sphénisciformes, Sphéniscidés	
Pinguin-Sturmtaucher	Common Diving Petrel		Новозеландский нырцовый буревестник
Pinselkormoran	Brandt's Cormorant		Кисточковый баклан
Pipile cumanensis	White-headed Piping Guan	Pénélope siffleuse	
Pithecophaga jefferyi	Monkey-eating Eagle		Обезьяноед
Platalea	Spoonbills	Spatules	
– *alba*	African Spoonbill	Spatule d'Afrique	Африканская белая колпица
– *leucorodia*	White Spoonbill	– blanche	Колпица
– *minor*	Black-faced Spoonbill	Petite Spatule	Малая колпица
Plataleinae	Spoonbills	Plataléinés	Колпицы
Plectropterus	Spur-winged Geese	Oies armées	Шпорцевые гуси
– *gambiensis*	Spur-wing	Oie de Gambie	Обыкновенный шпорцевый гусь
Plegadis chihi	White-faced Ibis	Ibis à face blanche	
– *falcinellus*	Glossy Ibis	– luisant	Каравайка
Plüschkopfente	Spectacled Eider	Eider de Fischer	Очковая гага
Podiceps	Grebes	Grèbes	Поганки
– *auritus*	Slavonian Grebe	Grèbe esclavon	Рогатая поганка
– *cristatus*	Great Crested Grebe	– huppé	Большая поганка
– *griseigena*	Red-necked Grebe	– jougris	Серощёкая поганка
– *nigricollis*	Black-necked Grebe	– à cou noir	Черношейная поганка
– *ruficollis*	Little Grebe	– castagneux	Малая поганка
Podicipedidae	Grebes	Podicipitidés	Поганки
Podicipediformes	Grebes	Podicipédiformes	Поганковые
Polemaëtus bellicosus	Martial Eagle	Aigle Martial	Орел-боец
Polihieracinae	Pigmy Falcons		Карликовые соколы
Polyborus lutosus	Guadalupe Caracara		Гваделупский каракара
– *plancus*	Crested Caracara		Каранхо
Polysticta stelleri	Steller's Eider	Eider de Steller	Сибирская гага
Prachteiderente	King Eider	– à tête grise	Гага-гребенутка
Prachtfregattvogel	Magnificent Man-o'-War Bird	Frégate superbe	Малый фрегат
Prachtschopfreiher	Javanese Pond Heron	Crabier malais	
Prachttaucher	Black-throated Diver	Plongeon lumme	Чернозобая гагара
Präriebussard	Swainson's Hawk	Buse de Swainson	
Präriefalk	Prairie Falcon	Faucon des prairies	Мексиканский сокол
Präriehuhn	Heath Hen	Poule des prairies	Большой луговой тетерев
Procellariidae	Procellariids	Procellariidés	Буревестники
Procellariiformes	Tube-Nosed Swimmers	Procellariiformes	Трубконосые
Pseudibis papillosa	Black Ibis	Ibis noir oriental	
Pseudogyps africanus	African White-backed Vulture	Griffon africain à dos blanc	Белоспинный сип
– *bengalensis*	Indian White-backed Vulture	– indien à dos blanc	Индийский гриф
Pterocnemia pennata	Darwin's Rhea	Nandou de Darwin	Нанду Дарвина
Pterodroma hasitata	Black-capped Petrel	Pétrel diablotin	Черношапочный тайфунник
– *leucoptera*	White-throated Petrel	– à ailes blanches	Белолобый тайфунник
– *macroptera*	Grey-winged Petrel		Длиннокрылый тайфунник
– *phaeopygia*	Hawaiian Petrel		Гавайский тайфунник
Pterodrominae	Gadfly Petrels		Тайфунники
Puffinus carneipes	Pale-footed Shearwater		Бледноногий буревестник
– *diomedea*	Mediterranean Shearwater	Puffin cendré	Большой белобрюхий буревестник
– *gravis*	Great Shearwater	– majeur	Большой пестробрюхий буревестник
– *griseus*	Sooty Shearwater	– fuligineux	Серый буревестник
– *puffinus*	Manx Shearwater	– des Anglais	Обыкновенный буревестник
– *tenuirostris*	Short-tailed Shearwater		Тонкоклювый буревестник
Purpurreiher	Purple Heron	Héron pourpre	Рыжая цапля
Pygopodes	Pygopodes	Pygopodes	Поганки и гагары
Pygoscelis adeliae	Adelie Penguin		Синий пингвин
– *papua*	Gentoo Penguin	Manchot Gentoo	Ослиный пингвин
Rabengeier	Black Vulture	Vautour noir	Американский черный гриф
Radjahgans	Radjah Shelduck		Пеганка-раджа
Rallenreiher	Squacco Heron	Héron crabier	Желтая цапля
Raubadler	Tawny Eagle	Aigle impérial	Африканский степной орел

GERMAN NAME	ENGLISH NAME	FRENCH NAME	RUSSIAN NAME
Rauhfußbussard	Northern Rough-legged Buzzard	Buse pattue	Мохноногий канюк
Rauhfußhühner	Grouse	Tétraoninés	
Rebhuhn	Grey Partridge	Perdrix grise	Серая куропатка
Regenstorch	Abdim's Stork	Cigogne d'Abdim	Абдимский аист
Regenwachtel	Rain Quail	Caille du Coromandel	Коромандельская перепелка
Reiher	Herons	Ardéidés	Цапли
Reiherente	Tufted Duck	Canard morillon	Хохлатая чернеть
Rhea americana	American Rhea	Nandou américain	Обыкновенный нанду
Rheae	Rheas	Nandous	Американские страусы
Rheidae	Rheas	Rhéidés	Нанду
Rhynchotus rufescens	Rufous Tinamou	Tinamou roussâtre	Бразильянский степной скрытохвост
Riesenibis	Giant Ibis	Ibis géant	
Riesenseeadler	Steller's Sea Eagle		Белоплечий орлан
Riesentafelente	Canvas-back	Milouin aux yeux rouges	
Riffreiher	Eastern Reef Heron	Aigrette des récifs	
Ringelgans	Brent Goose	Bernache cravant	Черная казарка
Rohrdommel	Bittern	Butor étoilé	Большая выпь
Röhrennasen	Tube-Nosed Swimmers	Procellariiformes	Трубконосые
Rohrweihe	Marsh Harrier	Busard harpaye	Камышевый лунь
Rollulus roulroul	Crowned Wood Partridge	Roulroul	Страусовый перепел
Rosaflamingo	European Flamingo	Flamant rose	Розово-красный фламинго
Rosapelikan	Eastern White Pelican	Pélican blanc	Розовый пеликан
Rosenroter Flamingo	European Flamingo	Flamant rose	Розово-красный фламинго
Rostgans	Ruddy Shelduck		Огарь
Rotbrusttao	Variegated Tinamou	Tinamou varié	
Rötelfalk	Lesser Kestrel	Faucon crécerellette	Степная пустельга
Rötelpelikan	Pink-backed Pelican	Pélican à dos rosé	Красноспинный пеликан
Roter Flamingo	American Flamingo	Flamant rouge	Красный фламинго
– Sichler	Scarlet Ibis	Ibis rouge	Красный ибис
Rotflügelguan	Chestnut-winged Chachalaca	Ortalide babillarde	Краснокрылый шуан
Rotfußfalk	Red-footed Falcon	Faucon kobez	Кобчик
Rotfußtölpel	– Booby	Fou aux pieds rouges	Красноногая олуша
Rothalsgans	Red-breasted Goose	Bernache à cou roux	Краснозобая казарка
Rothalstaucher	Red-necked Grebe	Grèbe jougris	Серощёкая поганка
Rothuhn	Red-legged Partridge	Perdrix rouge	Красная каменная куропатка
Rotkehlfrankolin	Angola Red-necked Partridge	Pterniste à cou nu	
Rotkehl-Karakara	Red-throated Caracara		Красношейный каракара
Rotkopfente	Red-head	Milouin à tête rousse	Краснолобый нырок
Rotkopfgans	Ruddy-headed Goose	Bernache à tête rousse	
Rotkopfmerlin	Red-headed Falcon	Faucon shikra	
Rotmilan	Red Kite	Milan royal	Красный коршун
Rot-Satyrhuhn	Satyr Tragopan	Tragopan satyre	
Rotschnabelente	Red-billed Teal	Canard à bec rouge	
Rotschnabel-Tropikvogel	– Tropic Bird	Paille-en-queue à bec rouge	Красноклювый фаэтон
Rotschulterbussard	Red-shouldered Hawk	Buse à épaulettes rousses	
Rotschwanzbussard	Red-tailed Hawk	– à queue rousse	Краснохвостый канюк
Rotschwanzguan	Jardine's Chachalaca		Краснохвостый шуан
Rotschwanz-Tropikvogel	Red-tailed Tropic Bird		Краснохвостый фаэтон
Ruderenten	Stiff-tailed Ducks	Erismatures	
– i. e. S.	Stiff-tails	Erismatures	
Ruderfüßer	Totipalmate Swimmers	Pélécaniformes	Веслоногие
Rundschwanzsperber	Cooper's Hawk	Épervier de Cooper	Ястреб Купера
Rußsturmtaucher	Sooty Shearwater	Puffin fuligineux	Серый буревестник
Saatgans	Bean Goose	Oie des moissons	Гуменник
Säger	Mergansers	Harles	Крохали
Sagittariidae	Secretary Birds	Sagittaridés	Африканские секретари
Sagittarius serpentarius	– Bird	Secrétaire	Секретарь
Samtente	Velvet Scoter	Macreuse brune	Черный турпан
Samthokko	Lesser Razor-billed Curassow	Petit Hocco à bec de rasoir	
Sandwichgans	Hawaiian Goose	Bernache des îles Sandwich	Гавайская казарка
Sarcidiornis melanotus	Comb Duck	Sarcidiorne à crête	Шишконосый гусь
Sarcogyps calvus	Pondicherry Vulture	Vautour de Pondichéry	
Sarcoramphus papa	King Vulture	– royal	Королевский гриф
Schakutinga	White-headed Piping Guan	Pénélope siffleuse	
Scharlachibis	Scarlet Ibis	Ibis rouge	Красный ибис
Schattenvogel	Hamerkop	Ombrette	Молотоглав
Scheckente	Steller's Eider	Eider de Steller	Сибирская гага
Schelladler	Greater Spotted Eagle	Aigle criard	Большой подорлик
Schellente	Common Golden-eye	Canard garrot	Обыкновенный гоголь
Schieferfalk	Sooty Falcon	Faucon concolore	
Schlangenadler	Bateleur, Harrier Eagles and Serpent Eagles, Eurasian Short-toed Eagle	Circaetinés, Circaète Jean-le-Blanc	Змееяд

GERMAN NAME	ENGLISH NAME	FRENCH NAME	RUSSIAN NAME
Schlangenhalsvögel	Darters	Anhingidés, Anhingas	Змеешейки, Анхинги
Schmalschnabel-Löffler	African Spoonbill	Spatule d'Afrique	Африканская белая колпица
Schmuckreiher	Snowy Egret	Aigrette neigeuse	
Schmutzgeier	Egyptian Vulture	Peronoptère d'Égypte	Стервятник
Schnatterente	Gadwall	Canard chipeau	Серая утка
Schneegans	Snow Goose	Oie des neiges	Белый гусь
Schneegeier	Himalayan Griffon		Снежный гималайский сип
Schneesturmvogel	Snow Petrel		Снежный буревестник
Schopfadler	Long-crested Eagle	Aigle huppé	Гребневый орел
Schopfhokko	Nocturnal Curassow		Хохлатый гокко
Schopfibis	Hermit Ibis	Ibis chauve	Горный ибис
Schopfkasarka	Korean Crested Shelduck		Хохлатая пеганка
Schopfmilan	Square-tailed Kite		Хохлатый коршун
Schreiadler	Lesser Spotted Eagle	Aigle pomarin	Малый подорлик
Schreiseeadler	African Sea Eagle	Aigle pêcheur	Крикливый орлан
Schuhschnäbel	Whale-headed Storks, Shoebills	Balaenicipitidés, Baleniceps	Китоглавы
Schuhschnabel	Shoebill	Bec-en-Sabot	Китоглав
Schuppensäger	Chinese Merganser	Harle chinois	Чемуйчатый крохаль
Schuppenwachtel	Mexican Scaled Quail	Quaglia azzurra	
Schwalbengleitaar	African Swallow-tailed Kite	Naucler d'Afrique	
Schwalbenweih	Swallow-tailed Kite	Milan à queue fourchue	Вилохвостый лунь
Schwäne	Swans	Cygnes	Лебеди
Schwanengans	Swan Goose	Oie cygnoïde	Сухонос
Schwarzachseladler	Cassin's Hawk Eagle	Aigle Autour de Cassin	
Schwarzachsel-Gleitaar	Letter-winged Kite	Élanion écrit	
Schwarzbauch-Sturmschwalbe	Black-bellied Storm Petrel	Pétrel des tropiques	Чернобрюхая качурка
Schwarzbrauenalbatros	Black-browed Albatross	Albatros à sourcil noir	Чернобровый альбатрос
Schwarzente	African Black Duck	Canard noir africain	
Schwarzflügel-Gleitaar	Black-winged Kite	Faucon blanc	Чернокрылый коршун
Schwarzfußalbatros	Black-footed Albatross	Albatros à pieds noirs	Черноногий альбатрос
Schwarzhalsreiher	Black-headed Heron	Héron à tête noire	
Schwarzhalsschwan	Black-necked Swan	Cygne à cou noir	Черношейный лебедь
Schwarzhalstaucher	– Grebe	Grèbe à cou noir	Черношейная поганка
Schwarzkappensturmtaucher	Black-capped Petrel	Pétrel diablotin	Черношапочный тайфунник
Schwarzkopfente	Black-headed Duck	Canard à tête noire de l'Argentine	Кукутковая утка
Schwarzkopfibis	Oriental Ibis	Ibis à tête noire	
Schwarzkopf-Moorente	Baer's Pochard		Нырок Бэра
Schwarzkopfruderente	Ruddy Duck	Erismature à tête noire	
Schwarzmilan	Black Kite	Milan noir	Черный коршун
Schwarzschnabelstorch	Oriental White Stork		Черноклювый аист
Schwarzschnabel-Sturmtaucher	Manx Shearwater	Puffin des Anglais	Обыкновенный буревестник
Schwarzschopfreiher	Tiger Bittern	Butor malais	
Schwarzstorch	Black Stork	Cigogne noire	Черный аист
Schwarzweißweihe	Pied Harrier	Busard pie	Чернопегий лунь
Schweif-Waldhuhn	Northern Sharp-tailed Grouse	Gelinotte à queue fine	Хвостатый тетерев
Scopidae	Hamerkops	Scopidés	Теневые птицы
Scopus	Hamerkops	Ombrettes	Молотоголовые цапли
– umbretta	Hamerkop	Ombrette	Молотоглав
Seeadler	White-tailed Sea Eagle	Pygargue à queue blanche	Орлан-белохвост
Seetaucher	Loons, Divers	Gaviiformes, Gaviidés, Plongeons	Гагаровые, гагары
Seidenreiher	Little Egret	Aigrette garzette	Малая белая цапля
Sekretär	Secretary Bird	Secrétaire	Секретарь
Sekretäre	– Birds	Sagittaridés	Африканские секретари
Sichelente	Falcated Teal	Sarcelle à faucilles	Касатка
Sichler	Ibises	Threskiornithinés	Ибисы
Silberadler	Wahlberg's Eagle	Aigle de Wahlberg	
Silberreiher	Great White Egret	Grande Aigrette	Большая белая цапля
Singhabicht	Pale Chanting Goshawk		Певчий ястреб
Singschwan	Whooper Swan	Cygne Sauvage	Лебедь-кликун
Somateria	Greater Eider Ducks	Eiders	Гаги
– fischeri	Spectecled Eider	Eider de Fischer	Очковая гага
– mollissima	Common Eider	– à duvet	Обыкновенная гага
– spectabilis	King Eider	– à tête grise	Гага-гребенутка
Spaltfußgans	Australian Magpie Goose	Oie semi-palmée	Расщепнолапый гусь
Spaltfußgänse	Magpie Geese	Anseranatinés, Oies semi-palmées	
Spatelente	Barrow's Golden-eye	Garrot islandais	Исландский гоголь
Sperber	Eurasian Sparrowhawk	Épervier d'Europe	Ястреб-перепелятник
Sperbergeier	Rüppell's Griffon	Vautour de Rüppell	
Spheniscidae	Penguins	Sphéniscidés	
Sphenisciformes	Penguins	Sphénisciformes	

GERMAN NAME	ENGLISH NAME	FRENCH NAME	RUSSIAN NAME
Spheniscus demersus	Black-footed Penguin	Manchot du Cap	Очковый пингвин
– humboldti	Peruvian Penguin		Пингвин Гумбольдта
– magellanicus	Magellanic Penguin	– de Magelhaen	
– mendiculus	Galapagos Penguin		Галапагосский пингвин
Spiegelgänse	South American Sheldgeese	Bernaches sud-américaines	
Spießente	Blue-billed Pintail	Canard pilet	Шилохвость
Spitzschwanzente	Yellow-billed Pintail	Pilet du Chili	
Spilornis cheela	Crested Serpent Eagle		Орел-хеела
– holospilus	Philippine Serpent Eagle		Филиппинский змеиный орел
Spizaëtus nipalensis	Mountain Hawk Eagle		Хохлатый орел
Sporengans	Spur-wing	Oie de Gambie	
Sporengänse	Spur-winged Geese	Oies armées	Шпорцевые гуси
Steinadler	Golden Eagle	Aigle doré	Беркут
Steinhuhn	Rock Partridge	Perdrix bartavelle	Кеклик
Steißfüße	Pygopodes	Pygopodes	Поганки и гагары
Steißhühner	Tinamous	Tinamous	Скрытохвости, скрытохвостые куры
Stellers Albatros	Stellers's Albatross	Albatros de Steller	Белоспинный альбатрос
Stephanoaëtus coronatus	Crowned Eagle	Blanchard	Венценосный орел
Steppenadler	Steppe Eagle		Восточный степной орел
Steppenweihe	Pallid Harrier	Busard pâle	Степной лунь
Sterntaucher	Red-throated Diver	Plongeon catmarin	Краснозобая гагара
Stockente	Northern Mallard	Canard colvert	Кряква
Störche	Storks	Ciconiidés	Аисты
Strauß	Ostrich	Autruche	Африканский страус
Strauße	Ostriches	Autruches, Struthionidés	Африканские страусы, страусы
Straußwachtel	Crowned Wood Partridge	Roulroul	Страусовый перепел
Streifengans	Bar-headed Goose	Oie à tête barrée	Индийский горный гусь
Streifenkiwi	Common Kiwi	Kiwi austral	Обыкновенный киви
Struthio	Ostriches	Autruches	Страусы
– camelus	Ostrich	Autruche	Африканский страус
Struthiones	Ostriches	Autruches	Африканские страусы
Struthionidae	Ostriches	Struthionidés	Страусы
Struthioniformes	Ratites	Ratites	Бегающие
Sturmschwalbe	Storm Petrel	Pétrel tempête	Малая прямохвостая качурка
Sturmschwalben	– Petrels	Hydrobatidés	Качурки
Sturmvögel	Procellariids	Procellariidés	Буревестники
Südamerikanische Löffelente	Red Shoveler	Souchet roux	Южноамериканская широконоска
Südlicher Karakara	Striated Caracara		Южный каракара
– Riesensturmvogel	Southern Giant Fulmar		Южный гигантский буревестник
– Rußalbatros	Light-mantled Sooty Albatross	Albatros fuligineux	Антарктический дымчатый альбатрос
Sula	Boobies	Fous	
– dactylatra	Masked Booby	Fou masqué	Голуболицая олуша
– leucogaster	Brown Booby	– à ventre blanc	
– sula	Red-footed Booby	– aux pieds rouges	Красноногая олуша
Sulidae	Boobies and Gannets	Sulidés	Олуши
Sumatrareiher	Dusky Grey Heron	Héron de Sumatra	Тифонова цапля
Sunda Marabu	Lesser Marabou	Petit Marabout	Яванский марабу
Tachyeres	Steamer Ducks	Canards-vapeurs	
– brachypterus	Falkland Flightless Steamer Duck	Canard à ailes courtes	
Tadorna	Shelducks	Tadornes et Casarcas	Пеганки
– cana	South African Shelduck	Casarca du Cap	Сероголовая пеганка
– cristata	Korean Crested Shelduck		Хохлатая пеганка
– ferruginea	Ruddy Shelduck		Огарь
– radjah	Radjah Shelduck		Пеганка-раджа
– tadorna	Common Shelduck	Tadorne de Belon	Пеганка
– tadornoides	Australian Shelduck	Casarca d'Australie	Австралийская пеганка
– variegata	Paradise Shelduck	– de Paradis	Райская пеганка
Tafelente	European Pochard	Canard milouin	Красноголовый нырок
Tanggans	Kelp Goose	Bernache antarctique	
Tannen-Waldhuhn	Hudsonian Spruce Grouse	Tétras des savanes	Канадский тетерев
Tataupa	Tataupa Tinamou	Tinamou tataupa	
Taucher	Grebes	Grèbes	Поганки
Tauchsturmvögel	Diving Petrels	Pélécanoïdidés, Pétrels plongeurs	Нырцовые буревестники
Temminck-Satyrhuhn	Temminck's Tragopan	Tragopan de Temminck	
Terathopius ecaudatus	Bateleur	Bateleur	Орел-скоморох
Tetraogallus caspicus	Caspian Snowcock	Perdrix des neiges caspienne	Каспийский улар
Tetraoninae	Grouse	Tétraoninés	
Tetrao urogallus	Capercaillie	Grand Tétras	Глухарь

GERMAN NAME	ENGLISH NAME	FRENCH NAME	RUSSIAN NAME
Tetrastes bonasia	Hazel Hen	Gelinotte des bois	Рябчик
Thalassoica antarctica	Antarctic Petrel		Антарктический буревестник
Thalassornis leuconotus	White-backed Duck	Canard à dos blanc	
Thaumatibis gigantea	Giant Ibis	Ibis géant	
Thermometerhuhn	Mallee Fowl	Leipoa ocellé	Лейпоа
Threskiornis	Ibises	Ibis	Ибисы
– aethiopica	Sacred Ibis	– sacré	Священный ибис
– melanocephala	Oriental Ibis	– à tête noire	
Threskiornithidae	Ibises and Spoonbills	Threskiornithidés	Ибисы
Threskiornithinae	Ibises	Threskiornithinés	Ибисы
Tigriornis leucolophus	White-crested Bittern	Butor à crête blanche	
Tinamidae	Tinamous	Tinamous	Скрытохвостые куры
Tinamiformes	Tinamous	Tinamous	Скрытохвосты
Tinaminae	Tinamous	Tinamous	Лесные скрытохвостые куры
Tinamus	Tinamous	Tinamous	Скрытохвостые куры
Tinamus major	Great Tinamou	Grand Tinamou	Большой тинаму
Tölpel	Boobies and Gannets	Sulidés	Олуши
– i. e. S.	Gannets	Fous	
Torgos tracheliotus	Lappet-faced Vulture	Vautour oricou	Ушастый гриф
Tragopan blythii	Blyth's Tragopan	Tragopan de Blyth	
– caboti	Cabot's Tragopan	– de Cabot	
– melanocephalus	Hasting's Tragopan	– de Hasting	
– satyrus	Satyr Tragopan	– satyre	
– temminckii	Temminck's Tragopan	– de Temminck	
Trauerente	Black Scoter	Macreuse noire	Синьга
Trauerenten	Scoters	Macreuses	Турпаны
Trauerhabicht	Black Goshawk	Épervier pie	
Trauerschwan	– Swan	Cygne noir	Черный лебедь
Trigonoceps occipitalis	White-headed Vulture	Vautour huppé	
Trompeterschwan	Trumpeter Swan	Cygne trompette	Лебедь-трубач
Tropikvögel	Tropic Birds	Phaëthontidés, Phaëthons	Фаэтоны
Tropische Tölpel	Boobies	Fous	
Truthahngeier	Turkey Vulture	Vautour à tête rouge	Индюшачий гриф
Tuberkelhokko	Mexican Curassow	Grand Hocco	
Tüpfelkormoran	Spotted Cormorant		Пятнистый баклан
Tüpfelpfeifgans	– Whistling Duck		Крапчатая древесная утка
Turmfalk	Old World Kestrel	Faucon crécerelle	Обыкновенная пустельга
Tympanuchus	Prairie Hens	Cupidons	
– cupido	Heath Hen	Poule des prairies	Большой луговой тетерев
– pallidicinctus	Lesser Prairie Hen		Малый луговой тетерев
Urotriorchis macrourus	Long-tailed Hawk	Autour à longue queue	Длиннохвостый ястреб
Virginiawachtel	Eastern Bob White	Colin de Virginie	Виргинская куропатка
Vögel	Birds	Oiseaux	Птицы
Vultur gryphus	Andean Condor	Condor des Andes	Андский кондор
Wachtel	Common Quail	Caille des blés	Обыкновенный перепел
Wachteln i. e. S.	Quails	Cailles	
Waldrapp	Hermit Ibis	Ibis chauve	Горный ибис
Wald-Steißhühner	Tinamous	Tinamous	Лесные скрытохвостые куры
Wanderalbatros	Wandering Albatross	Albatros hurleur	Странствующий альбатрос
Wanderfalk	Peregrine Falcon	Faucon pèlerin	Сокол-сапсан
Warzenibis	Black Ibis	Ibis noir oriental	
Weichnasen-Sturmvogel	Bulwer's Petrel	Pétrel de Bulwer	Тайфунник Бульвера
Weihen	Harriers and Crane Hawks	Circinés	Луни
Weißbauch-Fregattvogel	Christmas Frigate Bird		Белобрюхий фрегат
Weißbrauenhabicht	Variable Goshawk		Белый ястреб
Weißer Sichler	White Ibis	Ibis blanc	
Weißkehlente	Grey Teal	Sarcelle grise des Indes	
Weißkehl-Karakara	White-throated Caracara		Белогорлый каракара
Weißkopfruderente	White-headed Duck	Canard à tête blanche	Савка
Weißkopf-Seeadler	Bald Eagle	Aigle à tête blanche	Белоголовый орлан
Weißnackenreiher	White-crested Bittern	Butor à crête blanche	
Weißrückenente	White-backed Duck	Canard à dos blanc	
Weißschwanz-Tropikvogel	Yellow-billed Tropic Bird	Paille-en-queue à bec jaune	Белохвостый фаэтон
Weißstorch	White Stork	Cigogne blanche	Белый аист
Weißwangengans	Barnacle Goose	Bernache nonnette	Белощёкая казарка
Weißwangenwehrvogel	Crested Screamer	Chauna chavaria	Белощекая паламедея
Wehrvögel	Screamers	Anhimidés	Паламедеи
Wellenläufer	Leach's Petrel	Pétrel cul-blanc	Северная качурка
Wespenbussard	Honey Buzzard	Bondrée apivore	Осоед
West-Satyrhuhn	Hasting's Tragopan	Tragopan de Hasting	

GERMAN NAME	ENGLISH NAME	FRENCH NAME	RUSSIAN NAME
Wiesenweihe	Montagu's Harrier	Busard de Montagu	Луговой лунь
Witwenpfeifgans	White-faced Whistling Duck	Dendrocygne veuf	Монашенка
Wollhalsstörche	Bishop Storks	Cigognes épiscopales	Пуховые аисты
Wollkopfgeier	White-headed Vulture	Vautour huppé	
Würgfalk	Saker Falcon	Faucon sacre	Обыкновенный балобан
Xenorhynchus asiaticus	Black-necked Stork	Jabirou asiatique	Индийский исполинский аист
Zimtdommel	Cinnamon Bittern	Blongios cannelle	
Zimtente	– Teal	Sarcelle cannelle	
Zwergadler	Booted Eagle	Aigle botté	Орел-карлик
Zwergdommel	Little Bittern	Blongios nain	Малая выпь
Zwergfalken	Pigmy Falcons		Карликовые соколы
Zwergflamingo	Lesser Flamingo	Petit Flamant	Африканский карликовый фламинго
Zwergflamingos	– Flamingoes	Petits Flamants	
Zwerggans	– White-fronted Goose	Oie naine	Пискулька
Zwerggänsegeier	African White-backed Vulture	Griffon africain à dos Blanc	Белоспинный сип
Zwergglanzenten	Pygmy Geese		Карликовые утки
Zwergpinguin	Little Penguin		Карликовый пингвин
Zwergrohrdommel	– Bittern	Blongios nain	Малая выпь
Zwergsäger	Smew	Harle piette	Луток
Zwergscharbe	Pygmy Cormorant	Cormoran pygmée	Малый баклан
Zwergschneegans	Ross' Goose	Oie de Ross	Карликовый белый гусь
Zwergschwan	Bewick's Swan	Cygne de Bewick	Западный тундровый лебедь
Zwergtaucher	Little Grebe	Grèbe castagneux	Малая поганка
Zwergwachtel	Chinese Painted Quail	Caille peinte de Chine	

III. French—German—English—Russian

Dans la plupart des cas, les noms des sous-espèces sont formés en ajoutant au nom de l'espèce un adjectif ou une désignation géographique. Dans cette partie du dictionnaire zoologique, les noms français des sous-espèces formés de cette manière ne seront en général pas indiqués.

FRENCH NAME	GERMAN NAME	ENGLISH NAME	RUSSIAN NAME
Accipitridés	Habichtartige	Hawks, Old World Vultures and Harriers	
Aegypiinés	Altweltgeier	Old World Vultures	Настоящие грифы
Aigle à tête blanche	Weißkopf-Seeadler	Bald Eagle	Белоголовый орлан
– botté	Zwergadler	Booted Eagle	Орел-карлик
– criard	Schelladler	Greater Spotted Eagle	Большой подорлик
– de Bonelli	Habichtsadler	Bonelli's Eagle	Ястребиный орел
– de Verreaux	Kaffernadler	Verreaux's Eagle	Капский орел
– de Wahlberg	Silberadler	Wahlberg's Eagle	
– doré	Steinadler	Golden Eagle	Беркут
– huppé	Schopfadler	Long-crested Eagle	Гребневый орел
– impérial	Kaiseradler	Imperial Eagle	Орел-могильник
– Martial	Kampfadler	Martial Eagle	Орел-боец
– pêcheur	Schreiseeadler	African Sea Eagle	Крикливый орлан
– pomarin	Schreiadler	Lesser Spotted Eagle	Малый подорлик
Aigle Autour d'Ayres	Haubenzwergadler	Ayres' Eagle	
– – de Cassin	Schwarzachseladler	Cassin's Hawk Eagle	
Aigrette à bec jaune	Edelreiher	Yellow-billed Egret	Средняя белая цапля
– des récifs	Riffreiher	Eastern Reef Heron	
– garzette	Seidenreiher	Little Egret	Малая белая цапля
– neigeuse	Schmuckreiher	Snowy Egret	
Albatros à bec jaune	Gelbnasenalbatros	Yellow-nosed Albatross	Желтоклювый альбатрос
– à pieds noirs	Schwarzfußalbatros	Black-footed Albatross	Черноногий альбатрос
– à sourcil noir	Schwarzbrauenalbatros	Black-browed Albatross	Чернобровый альбатрос
– à tête grise	Graukopfalbatros	Grey-headed Albatross	Сероголовый альбатрос
– de Steller	Kurzschwanzalbatros	Steller's Albatross	Белоспинный альбатрос
– fuligineux	Südlicher Rußalbatros	Light-mantled Sooty Albatross	Антарктический дымчатый альбатрос
– hurleur	Wanderalbatros	Wandering Albatross	Странствующий альбатрос
Albatros royal	Königsalbatros	Royal Albatross	Королевский альбатрос
Anatidés	Entenvögel	Ducks and Geese	Утиные
Anatinés	Entenverwandte	Sheldgeese and Shelducks	
Anhimidés	Wehrvögel	Screamers	Паламедеи
Anhinga d'Amerique	Amerikanischer Schlangenhalsvogel	American Darter	Американская ахинга

FRENCH NAME	GERMAN NAME	ENGLISH NAME	RUSSIAN NAME
Anhingas	Schlangenhalsvögel	Darters	Анхинги
Anhingidés	Schlangenhalsvögel	Darters	Эмеешейки
Anseranatinés	Spaltfußgänse	Magpie Geese	
Ansériformes	Gänsevögel	Waterfowl and Screamers	Пластинчатоклювые
Ansérinés	Gänseverwandte	Whistling Ducks, Swans and Geese	
Aptérygidés	Kiwis	Kiwis	Киви
Ardéidés	Reiher	Herons	Цапли
Autour à longue queue	Langschwanzhabicht	Long-tailed Hawk	Длиннохвостый ястреб
– à ventre rouge	Australischer Habicht	Red Goshawk	Австралийский ястреб
– des palombes	Habicht	Goshawk	Ястреб-тетеревятник
– gabar	Gabar-Habicht	Gabar Goshawk	
– tachiro	Afrikanischer Habicht	African Sparrowhawk	
Autruche	Strauß	Ostrich	Африканский страус
Autruches	Strauße	Ostriches	Африканские страусы, Страусы
Balaenicipitidés	Schuhschnäbel	Shoebills	Китоглавы
Balbuzard fluvatile	Fischadler	Osprey	Скопа
Baleniceps	Schuhschnäbel	Shoebills	Китоглавы
Bateleur	Gaukler	Bateleur	Орел-скоморох
Bec-en-Sabot	Schuhschnabel	Shoebill	Китоглав
Bec-Ouvert	Afrika-Klaffschnabel	Open Bill	Африканский аист-разиня
– – asiatique	Indien-Klaffschnabel	Asian Open Bill	Индийский аист-разиня
Becs-Ouverts	Klaffschnäbel	Open Bills	Аисты-разини
Bernache à cou roux	Rothalsgans	Red-breasted Goose	Краснозобая казарка
– à crinière	Mähnengans	Maned Wood Duck	
– antarctique	Tanggans	Kelp Goose	
– à tête grise	Graukopfgans	Ashy-headed Goose	
– à tête rousse	Rotkopfgans	Ruddy-headed Goose	
– aux ailes bleues	Blauflügelgans	Abyssinian Blue-winged Goose	Голубокрылый гусь
– cravant	Ringelgans	Brent Goose	Черная казарка
– de Magellan	Magellangans	Magellan Goose	
– des Andes	Andengans	Andean Goose	
– des îles Sandwich	Hawaiigans	Hawaiian Goose	Гавайская казарка
– du Canada	Kanadagans	Canada Goose	Канадская казарка
– nonnette	Weißwangengans	Barnacle Goose	Белощёкая казарка
Bernaches	Meergänse	Brents	Казарки
– sud-américaines	Spiegelgänse	South American Sheldgeese	
Blanchard	Kronenadler	Crowned Eagle	Венценосный орел
Blongios cannelle	Zimtdommel	Cinnamon Bittern	
– chinois	Chinesendommel	Chinese Little Bittern	Китайский волгок
– de Schrenck	Mandschurendommel	Schrenck's Little Bittern	Амурский волгок
– nain	Zwergdommel	Little Bittern	Малая выпь
Bondrée apivore	Wespenbussard	Honey Buzzard	Осоед
Brante roussâtre	Kolbenente	Red-crested Duck	Красноносый нырок
Busard de Montagu	Wiesenweihe	Montagu's Harrier	Луговой лунь
– des sauterelles	Heuschreckenbussard	Grasshopper Buzzard	
– harpaye	Rohrweihe	Marsh Harrier	Камышевый лунь
– pâle	Steppenweihe	Pallid Harrier	Степной лунь
– pie	Schwarzweißweihe	Pied Harrier	Чернопегий лунь
– Saint-Martin	Kornweihe	Hen Harrier	Полевой лунь
Buse à épaulettes rousses	Rotschulterbussard	Red-shouldered Hawk	
– à poitrine noire	Haubenmilan	Black-breasted Buzzard Kite	
– à queue rousse	Rotschwanzbussard, Felsenbussard	Red-tailed Hawk, African Red-tailed Buzzard	Краснохвостый канюк
– de Swainson	Präriebussard	Swainson's Hawk	
– féroce	Adlerbussard	Long-legged Hawk	Канюк-курганник
– pattue	Rauhfußbussard	Northern Rough-legged Buzzard	Мохноногий канюк
– rouilleuse	Königsbussard	Ferruginous Rough-legged Hawk	
– unibande	Kehlstreifbussard	Lizard Buzzard	
– variable	Mäusebussard	Common Buzzard	Обыкновенный сарыч
Buses	Bussarde	Buzzards	Сарычи
Butor à crête blanche	Weißnackenreiher	White-crested Bittern	
– d'Amérique	Nordamerikanische Rohrdommel	American Bittern	Североамериканская выпь
– étoilé	Rohrdommel	Bittern	Большая выпь
– malais	Schwarzschopfreiher	Tiger Bittern	
Caille des blés	Wachtel	Common Quail	Обыкновенный перепел
– du Coromandel	Regenwachtel	Rain Quail	Коромандельская перепелка
– du Japon	Japanische Wachtel	Japanese Quail	Немой перепел
– harlequine	Harlekinwachtel	Harlequin Quail	
– peinte de Chine	Zwergwachtel	Chinese Painted Quail	
Cailles	Wachteln i. e. S.	Quails	
Canard à ailes courtes	Falkland-Dampfschiffente	Falkland Flightless Steamer Duck	

FRENCH NAME	GERMAN NAME	ENGLISH NAME	RUSSIAN NAME
— à bec jaune	Gelbschnabelente	Yellow-billed Duck	
— à bec rosé	Peposakaente	Rosy-billed Pochard	
— à bec rouge	Rotschnabelente	Red-billed Teal	
— à bec tacheté	Fleckschnabelente	Spot-billed Duck	Черная желтоносая кряква
— à dos blanc	Weißrückenente	White-backed Duck	
— à membrane	Lappenente	Australian Musk Duck	
— à sourcil blanc	Augenbrauenente	Grey Duck	
— à tête blanche	Weißkopfruderente	White-headed Duck	Савка
— à tête noire de l'Argentine	Kuckucksente	Black-headed Duck	Кукутковая утка
— Carolin	Brautente	Carolina Wood Duck	Американская брачная утка
— chipeau	Schnatterente	Gadwall	Серая утка
— colvert	Stockente	Northern Mallard	Кряква
— de Bahama	Bahama-Ente	Bahama Pintail	
— de Meller	Madagaskar-Ente	Meller's Duck	Мадагаскарская утка
— de Miquelon	Eisente	Long-tailed Duck	Морянка
— des Philippines	Philippinen-Ente	Philippine Duck	Филиппинская утка
— doméstique	Hausente	Domestic duck	Домашняя утка
— garrot	Schellente	Common Golden-eye	Обыкновенный гоголь
— mandarin	Mandarinente	Mandarin Duck	Мандаринка
— milouin	Tafelente	European Pochard	Красноголовый нырок
— milouinan	Bergente	Greater Scaup	Морская чернеть
— morillon	Reiherente	Tufted Duck	Хохлатая чернеть
— musqué	Moschusente	Moscovy Duck	Мускусная утка
— noir africain	Schwarzente	African Black Duck	
— noir de l'Amérique	Dunkelente	American Black Duck	
— nyroca	Moorente	Ferruginous White-eye	Белоглазый нырок
— pilet	Spießente	Blue-billed Pintail	Шилохвость
— siffleur	Pfeifente, Amerikanische Pfeifente	European Wigeon, American Wigeon	Свиязь, Американская свиязь
— souchet	Löffelente	Northern Shoveler	Широконоска
Canards siffleurs	Pfeifgänse	Whistling Ducks	Древесные утки
— vapeurs	Dampfschiffenten	Steamer Ducks	
Casarca d'Australie	Australische Kasarka	Australian Shelduck	Австралийская пеганка
— de Paradis	Paradieskasarka	Paradise Shelduck	Райская пеганка
— du Cap	Graukopfkasarka	South African Shelduck	Сероголовая пеганка
Casoar à casque	Helmkasuar	Australian Cassowary	Шлемоносный казуар
— de Bennett	Bennettkasuar	Bennett's Cassowary	Казуар Беннетта
— unicaronculé	Goldhalskasuar	One-wattled Cassowary	Лоскутный казуар
Casoars	Kasuare	Cassowaries	Казуары
Casuariidés	Kasuare	Cassowaries	Казуары
Céréopse de Nouvelle-Hollande	Hühnergans	Cape Barren Goose	Куриный гусь
Charognard	Kappengeier	Hooded Vulture	
Chauna chavaria	Weißwangenwehrvogel	Crested Screamer	Белощекая паламедея
Ciconiidés	Störche	Storks	Аисты
Cigogne américaine	Amerika-Nimmersatt	North American Wood Ibis	Американский ярибу
— blanche	Weißstorch	White Stork	Белый аист
— d'Abdim	Regenstorch	Abdim's Stork	Абдимский аист
— épiscopale	Afrikanisch-Indischer Wollhalsstorch	Bishop Stork	Африканско-индийский пуховый аист
— noire	Schwarzstorch	Black Stork	Черный аист
Cigognes	Eigentliche Störche	— Storks	
— épiscopales	Wollhalsstörche	Bishop Storks	Пуховые аисты
Circaète brun	Afrikanischer Schlangenadler	Brown Harrier Eagle	
— Jean-le-Blanc	Schlangenadler	Eurasian Short-toed Eagle	Змееяд
Circaetinés	Schlangenadler	Bateleur, Harrier Eagles and Serpent Eagles	
Circinés	Weihen	Harriers and Crane Hawks	Луни
Colin de Californie	Kalifornische Schopfwachtel	Valley Quail	Калифорнийский хохлатый перепел
— de Gambel	Helmwachtel	Gambel's Quail	Калифорнийский шлемоносный перепел
— de Masséna	Montezumawachtel	Montezuma's Quail	
— des montagnes	Berghaubenwachtel	Plumed Quail	
— de Virginie	Virginiawachtel	Eastern Bob White	Виргинская куропатка
Condor de Californie	Kalifornischer Kondor	California Condor	Калифорнийский кондор
— des Andes	Anden-Kondor	Andean Condor	Андский кондор
Cormoran africain	Gelbschnabel-Zwergscharbe	Long-tailed Cormorant	
— huppé	Krähenscharbe	Green Cormorant	Длинноносый баклан
— pygmée	Zwergscharbe	Pygmy Cormorant	Малый баклан
Crabier chinois	Bacchusreiher	Chinese Pond Heron	Белокрылая цапля
— malais	Prachtschopfreiher	Javanese Pond Heron	
Cracidés	Hokkos	Curassows, Guans and Chachalacas	Гокко
Crécerelle americaine	Buntfalk	American Sparrow Hawk	Американская пустельга

FRENCH NAME	GERMAN NAME	ENGLISH NAME	RUSSIAN NAME
Cupidons	Eigentliche Präriehühner	Prairie Hens	
Cygne à cou noir	Schwarzhalsschwan	Black-necked Swan	Черношеий лебедь
– américain	Pfeifschwan	Whistling Swan	
– coscoroba	Koskorobaschwan	Coscoroba Swan	Гигаитская утка коскороба
– de Bewick	Zwergschwan	Bewick's Swan	Западный тундровый лебедь
– muet	Höckerschwan	Mute Swan	Лепедь-шипун
– noir	Trauerschwan	Black Swan	Черный лебедь
– sauvage	Singschwan	Whooper, Whooper Swan	Лебедь-кликун
– trompette	Trompeterschwan	Trumpeter Swan	Лебедь-трубач
Cygnes	Schwäne	Swans	Лебеди
Dendrocygne siffleur	Indien-Pfeifgans	Indian Whistling Duck	Яванская древесная утка
veuf	Witwenpfeifgans	White-faced Whistling Duck	Монашенка
Dimorphe	Küstenreiher	Reef Heron	
Dinornithidés	Moas	Moas	
Diomédeidés	Albatrosse	Albatrosses and Mollymawks	Альбатросы
Dromicéiidés	Emus	Emus	Эму
Eider à duvet	Eiderente	Common Eider	Обыкновенная гага
– à tête grise	Prachteiderente	King Eider	Гага-гребенутка
– de Fischer	Plüschkopfente	Spectacled Eider	Очковая гага
– de Steller	Scheckente	Steller's Eider	Сибирская гага
Eiders	Eiderenten	Greater Eider Ducks	Гаги
Élaninés	Gleitaare	Kites	
Élanion écrit	Schwarzachsel-Gleitaar	Letter-winged Kite	
Emeu d'Australie	Emu	Emu	Эму
Épervier brun	Eckschwanzsperber	Shark-shinned Hawk	
– de Cooper	Rundschwanzsperber	Cooper's Hawk	Ястреб Купера
– d'Europe	Sperber	Eurasian Sparrowhawk	Ястреб-перепелятник
– pie	Trauerhabicht	Black Goshawk	
Éperviers	Habichte i. e. S.	Goshawks	Ястреба
Erismature à tête noire	Schwarzkopfruderente	Ruddy Duck	
– d'Argentine	Argentinische Schwarzkopfruderente	Argentine Ruddy Duck	
Erismatures	Ruderenten, Ruderenten i. e. S.	Stiff-tailed Ducks, Stiff-tails	
Falconidés	Falken	Falcons	
Falconiformes	Greifvögel	Diurnal Birds of Prey	Хищные птицы
Falconinés	Eigentliche Falken	Gyrfalcons, Falcons and Kestrels	Соколы
Faucon ardoisé	Graufalk	Grey Kestrel	
– blanc	Schwarzflügel-Gleitaar	Black-winged Kite	Чернокрылый коршун
– concolore	Schieferfalk	Sooty Falcon	
– crécerelle	Turmfalk	Old World Kestrel	Обыкновенная пустельга
– crécerellette	Rötelfalk	Lesser Kestrel	Степная пустельга
– de Feldegg	Feldeggsfalk	Feldegg's Falcon	Средиземноморский рыжеголовый балобан
– d'Eléonore	Eleonorenfalk	Eleonora's Falcon	Чеглок Элеоноры
– des chauves-souris	Fledermaus-Gleitaar	Bat Hawk	Индийский совиный сарыч
– des prairies	Präriefalk	Prairie Falcon	Мексиканский сокол
– émerillon	Merlin	Merlin	Дербник
– gerfaut	Gerfalk	Gyrfalcon	Кречет
– hobereau	Baumfalk	Hooby	Чеглок
– kobez	Rotfußfalk	Red-footed Falcon	Кобчик
– lanier	Lannerfalk	Lanner Falcon	Рыжеголовый балобан
– pèlerin	Wanderfalk	Peregrine Falcon	Сокол-сапсан
– Renard	Fuchsfalk	Fox Kestrel	
– sacre	Würgfalk	Saker Falcon	Обыкновенный балобан
– shikra	Rotkopfmerlin	Red-headed Falcon	
Fauconnet à collier	Indischer Zwergfalk	Rufous-thighed Falconet	Красноногий карликовый сокол
– à pattes jaunes	Langschwanz-Zwergfalk	Burmese Pigmy Falcon	
Faucons	Falken i. e. S.	Falcons	
Flamant rose	Flamingo, Rosaflamingo	Greater Flamingo, European Flamingo	Обыкновенный фламинго, Розово-красный фламинго
– rouge	Roter Flamingo	American Flamingo	Красный фламинго
Flamants	Flamingos	Flamingoes	Фламинго
Fou aux pieds rouges	Rotfußtölpel	Red-footed Booby	Красноногая олуша
– à ventre blanc	Brauntölpel	Brown Booby	
– de Bassan	Baßtölpel	Northern Gannet	Атлантическая олуша
– masqué	Maskentölpel	Masked Booby	Голуболицая олуша
Fous	Tropische Tölpel, Tölpel i. e. S.	Boobies, Gannets	
Francolin d'Europe	Halsbandfrankolin	Francolin	Турач
– du Mont Cameroun	Kamerunbergwald-Frankolin	Cameroun Mountain Francolin	
Francolins	Frankoline	Francolins	
Frégate aigle	Adlerfregattvogel	Ascension Frigate Bird	Орлиный фрегат
– superbe	Prachtfregattvogel	Magnificent Man-o'-War Bird	Малый фрегат
Frégates	Fregattvögel	Frigate Birds	Фрегаты

FRENCH NAME	GERMAN NAME	ENGLISH NAME	RUSSIAN NAME
Frégatidés	Fregattvögel	Frigate Birds	Фрегаты
Galliformes	Hühnervögel	Fowllike Birds	Куриные
Garrot albéole	Büffelkopfente	Bufflehead	Гоголь-головастик
Garrot harlequin	Kragenente	Harlequin Duck	Каменутка
– islandais	Spatelente	Barrow's Golden-eye	Исландский гоголь
Gaviidés	Seetaucher	Loons	Гагары
Gaviiformes	Seetaucher	Loons	Гагаровые
Gelinotte à queue fine	Schweif-Waldhuhn	Northern Sharp-tailed Grouse	Хвостатый тетерев
– des armoises	Beifußhuhn	Sage Grouse	Полынный тетерев
– des Bois	Haselhuhn	Hazel Hen	Рябчик
– huppée	Kragenhuhn	Eastern Ruffed Grouse	Воротничковый рябчик
Gorfou sauteur	Felsenpinguin	Rockhopper Penguin	
Grand Cormoran	Kormoran	Black Cormorant	Обыкновенный баклан
– Héron	Amerikanischer Graureiher	Great Blue Heron	
– Hocco	Tuberkelhokko	Mexican Curassow	
– Hocco à bec de rasoir	Mitu	Razor-billed Curassow	
– Marabout	Argala-Marabu	Greater Marabou	Индийский марабу-аргал
– Tétras	Auerhuhn	Capercaillie	Глухарь
– Tinamou	Großtao	Great Tinamou	Большой тинаму
Grande Aigrette	Silberreiher	Great White Egret	Большая белая цапля
Grèbe à cou noir	Schwarzhalstaucher	Black-necked Grebe	Черношейная поганка
– castagneux	Zwergtaucher	Little Grebe	Малая поганка
– esclavon	Ohrentaucher	Slavonian Grebe	Рогатая поганка
– huppé	Haubentaucher	Great Crested Grebe	Большая поганка
– jougris	Rothalstaucher	Red-necked Grebe	Серощёкая поганка
Grèbes	Taucher	Grebes	Поганки
Griffon african à dos blanc	Zweiggänsegeier	African White-backed Vulture	Белоспинный сип
– indien à dos blanc	Bengalgeier	Indian White backed Vulture	Индийский гриф
Gypaëte barbu	Bartgeier	Bearded Vulture	Бородач
Harle bièvre	Gänsesäger	Goosander	Большой крохаль
– chinois	Schuppensäger	Chinese Merganser	Чемуйчатый крохаль
– couronné	Kappensäger	Hooded Merganser	Американский крохаль
– huppé	Mittelsäger	Red-breasted Merganser	Средний крохаль
– piette	Zwergsäger	Smew	Луток
Harles	Säger	Mergansers	Крохали
Harpye	Harpyie	Harpy Eagle	
Héron à dos vert	Mangrovereiher	Green-backed Heron	Зеленая кваква
– ardoisé	Glockenreiher	Black Heron	
– à tête noire	Schwarzhalsreiher	Black-headed Heron	
– à ventre blanc	Dreifarbenreiher	Louisiana Heron	
– bihoreau à couronne noire	Nachtreiher	Black-crowned Night Heron	Кваква
– cendré	Graureiher	Grey Heron	Серая цапля
– crabier	Rallenreiher	Squacco Heron	Желтая цапля
– de Sumatra	Sumatrareiher	Dusky Grey Heron	Тифонова цапля
– garde-boeufs	Kuhreiher	Cattle Egret	Египетская цапля
– goliath	Goliathreiher	Goliath Heron	
– pourpré	Purpurreiher	Purple Heron	Рыжая цапля
– vert	Grünreiher	Green Heron	
Hérons bihoreaux	Nachtreiher	Night Herons	
Hobereau africain	Afrikanischer Baumfalk	African Hooby	
– à poitrine rousse	Indischer Baumfalk	Oriental Hooby	
Hocco de la Guiane	Glattschnabelhokko	Crested Curassow	Гладкоклювый гокко
Hydrobatidés	Sturmschwalben	Storm Petrels	Качурки
Ibis	Ibisse	Ibises	Ибисы
– à face blanche	Brillensichler	White-faced Ibis	
– à tête noire	Schwarzkopfibis	Oriental Ibis	
– blanc	Weißer Sichler	White Ibis	
– chauve	Waldrapp	Hermit Ibis	Горный ибис
– géant	Riesenibis	Giant Ibis	
– hagedash	Hagedasch	Hadada	
– luisant	Brauner Sichler	Glossy Ibis	Каравайка
– noir oriental	Warzenibis	Black Ibis	
– olivâtre	Olivgrüner Ibis	Olive Ibis	
– rouge	Roter Sichler	Scarlet Ibis	Красный ибис
– sacré	Heiliger Ibis	Sacred Ibis	Священный ибис
Ithagine sanguine	Blutfasan	Blood Pheasant	
Jabirou asiatique	Indien-Großstorch	Black-necked Stork	Индийский исполинский аист
– du Sénégal	Afrika-Sattelstorch	Seddle Bill	Африканский ярибу
Jabiru	Jabiru	Jabiru	
Kamichi cornu	Hornwehrvogel	Horned Screamer	Рогатая анхима
Kiwi austral	Streifenkiwi	Common Kiwi	Обыкновенный киви
– d'Owen	Fleckenkiwi	Owen's Kiwi	Киви Оуэна
Kiwis	Kiwivögel, Kiwis	Kiwis	Бескрылые, Киви
Lagopède à queue blanche	Amerikanisches Alpenschneehuhn	Canadian White-tailed Ptarmigan	Америсканская белая куропатка

FRENCH NAME	GERMAN NAME	ENGLISH NAME	RUSSIAN NAME
– des Saules	Moorschneehuhn	Willow Grouse	Белая куропатка
– muet	Alpenschneehuhn	Ptarmigan	Тундряная куропатка
Leipoa ocellé	Thermometerhuhn	Mallee Fowl	Лейпоа
Macreuse à lunettes	Brillenente	Surf Scoter	Пестроносый турпан
– brune	Samtente	Velvet Scoter	Черный турпан
– noire	Trauerente	Black Scoter	Синьга
Macreuses	Trauerenten	Scoters	Турпаны
Manchot de Magelhaen	Magellanpinguin	Magellanic Penguin	
– du Cap	Brillenpinguin	Blackfooted Penguin	Очковый пингвин
– Gentoo	Eselspinguin	Gentoo Penguin	Ослиный пингвин
– impérial	Kaiserpinguin	Emperor Penguin	Императорский пингвин
– royal	Königspinguin	King Penguin	Королевский пингвин
Marabout	Afrika-Marabu	Marabou	Африканский марабу
Marabouts	Marabus	Marabous	Марабу
Mégapodiidés	Großfußhühner	Moundfowl	Сорные куры
Milan à queue fourchue	Schwalbenweih	Swallow-tailed Kite	Вилохвостый лунь
– noir	Schwarzmilan	Black Kite	Черный коршун
– royal	Rotmilan	Red Kite	Красный коршун
Milouin à tête rousse	Rotkopfente	Red-head	Краснолобый нырок
– aux yeux rouges	Riesentafelente	Canvas-back	
– d'Australie	Australische Moorente	Australasian White-eye	
– de Madagascar	Madagaskar-Moorente	Madagascan White-eye	
Milvinés	Milane i. e. S.		Коршуны
Morillon à collier	Halsringente	Ring-necked Duck	
Nandou américain	Nandu	American Rhea	Обыкновенный нанду
– de Darwin	Darwin-Nandu	Darwin's Rhea	Нанду Дарвина
Nandous	Nandus	Rheas	Американские страусы
Naucler d'Afrique	Schwalbengleitaar	African Swallow-tailed Kite	
Oie à tête barrée	Streifengans	Bar-headed Goose	Индийский горный гусь
– cendrée	Graugans	Greylag Goose	Серый гусь
– cygnoïde	Schwanengans	Swan Goose	Сухонос
– – doméstique	Haus-Höckergans	Domestic Swan Goose	Домашний китайский гусь
– de Gambie	Sporengans	Spur-wing	Обыкновенный шпорцевый гусь
– d'Égypte	Nilgans	Egyptian Goose	Нильский гусь
– de l'Orinoque	Orinokogans	Orinoco Goose	
– de Ross	Zwergschneegans	Ross' Goose	Карликовый белый гусь
– des moissons	Saatgans	Bean Goose	Гуменник
– des neiges	Schneegans	Snow Goose	Белый гусь
– doméstique	Hausgans	Domestic goose	Домашний гусь
– empereur	Kaisergans	Emperor Goose	Белошей
– naine	Zwerggans	Lesser White-fronted Goose	Пискулька
– rieuse	Bleßgans	White-fronted Goose	Белолобый гусь
– semi-palmée	Spaltfußgans	Australian Magpie Goose	Расщепнолапый гусь
Oies armées	Sporengänse	Spur-winged Geese	Шпорцевые гуси
– semi-palmées	Spaltfußgänse	Magpie Geese	
Oiseau-Serpent	Altwelt-Schlangenhalsvogel	Darter	Ахинга старого света
Oiseaux	Vögel	Birds	Птицы
Ombrette	Hammerkopf	Hamerkop	Молотоглав
Ombrettes	Hammerköpfe	Hamerkops	Молотоголовые цапли
Ortalide babillarde	Rotflügelguan	Chestnut-winged Chachalaca	Краснокрылый шуан
– du Mexique	Braunflügelguan	Lesser Grey-headed Chachalaca	Бурокрылый шуан
Paille-en-queue à bec jaune	Weißschwanz-Tropikvogel	Yellow-billed Tropic Bird	Белохвостый фаэтон
– – – à bec rouge	Rotschnabel-Tropikvogel	Red-billed Tropic Bird	Красноклювый фаэтон
Pauxi pierre	Helmhokko	Helmeted Curassow	Шлемоносный гокко
Pélican à dos rosé	Rötelpelikan	Pink-backed Pelican	Красноспинный пеликан
– blanc	Rosapelikan	Eastern White Pelican	Розовый пеликан
– frisé	Krauskopfpelikan	Dalmatian Pelican	Кудявый пеликан
Pélécanidés	Pelikane	Pelicans	Пеликаны
Pélécaniformes	Ruderfüßer	Totipalmate Swimmers	Веслоногие
Pélécanoïdidés	Tauchsturmvögel	Diving Petrels	Нырцовые буревестники
Pélicans	Pelikane	Pelicans	Пеликаны
Pénélope siffleuse	Schakutinga	White-headed Piping Guan	
Percnoptère d'Égypte	Schmutzgeier	Egyptian Vulture	Стверятник
Perdricinés	Feldhühner	Old World Quails	
Perdrix bartavelle	Steinhuhn	Rock Partridge	Кеклик
– de Barbarie	Felsenhuhn	Barbary Partridge	Берберская каменная куропатка
– de Hay	Arabisches Sandhuhn	Sand Partridge	Аравийская пустынная курочка
– des neiges caspienne	Kaspisches Königshuhn	Caspian Snowcock	Каспийский улар
– grise	Rebhuhn	Grey Partridge	Серая куропатка
– lerwa	Haldenhuhn	Snow Partridge	Гималайская куропатка
– percheuse à collier	Hügelhuhn		Древесная куропатка
– rouge	Rothuhn	Red-legged Partridge	Красная каменная куропатка

FRENCH NAME	GERMAN NAME	ENGLISH NAME	RUSSIAN NAME
Petit Butor	Indianerdommel	Lesser Bittern	
– Flamant	Zwergflamingo	Lesser Flamingo	Африканский карликовый фламинго
– Héron bleu	Blaureiher	Little Blue Heron	
– Hocco à bec de rasoir	Samthokko	Lesser Razor-billed Curassow	
– Marabout	Sunda-Marabu	Lesser Marabou	Яванский марабу
Petite Buse	Breitschwingenbussard	Broad-winged Hawk	Ширококрылый канюк
– Spatule	Kleiner Löffler	Black-faced Spoonbill	Малая колпица
Petits Flamants	Zwergflamingos	Lesser Flamingoes	
Pétrel à ailes blanches	Brustbandsturmtaucher	White-throated Petrel	Белолобый тайфунник
– cul-blanc	Wellenläufer	Leach's Petrel	Северная качурка
– damier	Kapsturmvogel	Pintado Petrel	Капский буревестник
– de Bulwer	Weichnasen-Sturmvogel	Bulwer's Petrel	Тайфунник Бульвера
– des tropiques	Schwarzbauch-Sturmschwalbe	Black-bellied Storm Petrel	Чернобрюхая качурка
– diablotin	Schwarzkappensturmtaucher	Black-capped Petrel	Черношапочный тайфунник
– frégate	Fregattensturmschwalbe	Frigate Petrel	Морская качурка
– glacial	Eissturmvogel	Arctic Fulmar	Атлантический глупыш
– océanite	Buntfüßige Sturmschwalbe	Wilson's Petrel	Качурка Вильсона
– tempête	Sturmschwalbe	Storm Petrel	Малая прямохвостая качурка
Pétrels plongeurs	Tauchsturmvögel	Diving Petrels	
Phaëthons	Tropikvögel	Tropic Birds	Фаэтоны
Phaëthontidés	Tropikvögel	– Birds	Фаэтоны
Phalacrocoracidés	Kormorane	Cormorants	Бакланы
Phalacrocoracinés	Kormorane	Cormorants	Бакланы
Phasianidés	Fasanenartige	Pheasants, Quails and Peacocks	
Phoenicoptéridés	Flamingos	Flamingoes	Фламинго
Pilet du Chili	Spitzschwanzente	Yellow-billed Pintail	
Plataléinés	Löffler	Spoonbills	Колпицы
Plongeon à bec blanc	Gelbschnabel-Eistaucher	White-billed Diver	Белоносая гагара
– catmarin	Sterntaucher	Red-throated Diver	Краснозобая гагара
– imbrin	Eistaucher	Great Northern Diver	Полярная гагара
– lumme	Prachttaucher	Black-throated Diver	Чернозобая гагара
Plongeons	Seetaucher	Divers	Гагары
Podicipédiformes	Lappentaucher	Grebes	Поганковые
Podicipitidés	Lappentaucher	Grebes	Поганки
Poule des prairies	Präriehuhn	Heath Hen	Большой луговой тетерев
Procellariidés	Sturmvögel	Procellariids	Буревестники
Procellariiformes	Röhrennasen	Tube-Nosed Swimmers	Трубконосые
Pterniste à cou nu	Rotkehlfrankolin	Angola Red-necked Partridge	
Puffin cendré	Gelbschnabel-Sturmtaucher	Mediterranean Shearwater	Большой белобрюхий буревестник
– des Anglais	Schwarzschnabel-Sturmtaucher	Manx Shearwater	Обыкновенный буревестник
– fuligineux	Rußsturmtaucher	Sooty Shearwater	Серый буревестник
– majeur	Großer Sturmtaucher	Great Shearwater	Большой пестробрюхий буревестник
Pygargue à queue blanche	Seeadler, Bandseeadler	White-tailed Sea Eagle, Pallas' Sea Eagle	Орлан-белохвост, Орлан-долгохвост
Pygopodes	Steißfüße	Pygopodes	Поганки и гагары
Quaglia azzurra	Schuppenwachtel	Mexican Scaled Quail	
Ratites	Laufvögel	Ratites	Бегающие
Rhéidés	Nandus	Rheas	Нанду
Roulroul	Straußwachtel	Crowned Wood Partridge	Страусовый перепел
Sagittaridés	Sekretäre	Secretary Birds	Африканские секретари
Sarcelle à faucilles	Sichelente	Falcated Teal	Касатка
– cannelle	Zimtente	Cinnamon Teal	
– d'Australie	Kastanienente	Chestnut Teal	
– de Coromandel	Indische Zwergglanzente	Cotton Pygmy Goose	Индийская карликовая утка
– de Madagascar	Afrikanische Zwergglanzente	African Pygmy Goose	
– d'été	Knäkente	Garganey	Чирок-трескунок
– d'hiver	Krickente	Teal	Чирок-свистунок
– du Brésil	Amazonas-Ente	Brazilian Teal	
– du Cap	Kapente	Cape Teal	
– du Chili	Chile-Krickente	South American Green-winged Teal	Чилийский чирок
– élégante	Gluckente	Baikal Teal	Чирок-клоктун
– grise des Indes	Weißkehlente	Grey Teal	
– Hottentote	Hottentotten-Ente	Hottentot Teal	
– marbrée	Marmelente	Marbled Teal	Мраморный чирок
– soucrourou	Blauflügelente	Blue-winged Teal	Синекрылый чирок
– versicolore	Kappenente	Silver Teal	
– verte d'Australie	Grüne Zwergglanzente	Green Pygmy Goose	
Sarcidiorne à crête	Höckerglanzente	Comb Duck	Шишконосый гусь

FRENCH NAME	GERMAN NAME	ENGLISH NAME	RUSSIAN NAME
Savacou huppé	Kahnschnabel	Boat-billed Heron	Цапля-челноклюв
Scopidés	Hammerköpfe	Hamerkops	Теневые птицы
Secrétaire	Sekretär	Secretary Bird	Секретарь
Siffleur du Brésil	Chile-Pfeifente	Chiloë Wigeon	Чилийская свиязь
Souchet du Cap	Kap-Löffelente	Cape Shoveler	Капская широконоска
– roux	Südamerikanische Löffelente	Red Shoveler	Южноамериканская широконоска
Soui	Brauntao	Little Tinamou	
Spatule blanche	Löffler	White Spoonbill	Колпица
– d'Afrique	Schmalschnabel-Löffler	African Spoonbill	Африканская белая колпица
Spatules	Löffler	Spoonbills	
Sphéniscidés	Pinguine	Penguins	
Sphénisciformes	Pinguine	Penguins	
Struthionidés	Strauße	Ostriches	Страусы
Sulidés	Tölpel	Boobies and Gannets	Олуши
Tadorne de Belon	Brandgans	Common Shelduck	Пеганка
Tadornes et Casarcas	Halbgänse	Shelducks	Пеганки
Tantale blanc	Malaien-Nimmersatt	Malayan Wood Ibis	Малайский тантал
– ibis	Afrika-Nimmersatt	Wood Ibis	Африканский тантал
– indien	Indien-Nimmersatt	Painted Stork	Индийский тантал
Tantales	Nimmersatte	Wood Ibises	Танталы
Tétraoninés	Rauhfußhühner	Grouse	
Tétras des savanes	Tannen-Waldhuhn	Hudsonian Spruce Grouse	Канадский тетерев
– lyre	Birkhuhn	Black Grouse	Тетерев
– – de Géorgie	Kaukasisches Birkhuhn	Caucasian Black Grouse	Кавказский тетерев
– sombre	Felsengebirgshuhn	Dusky Grouse	Дымчатый тетерев
Threskiornithidés	Ibisvögel	Ibises and Spoonbills	Ибисы
Threskiornithinés	Sichler	Ibises	Ибисы
Tinamou huppé	Perl-Steißhuhn	Crested Tinamou	
– roussâtre	Pampashuhn	Rufous Tinamou	Бразильянский степной скрытохвост
– tataupa	Tataupa	Tataupa Tinamou	
– varié	Rotbrusttao	Variegated Tinamou	
Tinamous	Steißhühner, Wald-Steißhühner	Tinamous	Скрытохвосты, Скрытохвостые куры, Лесные скрытохвостые куры
Tragopan de Blyth	Blyth-Satyrhuhn	Blyth's Tragopan	
– de Cabot	Cabot-Satyrhuhn	Cabot's Tragopan	
– de Hasting	West-Satyrhuhn	Hasting's Tragopan	
– de Temminck	Temminck-Satyrhuhn	Temminck's Tragopan	
– satyre	Rot-Satyrhuhn	Satyr Tragopan	
Vautour à bec long	Indischer Geier	Long-billed Vulture	Индийский белоголовый сип
– à tête rouge	Truthahngeier	Turkey Vulture	Индюшачий гриф
– de Pondichéry	Lappengeier	Pondicherry Vulture	
– de Rüppell	Sperbergeier	Rüppell's Griffon	
– fauve	Gänsegeier	Griffon Vulture	Белоголовый сип
– huppé	Wollkopfgeier	White-headed Vulture	
– moine	Mönchsgeier	Black Vulture	Чёрный гриф
– noir	Rabengeier	Black Vulture	Американский черный гриф
– oricou	Ohrengeier	Lappet-faced Vulture	Ушастый гриф
– palmiste	Palmgeier	Palm-nut Vulture	Грифовый орлан
– royal	Königsgeier	King Vulture	Королевский гриф
Vulturidés	Neuweltgeier	Cathartines	Американские грифы

IV. Russian—German—English—French

Названия подвидов отличаются от видовых чаще всего лишь дополнительным прилагательным, главным образом географического характера. Такие русские названия подвидов как правило не включены в данную часть зоологического словаря.

RUSSIAN NAME	GERMAN NAME	ENGLISH NAME	FRENCH NAME
Абдисмкий аист	Regenstorch	Abdim's Stork	Cigogne d'Abdim
Австралийская выпь	Australische Rohrdommel	Brown Bittern	
Австралийская олуша	Australischer Tölpel	Australasian Gannet	
Австралийская пеганка	Australische Kasarka	Australian Shelduck	Casarca d'Australie
Австралийская пустельга	Australischer Turmfalk	Naukeen Kestrel	
Австралийские страусы	Kasuarvögel	Cassowaries and Emus	Casoars et Emeus

RUSSIAN NAME	GERMAN NAME	ENGLISH NAME	FRENCH NAME
Австралийский малый баклан	Australische Zwergscharbe	Little Pied Cormorant	
Австралийский пеликан	Brillenpelikan	Austialian Pelican	
Австралийский ястреб	Australischer Habicht	Red Goshawk	Autour à ventre rouge
Аисты	Störche	Storks	Ciconiidés
Аисты-разини	Klaffschnäbel	Open Bills	Becs-Ouverts
Альбатрос Буллера	Bullers Albatros	Buller's Albatross	
Альбатросы	Albatrosse	Albatrosses and Mollymawks	Diomédeidés
Американская анхинга	Amerikanischer Schlangenhalsvogel	American Darter	Anhinga d'Amerique
Американская белая куропатка	Amerikanisches Alpenschneehuhn	Canadian White-tailed Ptarmigan	Lagopède à queue blanche
Американская брачная утка	Brautente	Carolina Wood Duck	Canard Carolin
Американская пустельга	Buntfalk	American Sparrow Hawk	Crécerelle americaine
Американская свиязь	Amerikanische Pfeifente	American Wigeon	Canard siffleur
Американские грифы	Neuweltgeier	Cathartines	Vulturidés
Американские страусы	Nandus	Rheas	Nandous
Американский крохаль	Kappensäger	Hooded Merganser	Harle couronné
Американский черный гриф	Rabengeier	Black Vulture	Vautour noir
Американский ярибу	Amerika-Nimmersatt	North American Wood Ibis	Cigogne américaine
Амурский волгок	Mandschurendommel	Schrenck's Little Bittern	Blongios de Schrenck
Андский каракара	Anden-Karakara	Mountain Caracara	
Андский кондор	Anden-Kondor	Andean Condor	Condor des Andes
Андский фламинго	Andenflamingo	– Flamingo	
Антарктический буревестник	Antarktissturmvogel	Antarctic Petrel	
Антарктический глупыш	Antarktischer Eissturmvogel	– Fulmar	
Антарктический дымчатый альбатрос	Südlicher Rußalbatros	Light-mantled Sooty Albatross	Albatros fuligineux
Анхинга старого света	Altwelt-Schlangenhalsvogel	Darter	Oiseau-Serpent
Анхинги	Schlangenhalsvögel	Darters	Anhingas
Аравийская пустынная курочка	Arabisches Sandhuhn	Sand Partridge	Perdrix de Hay
Атлантическая олуша	Baßtölpel	Northern Gannet	Fou de Bassan
Атлантический глупыш	Eissturmvogel	Arctic Fulmar	Pétrel glacial
Африканская белая колпица	Schmalschnabel-Löffler	African Spoonbill	Spatule d'Afrique
Африканские секретари	Sekretäre	Secretary Birds	Sagittaridés
Африканские страусы	Strauße	Ostriches	Autruches
Африканский аист-разиня	Afrika-Klaffschnabel	Open Bill	Bec-Ouvert
Африканский карликовый фламинго	Zwergflamingo	Lesser Flamingo	Petit Flamant
Африканский марабу	Afrika Marabu	Marabou	Marabout
Африканский страус	Strauß	Ostrich	Autruche
Африканский тантал	Afrika-Nimmersatt	Wood Ibis	Tantale ibis
Африканский ярибу	Afrika-Sattelstorch	Seddle Bill	Jabirou du Sénégal
Африканско-индийский пуховый аист	Afrikanisch-Indischer Wollhalsstorch	Bishop Stork	Cigogne épiscopale
Бакланы	Kormorane	Cormorants	Phalacrocoracidés, Cormorans
Бегающие	Laufvögel	Ratites	Ratites
Белобокий фрегат	Kleiner Fregattvogel	Lesser Frigate Bird	
Белобрюхий фрегат	Weißbauch-Fregattvogel	Christmas Frigate Bird	
Белоглазый нырок	Moorente	Ferruginous White-eye	Canard nyroca
Белоголовый орлан	Weißkopf-Seeadler	Bald Eagle	Aigle à tête blanche
Белоголовый сип	Gänsegeier	Griffon Vulture	Vautour fauve
Белогорлый каракара	Weißkehl-Karakara	White-throated Caracara	
Белокрылая цапля	Bachusreiher	Chinese Pond Heron	Crabier chinois
Белолобый гусь	Bleßgans	White-fronted Goose	Oie rieuse
Белолобый тайфунник	Brustbandsturmtaucher	White-throated Petrel	Pétrel à ailes blanches
Белоносая гагара	Gelbschnabel-Eistaucher	White-billed Diver	Plongeon à bec blanc
Белоплечий орлан	Riesenseeadler	Steller's Sea Eagle	
Белоспинный альбатрос	Kurzschwanzalbatros	– Albatross	Albatros de Steller
Белоспинный сип	Zwerggänsegeier	African White-backed Vulture	Griffon africain à dos blanc
Белохвостый фаэтон	Weißschwanz-Tropikvogel	Yellow-billed Tropic Bird	Paille-en-queue à bec jaune
Белошей	Kaisergans	Emperor Goose	Oie empereur
Белощёкая казарка	Weißwangengans	Barnacle Goose	Bernache nonnette
Белощекая паламедея	Weißwangenwehrvogel	Crested Screamer	Chauna chavaria
Белая куропатка	Moorschneehuhn	Willow Grouse	Lagopède des Saules
Белый аист	Weißstorch	White Stork	Cigogne blanche
Белый гусь	Schneegans	Snow Goose	Oie des neiges
Белый ястреб	Weißbrauenhabicht	Variable Goshawk	
Берберская каменная куропатка	Felsenhuhn	Barbary Partridge	Perdrix de Barbarie
Берингов баклан	Nordpazifischer Kormoran	Pelagic Cormorant	

RUSSIAN NAME	GERMAN NAME	ENGLISH NAME	FRENCH NAME
Беркут	Steinadler	Golden Eagle	Aigle doré
Бледноногий буревестник	Blaßfußsturmtaucher	Pale-footed Shearwater	
Большая белая цапля	Silberreiher	Great White Egret	Grande Aigrette
Большая выпь	Rohrdommel	Bittern	Butor étoilé
Большеног Фрейсинэ	Freycinet-Großfußhuhn	Scrub Fowl	
Большой белобрюхий буревестник	Gelbschnabel-Sturmtaucher	Mediterranean Shearwater	Puffin cendré
Большой крохаль	Gänsesäger	Goosander	Harle bièvre
Большой луговой тетерев	Präriehuhn	Heath Hen	Poule des prairies
Большой пестробрюхий буревестник	Großer Sturmtaucher	Great Shearwater	Puffin majeur
Большая поганка	Haubentaucher	– Crested Grebe	Grèbe huppé
Большой подорлик	Schelladler	Greater Spotted Eagle	Aigle criard
Большой тинаму	Großtao	Great Tinamou	Grand Tinamou
Большой фрегат	Bindenfregattvogel	– Frigate Bird	
Бородач	Bartgeier	Bearded Vulture	Gypaëte barbu
Бразильянский степной скрытохвост	Pampashuhn	Rufous Tinamou	Tinamou roussâtre
Браминский коршун	Brahminenweih	Brahming Kite	
Буревестник-горлица	Feenwalvogel	Fairy Prion	
Буревестники	Sturmvögel	Procellariids	Procellariidés
Бурокрылый шуан	Braunflügelguan	Lesser Grey-headed Chachalaca	Ortalide du Mexique
Бурый пеликан	Brauner Pelikan	Brown Pelican	
Веерохвостые	Neuvögel	Typical birds	
Венценосный орел	Kronenadler	Crowned Eagle	Blanchard
Веслоногие	Ruderfüßer	Totipalmate Swimmers	Pélécaniformes
Вилохвостый лунь	Schwalbenweih	Swallow-tailed Kite	Milan à queue fourchue
Виргинская куропатка	Virginiawachtel	Eastern Bob White	Colin de Virginie
Воротничковый рябчик	Kragenhuhn	– Ruffed Grouse	Gelinotte huppée
Восточный степной орел	Steppenadler	Steppe Eagle	
Восточный тундровый лебедь	Jankowski-Schwan	Jankowski's Swan	
Гавайская казарка	Hawaiigans	Hawaiian Goose	Bernache des îles Sandwich
Гавайский тайфунник	Hawaiisturmvogel	– Petrel	
Гага-гребенутка	Prachteiderente	King Eider	Eider à tête grise
Гагаровые	Seetaucher	Loons	Gaviiformes
Гагары	Seetaucher	Loons, Divers	Gaviidés, Plongeons
Гаги	Eiderenten	Greater Eider Ducks	Eiders
Галапагосский пингвин	Galapagospinguin	Galapagos Penguin	
Гваделупский каракара	Guadalupe-Karakara	Guadalupe Caracara	
Гигаитская утка коскороба	Koskorobaschwan	Coscoroba Swan	Cygne coscoroba
Гималайская куропатка	Haldenhuhn	Snow Partridge	Perdrix lerwa
Гладкоклювый гокко	Glattschnabelhokko	Crested Curassow	Hocco de la Guiane
Глупыши	Möwensturmvögel	Fulmars	
Глухарь	Auerhuhn	Capercaillie	Grand Tétras
Гоголь-головастик	Büffelkopfente	Bufflehead	Garrot albéole
Гокко	Hokkos	Curassows, Guans and Chachalacas	Cracidés
Голубой тайфунник	Blausturmvogel	Blue Petrel	
Голубокрылый гусь	Blauflügelgans	Abyssinian Blue-winged Goose	Bernache aux ailes bleues
Голуболицая олуша	Maskentölpel	Masked Booby	Fou masqué
Горный ибис	Waldrapp	Hermit Ibis	Ibis chauve
Горный шуан	Bergguan	Derby's Mountain Pheasant	
Гребневый орел	Schopfadler	Long-crested Eagle	Aigle huppé
Грифовый орлан	Palmgeier	Palm-nut Vulture	Vautour palmiste
Гуменник	Saatgans	Bean Goose	Oie des moissons
Дербник	Merlin	Merlin	Faucon émerillon
Длиннокрылый тайфунник	Langflügelsturmtaucher	Grey-winged Petrel	
Длинноносый баклан	Krähenscharbe	Green Cormorant	Cormoran huppé
Длиннохвостый ястреб	Langschwanzhabicht	Long-tailed Hawk	Autour à longue queue
Домашний гусь	Hausgans	Domestic goose	Oie doméstique
Домашний китайский гусь	Haus-Höckergans	– swan goose	– cygnoïde doméstique
Домашняя утка	Hausente	– duck	Canard doméstique
Древесная куропатка	Hügelhuhn		Perdrix percheuse à collier
Древесные утки	Pfeifgänse	Whistling Ducks	Canards siffleurs
Дымчатый тетерев	Felsengebirgshuhn	Dusky Grouse	Tétras sombre
Египетская цапля	Kuhreiher	Cattle Egret	Héron garde-boeufs
Желтая цапля	Rallenreiher	Squacco Heron	– crabier
Желточубый пингвин	Gelbschopfpinguin	Erect-crested Penguin	
Желтозобый каракара	Gelbkinn-Karakara	Yellow-throated Caracara	
Желтоклювый альбатрос	Gelbnasenalbatros	Yellow-nosed Albatross	Albatros à bec jaune
Желтоклювая чепура-нужда	China-Seidenreiher	Chinese Egret	
Западносибирский тетеревятник	Sibirischer Habicht	Siberian Goshawk	
Западный тундровый лебедь	Zwergschwan	Bewick's Swan	Cygne de Bewick

RUSSIAN NAME	GERMAN NAME	ENGLISH NAME	FRENCH NAME
Зеленая кваква	Mangrovereiher	Green-backed Heron	Héron à dos vert
Змееяд	Schlangenadler	Eurasian Short-toed Eagle	Circaète Jean-le-Blanc
Золоточубый пингвин	Goldschopfpinguin	Macaroni Penguin	
Ибисы	Ibisvögel, Sichler, Ibisse	Ibises (and Spoonbills)	Threskiornithidés, Threskiornithinés, Ibis
Императорский пингвин	Kaiserpinguin	Emperor Penguin	Manchot impérial
Индийская карликовая утка	Indische Zwergglanzente	Cotton Pygmy Goose	Sarcelle de Coromandel
Индийский аист-разиня	Indien-Klaffschnabel	Asian Open Bill	Bec-Ouvert asiatique
Индийский балобан-лаггар	Laggarfalk	Laggar Falcon	
Индийский белоголовый сип	Indischer Geier	Long-billed Vulture	Vautour à bec long
Индийский горный гусь	Streifengans	Bar-headed Goose	Oie à tête barrée
Индийский гриф	Bengalgeier	Indian White-backed Vulture	Griffon indien à dos blanc
Индийский исполинский аист	Indien-Großstorch	Black-necked Stork	Jabirou asiatique
Индийский марабу-аргал	Argala-Marabu	Greater Marabou	Grand Marabout
Индийский совиный сарыч	Fledermaus-Gleitaar	Bat Hawk	Faucon des chauves-souris
Индийский тантал	Indien-Nimmersatt	Painted Stork	Tantale indien
Индюшачий гриф	Truthahngeier	Turkey Vulture	Vautour à tête rouge
Исландский гоголь	Spatelente	Barrow's Golden-eye	Garrot islandais
Кавказский тетерев	Kaukasisches Birkhuhn	Caucasian Black Grouse	Tétras lyre de Géorgie
Казарки	Meergänse	Brents	Bernaches
Казуар Беннетта	Bennettkasuar	Bennett's Cassowary	Casoar de Bennett
Казуары	Kasuare	Cassowaries	Casuariidés, Casoars
Калифорнийский кондор	Kalifornischer Kondor	California Condor	Condor de Californie
Калифорнийский хохлатый перепел	Kalifornische Schopfwachtel	Valley Quail	Colin de Californie
Калифорнийский шлемоносный перепел	Helmwachtel	Gambel's Quail	– de Gambel
Каменутка	Kragenente	Harlequin Duck	Garrot harlequin
Камышевый лунь	Rohrweihe	Marsh Harrier	Busard harpaye
Канадская казарка	Kanadagans	Canada Goose	Bernache du Canada
Канадский тетерев	Tannen-Waldhuhn	Hudsonian Spruce Grouse	Tétras des savanes
Канюк-курганник	Adlerbussard	Long-legged Hawk	Buse féroce
Капская олуша	Kaptölpel	Cape Gannet	
Капская широконоска	Kap-Löffelente	Cape Shoveler	Souchet du Cap
Капский буревестник	Kapsturmvogel	Pintado Petrel	Pétrel damier
Капский орел	Kaffernadler	Verreaux's Eagle	Aigle de Verreaux
Каравайка	Brauner Sichler	Glossy Ibis	Ibis luisant
Каранхо	Carancho	Crested Caracara	
Карликовые соколы	Zwergfalken	Pigmy Falcons	
Карликовые утки	Zwergglanzenten	– Geese	
Карликовый белый гусь	Zwergschneegans	Ross' Goose	Oie de Ross
Карликовый лунь	Perlenweih	Pearl Kite	
Карликовый пингвин	Zwergpinguin	Little Penguin	
Касатка	Sichelente	Falcated Teal	Sarcelle à faucilles
Каспийский улар	Kaspisches Königshuhn	Caspian Snowcock	Perdrix des neiges caspienne
Качурка Вильсона	Buntfüßige Sturmschwalbe	Wilson's Petrel	Pétrel océanite
Качурки	Sturmschwalben	Storm Petrels	Hydrobatidés
Кваква	Nachtreiher	Black-crowned Night Heron	Héron bihoreau à couronne noire
Кеклик	Steinhuhn	Rock Partridge	Perdrix bartavelle
Киви	Kiwis	Kiwis	Aptérygidés, Kiwis
Киви Оуэна	Fleckenkiwi	Owen's Kiwi	Kiwi d'Owen
Кисточковый баклан	Pinselkormoran	Brandt's Cormorant	
Китайский волгок	Chinesendommel	Chinese Little Bittern	Blongios chinois
Китайский красноногий ибис	Japanischer Ibis	Japanese Crested Ibis	
Китоглав	Schuhschnabel	Shoebill	Bec-en-Sabot
Китоглавы	Schuhschnäbel	Shoebills	Balaenicipitidés, Baleniceps
Клинохвостый орел	Keilschwanzadler	Wedge-tailed Eagle	
Кобчик	Rotfußfalk	Red-footed Falcon	Faucon kobez
Колпица	Löffler	White Spoonbill	Spatule blanche
Колпицы	Löffler	Spoonbills	Plataléinés
Королевский альбатрос	Königsalbatros	Royal Albatross	Albatros royal
Королевский гриф	Königsgeier	King Vulture	Vautour royal
Королевский пингвин	Königspinguin	– Penguin	Manchot royal
Коромандельская перепелка	Regenwachtel	Rain Quail	Caille du Coromandel
Короткоклювый фламинго	James-Flamingo	James' Flamingo	
Коршуны	Milane i. e. S.		Milvinés
Крапчатая древесная утка	Tüpfelpfeifgans	Spotted Whistling Duck	
Красная каменная куропатка	Rothuhn	Red-legged Partridge	Perdrix rouge

RUSSIAN NAME	GERMAN NAME	ENGLISH NAME	FRENCH NAME
Красноголовый нырок	Tafelente	European Pochard	Canard milouin
Краснозобая гагара	Sterntaucher	Red-throated Diver	Plongeon catmarin
Краснозобая казарка	Rothalsgans	Red-breasted Goose	Bernache à cou roux
Краснокрылый шуан	Rotflügelguan	Chestnut-winged Chachalaca	Ortalide babillarde
Красноклювый фаэтон	Rotschnabel-Tropikvogel	Red-billed Tropic Bird	Paille-en-queue à bec rouge
Краснолицый баклан	Aleuten-Kormoran	Red-faced Cormorant	
Краснолобый нырок	Rotkopfente	Red-head	Milouin à tête rousse
Красноногая олуша	Rotfußtölpel	Red-footed Booby	Fou aux pieds rouges
Красноногий карликовый сокол	Indischer Zwergfalk	Rufous-thighed Falconet	Fauconnet à collier
Красноносый нырок	Kolbenente	Red-crested Duck	Brante roussâtre
Красноспинный пеликан	Rötelpelikan	Pink-backed Pelican	Pélican à dos rosé
Краснохвостый канюк	Rotschwanzbussard	Red-tailed Hawk	Buse à queue rousse
Краснохвостый фаэтон	Rotschwanz-Tropikvogel	– – Tropic Bird	
Краснохвостый шуан	Rotschwanzguan	Jardine's Chachalaca	
Красношеий каракара	Rotkehl-Karakara	Red-throated Caracara	
Красный ибис	Roter Sichler	Scarlet Ibis	Ibis rouge
Красный коршун	Rotmilan	Red Kite	Milan royal
Красный фламинго	Roter Flamingo	American Flamingo	Flamant rouge
Кречет	Gerfalk	Gyrfalcon	Faucon gerfaut
Крикливый орлан	Schreiseeadler	African Sea Eagle	Aigle pêcheur
Крохали	Säger	Mergansers	Harles
Кряква	Stockente	Northern Mallard	Canard colvert
Кубинская древесная утка	Kuba-Pfeifgans	Black-billed Whistling Duck	
Кудрявый пеликан	Krauskopfpelikan	Dalmatian Pelican	Pélican frisé
Кукутковая утка	Kuckucksente	Black-headed Duck	Canard à tête noire de l'Argentine
Куриные	Hühnervögel	Fowllike Birds	Galliformes
Куриный гусь	Hühnergans	Cape Barren Goose	Céréopse de Nouvelle-Hollande
Кустовая курица	Buschhuhn	Brush Turkey	
Лайсанская кряква	Laysan-Stockente	Laysan Duck	
Лебеди	Schwäne	Swans	Cygnes
Лебедь-кликун	Singschwan	Whooper, Whooper-Swan	Cygne sauvage
Лебедь-трубач	Trompeterschwan	Trumpeter Swan	– trompette
Лебедь-шипун	Höckerschwan	Mute Swan	– muet
Лейпоа	Thermometerhuhn	Mallee Fowl	Leipoa ocellé
Лесные скрытохвостые куры	Wald-Steißhühner	Tinamous	Tinamous
Лоскутный казуар	Goldhalskasuar	One-wattled Cassowary	Casoar unicaronculé
Луговой лунь	Wiesenweihe	Montagu's Harrier	Busard de Montagu
Луни	Weihen	Harriers and Crane Hawks	Circinés
Луток	Zwergsäger	Smew	Harle piette
Мадагаскарская утка	Madagaskar-Ente	Meller's Duck	Canard de Meller
Малайский тантал	Malaien-Nimmersatt	Malayan Wood Ibis	Tantale blanc
Малая белая цапля	Seidenreiher	Little Egret	Aigrette garzette
Малая выпь	Zwergdommel	Little Bittern	Blongios nain
Малая канадская казарка	Dunkle Zwergkanadagans	Cackling Canada Goose	
Малая колпица	Kleiner Löffler	Black-faced Spoonbill	Petite Spatule
Малая поганка	Zwergtaucher	Little Grebe	Grèbe castagneux
Малая прямохвостая качурка	Sturmschwalbe	Storm Petrel	Pétrel tempête
Малый баклан	Zwergscharbe	Pygmy Cormorant	Cormoran pygmée
Малый луговой тетерев	Kleines Präriehuhn	Lesser Prairie Hen	
Малый подорлик	Schreiadler	– Spotted Eagle	Aigle pomarin
Малый фрегат	Prachtfregattvogel	Magnificent Man-o'-War Bird	Frégate superbe
Мандаринка	Mandarinente	Mandarin Duck	Canard mandarin
Марабу	Marabus	Marabous	Marabouts
Мексиканский сокол	Präriefalk	Prairie Falcon	Faucon des prairies
Молотоглав	Hammerkopf	Hamerkop	Ombrette
Молотоголовые цапли	Hammerköpfe	Hamerkops	Ombrettes
Монашенка	Witwenpfeifgans	White-faced Whistling Duck	Dendrocygne veuf
Морская качурка	Fregattensturmschwalbe	Frigate Petrel	Pétrel frégate
Морская чернеть	Bergente	Greater Scaup	Canard milouinan
Морянка	Eisente	Long-tailed Duck	– de Miquelon
Мохноногий канюк	Rauhfußbussard	Northern Rough-legged Buzzard	Buse pattue
Мохноногий курганник	Hochlandbussard	Upland Buzzard	
Мраморный чирок	Marmelente	Marbled Teal	Sarcelle marbrée
Мускусная утка	Moschusente	Moscovy Duck	Canard musqué
Нанду	Nandus	Rheas	Rhéidés
Нанду Дарвина	Darwin-Nandu	Darwin's Rhea	Nandou de Darwin
Настоящие грифы	Altweltgeier	Old World Vultures	Aegypiinés
Немой перепел	Japanische Wachtel	Japanese Quail	Caille du Japon
Нильский гусь	Nilgans	Egyptian Goose	Oie d'Égypte
Новозеландский нырцовый буревестник	Pinguin-Sturmtaucher	Common Diving Petrel	
Нырок Бэра	Schwarzkopf-Moorente	Baer's Pochard	

RUSSIAN NAME	GERMAN NAME	ENGLISH NAME	FRENCH NAME
Нырцовые буревестники	Tauchsturmvögel	Diving Petrels	Pélécanoïdidés
Обезьяноед	Affenadler	Monkey-eating Eagle	
Обыкновенная гага	Eiderente	Common Eider	Eider à duvet
Обыкновенная пустельга	Turmfalk	Old World Kestrel	Faucon crécerelle
Обыкновенный баклан	Kormoran	Black Cormorant	Grand Cormoran
Обыкновенный балобан	Würgfalk	Saker Falcon	Faucon sacre
Обыкновенный буревестник	Schwarzschnabel-Sturmtaucher	Manx Shearwater	Puffin des Anglais
Обыкновенный гоголь	Schellente	Common Golden-eye	Canard garrot
Обыкновенный киви	Streifenkiwi	– Kiwi	Kiwi austral
Обыковенный нанду	Nandu	American Rhea	Nandou américain
Обыкновенный перепел	Wachtel	Common Quail	Caille des blés
Обыкновенный сарыч	Mäusebussard	– Buzzard	Buse variable
Обыкновенный фламинго	Flamingo	Greater Flamingo	Flamant rose
Обыкновенный шпорцевый гусь	Sporengans	Spur-wing	Oie de Gambie
Огарь	Rostgans	Ruddy Shelduck	
Олуши	Tölpel	Boobies and Gannets	Sulidés
Орел-боец	Kampfadler	Martial Eagle	Aigle Martial
Орел-карлик	Zwergadler	Booted Eagle	– botté
Орел-могильник	Kaiseradler	Imperial Eagle	– impérial
Орел-скоморох	Gaukler	Bateleur	Bateleur
Орел-хеела	Indischer Schlangenhabicht	Crested Serpent Eagle	
Орлан-белохвост	Seeadler	White-tailed Sea Eagle	Pygargue à queue blanche
Орлан-долгохвост	Bandseeadler	Pallas' Sea Eagle	Pygargue à queue blanche
Орлиный фрегат	Adlerfregattvogel	Ascension Frigate Bird	Frégate aigle
Ослиный пингвин	Eselspinguin	Gentoo Penguin	Manchot Gentoo
Осоед	Wespenbussard	Honey Buzzard	Bondrée apivore
Очковая гага	Plüschkopfente	Spectacled Eider	Eider de Fischer
Очковый пингвин	Brillenpinguin	Blackfooted Penguin	Manchot du Cap
Очковый баклан	Brillenkormoran	Pallas' Cormorant	
Паламедеи	Wehrvögel	Screamers	Anhimidés
Певчий ястреб	Singhabicht	Pale Chanting Goshawk	
Пеганка	Brandgans	Common Shelduck	Tadorne de Belon
Пеганка-раджа	Radjahgans	Radjah Shelduck	
Пеганки	Halbgänse	Shelducks	Tadornes et Casarcas
Пегая древесная утка	Fahlpfeifgans	Fulvous Whistling Duck	
Пеликан-носорог	Nashornpelikan	American White Pelican	
Пеликаны	Pelikane	Pelicans	Pélécanidés, Pélicans
Пестроносый турпан	Brillenente	Surf Scoter	Macreuse à lunettes
Пингвин Гумбольдта	Humboldtpinguin	Peruvian Penguin	
Пискулька	Zwerggans	Lesser White-fronted Goose	Oie naine
Пластинчатоклювые	Gänsevögel	Waterfowl and Screamers	Ansériformes
Поганки	Lappentaucher	Grebes	Podicipitidés, Grèbes
Поганки и гагары	Steißfüße	Pygopodes	Pygopodes
Поганковые	Lappentaucher	Grebes	Podicipédiformes
Полевой лунь	Kornweihe	Hen Harrier	Busard Saint-Martin
Полынный тетерев	Beifußhuhn	Sage Grouse	Gelinotte des armoises
Полярная гагара	Eistaucher	Great Northern Diver	Plongeon imbrin
Птицы	Vögel	Birds	Oiseaux
Пуховые аисты	Wollhalsstörche	Bishop Storks	Cigognes épiscopales
Пятнистый баклан	Tüpfelkormoran	Spotted Cormorant	
Райская пеганка	Paradieskasarka	Paradise Shelduck	Casarca de Paradis
Расщепнолапый гусь	Spaltfußgans	Australian Magpie Goose	Oie semi-palmée
Рогатая анхима	Hornwehrvogel	Horned Screamer	Kamichi cornu
Рогатая поганка	Ohrentaucher	Slavonian Grebe	Grèbe esclavon
Розово-красный фламинго	Rosaflamingo	European Flamingo	Flamant rose
Розовый пеликан	Rosapelikan	Eastern White Pelican	Pélican blanc
Рыжая цапля	Purpurreiher	Purple Heron	Héron pourpré
Рыжеголовый балобан	Lannerfalk	Lanner Falcon	Faucon lanier
Рябчик	Haselhuhn	Hazel Hen	Gelinotte des Bois
Савка	Weißkopfruderente	White-headed Duck	Canard à tête blanche
Сарычи	Bussarde	Buzzards	Buses
Свиязь	Pfeifente	European Wigeon	Canard siffleur
Священный ибис	Heiliger Ibis	Sacred Ibis	Ibis sacré
Северная качурка	Wellenläufer	Leach's Petrel	Pétrel cul-blanc
Северный гигантский буревестник	Nördlicher Riesensturmvogel	Northern Giant Fulmar	
Северный дымчатый альбатрос	– Rußalbatros	Sooty Albatross	
Североамериканская выпь	Nordamerikanische Rohrdommel	American Bittern	Butor d'Amérique
Секретарь	Sekretär	Secretary Bird	Secrétaire
Серая качурка	Gabelschwanz-Wellenläufer	Fork-tailed Storm Petrel	
Серая куропатка	Rebhuhn	Grey Partridge	Perdrix grise
Серая утка	Schnatterente	Gadwall	Canard chipeau
Серая цапля	Graureiher	Grey Heron	Héron cendré

RUSSIAN NAME	GERMAN NAME	ENGLISH NAME	FRENCH NAME
Сероголовая пеганка	Graukopfkasarka	South African Shelduck	Casarca du Cap
Сероголовый альбатрос	Graukopfalbatros	Grey-headed Albatross	Albatros à tête grise
Серощёкая поганка	Rothalstaucher	Red-necked Grebe	Grèbe jougris
Серый буревестник	Rußsturmtaucher	Sooty Shearwater	Puffin fuligineux
Серый гусь	Graugans	Greylag Goose	Oie cendrée
Серый пеликан	Graupelikan	Grey Pelican	
Сибирская гага	Scheckente	Steller's Eider	Eider de Steller
Синекрылый чирок	Blauflügelente	Blue-winged Teal	Sarcelle soucrourou
Синий пингвин	Adeliepinguin	Adelie Penguin	
Синьга	Trauerente	Black Scoter	Macreuse noire
Скопа	Fischadler	Osprey	Balbuzard fluvatile
Скрытохвостые куры	Steißhühner	Tinamous	Tinamous
Снежный буревестник	Schneesturmvogel	Snow Petrel	
Снежный гималайский сип	Schneegeier	Himalayan Griffon	
Сокол-сапсан	Wanderfalk	Peregrine Falcon	Faucon pèlerin
Соколы	Eigentliche Falken	Gyrfalcons, Falcons and Kestrels	Falconinés
Сорные куры	Großfußhühner	Moundfowl	Mégapodiidés
Средиземноморский рыжеголовый балобан	Feldeggsfalk	Feldegg's Falcon	Faucon de Feldegg
Средний крохаль	Mittelsäger	Red-breasted Merganser	Harle huppé
Средняя белая цапля	Edelreiher	Yellow-billed Egret	Aigrette à bec jaune
Степная пустельга	Rötelfalk	Lesser Kestrel	Faucon crécerellette
Степной лунь	Steppenweihe	Pallid Harrier	Busard pâle
Стервятник	Schmutzgeier	Egyptian Vulture	Percnoptère d'Égypte
Странствующий альбатрос	Wanderalbatros	Wandering Albatross	Albatros hurleur
Страусовый перепел	Straußwachtel	Crowned Wood Partridge	Roulroul
Страусы	Strauße	Ostriches	Struthionidés, Autruches
Сухонос	Schwanengans	Swan Goose	Oie cygnoïde
Тайфунник Бульвера	Weichnasen-Sturmvogel	Bulwer's Petrel	Pétrel de Bulwer
Тайфунники	Hakensturmtaucher	Gadfly Petrels	
Танталы	Nimmersatte	Wood Ibises	Tantales
Теневые птицы	Hammerköpfe	Hamerkops	Scopidés
Тетерев	Birkhuhn	Black Grouse	Tétras lyre
Тифонова цапля	Sumatrareiher	Dusky Grey Heron	Héron de Sumatra
Толстоклювый пингвин	Dickschnabelpinguin	Fiordland Penguin	
Тонкоклювый буревестник	Millionensturmtaucher	Short-tailed Shearwater	
Трубконосые	Röhrennasen	Tube-nosed Swimmers	Procellariiformes
Тундряная куропатка	Alpenschneehuhn	Ptarmigan	Lagopède muet
Турач	Halsbandfrankolin	Francolin	Francolin d'Europe
Турпаны	Trauerenten	Scoters	Macreuses
Утиные	Entenvögel	Ducks and Geese	Anatidés
Ушастый гриф	Ohrengeier	Lappet-faced Vulture	Vautour oricou
Фаэтоны	Tropikvögel	Tropic Birds	Phaëthontidés, Phaëthons
Филиппинская утка	Philippinen-Ente	Philippine Duck	Canard des Philippines
Фламинго	Flamingos	Flamingoes	Flamants, Phoenicoptéridés
Фрегаты	Fregattvögel	Frigate Birds	Frégatidés, Frégates
Хвостатый тетерев	Schweif-Waldhuhn	Northern Sharp-tailed Grouse	Gelinotte à queue fine
Химанго	Chimango	Chimango Caracara	
Хищные птицы	Greifvögel	Diurnal Birds of Prey	Falconiformes
Хохлатая пеганка	Schopfkasarka	Korean Crested Shelduck	
Хохлатая чернеть	Reiherente	Tufted Duck	Canard morillon
Хохлатый гокко	Schopfhokko	Nocturnal Curassow	
Хохлатый коршун	Schopfmilan	Square-tailed Kite	
Хохлатый орел	Nepal-Haubenadler	Mountain Hawk Eagle	
Хохлатый осоед	Malayen-Wespenbussard	Crested Honey Buzzard	
Цапли	Reiher	Herons	Ardéidés
Цапля-челноклюв	Kahnschnabel	Boat-billed Heron	Savacou huppé
Чеглок	Baumfalk	Hooby	Faucon hobereau
Чеглок Элеоноры	Eleonorenfalk	Eleonora's Falcon	Faucon d'Eléonore
Чемуйчатый крохаль	Schuppensäger	Chinese Merganser	Harle chinois
Черная желтоносая кряква	Fleckschnabelente	Spot-billed Duck	Canard à bec tacheté
Черная казарка	Ringelgans	Brent Goose	Bernache cravant
Чернобровый альбатрос	Schwarzbrauenalbatros	Black-browed Albatross	Albatros à sourcil noir
Чернобрюхая качурка	Schwarzbauch-Sturmschwalbe	Black-bellied Storm Petrel	Pétrel des tropiques
Чернозобая гагара	Prachttaucher	Black-throated Diver	Plongeon lumme
Черноклювый аист	Schwarzschnabelstorch	Oriental White Stork	
Чернокрылый коршун	Schwarzflügel-Gleitaar	Black-winged Kite	Faucon blanc
Черноногий альбатрос	Schwarzfußalbatros	Black-footed Albatross	Albatros à pieds noirs
Чернопегий лунь	Schwarzweißweihe	Pied Harrier	Busard pie
Черношапочный тайфунник	Schwarzkappensturmtaucher	Black-capped Petrel	Pétrel diablotin
Черношеий лебедь	Schwarzhalsschwan	Black-necked Swan	Cygne à cou noir
Черношейная поганка	Schwarzhalstaucher	– – Grebe	Grèbe à cou noir
Черный аист	Schwarzstorch	Black Stork	Cigogne noire
Чёрный гриф	Mönchsgeier	– Vulture	Vautour moine
Черный коршун	Schwarzmilan	– Kite	Milan noir

RUSSIAN NAME	GERMAN NAME	ENGLISH NAME	FRENCH NAME
Черный лебедь	Trauerschwan	– Swan	Cygne noir
Черный турпан	Samtente	Velvet Scoter	Macreuse brune
Черный шуан	Mohrenguan	Guatemalan Black Chachalaca	
Чилийская свиязь	Chile-Pfeifente	Chiloë Wigeon	Siffleur du Brésil
Чилийский фламинго	Chilenischer Flamingo	Chilean Flamingo	
Чилийский чирок	Chile-Krickente	South American Green-winged Teal	Sarcelle du Chili
Чирок-клоктун	Gluckente	Baikal Teal	– élégante
Чирок-свистунок	Krickente	Teal	– d'hiver
Чирок-трескунок	Knäkente	Garganey	– d'été
Шилохвость	Spießente	Blue-billed Pintail	Canard pilet
Ширококрылый канюк	Breitschwingenbussard	Broad-winged Hawk	Petite Buse
Широконоска	Löffelente	Northern Shoveler	Canard souchet
Шишконосый гусь	Höckerglanzente	Comb Duck	Sarcidiorne à crète
Шлемоносный гокко	Helmhokko	Helmeted Curassow	Pauxi pierre
Шлемоносный казуар	Helmkasuar	Australian Cassowary	Casoar à casque
Шпорцевые гуси	Sporengänse	Spur-winged Geese	Oies armées
Змеешейки	Schlangenhalsvögel	Darters	Anhingidés
Эму	Emu	Emu	Émeu d'Australie
Южноамериканская широконоска	Südamerikanische Löffelente	Red Shoveler	Souchet roux
Южный гигантский буревестник	Südlicher Riesensturmvogel	Southern Giant Fulmar	
Южный каракара	– Karakara	Striated Caracara	
Яванская древесная утка	Indien-Pfeifgans	Indian Whistling Duck	Dendrocygne siffleur
Яванский марабу	Sunda-Marabu	Lesser Marabou	Petit Marabout
Японский баклан	Japanischer Kormoran	Japanese Cormorant	
Ястреба	Habichte i. e. S.	Goshawks	Éperviers
Ястребиный орел	Habichtsadler	Bonelli's Eagle	Aigle de Bonelli
Ястреб Купера	Rundschwanzsperber	Cooper's Hawk	Épervier de Cooper
Ястреб-перепелятник	Sperber	Eurasian Sparrowhawk	– d'Europe
Ястреб-тетеревятник	Habicht	Goshawk	Autour des palombes

Conversion Tables of Metric to U.S. and British Systems

U.S. Customary to Metric		*Metric to U.S. Customary*	
	Length		
To convert	*Multiply by*	*To convert*	*Multiply by*
in. to mm.	25.4	mm. to in.	0.039
in. to cm.	2.54	cm. to in.	0.394
ft. to m.	0.305	m. to ft.	3.281
yd. to m.	0.914	m. to yd.	1.094
mi. to km.	1.609	km. to mi.	0.621
	Area		
sq. in. to sq. cm.	6.452	sq. cm. to sq. in.	0.155
sq. ft. to sq. mi.	0.093	sq. m. to sq. ft.	10.764
sq. yd. to sq. m.	0.836	sq. m. to sq. yd.	1.196
sq. mi. to ha.	258.999	ha. to sq. mi.	0.004
	Volume		
cu. in. to cc.	16.387	cc. to cu. in.	0.061
cu. ft. to cu. m.	0.028	cu. m. to cu. ft.	35.315
cu. yd. to cu. m.	0.765	cu. m. to cu. yd.	1.308
	Capacity (liquid)		
fl. oz. to liter	0.03	liter to fl. oz.	33.815
qt. to liter	0.946	liter to qt.	1.057
gal. to liter	3.785	liter to gal.	0.264
	Mass (weight)		
oz. avdp. to g.	28.35	g. to oz. avdp.	0.035
lb. avdp. to kg.	0.454	kg. to lb. avdp.	2.205
ton to t.	0.907	t. to ton	1.102
l. t. to t.	1.016	t. to l. t.	0.984

Abbreviations

U.S. Customary

avdp.—avoirdupois
ft.—foot, feet
gal.—gallon(s)
in.—inch(es)
lb.—pound(s)
l. t.—long ton(s)
mi.—mile(s)
oz.—ounce(s)
qt.—quart(s)
sq.—square
yd.—yard(s)

Metric

cc.—cubic centimeter(s)
cm.—centimeter(s)
cu.—cubic
g.—gram(s)
ha.—hectare(s)
kg.—kilogram(s)
m.—meter(s)
mm.—millimeter(s)
t.—metric ton(s)

By kind permission of Walker: Mammals of the World
©1968 Johns Hopkins Press, Baltimore, Md., U.S.A.

TEMPERATURE

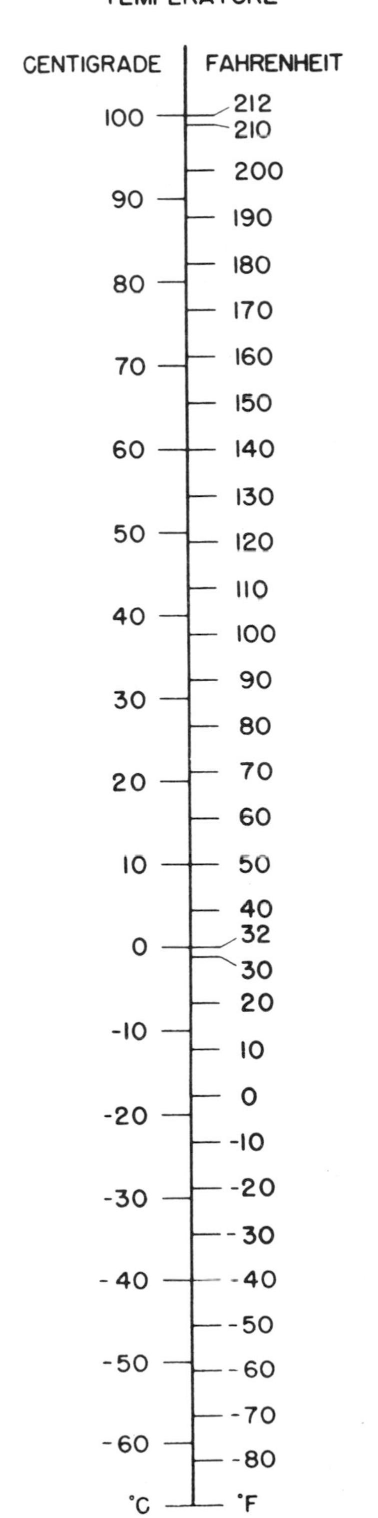

AREA

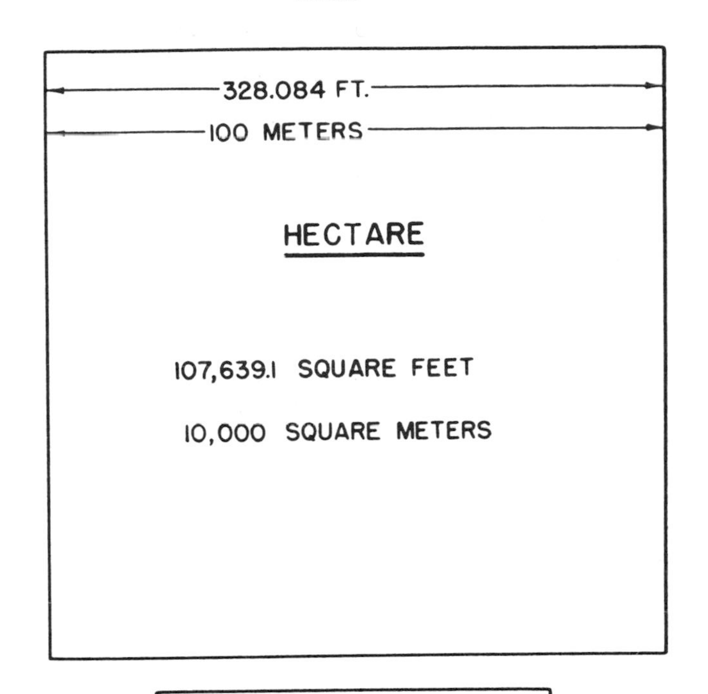

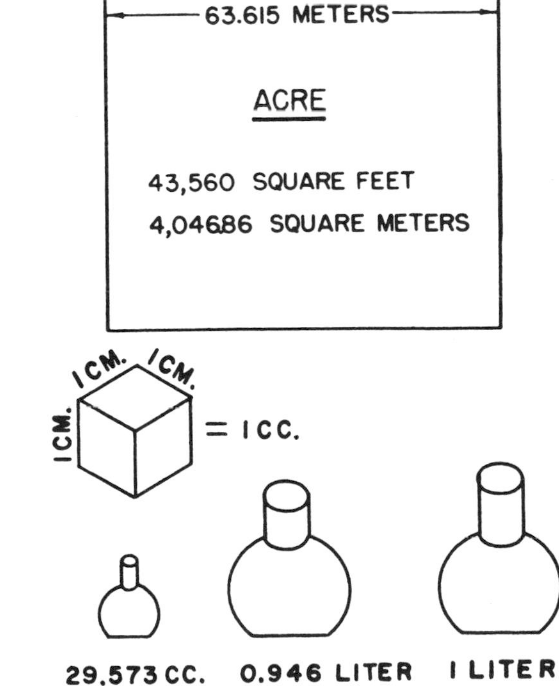

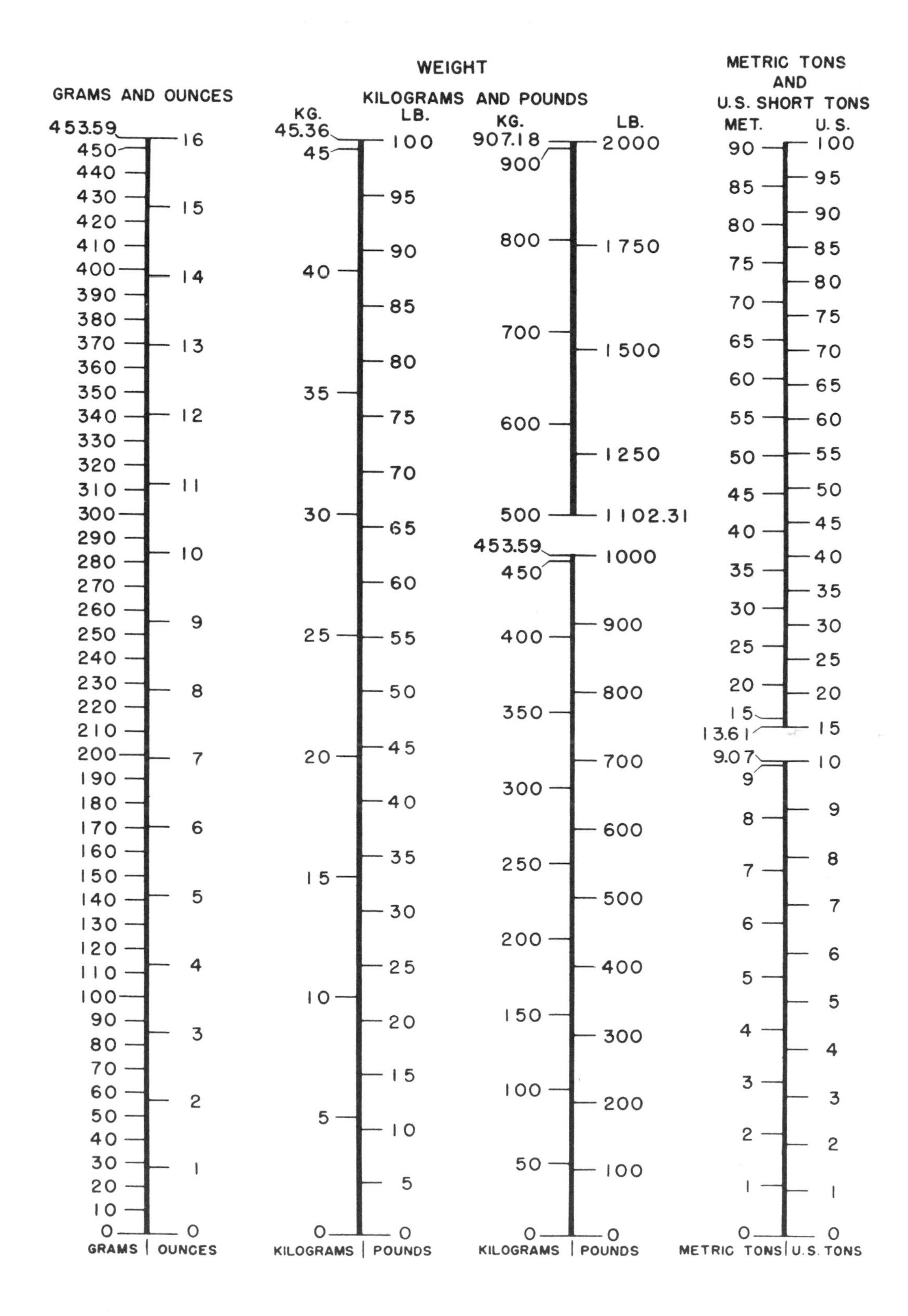
WEIGHT
GRAMS AND OUNCES
KILOGRAMS AND POUNDS
METRIC TONS AND U.S. SHORT TONS
KG.
LB.
MET.
U.S.
GRAMS | OUNCES
KILOGRAMS | POUNDS
KILOGRAMS | POUNDS
METRIC TONS | U.S. TONS

LENGTH: MILLIMETERS AND INCHES

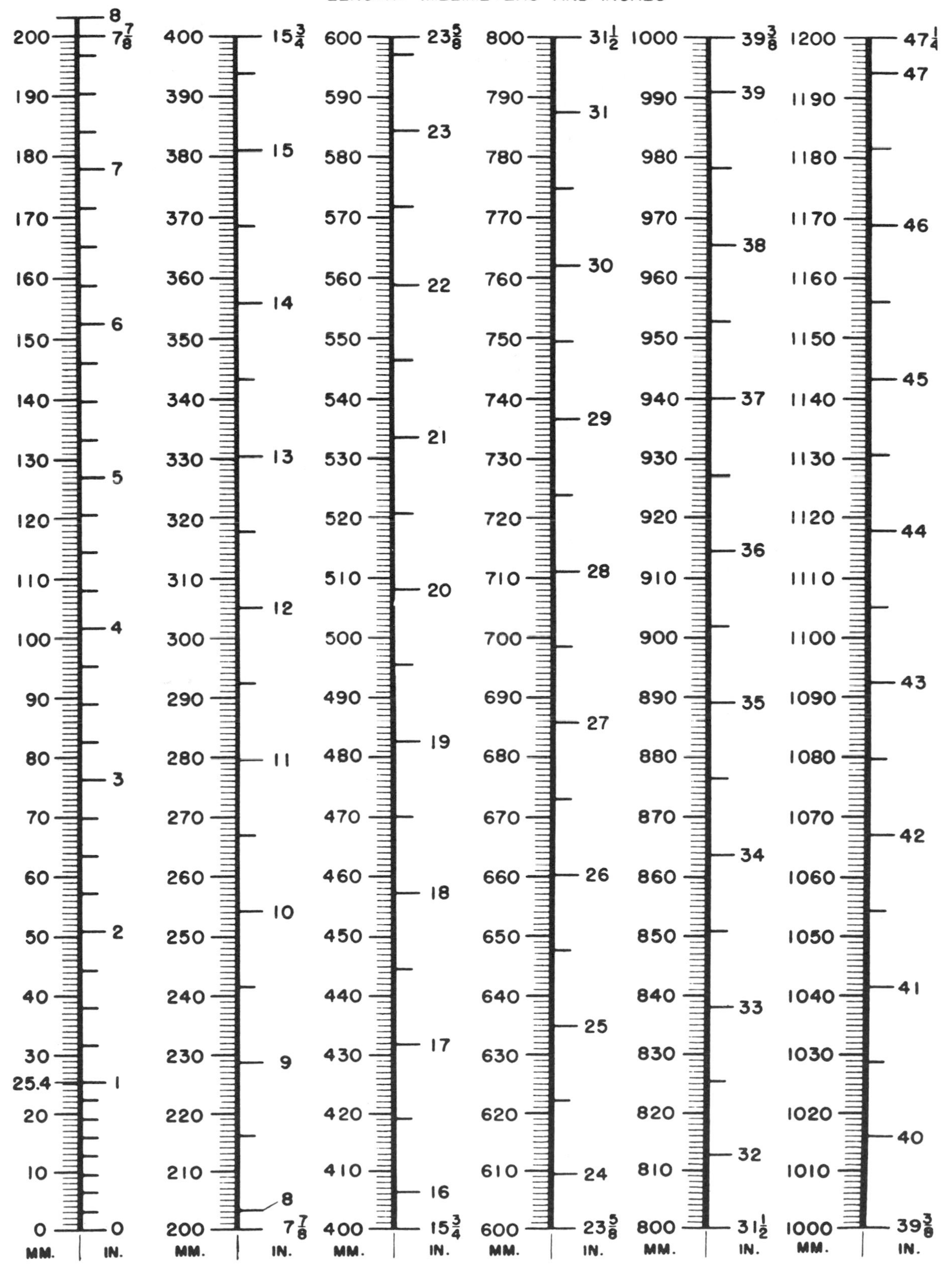

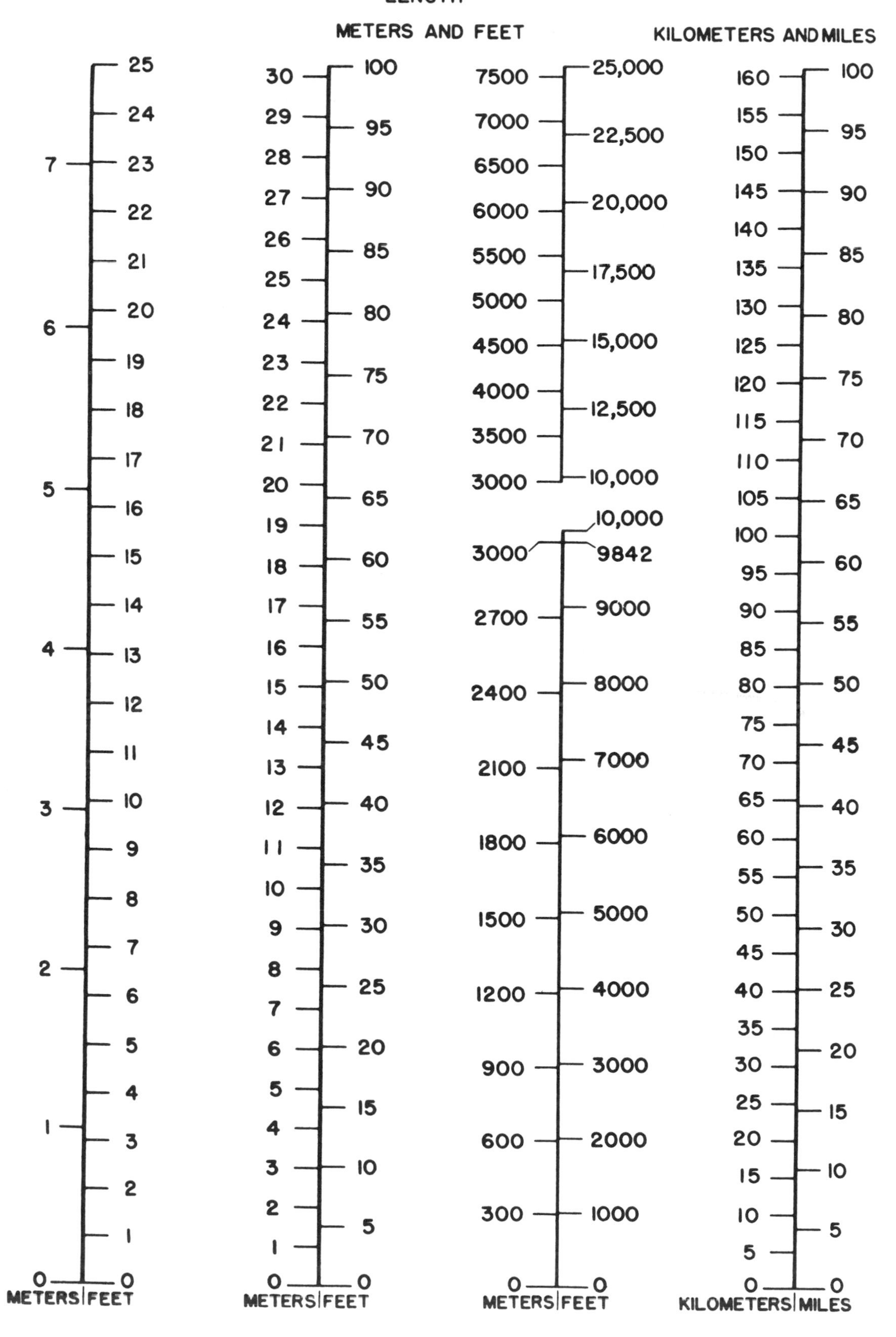
LENGTH
METERS AND FEET
KILOMETERS AND MILES
METERS
FEET
KILOMETERS
MILES
3000
10,000
9842
25,000
7500

Supplementary Readings

These references of books and articles published in scientific journals deal with animals and topics that are covered in this volume. Some of these were the original sources on which the content of this book is based. These titles are intended as an aid to readers who are interested in additional information and more detailed coverage of the subjects contained in this book*.

BIRDS—GENERAL BOOKS:

Alexander, W. B. 1963. *Birds of the Ocean.* G. P. Putnam's Sons, New York.

Allen, A. A. 1934. *American Bird Biographies.* Sutton, Ithaca, New York.

—. 1961. *The Book of Bird Life.* D. Van Nostrand Company, New York.

Allen, G. M. 1925. *Birds and their Attributes.* Marshall Jones Company, Boston.

Amadon, D. 1966. *Birds around the World: A Geographical Look at Evolution and Birds.* Natural History Press, Garden City, New York.

Armstrong, E. A. 1942. *Bird Display and Behavior.* Lindsay Drummond Ltd., London. (Reissued in paperback by Dover Publications.)

Austin, O. L., Jr. and A. Singer. 1961. *Birds of the World.* Golden Press, New York.

Bannerman, D. A. 1963-1968. *Birds of the Atlantic Islands.* 4 vols. Oliver and Boyd, Edinburgh.

Barruel, P. 1954. *Birds of the World: Their Life and Habits.* Oxford University Press, New York.

Bent, A. C. 1919-1958. *Life Histories of North American Birds.* 20 vols. Bulletin of the United States National Museum. (Reissued in paperback by Dover Publications.)

Berger, A. J. 1961. *Bird Study.* John Wiley and Sons, New York.

Darling, L. and L. Darling. 1962. *Bird.* Houghton Mifflin Company, Boston.

Falla, R. A., R. B. Sibson and E. G. Turbott. 1967. *A Field Guide to the Birds of New Zealand.* Houghton Mifflin Company, Boston.

Farner, D. S., J. R. King and K. C. Parkes, eds. 1971. *Avian Biology.* Academic Press, New York.

Fisher, J. 1954. *A History of Birds.* Houghton Mifflin Company, Boston.

— and R. M. Lockley. 1954. *Seabirds: An Introduction to the Natural History of the Seabirds of the North Atlantic.* Collins, Sons and Company, London.

— and R. T. Peterson. 1963. *The World of Birds.* Doubleday and Company, New York.

Gilliard, E. T. 1958. *Living Birds of the World.* Doubleday and Company, New York.

Greenewalt, C. H. 1968. *Bird Song: Acoustics and Physiology.* Smithsonian Institution Press, Washington, D.C.

Greenway, J. C., Jr. 1958. *Extinct and Vanishing Birds of the World.* American Committee for International Wildlife Protection. Special Publication No. 13.

Heinroth, O. and K. Heinroth. 1958. *The Birds.* University of Michigan Press, Ann Arbor.

Hess, G. 1951. *The Bird: Its Form and Structure.* Greenberg, New York.

Hill, R. 1967. *Australian Birds.* Thomas Nelson (Australia), Melbourne.

Howard, H. E. 1948. *Territory in Bird Life.* Collins, London.

Lack, D. 1971. *Ecological Isolation in Birds.* Harvard University Press, Cambridge.

Lanyon, W. E. 1963. *Biology of Birds.* Natural History Press, Garden City, New York.

Mackworth-Praed, C. W. and C. H. B. Grant. 1952-1955. *Birds of Eastern and Northeastern Africa.* 2 vols. Longmans, Green and Company, London.

— and —. 1962-1963. *Birds of the Southern Third of Africa.* 2 vols. Longmans, Green and Company, London.

Marshall, A. J. 1960-1061. *Biology and Comparative Physiology of Birds.* 2 vols. Academic Press, New York.

Matthews, G. V. T. 1955. *Bird Navigation.* Cambridge University Press, Cambridge.

Meyer de Schauensee, R. 1966. *The Species of Birds of South America and Their Distribution.* Academy of Natural Sciences of Philadelphia.

Murphy, R. C. and D. Amadon. 1953. *Land Birds of America.* McGraw-Hill Book Company, New York.

Pearson, T. G. 1936. *Birds of America.* Garden City Books, Garden City, New York.

Peters, J. L. 1931-1960. *Checklist of Birds of the World.* 9 vols. Harvard University Press, Cambridge, Massachusetts.

Peterson, R. T. 1947. *A Field Guide to the Birds.* Houghton Mifflin Company, Boston.

—. 1963. *The Birds.* Time, Inc., New York.

. 1964. *Birds over America.* Dodd, Mead and Company, New York.

—, G. Mountfort and P. A. D. Hollom. 1966. *A Field Guide to the Birds of Britain and Europe.* Houghton Mifflin Company, Boston.

Pettingill, O. S., Jr., ed. 1965. *The Bird Watcher's America.* McGraw-Hill Book Company, New York.

—. 1970. *Ornithology in Laboratory and Field.* Burgess Publishing Company, Minneapolis.

Pickwell, G. B. 1939. *Birds.* McGraw-Hill Book Company, New York.

Pough, R. H. 1946-1951. *Audubon Bird Guide: Eastern Landbirds, Waterbirds.* 2 vols. Doubleday and Company, New York.

Pycraft, W. P. 1910. *A History of Birds.* Methuen and Company, London.

Rand, A. L. 1967. *Ornithology: An Introduction.* W. W. Norton and Company, New York.

Robbins, C. S., B. Bruun, H. S. Zim and A. Singer. 1966. *Birds of North America.* Golden Press, New York.

Saunders, A. A. 1954. *The Lives of Wild Birds.* Doubleday and Company, Garden City, New York.

Stefferud, A. and A. L. Nelson. 1966. *Birds in Our Lives.* U.S. Department of the Interior, Washington, D.C.

Thomson, A. L., ed. 1964. *A New Dictionary of Birds.* McGraw-Hill Book Company, New York.

Van Tyne, J. and A. Berger. 1959. *Fundamentals of Ornithology.* John Wiley and Sons, New York. (Reissued in paperback by Dover Publications.)

Vaurie, C. 1965. *The Birds of the Palearctic Fauna: A Systematic Reference. Non Passeriformes.* H. F. and G. Witherby, London.

Wallace, G. J. 1963. *An Introduction to Ornithology.* The Macmillan Company, New York.

Watson, G. E. 1966. *Seabirds of the Tropical Atlantic Ocean,* Smithsonian Publication 4680. Smithsonian Press, Washington, D.C.

Welty, J. C. 1962. *The Life of Birds.* W. B. Saunders Company, Philadelphia and London.

Wetmore, A. 1931. *Birds.* Smithsonian Scientific Series, Vol. 9, Washington, D.C.

— and others. 1965. *Water, Prey and Game Birds of North America.* National Geographic Society, Washington, D.C.

Wing, L. W. 1956. *Natural History of Birds: A Guide to Ornithology.* Ronald Press Company, New York.

Witherby, H. F. and others. 1938-1941. *The Handbook of British Birds.* 5 vols. H. F. and G. Witherby, London.

*Supplementary Readings prepared by John B. Brown.

SCIENTIFIC JOURNALS:

Greenewalt, C. H. 1969. How Birds Sing. *Scientific American* 221(5):126-139.

Mayr, E. and D. Amadon. 1951. A Classification of Recent Birds. *American Museum Novitiates* 1496: 1-42.

Nisbet, I. C. T. and W. H. Drury, Jr. 1967. Scanning the Sky: Birds on Radar. *Massachusetts Audubon* 51:166-174.

Parkes, K. C. 1966. Speculations on the Origins of Feathers. *The Living Bird* 5:77-86.

Schmidt-Nielson, Knut. 1959. Salt Glands. *Scientific American* 200(1):109-116.

—. 1971. How Birds Breathe. *Scientific American* 225(6):72-79.

Short, L. L. 1969. Taxonomic Aspects of Avian Hybridization. *Auk* 86(1):84-105.

Stager, K. 1967. Avian Olfaction. *American Zoologist* 7(3):415-419.

Taylor, T. G. 1970. How an Eggshell is Made. *Scientific American* 222(3):88-97.

Tucker, Vance A. 1969. The Energetics of Bird Flight. *Scientific American* 220(5):70-81.

Wetmore, A. 1960. A Classification of Birds of the World. *Smithsonian Miscellaneous Collections* 139(11):1-37.

BIRDS COVERED IN THIS VOLUME BOOKS:

Allen, D. L. 1956. *Pheasants in North America.* Stackpole Company, Harrisburg, Pennsylvania, and Wildlife Management Institute, Washington.

Allen, R. P. 1942. *The Roseate Spoonbill.* National Audobon Research Report No. 2, New York. (Dover reprint available.)

—. 1956. *The Flamingos: Their Life History and Survival, with Special Reference to the American or West Indian Flamingo (Phoenicopterus ruber).* National Audubon Society Research Report No. 5, New York.

Ammann, G. A. 1957. *The Prairie Grouse of Michigan.* Game Division, Department of Conservation.

Austing, G. R. 1964. *The World of the Red-tailed Hawk.* J. B. Lippincott Company, Philadelphia.

Banko, W. E. 1960. *The Trumpeter Swan: Its History, Habits, and Population in the United States.* North American Fauna No. 63:i-x;1-214. U.S. Fish and Wildlife Service, Washington, D.C.

Beebe, W. 1926. *Pheasants, Their Lives and Homes.* 2 vols. Doubleday, Page and Company, Garden City, New York.

Bennett, L. J. 1938. *The Blue-winged Teal: Its Ecology and Management.* Collegiate Press, Ames, Iowa.

Bent, A. C. 1922. *Life Histories of North American Petrels and Pelicans and Their Allies.* U.S. National Museum Bulletin 121.

—. 1923-1925. *Life Histories of North American Wild Fowl,* Parts I and II. U.S. National Museum Bulletin 126, 130.

—. 1926. *Life Histories of North American Marsh Birds.* U.S. National Museum Bulletin 135.

—. 1937-1938. *Life Histories of North American Birds of Prey,* Parts I and II. U.S. National Museum Bulletin 167, 170.

Brown, L. and D. Amadon. 1968. *Eagles, Hawks and Falcons of the World.* 2 vols. McGraw-Hill Book Company, New York.

Bump, G. and others. 1947. *The Ruffed Grouse: Life History, Propogation, Management.* New York State Conservation Department, Albany.

Cottrille, W. P. and B. D. Cottrille. 1958. *Great Blue Heron: Behavior at the Nest.* University of Michigan Museum of Zoology Miscellaneous Publication No. 102.

Craighead, J. and F. Craighead. 1956. *Hawks, Owls and Wildlife.* Stackpole Company, Harrisburg, Pennsylvania. (Dover reprint available.)

Delacour, J. 1951. *The Pheasants of the World.* Country Life, London.

—. 1954-1964. *The Waterfowl of the World.* 4 vols. Country Life, London.

Edminister, F. C. 1947. *The Ruffed Grouse: Its Life History, Ecology and Management.* The Macmillan Company, New York.

Einarsen, A. S. 1965. *Black Brant: Sea Goose of the Pacific Coast.* University of Washington Press, Seattle.

Fisher, J. 1952. *The Fulmar.* Collins, London.

Girard, G. I. 1937. *Life History, Habits and Food of the Sage Grouse, Centrocercus urophasianus* Bonaparte. University of Wyoming Publications 3:1-56.

Gordon, S. 1955. *The Golden Eagle: King of Birds.* Collins, London.

Grice, D. and J. P. Rogers. 1965. *The Wood Duck in Massachusetts.* Massachusetts Division of Fisheries and Game, Boston.

Grossman, M. L. and J. Hamlet, 1964. *Birds of Prey of the World.* Clarkson S. Potter, New York.

Gurney, J. H. 1913. *The Gannet: A Bird with a History.* Witherby and Company, London.

Hanson, H. C. 1965. *The Giant Canada Goose.* Southern Illinois University Press, Carbondale.

Haverschmidt, F. 1949. *The Life of the White Stork.* E. J. Brill, Leiden.

Herrick, F. H. 1934. *The American Eagle.* D. Appleton-Century Company, New York.

Hewitt, O. H. 1967. *The Wild Turkey and Its Management.* The Wildlife Society, Washington, D.C.

Hickey, J. J., ed. 1969. *Peregrine Falcon Populations: Their Biology and Decline.* University of Wisconsin Press, Madison.

Hochbaum, H. A. 1959. *The Canvasback on a Prairie Marsh.* Stackpole Company, Harrisburg, Pennsylvania, and Wildlife Management Institute, Washington, D.C.

Johnsgard, P. A. 1965. *Handbook of Waterfowl Behavior.* Cornell University Press, Ithaca, New York.

Koford, C. B. 1953. *The California Condor.* National Audubon Society Research Report No. 4, New York. (Dover reprint available.)

Kortright, F. H. 1942. *The Ducks, Geese and Swans of North America.* American Wildlife Institute, Washington, D.C.

Latham, R. M. 1956. *Complete Book of the Wild Turkey.* Stackpole Company, Harrisburg, Pennsylvania.

Lewis, H. F. 1929. *The Natural History of the Double-crested Cormorant (Phalacrocorax auritus auritus).* Ru-Mi-Lou Books, Ottawa, Canada.

Lockley, R. M. 1961. *Shearwaters.* Doubleday and Company, Garden City, New York.

—. 1962. *Puffins.* Doubleday and Company, Garden City, New York.

Lowe, F. A. 1955. *The Heron.* Collins, London.

MacPherson, H. B. 1909. *The Home Life of a Golden Eagle.* Witherby and Company, London.

Mendall, H. L. 1958. *The Ring-necked Duck in the Northeast.* University of Maine Bulletin, 60:i-xvi:1-317.

Meyerriecks, A. J. 1960. *Comparative Breeding Behavior of Four Species of North American Herons.* Publication of the Nuttall Ornithology Club No. 2, Cambridge, Massachusetts.

Millais, J. G. 1902. *The Natural History of British Surface-feeding Ducks.* Longmans, Green and Company, London.

—. 1918. *British Diving Ducks.* 2 vols. Longmans, Green and Company, London.

Olson, S. T. and W. H. Marshall. 1952. *The Common Loon in Minnesota.* Minnesota Museum of Natural History Occasional Papers No. 5.

Palmer, R. S., ed. 1962. *Handbook of North American Birds. Volume 1. Loons through Flamingos.* Yale University Press, New Haven, Connecticut.

Phillips, J. C. 1922-1926. *A Natural History of the Ducks.* 4 vols. Houghton Mifflin Company, Boston.

Rosene, W., Jr. 1969. *The Bobwhite Quail: Its Life and Management.* Rutgers University Press, New Brunswick, New Jersey.

Sanger, M. B. 1967. *World of the Great White Heron: A Saga of the Florida Keys.* Devin-Adair Company, New York.

Savage, C. 1952. *The Mandarin Duck.* Adam and Charles Black, London.

Soper, J. D. 1930. *The Blue Goose.* Department of the Interior, Ottawa, Canada.

Sowls, L. K. 1955. *Prairie Ducks: A Study of Their Behavior, Ecology, and Management.* Stackpole Company, Harrisburg, Pennsylvania, and Wildlife Management Institute, Washington, D.C.

Sparks, J. and T. Soper. 1967. *Penguins.* David and Charles, Newton Abbot.

Wormer, J. van. 1968. *The World of the Canada Goose.* J. B. Lippincott Company, Philadelphia.

Wright, B. S. 1954. *High Tide and East Wind: The Story of the Black Duck.* Stackpole Company, Harrisburg, Pennsylvania, and Wildlife Management Institute, Washington, D.C.

SCIENTIFIC JOURNALS:

Allen, R. P. and F. P. Mangels. 1940. Studies of the Nesting Behavior of the Black-crowned Night Heron. *Proceedings of the Linnaean Society of New York,* Nos. 50-51:1-28.

Chapman, F. M. 1905. A Contribution to the Life History of the American Flamingo *(Phoenicopterus ruber),* with remarks upon specimens. *Bulletin of the American Museum of Natural History* 21:53-77.

Low, J. B. 1945. Ecology and Management of the Redhead, *Nyroca americana,* in Iowa. *Ecology Monographs* 15:35-69.

Picture Credits

Artists: Z. Burian (p. 26). W. Ebke (p. 88 upper left, lower left, lower right). W. Eigener p. 88 upper right, 121, 216, 217, 345, 365). G. Kapitzke (p. 35, 36). F. Reimann (p. 61/62, 255/256, 334, 346, 355, 356, 366, 375, 376, 385/386, 397, 398, 403, 404, 417, 418, 423). J. Ritter (p. 87, 115, 116, 124, 138, 147, 148, 157, 158, 167, 168, 177, 195, 196, 197, 198, 215, 222, 237, 238, 267, 268, 273, 274, 281, 282, 283, 284, 294, 295, 296, 303, 304, 305, 306, 307, 308, 309, 310, 317, 318, 319, 320, 424, 435, 436, 445, 446, 470, 479, 480, 481, 482).

Scientific advisors to the artists: Prof. Dr. J. Augusta (Burian), Dr. D. Heinemann (Eigener), Prof. Dr. H. Dathe (Reimann), Prof. Dr. G. Niethammer (Ebke), Dr. J. Steinbacher (Ritter), Prof. Dr. H. Wilkens (Kapitzke). Color photographs: Aichhorn (p. 328). Barnfather/ Photo Researchers (p. 332 left and right). Des Bartlett/ Photo Researchers (p. 221). Costa/Pip (p. 243). Friedmann/Pip (p. 184/185). Grzimek/Okapia (p. 327 below, 329). Hill/Pip (p. 183). Kapocsy (p. 25). Kraft/Bavaria (p. 97). van Lawick-Goodall (p. 330/331). Layer (p. 244). Quedens (p. 332 upper left, 333 above, 467, 468, 469). Scheer (p. 178 below). Schuhmacher/Barth (p. 122/123). Sielmann (p. 137). Tomisch/Bavaria (p. 327 above). Trevor/Pip (p. 98, 220). V-Dia-Verlag (p. 218/219). Zellmann (p. 178 below). ZFA (p. 186, 133 below). Zingel (p. 293). Line drawings: J. Kühn (Distribution maps and Fig. 12-3, p. 279). From Berndt/ Meise, Natural History of Birds, with the permission of Franck'sche Verlagshandlung in Stuttgart (Fig. 1-2, p. 20; Fig. 1-4, p. 22; Fig. 1-5 and 1-6, p. 33; p. 34, lower two drawings; Fig. 1-9 and 1-10, p. 39; Fig. 1-11, p. 40; Fig. 1-12, p. 41; Fig. 1-15, p. 49; Fig. 1-16, p. 50: Fig. 1-25 and 1-26, p. 74 and 75; Fig. 1-27 and 1-28, p. 78; Fig. 1-29, p. 79; Fig. 2-2, p. 85; Fig. 4-3, p. 111; Fig. 5-2, p. 130; Fig. 5-5, p. 134; Fig. 5-7, p. 135; Fig. 6-1, p. 141; Fig. 6-5, p. 146; Fig. 7-1, p. 164; Fig. 7-9, p. 169; Fig. 7-26, p. 182; Fig. 9-4, p. 227; Fig. 9-7, p. 233; Fig. 9-9, p. 240; Fig. 11-5 and 11-6, p. 262; Fig. 13-9, p. 313; Fig. 13-10, p. 314; Fig. 13-19, p. 322; Fig. 15-10 and 15-11 top, p. 350; Fig. 15-22, 15-23 top, and 15-24, p. 358; Fig. 15-25 and 15-27, p. 359; Fig. 15-45, p. 369; Fig. 15-46, p. 370; Fig. 15-65 top, p. 381; Fig. 15-71, p. 389; Fig. 15-83 top, p. 406; Fig. 15-92, p. 410; Fig. 16-1, p. 416; Fig. 16-6, p. 422; Fig. 16-8, p. 425; Fig. 16-10, p. 426; Fig. 16-15, p. 430; Fig. 18-10, p. 464; Fig. 18-14 and 18-15, p. 471). H. Brüll (Fig. 14-1 and 14-2, p. 337; Fig. 14-5 and 14-6, p. 339; Fig. 14-7, p. 340; Fig. 15-11 bottom, p. 350; Fig. 15-23 bottom, p. 358; Fig. 5-26, p. 359; Fig. 15-28, p. 360; Fig. 15-65 bottom, p. 381; Fig. 15-72, p. 389; Fig. 15-82 and 15-83 bottom, p. 406). A. Festetics (Fig. 8-6, p. 193; Fig. 8-10, p. 202). S. Raethel (Fig. 19-14, p. 488; Fig. 19-15 and 19-16, p. 489; Fig. 19-17, 19-18, 19-19 and 19-20, p. 490; Fig. 19-22, p. 493). E. Diller (all others, including: according to Baur-Glutz, Fig. 8-1 and 8-2, p. 190; Fig. 8-19, p. 212; Fig. 9-14, p. 246; Fig. 9-15, p. 247; according to Berndt/Meise, Fig. 1-13, p. 42; according to Blume, Fig. 1-17 and 1-18, p. 54; Fig. 1-20, p. 59; Fig. 1-22, p. 67; Fig. 4-8, p. 112; Fig. 7-16, 7-17, 7-18 and 7-19, p. 175; Fig. 8-4, p. 191; Fig. 8-5, p. 192; Fig. 9-2, p. 224; Fig. 9-3, p. 225; Fig. 11-7, p. 265; Fig. 11-8, p. 266; Fig. 11-9, p. 269; Fig. 11-10, p. 270; Fig. 14-3 and 14-4, p. 338; Fig. 18-5, p. 460; Fig. 18-6 and 18-7, p. 461; according to Boetticher/Grummet, Fig. 11-4, p. 261; according to Eibl-Eibesfeldt, Fig. 6-2, p. 144; Fig. 12-12, p. 289; according to photographs, Fig. 7-27, p. 182; Fig. 15-91, p. 410; according to Grassé, Fig. 1-3, p. 21; according to O. Heinroth, Fig. 1-19, p. 56; according to Hornberger-Kacher, Fig. 9-5, p. 229; according to Johnsgard, Fig. 11-2, p. 258; Fig. 12-2, p. 279; Fig. 12-4, p. 280; Fig. 12-9, p. 286; Fig. 13-4, 13-5 and 13-6, p. 301; according to Krösche, Fig. 3-6, p. 107; according to Krumbiegel, Fig. 3-3, p. 93; Fig. 3-4, p. 95; according to Life "Wonders of Nature," Fig. 1-30, p. 80; according to Lorenz, Fig. 15-66, p. 382; according to Niethammer, Fig. 1-1, p. 18; according to Scott/Delacour, Fig. 11-3, p. 259; according to Sigmund, Fig. 1-14, p. 47; according to Studer-Thiersch, Fig. 10-2, p. 250; according to van Tets, Fig. 7-20 and 7-21, p. 176; Fig. 7-22, p. 179; according to Tinbergen/Lorenz, Fig. 12-13, p. 290; according to Warham, Fig. 6-3 and 6-4, p. 145; Fig. 7-24, p. 180 and 181; according to Wobus, Fig. 4-9, p. 113; Fig. 4-10, p. 114; Fig. 4-11, p. 117).

Index

Heavy type indicates the main entry, an asterisk indicates an illustration, and m indicates a distribution map.*

Abbreviations and Symbols

C, °C ...... Celsius, degrees centigrade

C.S.I.R.O. .. Commonwealth Scientific and Industrial Res. Org. (Australia)

f .... following (page)

ff .... following (pages)

L .... total length (from tip of bill to end of tail)

I.R.S.A.C.... Institute for Scientific Res. in Central Africa, Congo

I.U.C.N. .... Intern. Union for Conserv. of Nature and Natural Resources

BH .... body height

TL .... tail length

i.n.s. .... in a narrower sense

WL .... wing length

BL .... body length

♂ .... male

♂♂ .... males

♀ .... female

♀♀ .... females

♂♀ .... pair

+ .... extinct

▷ .... following (opposite page) color plate

▷▷ .... Color plate or double color plate on the page following the next

▷▷▷ .... Third color plate or double color plate (etc.)

⟠ .... Endangered species and subspecies